HF Radio
Systems & Circuits

HF Radio Systems & Circuits, formerly *Single Sideband Systems & Circuits*, is a member of the Noble Publishing Classic Series. Titles selected for the Classic Series are important works which have stood the test of time and remain significant today. While many publishers lose interest in past titles, Noble Publishing is proud to recognize classic works and to revive their contribution.

HF Radio
Systems & Circuits

William E. Sabin and Edgar O. Schoenike
Editors

Written by members of the Engineering Staff,
Collins Divisions, Rockwell Corporation

N⊕BLE

Noble Publishing Corporation
Atlanta

Standard Cataloging-in-Publication Data

Sabin, William E. and Edgar O. Schoenike, eds.
    HF Radio Systems & Circuits—Rev. 2nd Ed.

    Originally published as *Single Sideband Systems
        and Circuits, Second Edition*:
        New York: McGraw-Hill, Inc., 1995.
    Includes bibliographical references and index.
    1. Radio, Single-sideband. 2. Radio circuits. I. Sabin, William
E. II. Schoenike, Edgar O.
TK6562.S54S56   1998   621.384'153—dc20
ISBN 1-884932-04-5

N⊕BLE

International Standard Book Number 1-884932-04-5

Contents

Chapter 17. Software for SSB

William E. Sabin

Preface to the Revised Second Edition

HF Radio Systems & Circuits is a unique reference book, one which contains information not easily found elsewhere. Its subject is radio communications, which is a rapidly-growing technology at all frequencies of operation, reaching new markets with wireless products that offer remarkable convenience, excellent performance and lower cost.

This book on high-frequency (2 to 30 MHz) communications serves two purposes for radio engineers. Most obvious is its thorough coverage of HF single sideband communications systems and the circuits they contain. Domestic and international use of SSB and other transmission modes in this frequency range is increasing as various government and private agencies seek to reach a common set of interoperability protocols. This book's original title, *Single Sideband Systems & Circuits*, indicates its emphasis in this area.

The second purpose of *HF Radio Systems & Circuits* is for more broadly-based reference data. Most of the topics included in this book are also valuable for applications other than SSB communications. Data on speech characteristics and speech processing can be applied to any type of voice communications. Design guidelines for amplifiers, mixers, oscillators, synthesizers and other key circuits are useful to any RF or microwave design engineer. The test procedures and instrumentation data presented here has wide applicability, as well. This wide applicability is the reason for changing the book's title.

In short, this is an excellent radio-frequency design reference book. Its thorough coverage of basic communication principles and circuit building blocks clearly warrants its continued publication as part of Noble Publishing's Classic Series. This edition has a modest number of corrections and revisions to the second edition, including updated software utilities on diskette.

Gary Breed
President
Noble Publishing Corporation

Preface to the Second Edition

Since the publication of the first edition, several important advances in SSB technology have occurred:

1. The technology of link establishment has become a major topic in HF SSB communications. A new chapter is dedicated to that subject.

2. The use of digital signal processing has advanced. Integrated circuits and design algorithms have made great strides.

3. The use of MOSFETs in power amplifier design has become widespread.

4. The use of the personal computer for the design and simulation of circuits and for systems analysis has become a fact of life for design engineers.

5. There is an increasing interest in the use of pilot carrier SSB in various applications. The widespread manufacture of special integrated circuits and design methods for personal communication systems, such as cellular telephones, has implications for low-cost SSB equipment.

6. Recent equipment designs such as the Spectrum 2000 are excellent examples of the union of the SSB radio with the personal computer.

7. Recent receiver design methods, such as direct conversion, are becoming more useful as digital processing becomes more cost-effective and more efficient.

8. The embedding of digital circuitry and sensitive analog receiver circuits in the same framework has created new philosophies regarding electromagnetic compatibility.

This second edition provides coverage of all these topics in sufficient detail to enable the design engineer and the advanced amateur experimenter to get started in these various subjects. Many other subjects in the various chapters have been expanded and modernized.

William E. Sabin
1400 Harold Dr., SE
Cedar Rapids, IA 52403

Edgar O. Schoenike
325 Rockledge Trail
Floral, AR 72534

Preface to the First Edition

It was 23 years ago that McGraw-Hill published *Single Sideband Principles and Circuits*, by E.W. Pappenfus, W.B. Bruene, and E.O. Schoenike. Since then many changes have occurred in the components, circuits, and systems analysis used in SSB technology. In 1964 transistors were just beginning to become dominant in low-power amplifiers and oscillators, while vacuum tubes still reigned supreme for medium- and high-power amplifiers. Integrated circuits were just beginning to be used in SSB equipment, and microprocessors were unknown. The development of these new electronic components has led to new circuits, greater flexibility in design, and more sophisticated equipment control.

One aim of this book is to incorporate an explanation of the developments that have taken place in the design of SSB equipment while retaining explanations of those techniques which have withstood the test of time. Thus, solid-state power amplifiers and power supplies are discussed in detail, but advances in high-power vacuum tube amplifiers are not overlooked. Similarly, balanced diode mixers and modulators, instead of being superseded, are used today perhaps more widely than ever before, and so are also fully covered.

This book was written at the level of a practicing engineer, although it will be appreciated by the engineering student and advanced amateur as well. Most explanations are intended to be practical in nature, but the theoretical basis of SSB is treated in some detail, design principles are not overlooked, and, when relevant, performance trade-offs are discussed. However, only amplitude modulation SSB is discussed. Angle-modulated single sideband is beyond the scope of this book.

Special emphasis is placed on the system analysis and system design of the SSB communications link. The cost and complexity of modern communications equipment and systems are such that accurate estimates of performance, prior to commitment of resources, are essential.

Greater sophistication in design has led to specialization, and it is therefore appropriate that a large number of experts in specialized areas should contribute to this book. It is almost impossible for one person to completely master all of the disciplines involved; therefore the editors have gathered together some of the leading equipment designers and analysts from the Collins Division of the Rockwell Corporation to contribute chapters in their fields of specialization.

Besides the chapter authors, the editors give special thanks to our engineering colleagues for their ideas and contributions to this book, and to Rockwell International for its permission to publish.

We also owe a special debt to the many secretaries who contributed their spare time to the word processor typing chores.

In chapters where multiple authors appear, names are listed in alphabetical order.

One final note: Recognizing as we do the important and growing role of women in the sciences, every effort has been made to use gender-neutral language in the writing of this book. In the single instance of "specsmanship," however, no generally recognized gender-neutral equivalent exists; thus the term should be taken in a purely generic sense, intended to apply to both women and men.

William E. Sabin
Edgar O. Schoenike

This book is the result of a difficult and comprehensive team effort by many members of the Engineering Staff of the Collins Division of Rockwell Corporation. It is dedicated to them and to all the employees, past and present, of the Collins Division of the Rockwell Corporation on the occasion of the fiftieth anniversary of the founding of the Collins Radio Company.

A

B

Figure I.1 The Rockwell HF SSB Comm Central station in Cedar Rapids, Iowa. (A) The operating console with Collins 651S SSB receivers and Rockwell computers, attended by two operators, 24 hours a day. (B) The SSB transmitter/receiver rack consists of four 10-kW and three 1-kW units. A remote-controlled, unattended station in Newport Beach, California, is accessed by a dedicated telephone line. Automatic link establishment (ALE) computer programs are used extensively. A 20-acre antenna farm contains 13 HF antennas.(*David G. Berner*)

1

Overview of Single Sideband

William E. Sabin

1.1 The Radio Link

The principal task that confronts a radio communications link between two or many mutually distant points is to provide, within the framework of a limited available transmitter power, reliable, high-quality communication. Very often, real-time voice contact is desired. Opposing this goal are the inimical characteristics of the radio frequency (RF) spectrum. Among these are noise from the ionosphere and the galaxy, artificially produced electrical noise, severe variations in the received signal strength (fading) that are observed over time spans from milliseconds to hours or days, propagation disturbances, interference from other users of the spectrum, and multiple arrivals of the signal along different paths. Interference caused to other users is aggravated by technical limitations in transmitter spectral purity and directional antenna design. Interference experienced from other users is increased by deficiencies in receiver design and receiving antennas. One especially difficult mode of interference is between transmitters and receivers that are in close proximity (collocated). Also involved here is the creation of false signals (intermodulation or IM) due to nonlinearities within the collocated environment.

The approach to reliable communication used by radio engineers is to obtain from a given amount of available transmitter power the maximum amount of intelligibility of a speech signal or the minimum error rate of a digital signal (at the distant receiver) under the conditions described above. Two important constraints in this design are the conservation of bandwidth and time. That is, the spectrum in use very often requires a small ratio of RF bandwidth to baseband message bandwidth in order to accommodate a large number of users. Also, the time used to transmit a message is very often required to be nearly the same as the duration of the original message; in this situation the amounts of redundancy and encoding available must be small. An equivalent statement is that the ideal communication channel, in the sense described by Shannon [1], is, in most near-real-time situations, not approached. In practice, voice or digital messages usually contain inherently high levels of redundancy or predictability, except for key elements that may be repeated several times by a good operator.

One of the principal system design approaches used is to select a method of modulation that is optimal within the environment and the constraints described above. For real-time speech communication, from 10 kHz to 250 MHz, and in recent times up to as high as 10 GHz, the use of single-sideband (SSB) suppressed-carrier (or quite often reduced-carrier) modulation has provided a very satisfactory answer. A long period (75 years) of analytical and experimental investigation has proven the efficacy of this method, especially in the high-frequency (HF) band, which is a difficult arena.

A further consideration in situations where the volume and weight of the transmitter are critical is that SSB is competitive with narrowband frequency modulation (FM) in terms of communications effectiveness for a given weight and size. The results of recent studies will be considered in later sections. Single sideband also is a decisive improvement, in nearly all respects, over high-level, double-sideband amplitude modulation (AM).

One of the costs involved in SSB, as compared with AM, is the additional complexity of the receiver versus the conventional low-cost AM broadcast receiver. Various responses to this will be considered in this book. In the transmitter, the need for large amounts of linear amplification of the RF signal is a technical and economic burden.

The development of phase-lock loop (PLL) techniques has opened up many uses of reduced-carrier SSB, where the reduced pilot carrier provides frequency-locked and phase-locked reception and serves other functions. The improvements in frequency synthesizer design and low-cost, portable frequency standards have made SSB practical at much higher frequencies than were possible a few years ago.

A further enhancement of SSB has been the development of speech processors that utilize the peak power capabilities of the transmitter more effectively by compressing the dynamic range of human speech, thereby increasing the average power. The LINCOMPEX system and other companders have the ability to restore the original dynamic range at the receiver, providing a telephone-grade signal. Vocoder and other techniques that help to reduce bandwidth with no appreciable loss of intelligibility have been undergoing continuous development.

In long-distance HF (2- to 30-MHz) communication, a major problem is to locate a favorable frequency very quickly and automatically tune to it (automatic link establishment or ALE). The extensive application of microprocessor technology has produced highly programmable, remotely controllable radios that combine with recently perfected techniques to produce orders of magnitude of improvement in HF link reliability. Link quality analysis (LQA) quickly determines the ability of the selected channel to support message transfer.

In very recent years the revolution in digital signal processing technologies has been applied to SSB receivers and transmitters to produce levels of performance and flexibility (programmability) that are setting new standards in radio design. Also, developments in high-power, solid-state transistors and circuit design have revolutionized linear power amplifier (LPA) design. The development of ultralinear power amplifiers using feedforward techniques promises to reduce transmitter distortion products by two or three orders of magnitude. Improvements in high-power, fast-tuning vacuum tube amplifiers are described in this book. Advanced measurement techniques for high-performance receivers and transmitters have been developed. The design of antenna couplers used in SSB transmitters has been elevated to a high level of sophistication. To complete the discussion, recent developments in intermediate frequency (IF) filters and tunable bandpass filters for SSB equipment are covered.

1.2   Overview of SSB Equipment

The transmitter

Figure 1.1C shows a block diagram for a complete SSB transmitter that uses the filter method of generating the desired wave. A baseband signal $f(t)$ has a spectrum $F(\omega)$ which is shown in Figure 1.1A. For the purpose of this discussion, it is often preferred, but not essential, to use the concept of a *two-sided* spectrum shown in Figure 1.1B, where the physical signal is decomposed mathematically into two coherent, complex conjugate segments, one of which is at a fictitious "negative frequency."

This $f(t)$ is multiplied in a balanced mixer by the local oscillator (LO) wave cos $(2\pi f_0 t)$. The spectrum at the mixer output, Figure 1.1D, is therefore the convolution of $F(\omega)$, the spectrum of $f(t)$, and the spectrum of the LO, shown at Figure 1.1E. Observe that although the balanced mixer output contains very little carrier, the phase noise impurities in the LO are transferred to the output signal in the form of amplitude sidebands, as shown. These often tend to establish the maximum in-band signal-to-noise (S/N) ratio of the desired transmitter signal.

The narrowband filter passes accurately one of the sidebands and sharply attenuates the other sideband, the carrier frequency, high and low speech frequencies, wideband LO noise and other noise, and spurious emissions from the low-level stages of the transmitter. Therefore this filter is a critical item in the design. Alternative approaches that attempt to eliminate this filter are considered in Chapter 2.

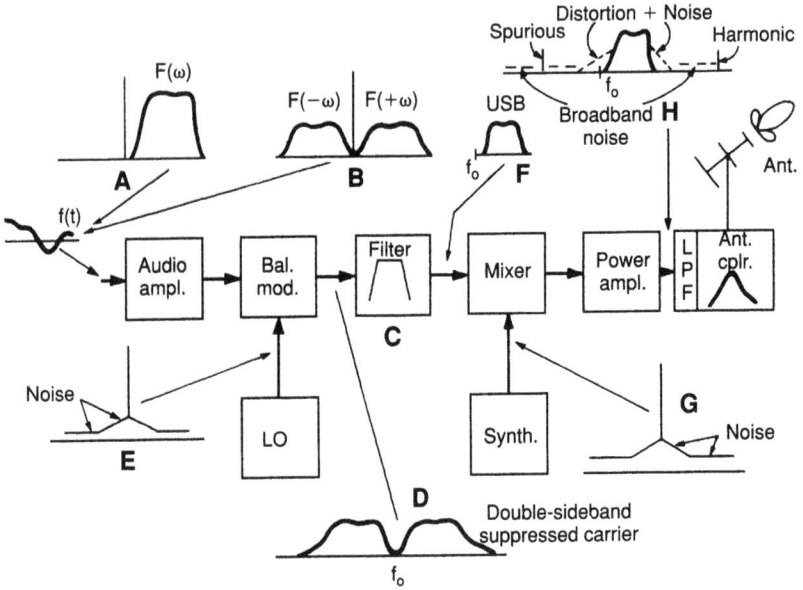

Figure 1.1 Block diagram of an SSB transmitter. Parts (A) and (B) show the baseband signal as one-sided or two-sided spectra; (D) is the double-sideband suppressed-carrier signal; (F) is the filtered SSB signal which is amplified, contaminated, and filtered as it moves to the antenna; (H) is the output of the PA; (E) and (G) show the contaminated local oscillator signals.

Following the filter, amplification, frequency translation, speech processing (to increase the average power), and power amplification occur. During these processes, LO contamination, amplifier noise, out-of-band and in-band distortion products, and discrete spurious frequencies and harmonics are added to the desired signal. These spurious emissions either degrade the desired signal or are a source of interference to other users, especially those who are collocated or on closely adjacent or certain other frequencies. At the output of the transmitter, the impedance of the antenna is transformed to the desired power amplifier (PA) load impedance, and band filtering is also performed. Adjacent transmitters must be electrically isolated from each other to prevent interactions between their outputs. The antenna tries to focus the energy on the desired destination.

The design challenge for the transmitter can thus be defined: to provide accurately the required power output with levels of fidelity, frequency stability, and undesired emissions that are acceptable to the user, and to provide as much communications effectiveness as possible. The desired usage may require that a carrier signal be wholly or partially inserted if needed.

Other aspects of the design are weight and size restrictions, temperature rise, reliability, and duty cycle. Furthermore, the equipment may have to perform Morse code and data transmission functions in addition to voice. In some cases the desired output may be two or four independent adjacent channels with low levels of cross-channel interference. The radio may require some kind of remote control, automatic tuning, or unattended operation. Finally, there are always cost constraints imposed by the marketplace. The design process, then, consists of finding a set of technical and economic compromises that satisfy the user in all these respects.

SSB power relationships

As an aid to understanding later discussions, consider Figure 1.2. The RF signal voltage, with carrier absent, varies in amplitude in accordance with the modulation. The line joining the tips of the RF cycles is the envelope. At the highest point of the envelope the instantaneous voltage reaches its highest value. The peak instantaneous power (or simply peak power) is determined at this point. Calculated over one cycle of the RF voltage at this highest point, the resulting average power is called the peak envelope power (PEP). Over a period of time that is much longer than the modulating waveforms, the average power is determined from the average value of the square of the rms voltage, as shown.

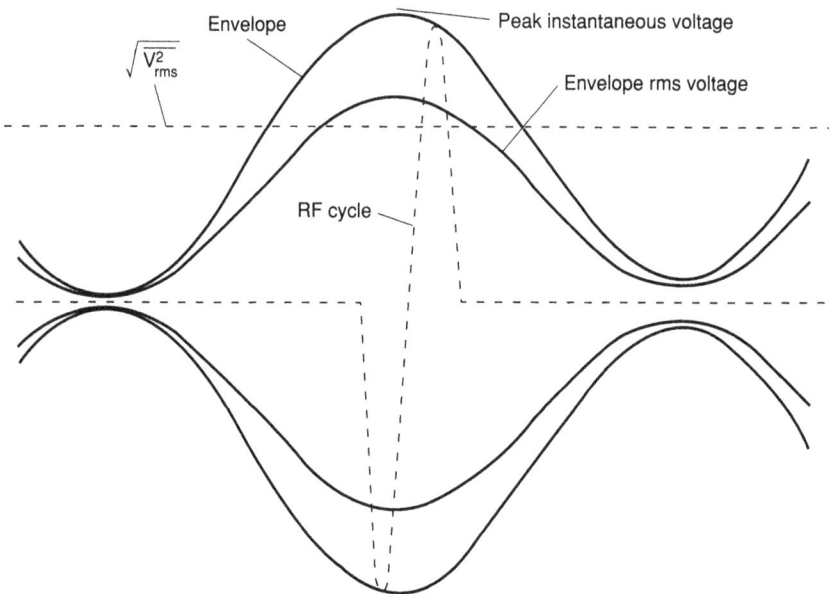

Figure 1.2 The RF output voltage of the transmitter for a modulated SSB signal is used to determine the peak power, the peak envelope power, and the average power.

The receiver

Figure 1.3 is a block diagram that exemplifies the design problems for an SSB receiver. The desired signal frequency is translated to the passband of a narrowband filter. The circuitry to the left of the second mixer, called the front end, or translator, is vulnerable to undesired signals that produce intermodulation distortion (IMD), desensitization, reciprocal mixing (noise modulation) caused by synthesizer phase noise, spurious responses due to undesired mixer products and synthesizer contamination, and damage due to nearby transmitters. The antenna lead also can conduct undesired receiver emissions, which are radiated and produce local interference. The reception of weak signals may be restricted by in-band noise generated by the receiver, but at lower frequencies noise picked up by the antenna usually sets the limit of sensitivity unless the receiving antenna is very inefficient. A design goal is to make internal noise negligible in comparison, but this goal conflicts with the need for immunity to undesired signals. A directional antenna augments the intended signal-to-interference ratios.

The block diagram in Figure 1.3 shows also that selectivity may precede the receiver input. Increased immunity to interference is then possible, but at considerably greater cost. A goal is to minimize this selectivity cost. This desire has led to the development of receiver front ends having very high

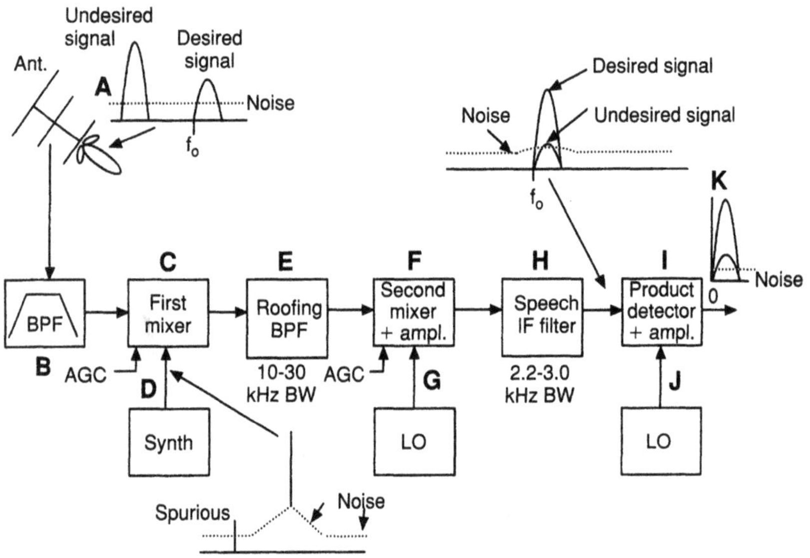

Figure 1.3 Block diagram of an SSB receiver. The desired signal at (A) is accompanied by noise and undesired signals. Local oscillator noise and spurious responses at (D) further contaminate the desired signal. The narrow filter at (H) rejects out-of-band interference and noise. The desired output at (I) is degraded by noise and spurious responses.

dynamic range. In many systems multiple receivers must operate effectively, using diplexing networks, from the same antenna. *Active* receiving antenna arrays having electronically steerable directivity are coming into use in some systems.

A high order of frequency stability and precise frequency control are needed in an SSB receiver. Frequency synthesis, controlled by a stable frequency standard, is ubiquitous in high-quality SSB equipment. Following the roofing filter, more amplification and conversion take place. Further narrowband filtering at a lower IF, closely conforming to the desired SSB frequency band, is needed. The circuits between these two filters are subject to overload by strong signals on closely adjacent frequencies that are within the passband of the wider filter. These stages and also those after the narrowband filter are sources of noise and distortion added to the desired signal. After the final filter a product detector having very desirable weak-signal properties translates to baseband. The output level is accurately regulated by an automatic gain control (AGC) that is operated by the speech sideband energy. The AGC also protects the IF and RF stages from overload by the desired signal and must have excellent transient response since it is driven by the rapidly fluctuating speech signal and by undesired impulses. An optional noise blanker removes these impulses.

The receiver may also utilize a reduced pilot carrier for frequency/phase-locked reception and for AGC and squelch purposes. Coherent phase-locked AM reception is also optional. The same receiver is also usually used for continuous wave (CW), radio teletype (RTTY), and conventional AM modes. Additional features are voice-actuated squelch, computer control of frequency and other functions, speech companding, and telephone line and IF outputs.

1.3   Attributes of SSB

There are two principal advantages to SSB: its narrow bandwidth, both in transmission and reception (these enhance each other), and its efficient usage of the transmitter's primary power source. This latter point translates also into weight and volumetric efficiencies that are crucial in many situations. It used to be that a 100-W transmitter stood in a 5-ft rack and weighed hundreds of pounds. Now a vastly superior transmitter (plus a receiver) can be hand-carried in a small luggage case. The absence of a carrier means that the entire peak power capability is devoted to the information-bearing content of the signal. There is, however, an anomaly inherent in the SSB wave, which Chapter 2 evaluates. Under certain conditions the baseband waveform can produce excessive peaks in the RF wave that result in distortion in the power amplifiers. A comparable overload effect can occur in the receiver and has limited the use of SSB in certain data applications.

The narrow bandwidth is achieved only when spurious emissions are tightly controlled; therefore SSB logically demands an extremely high signal purity. This reduces somewhat the efficiencies mentioned above, since the design must be more conservative than would otherwise be needed. Despite this, and because of technical developments, the choice between double-sideband suppressed carrier (DSBSC) and SSB has, for nearly all systems, been resolved in favor of the reduced spectral usage of a clean SSB signal. Also, a possible signal-to-noise ratio improvement of 3 dB exists for SSB, relative to DSBSC, when compared on the basis of equal peak envelope power.

At the receiver, the product detection and narrow bandwidth imply that the signal-to-noise ratio (for a perfect receiver) remains constant along the signal path. The quieting effect associated with FM, for example, is normally absent in SSB. However, an equivalent effect can be synthesized, as discussed in Chapter 3. The use of product detection also means that the severe distortion caused by carrier dropout, which is so well known in conventional AM, is eliminated. This kind of detection, when applied to AM receivers, is called *exalted-carrier* reception.

Frequency differences between signal and receiver are detrimental to SSB and reduce the intelligibility and musical quality considerably. This is discussed in Chapter 2, and it turns out that for speech some offset is tolerable, which makes SSB for speech much easier to implement since the pilot carrier is often (but not always) unnecessary.

In an SSB receiver without pilot carrier the AGC is derived from the speech content. During a pause the noise level increase can sometimes be annoying in a commercial or military link (amateurs and some others do not seem to care as much). Therefore SSB AGC and squelch designs are needed that circumvent this natural problem; or as mentioned before, the pilot carrier can be used. In SSB the use of voice-actuated transmit/receive (T/R) switching has been widely adopted, perhaps coincidentally. This practice has led to more efficient channel usage and better communication practice.

1.4 Threshold SSB Signal Reception

The evaluation of the reception of SSB signals that are near the antenna noise level helps to put the receiver and antenna design requirements into perspective.

Atmospheric (terrestrial) noise, including man-made noise, is maximum at frequencies below 10 MHz, where its value is, on the average, 40 dB above thermal (−174 dBm/Hz) and diminishes at 20 dB/*octave*. At VHF and above it is of little importance. Galactic (outer space) noise averages about 20 dB above thermal at 20 MHz and decreases at about 20 dB/*decade*. But at certain locations and times and in certain directions (using directional antennas) the composite noise can be much less than this even at lower HF.

Very quiet conditions, along with very weak signals, are frequently reported even at 4 or 5 MHz. Therefore antenna efficiency and receiver noise figures cannot always be neglected for reliable reception at these frequencies. This conclusion may be modified somewhat by the fact that a perfectly efficient dipole antenna gathers about 12 dB more signal and noise power at 1.6 MHz than it does at 30 MHz [2]. If the noise is uniform from all directions, a directional antenna receives the same amount of noise as an omnidirectional antenna, because of its gain, but the signal is also increased by the antenna gain. So the signal-to-noise ratio is improved.

System performance is often better understood in terms of the *noise temperature* concept. We wish to relate this concept to a *noise figure* concept, which is more familiar to many HF SSB engineers. The *effective noise temperature* of a device is defined as

$$T_E = 290 \ (F - 1)$$

F is the noise factor, which is specified or measured in the lab at 290 K, according to an IEEE standard [3]. $F - 1$ is the excess noise, relative to thermal noise; so an active device that contributes no excess noise has an equivalent noise temperature of 0 K. For a purely passive device that attenuates the signal we have

$$T_E = 290 \ (L - 1)$$

where L is the loss factor, which is the same as its noise factor and the reciprocal of its gain $G < 1$.

For a cascaded system the equivalent system noise temperature is given by

$$T_S = T_G + T_{E1} + \frac{T_{E2}}{G_1} + \frac{T_{E3}}{G_1 G_2} + \frac{T_{E4}}{G_1 G_2 G_3} + \dots \tag{1.1}$$

where T_G is the generator (antenna) temperature, the T_E numbers are the receiver component effective noise temperatures, and the G numbers are the gain values. The system noise floor is related to T_S by the relation $N = kT_S B_N$, where B_N is noise bandwidth and k is Boltzmann's constant. The ratio T_S/T_G can be correctly thought of as a system noise factor, F_S, which is not pegged to 290 K but depends in part on the antenna temperature and can range from a few degrees Kelvin to several thousand degrees Kelvin. F_S is the ratio of total noise to generator noise, both observed at the receiver output.

As one example, if a receiver has a noise temperature of 1000 K (noise factor 4.45 or 6.5 dB) and the antenna temperature (often called the sky temperature) is 1500 K (-167 dBm/Hz), then

$$T_S = 1000 \text{ K} + 1500 \text{ K} = 2500 \text{ K}$$

and the value of F_S would be

$$2500 \text{ K}/1500 \text{ K} = 1.67 \text{ (or 2.2 dB)}.$$

From this example, we see that a reduction of receiver noise factor, F, would not be very helpful, unless much quieter sky conditions prevailed. If we place a 10-dB ($F = 10$, $G = 0.1$) attenuator ($T_E = 2610$ K) in the antenna lead, then

$$T_S = 1500 + 2610 + 1000/0.1 = 14110 \text{ K}$$

and $F_S = 9.407$ (or 9.7 dB). Note that the attenuator has a doubly degrading effect, once in the second term (2610) and again in the third term (0.1). Compare this 9.7-dB value of F_S with the noise figure, F, of the receiver plus attenuator, $6.5 + 10 = 16.5$ dB. The meaning of these numbers is the following:

1. At 1500 K sky temperature the 16.5-dB receiver (including the attenuator) has a 9.7-dB ratio of total noise to antenna noise.
2. At 290 K sky temperature a 9.7-dB receiver also has the same ratio of total noise to antenna noise.
3. The two situations are equivalent as far as system performance is concerned.

The attenuator therefore does not affect system noise performance drastically but it might reduce a tendency toward receiver overload caused by strong undesired signals. A receiver using a large high-gain antenna at low frequencies may need this help at times.

This approach can be used to determine various useful system parameters of interest to the SSB system designer.

1.5 Envelope Elimination and Restoration

As pointed out previously, one of the cost and complexity burdens of SSB is that the transmitter requires a large amount of highly linear amplification. The concept of envelope elimination and restoration has been developed as an alternative approach that reduces this requirement. The method is shown in Figure 1.4.

The SSB signal is amplitude limited at a very low level and this limited signal is amplified to a high level by a string of amplifiers that can be nonlinear. The envelope of the low-level SSB signal is amplified independently by a highly efficient switching-mode amplifier. The output power amplifier is then high-level modulated by the envelope. The output of the power amplifier is the desired high-power SSB signal.

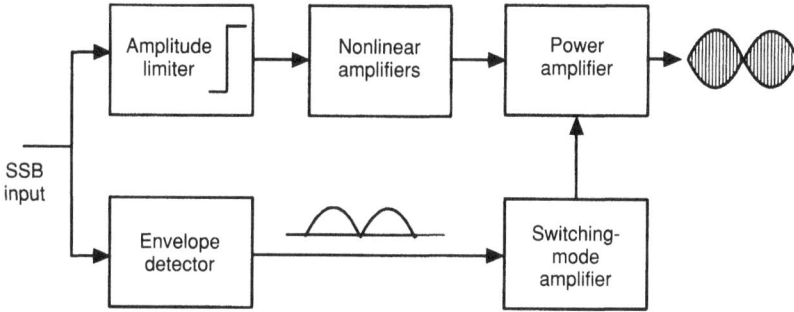

Figure 1.4 Block diagram of an envelope elimination and restoration method.

The two main requirements for this system are that the envelope be reconstructed very accurately and that the time delays in the two channels match very closely at the point of recombination. The penalties in these respects are poor intermodulation performance and a broad output spectrum. The potential advantages are a simplification of the signal path design and a reduction of primary power consumption. References 4 and 5 and section 2.1 of this book give more details.

References

1. C. E. Shannon, "Communications in the Presence of Noise," *Proc. of IRE* 37 (January 1949): 10–21.

2. H. A. Haus et al., "Description of the Noise Performance of Amplifiers and Receiving Systems," *Proc. of IEEE* (March 1963): 436–442.

3. J. D. Kraus and K. R. Carver, "Electromagnetics," 2nd ed. Sec. 14-5. (New York: McGraw-Hill, 1973).

4. F. H. Raab, "Class-S High-Efficiency Amplitude Modulator," *RF Design* (May 1994): 70–74. See also references in this article.

5. L. R. Kahn, "Single-Sideband Transmission by Envelope Elimination and Restoration," *Proc. of IRE* 40 (1952): 803–806.

2

System Design Considerations

Robert L. Craiglow (Section 2.1)
Edgar O. Schoenike (Sections 2.2, 2.3)

As pointed out in Chapter 1, the basic building blocks of an SSB communication system include the transmitter, the receiver, and the antenna. For many purposes, these are adequate for satisfactory communication. For more complex systems, additional specialized equipment supplement these basic components. Following this chapter, we cover specialized topics in transmitter and receiver design, such as transceivers, exciters, power amplifiers, filters, speech processing, and frequency standards, as well as other system elements such as preselectors, postselectors, and antenna couplers. When other than analog speech is to be transmitted, a modulator/demodulator (modem) is required as well, but this subject has become quite specialized and is beyond the scope of this book.

Preselectors and postselectors, discussed in Chapter 9, are often required in a system when simultaneous operation (SIMOP) is required of a collocated transmitter and receiver. A preselector may also be necessary if receiver operation is required in close proximity to an unrelated nearby transmitter operating in the same frequency range. When the antenna impedance varies widely over the operating band, a transmitting antenna coupler is required to present a more constant load impedance to the power amplifier.

The purpose of this chapter is not to discuss the system elements, but to describe important topics in system design and analysis. Section 2.1 starts with the basic voice signal that is to be transmitted. The nature of speech and hearing is reviewed, and speech intelligibility and the articulation index (AI) are given precise definitions. The effect of speech clipping on these characteristics is discussed. Section 2.1 continues with an exposition of signal representations and Hilbert transforms, leading into the various forms of AM, including SSB. For completeness, and to form a basis of comparison with SSB, angle modulation is also discussed. Finally, a comparison is made between the various forms of AM and FM.

Section 2.2 covers collocated system designs. The various factors that cause problems in simultaneous operation of receiver and transmitter are discussed. These include transmitter out-of-band noise and distortion, receiver distortion and reciprocal mixing, and transmitter back-IM distortion.

One of the reasons for a resurgence of interest in HF radio transmission by the military services is its use as a backup to satellite communications. To meet modern needs for communication in a potential jamming environment, such equipment must be designed to incorporate electronic countercountermeasures (ECCM). Those elements of ECCM design that affect the SSB receiver and transmitter are discussed in section 2.3.

2.1 Analog Voice Modulation

In this section we introduce the tools necessary to evaluate and compare various analog voice modulation systems. In order to better understand the factors affecting speech intelligibility, we briefly review the processes of speech and hearing and present a method for the calculation of the AI, an empirical measure of intelligibility. The effects of speech processing techniques are evaluated and the common voice modulation systems analyzed and compared.

Speech intelligibility

Speech sounds are controlled by the position of lips, teeth, tongue, and velum, which establish a set of acoustic resonant frequencies that characterize the vocal tract. These resonances are excited by harmonic-rich, quasi-periodic pulses of air from the vocal chords for voiced sounds and by the hiss of turbulent air passing through a constriction in the vocal tract for unvoiced sounds. The resultant speech spectrum shows these characteristic resonances or formant frequencies.

The ear performs a short-term spectral analysis of the speech sounds over an interval of some 1/8 sec with a frequency resolution of 50 to 500 Hz, depending on frequency, thus determining the characteristic resonant or formant frequencies of the vocal tract which are deciphered by the brain as a particular speech sound or phoneme.

Intelligibility can be measured either experimentally in terms of syllable, word, or sentence test scores or in terms of a calculable empirical measure known as the *articulation index* (AI). Test scores can be obtained only after the communication system has been built or simulated in detail. The AI, on the other hand, can be readily calculated from known characteristics of a proposed system and has been shown to be a reliable indicator of intelligibility for a wide range of system characteristics and test procedures [1]. The AI is therefore used here as the measure of intelligibility for the evaluation and comparison of speech processing and modulation systems. The AI varies from zero for a completely unintelligible system to one for a system giving the maximum possible intelligibility. Typical relationships between syllable, word, and sentence test scores and the AI are shown in Figure 2.1 [2].

There is a critical level of intelligibility for every task below which satisfactory performance can no longer be achieved. It occurs near an AI of 0.3 for a wide variety of applications, although trained communicators using a limited vocabulary and appropriate protocol can communicate with AIs as low as 0.2 [3, 4]. The threshold will be taken here as 0.3.

While an AI of 0.3 provides a usable communications link, fatigue is high and user acceptance low. For day-to-day operations, therefore, the AI should be 0.5 or higher. Of course, the higher the intelligibility and quality, the higher the customer satisfaction.

Articulation index

A speech sound or phoneme can be characterized by its short-term power spectral density. Low-frequency resonances in the vocal tract have narrower bandwidths than high-frequency resonances. Similarly, the spectral resolution bandwidth of the ear is narrower at lower frequencies. One might suspect, therefore, that lower speech frequencies would contribute more to

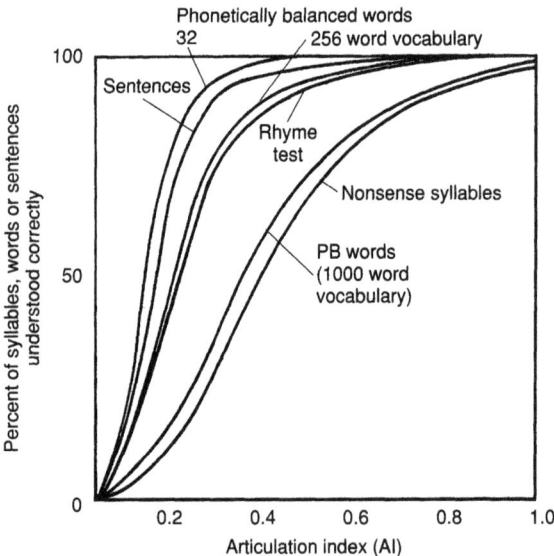

Figure 2.1 Relation between AI and various measures of speech intelligibility.

intelligibility per hertz of bandwidth than higher frequencies. Indeed, this has been verified experimentally [5, 6]. A frequency weighting function $W(f)$ is used to give different weightings or importance to various portions of the spectrum. The short-term spectral density of speech $S(f,t)$ at any given frequency f varies with time t over a range of some 30 dB and contributes to intelligibility in noise only when it is greater than the noise power spectral density $N(f)$ at the same frequency. The speech spectrum $S(f,t)$ at frequency f contributes to intelligibility in direct proportion to the percentage of time that it is greater than $N(f)$ and in direct proportion to $W(f)$ for that frequency. These observations form the basis for calculating the AI.

Several methods for the calculation of the AI have been developed by different investigators [1–3, 5–7]. The details of these methods differ, but the principles and results are essentially the same. The integral form of the AI will be presented here.

Frequencies in the range of 200 to 6100 Hz contribute to intelligibility in direct proportion to the empirical frequency weighting function given by

$$W(f) = 5.0 \times 10^{-4} \exp\left[-4.2 \times 10^{-4} f\right] \qquad 200 \leq f \leq 6100 \text{ Hz} \qquad (2.1)$$

In the absence of noise, the AI for speech which is perfectly bandwidth-limited between f_a and f_b is given by

$$\text{AI} = \int_{f_a}^{f_b} W(f)\,df \qquad (2.2)$$

Let $P(f)$ be the portion of the time or probability that $S(f,t)/N(f)$ is greater than 1. The AI in noise is thus given by

$$\text{AI} = \int_{f_a}^{f_b} W(f)\,P(f)\,df \qquad (2.3)$$

The short-term power spectral density $S(f,t)$ is distributed uniformly in decibels over a range from 18 dB below the long-term average spectral density $S(f)$ to 12 dB above $S(f)$. Therefore, if $S(f)/N(f)$ is 18 dB or greater, $P(f)$ is 1, and if it is –12 dB or lower, $P(f)$ is 0. Within the range of $-12 \text{ dB} < S(f)/N(f) < 18 \text{ dB}$, the probability $P(f)$ is directly proportional to $S(f)/N(f)$ (in dB), or

$$P(f) = \frac{12 + 10 \log_{10}\left[S(f)/N(f)\right]}{30} \qquad -12 \text{ dB} \leq \frac{S(f)}{N(f)} \leq +18 \text{ dB} \qquad (2.4)$$

In order to calculate the AI, we must know the long-term speech spectral density $S(f)$ and the noise spectral density $N(f)$. The long-term voice spectrum for an adult male drops off at 6 to 12 dB/octave above 500 Hz with an average spectrum approximated by

$$V(f) = \frac{1 + (f/4000)^4}{[1 + (200/f)^4][1 + (f/500)^{3.7}][1 + (f/8000)^4]} \qquad (2.5)$$

If $H(f)$ is the voltage-frequency response of the overall speech communications system, then the filtered speech spectral density is given by

$$S(f) = V(f)|H(f)|^2 \qquad (2.6)$$

The noise power spectral density is determined by the nature of the system. On radio links, background noise is generally white over the band of interest, and this will be assumed here.

Figure 2.2 shows the AI as a function of the long-term average audio signal-to-noise ratio $\overline{S}/\overline{N}$ for normal bandwidth-limited speech in white noise where the average speech power $\overline{S}$ is given by

$$\overline{S} = \int_{f_a}^{f_b} S(f)\,df \qquad (2.7)$$

and the average noise power $\overline{N}$ is given by

$$\overline{N} = \int_{f_a}^{f_b} N(f)\,df \qquad (2.8)$$

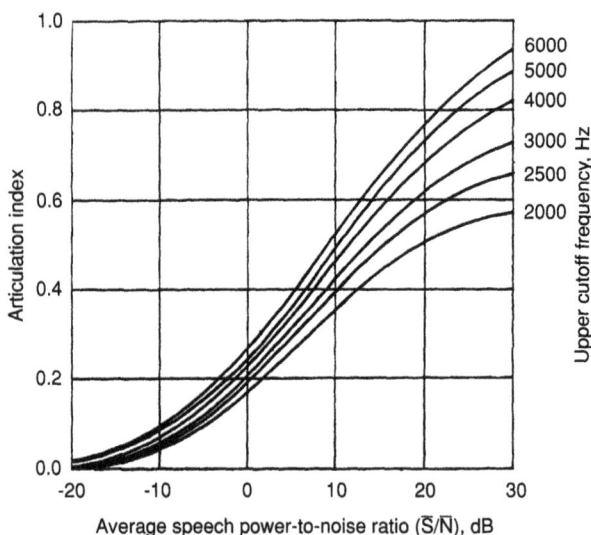

Figure 2.2 Intelligibility of normal speech in white noise with 200-Hz lower cutoff frequency versus signal-to-noise ratio.

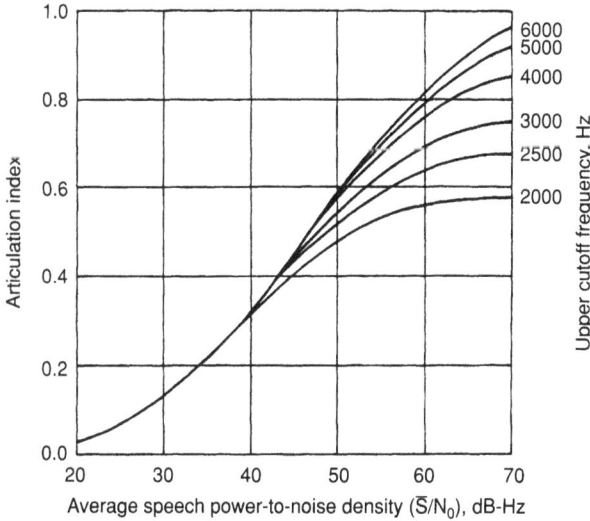

Figure 2.3 Intelligibility of normal speech in white noise with 200-Hz lower cutoff frequency versus signal-to-noise density ratio.

In this figure, the lower cutoff frequency f_a is 200 Hz and the upper cutoff frequency f_b is between 2000 and 6000 Hz.

For white noise, it is generally more useful to give the AI as a function of the ratio of the average signal power $\overline{S}$ to the noise density N_0, or $\overline{S}/N_0$, as shown in Figure 2.3, where N_0 is the noise power per hertz of bandwidth. The curves of Figures 2.2 and 2.3 are for ideal bandpass filters.

Preemphasis

Since higher-frequency components of normal speech are very weak they are readily lost in noise. Therefore these components are often boosted or preemphasized. The preemphasis that maximizes the AI when the average speech power $\overline{S}$ and the noise power spectral density $N(f)$ are fixed can be found using variational calculus.

Assume that $\overline{S}$, given by Equation 2.7, and $N(F)$ are fixed and that we wish to maximize the AI given by Equation 2.3. In the region in which $S(F)/N(F)$ is between −12 and +18 dB, $P(f)$ is given by

$$P(f) = \frac{12 + 10 \log_{10}\left[\dfrac{S(f)}{N(f)}\right]}{30} = K_1 + K_2 \ln\left[\frac{S(f)}{N(f)}\right] \qquad (2.9)$$

where K_1 and K_2 are constants. The integral to be maximized therefore becomes

$$\text{AI} = \int_{f_a}^{f_b} W(f)K_2 \ln[S(f)/N(f)]df + K_3 \tag{2.10}$$

The details of the derivation are omitted, but the optimum spectrum is

$$S_0(f) = K_4 W(f) \tag{2.11}$$

where K_4 is a constant. Surprisingly, this optimum speech spectrum is independent of $N(f)$ within the range of $S(f)/N(f)$ from -12 to $+18$ dB.

If $S(f)/N(f)$ is above $+18$ dB at some frequency, the useless excess speech power should be redistributed throughout the rest of the frequency range according to Equation 2.11. Similarly, if $S(f)/N(f)$ is below -12 dB at some frequency, then this wasted power should be completely removed and redistributed throughout the remaining frequency range according to Equation 2.11. If the initial bandwidth is 200 to 6100 Hz, the resulting speech spectrum will have both the optimum preemphasis and bandwidth.

In a typical communications system, $\overline{S/N}(f)$ at the receiver is not known a priori and the design must be optimized for some assumed value. In most practical situations the speech spectral density given by Equation 2.11 is nearly optimum without power redistribution. The corresponding preemphasis is thus

$$|H(f)|^2 = \frac{W(f)}{V(f)} \tag{2.12}$$

where $V(f)$ is given by Equation 2.5 and $W(f)$ is given by Equation 2.1. The exact preemphasis is not critical, and an adequate approximation to the optimum is a preemphasis of 6 dB/octave above 500 Hz.

Figure 2.4 shows the AI versus average signal-to-noise ratio $\overline{S/N}$ for speech with optimum preemphasis in white noise, while Figure 2.5 shows the AI versus $\overline{S/N_0}$. It will be noted that an AI of 0.3 is obtained with a 2- to 5-dB lower signal-to-noise ratio than for unpreemphasized speech.

Nonwhite noise. In the case of FM systems operating above the FM improvement threshold, the audio output noise power spectral density is not white but rather increases as the square of the audio frequency (f^2-type noise). Articulation index curves for speech in f^2-type noise are shown with and without preemphasis in Figures 2.6 and 2.7.

Peak speech power. Up to this time, we have related intelligibility to the long-term average speech power for continuous speech $\overline{S}$. However, most

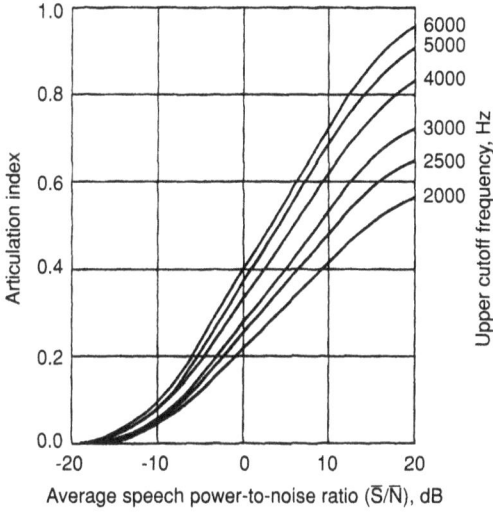

Figure 2.4 Intelligibility of preemphasized speech in white noise with 200-Hz lower cutoff frequency versus signal-to-noise ratio.

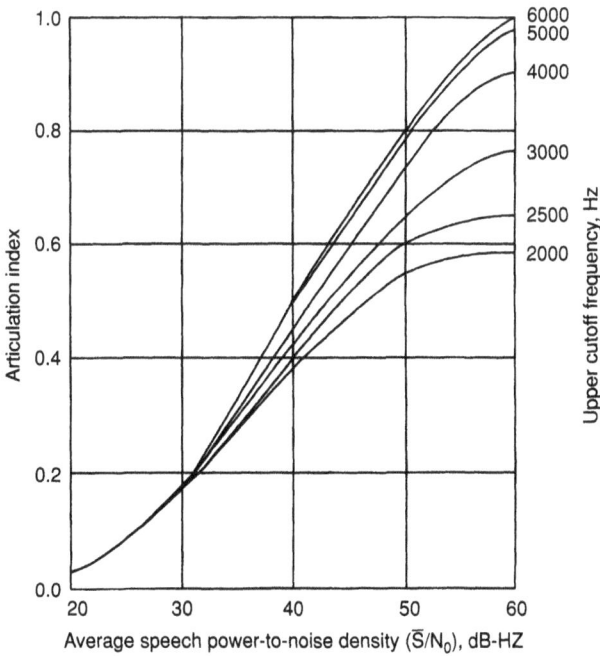

Figure 2.5 Intelligibility of preemphasized speech in white noise with 200-Hz lower cutoff frequency versus signal-to-noise density ratio.

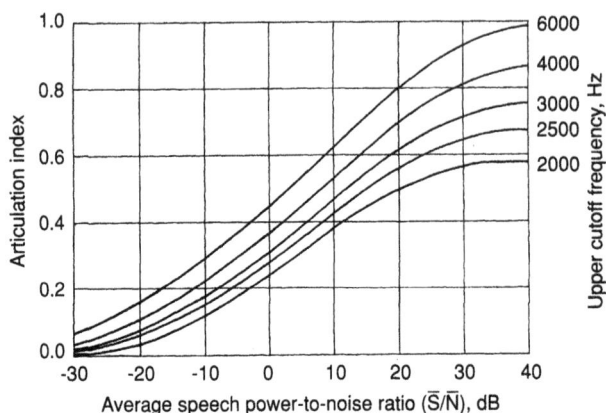

Figure 2.6 Intelligibility of normal speech in f^2 noise with 200-Hz lower cutoff frequency versus signal-to-noise ratio.

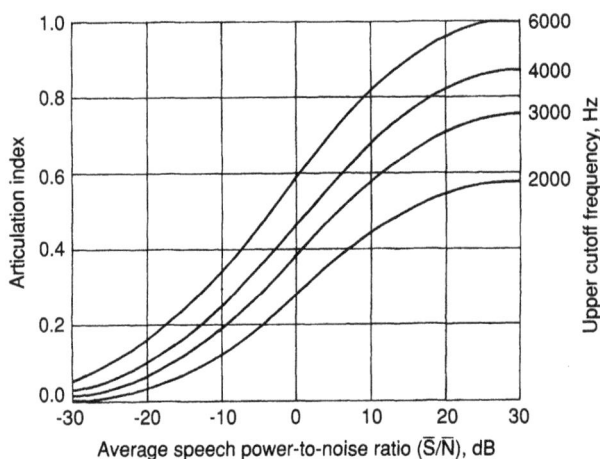

Figure 2.7 Intelligibility of preemphasized speech in f^2 noise with 200-Hz lower cutoff frequency versus signal-to-noise ratio.

modulation systems are peak power-limited. The results obtained for average speech power $\overline{S}$ can be extended to peak instantaneous speech power S_p by means of the peak-to-average power ratio given by

$$R = \frac{S_p}{\overline{S}} \tag{2.13}$$

Since it is difficult to define an absolute instantaneous peak power, we use the value that is exceeded only 0.01% of the time. The ratio R for normal speech is 14.5 dB. Optimum preemphasis increases the peak-to-average power ratio by 4.5 dB, thus generally offsetting the 2- to 5-dB reduction in the required average power. Preemphasis alone therefore does not generally improve communications efficiency for peak power-limited systems.

Peak clipping. The peak-to-average power ratio of speech can be reduced by symmetrical clipping of the positive and negative voltage peaks. In the limit of infinite clipping, speech is reduced to square waves with a peak-to-average power ratio of 1 (0 dB). Clipping causes an increase in signal bandwidth because of the generation of distortion products. It is generally necessary, therefore, to refilter speech after clipping. This increases R somewhat, but there is still a significant overall reduction. Typical reductions in R with clipping and refiltering are shown in Figure 2.8. The amount of clipping is taken as the ratio of the power peaks before clipping to the instantaneous power level at which peak clipping commences. The amount of repeaking that occurs depends on the filter bandwidth, cutoff rate, and delay distortion. The results will therefore depend heavily on the filter used.

Surprisingly, the distortion introduced by clipping has little effect on intelligibility [8]. Although it reduces the power level fluctuation with time, it has little effect on the shape of the short-term power spectrum or the location of the formant frequencies. It is generally assumed that peak clipping does not affect the intelligibility versus signal-to-noise ratio curves (Figures 2.2 through 2.7) for a range of preemphasis of from 0 to 12 dB/octave [8]. Beyond these limits the attenuated high- or low-frequency portions of the spectrum will be masked by clipper distortion products, and the intelligibil-

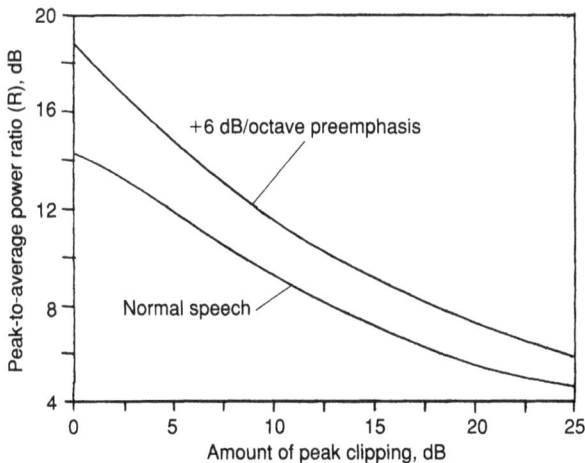

Figure 2.8 Peak-to-average power ratio for speech versus the amount of symmetrical audio peak clipping, with postclipping refiltering to bandwidth of 50 Hz to 12 kHz.

ity is therefore degraded. This masking effect is minimized by preemphasis of 6 to 12 dB/octave above 500 Hz. This means that the optimum preemphasis to minimize distortion after clipping is approximately the same as that for average power-limited systems without clipping.

Peak clipping degrades the subjective quality of speech, primarily because of the increase in background noise between syllables and words. While infinite clipping greatly increases communications power efficiency for peak power-limited systems, the intersyllable noise is annoying and fatiguing. In general, clipping levels significantly above 24 dB result in unacceptable levels of intersyllable noise. Extreme clipping is acceptable only for emergency communications where peak power efficiency is the only criterion. For most applications, optimum preemphasis, followed by a modest amount of some 12 to 24 dB of symmetrical peak clipping, will provide a good compromise between high communications efficiency and fair subjective quality for peak power-limited systems.

For peak power-limited systems, optimum preemphasis followed by 21 dB of audio peak clipping will reduce the power required to achieve an AI of 0.3 by 9 to 12 dB. By contrast, for average power-limited systems, optimum preemphasis will reduce the power required to obtain an AI of 0.3 by 2 to 5 dB, while peak clipping will have no beneficial effects.

In conclusion, peak clipping can improve the intelligibility of peak power-limited speech at high noise levels but only at a sacrifice in subjective speech quality. It is therefore recommended only in cases where communication is likely to be marginal and is vital regardless of quality. Clipping will not improve intelligibility or quality in average power-limited systems or if the clipping is done after the point in the system where the noise is introduced. Therefore, if excessive noise is present at the transmitter audio input, then peak clipping should not be used at the transmitter. Also, since most of the end-to-end link noise is usually added prior to the receiver audio output, peak clipping should not normally be used at the receiver.

Volume compression

A volume compressor is a gain-controlled amplifier that measures the short-term average output signal envelope and adjusts the gain in an attempt to keep that level constant. Typically the gain reduction or attack time constant is on the order of tens to hundreds of milliseconds and the gain increase or release time constant is on the order of hundreds of milliseconds. Peak limiters are similar except that the attack time constant is no more than a few milliseconds. Volume compressors and peak limiters do not generate significant distortion, degrade speech quality, or reduce the peak-to-average power ratio of their output significantly. They are used to compensate for differences in speech level inputs of different speakers and to avoid transmitter overload by large peaks in the audio input signal. The only subjective deleterious effect is that they will bring the background noise up during prolonged periods of silence. The shorter the release time constant the faster the background noise level comes up.

Compandoring

A compandor is a volume compressor and expandor that compresses the speech volume range into the transmitter and then expands the volume range out of the receiver in order to reestablish the original range of volume [9]. The compressor will reduce the peak-to-average power ratio significantly if the attack and release time constants are no more than a few milliseconds. In the limit, as the attack and release time approach zero, the compressor becomes a clipper. The expandor restores the original range of volume, thus quieting the output between syllables and providing high subjective speech quality. However, to achieve this high quality the signals used for the control of the expandor must have a good signal-to-noise ratio.

The compressors use a gain control device whose voltage gain A is inversely proportional to a control voltage or current I so that

$$A = K/I \qquad (2.14)$$

where K is a constant. If the rectified output envelope E is used to control the device gain, then $I = E$ and the output voltage envelope, under steady-state conditions, is given by

$$E = eA = e\left(\frac{K}{I}\right) = e\left(\frac{K}{E}\right) \text{ or } E^2 = eK \text{ or } E = \sqrt{eK} \qquad (2.15)$$

where e is the input envelope level. The output envelope changes (in decibels) are halved. In other words, if there is a 2-dB change in the input envelope level there will only be a 1-dB change in the output envelope level. This is called a 2:1 compression ratio. By cascading two compressors a 4:1 compression ratio is obtained.

The gain control device for an expandor is usually made to have a gain that is directly proportional to the control voltage or current I so that

$$A = IK \qquad (2.16)$$

and the output envelope E is related to the input envelope e by

$$E = eA = eKI = eKe = e^2K \qquad (2.17)$$

The output envelope level changes (in decibels) are doubled. In other words, a 1-dB change in the input envelope will cause a 2-dB change in the output envelope. Cascading two such expandors provides a 4:1 expansion ratio.

Thus compressors at the transmitter and expandors at the receiver will provide an output with the same envelope level variations as the input, while still reducing the peak-to-average power of the modulating signal.

Figure 2.9 shows a system using the 2:1 compandor described above. Such systems are used on FM links but have yet to be proven on SSB links because the link gain fluctuations due to fading and imperfect receiver AGC are exaggerated by the expandor. In addition, it would be desirable to suppress the output level variations completely. Chapter 7 gives further discussion regarding the implementation of an SSB compandor.

If the gain of the compressors is controlled by the input envelope as shown in Figure 2.10, then the final output envelope V is given by

$$V = Ae = \left(\frac{K}{I}\right) e = \left(\frac{K}{e}\right) e = K \qquad (2.18)$$

Thus the output envelope is a constant K regardless of the input magnitude. Of course, since the gain of the gain control device cannot go to infinity, there is a lower input signal limit below which the output level will change

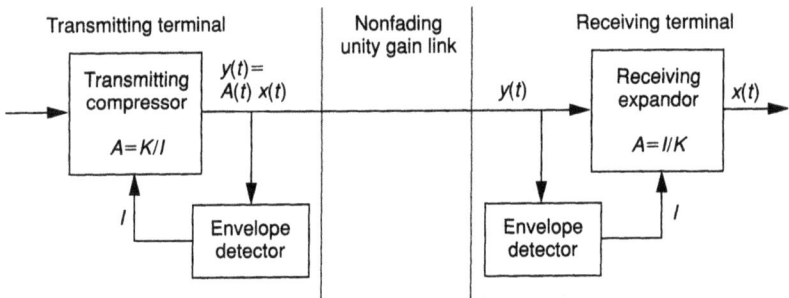

Figure 2.9 Compandor link with single-stage compressor and expandor.

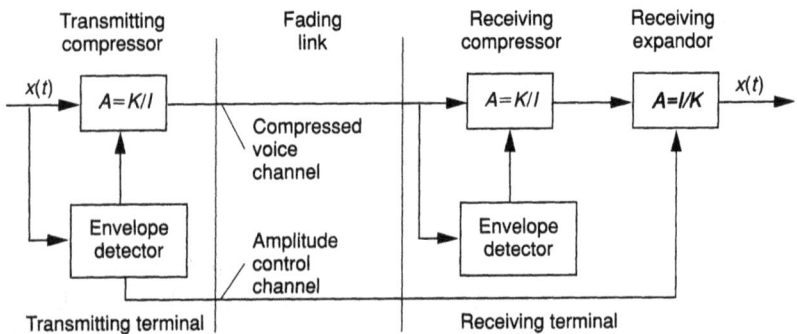

Figure 2.10 Compandored link with complete envelope compression.

in direct proportion to the input signal level. The range of compression and expansion can be extended by placing 2:1 compressors and expandors in front of the compressors and expandors in Figure 2.10.

Volume expansion at the output of the receiver is achieved by using an expandor as shown in Figure 2.10. Since the transmitted signal has no level fluctuations its envelope can no longer be used to control the gain of the expandors. It is necessary therefore to send a separate signal to control the voltage gain of the expandor. Before volume expansion, the received signal envelope into the expandor must be made a constant K so that fading will not introduce volume fluctuations. This is achieved by a fading compressor at the receiver, preceding the expandor.

Representation of signals

Modern signal processing systems often use 2-wire, 2-phase circuits and processor algorithms. A familiarity with 2-phase methods is therefore essential.

A sine-wave signal can be represented by the projection of a rotating vector on the horizontal, in-phase or I axis (see Figure 2.11) and is given by

$$i(t) = \rho\cos(\omega t + \theta) \tag{2.19}$$

where ρ is peak amplitude, ω is frequency, t is time and θ is initial phase. The rotating vector's projection on the vertical, quadrature-phase, or Q axis is

$$q(t) = \rho\sin(\omega t + \theta) \tag{2.20}$$

As time progresses this Cartesian vector $[i(t),q(t)]$ rotates counterclockwise at an angular velocity ω. The signals $i(t)$ and $q(t)$ are on the I and Q axis or wires, respectively. This rotating vector can be written in real Cartesian coordinates or in complex form, where the real part designates the in-phase or I signal and the imaginary part designates the quadrature, or Q, signal. The rotating vector's various forms are:

Cartesian Vector: $\mathbf{X}(t) = [i(t),q(t)] = [\rho\cos(\omega t + \theta) , \rho\sin(\omega t + \theta)]$

Complex Cartesian: $x(t) = \rho\cos(\omega t + \theta) + j\rho\sin(\omega t + \theta)$ $\qquad$ (2.21)

Complex Polar: $x(t) = \rho e^{j(\omega t + \theta)}$ (by Euler's Theorem)

where bold letters are Cartesian vectors and italicized letters are complex vectors. The imaginary multiplier j in the complex Cartesian form has no physical significance except as a label for the quadrature wire or term. Phasors, as used in AC circuit theory, and complex 2-phase notation are

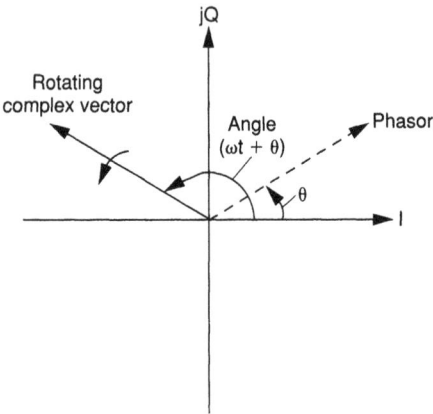

Figure 2.11 Rotating complex vector representation of a sine wave.

identical in form but have important differences in interpretation since phasors apply *only* to 1-wire circuits. Phasor notation will *not* be used here.

A counterclockwise rotation is called a "positive frequency." A "negative-frequency," clockwise-rotating vector, in Cartesian coordinates, is

$$\mathbf{Y}(t) = [\rho\cos(-\omega t - \theta), \rho\sin(-\omega t - \theta)] = [\rho\cos(\omega t + \theta), -\rho\sin(\omega t + \theta)] \quad (2.22)$$

The I and Q signals exist on separate wires and both have real physical significance. Positive- and negative-frequency signal components are distinct, independent and separable and any $[i(t), q(t)]$ signal can be resolved into the sum of separate rotating vectors, representing the different frequency components in the 2-phase signal.

A 1-wire, 2-phase sine-wave signal is obtained by adding the two counter-rotating vectors $\mathbf{X}(t)$ and $\mathbf{Y}(t)$, so that the quadrature components cancel as shown in Figure 2.12 and Equation 2.23.

$$\mathbf{Z}(t) = \mathbf{X}(t) + \mathbf{Y}(t) = [2\rho\cos(\omega t + \theta), 0] \quad (2.23)$$

The quadrature component for 1-wire, 1-phase signals must of course be zero. It follows that any 1-wire signal decomposes into equal postive- and negative-frequency components so as to cancel the quadrature term. However, when this signal is decomposed into a sum of rotating vectors, each rotating vector has a quadrature component. Therefore the quadrature term must be retained during mathematical analysis even though $q(t)$ is zero. There is no need to associate complex numbers with 2-phase signals since both wires are real, with real physical signals on them. However it is occasionally convenient to represent signals in complex form by associating the Q wire with the imaginary or j part of a complex number. This association will be used here in modeling frequency translation.

In 1-wire AC circuit analysis using phasors, sine-wave signals are represented by positive-frequency, rotating complex vectors. After analysis the signals are converted back to real, 1-wire signals by taking the real part of the result. By contrast, in the complex 2-phase method and in Fourier transform analysis a 1-wire, sine-wave signal is represented by a pair of counterrotating complex vectors giving the output directly.

Hilbert transform

The Hilbert transform performs a broad-band 90° phase lag to each frequency component of the input signal $i(t)$. The output $q(t)$ is

$$q(t) = \frac{1}{\pi} \int_{-\infty}^{+\infty} \frac{i(t-\tau)}{\tau} d\tau = \lim_{\varepsilon \to 0} \left[\frac{1}{\pi} \int_{-\infty}^{-\varepsilon} \frac{i(t-\tau)}{\tau} d\tau + \frac{1}{\pi} \int_{+\varepsilon}^{+\infty} \frac{i(t-\tau)}{\tau} d\tau \right] \qquad (2.24)$$

This operation corresponds to passing the input signal through a linear filter with an impulse response $1/t$. The 1-wire input signal $i(t)$ of course has matching positive and negative-frequency components. The 2-wire signal vector $[i(t),q(t)]$ however has only positive frequency components since for every frequency component in $i(t)$, $q(t)$ contains a component of the same frequency and amplitude but lagging by 90°. Such special signals containing only positive frequency components are called *analytic signals*.

Frequency translation

Frequency translation corresponds to multiplication of the signals $x(t)$ and $y(t)$ expressed in complex notation. If both of these are rotating vectors, this product is a single frequency rotating complex vector $z(t)$

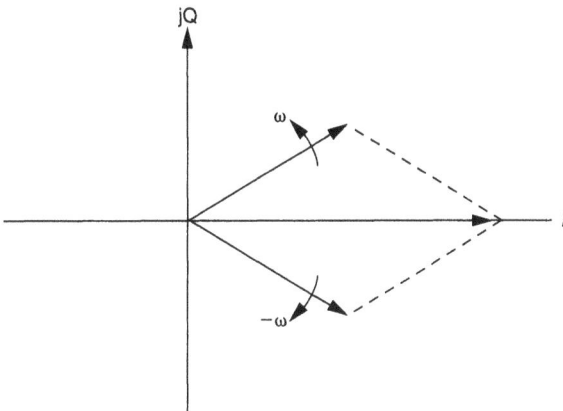

Figure 2.12 A purely in-phase signal generated by equal counterrotating phasors.

$$z(t) = x(t)y(t) = [i_x(t) + jq_x(t)][i_y(t) + jq_y(t)]$$
$$= [\cos(\omega_x t) + j\sin(\omega_x t)][\cos(\omega_y t) + j\sin(\omega_y t)] \qquad (2.25)$$
$$= \cos(\omega_x + \omega_y)t + j\sin(\omega_x + \omega_y)t = i_z(t) + jq_z(t)$$

This complex multiplication produces only the sum frequency output. An electrical implementation is shown in Figure 2.13 where the real and imaginary components correspond to I and Q wires respectively for the input, injection and output.

For frequency translation the output spectrum is the convolution of the signal and injection spectra. The convolution can be performed graphically by offsetting the signal spectrum by the amount of the injection frequency. If the injection is 1-wire, 1-phase the ouput spectrum is the sum of the signal spectra for both positive and negative offsets as shown in figure 2.20.

SSB filters

A single sideband filter passes only positive frequency signals. It is implemented as shown in Figure 2.14. The performance can be readily verified by inputting a positive-frequency vector [I=cos(ωt), Q=sin(ωt)] and a negative frequency vector [I=cos(-ωt), Q=sin(-ωt)] as test signals. Carrying out the trigonometric operations, the positive-frequency input produces the output vector [I=2cos(ωt), Q=2sin(ωt)] and the negative frequency input produces the output vector [I=0, Q=0]. To pass only negative-frequency signals, change the signs into the adders as shown in   Figure 2.14. Since any physical

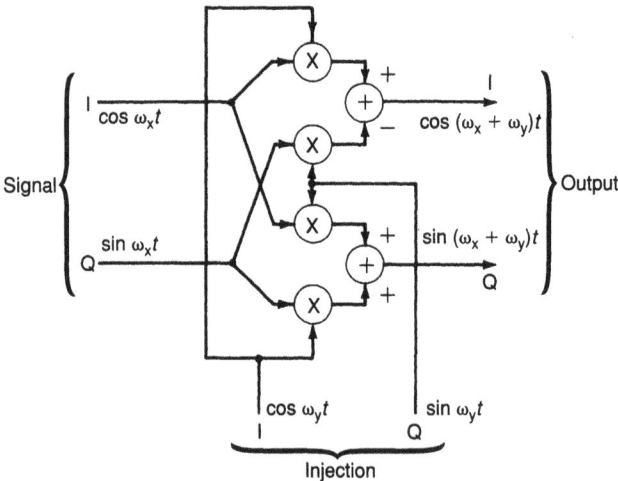

Figure 2.13 Implementation of a full complex mixer or frequency translator.

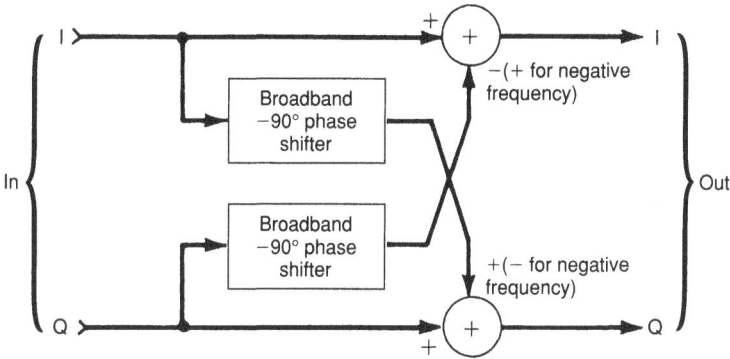

Figure 2.14 Filter that passes only positive-frequency components.

-90° phase shifter will delay the signals, matching delays, not shown in the figure, must be inserted in the straight-through I and Q paths.

AM family

The AM family includes full-carrier AM, double-sideband suppressed-carrier (DSBSC) modulation, single-sideband (SSB) modulation, and amplitude equivalent modulation (AME). We will discuss each of these forms of modulation and demodulation next.

Amplitude modulation. In full-carrier AM, the amplitude of an RF sine-wave carrier is made to vary in direct proportion to the modulating function, as shown in Figure 2.15. Amplitude modulation may be produced by multiplying a sine-wave carrier $A \cos (2\pi f_c t)$ by a modulating function $m(t)$ plus a direct current (dc) offset:

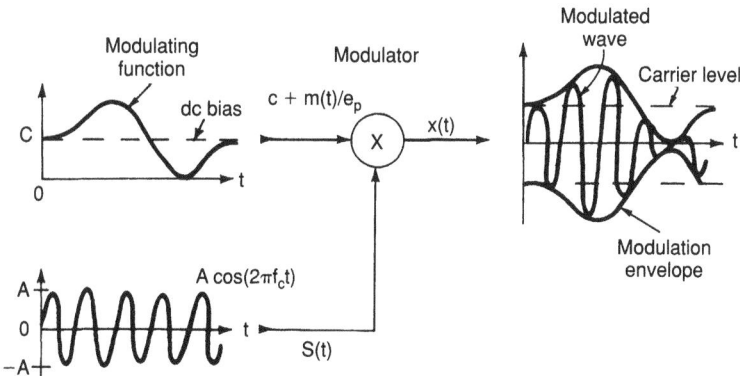

Figure 2.15 AM modulator block diagram.

$$x(t) = \underbrace{\left[1 + \frac{m(t)}{e_p}\right]}_{\substack{\text{Modulation}\\\text{envelope}}} A \cos(2\pi f_c t)$$

(2.26)

$$= \underbrace{A \cos(\omega_c t)}_{\text{Carrier}} + \underbrace{A \frac{m(t)}{e_p} \cos(\omega_c t)}_{\text{Sidebands}}$$

where A = peak amplitude of the sine wave carrier
$m(t)$ = modulating function
e_p = peak value
f_c = carrier frequency
ω_c = radian carrier frequency

Generally, AM is demodulated by a rectifier or envelope detector. Since the output of such a demodulator depends only on the absolute voltage of the modulated signal, severe output distortion will result if the modulating voltage into the multiplier goes negative. It is therefore necessary to add a dc component to the modulating signal.

Multiplication of the modulating waveform by a sine-wave carrier is equivalent to frequency translation of the modulating frequency by an amount f_c.

If the modulating waveform is represented by the Fourier series

$$\sum_{i=1}^{N} = a_i \cos(\omega_i t + \theta)$$

the modulated waveform is

$$x(t) = A\left(1 + \sum_{i=1}^{N} a_i \cos(\omega_i t + \theta)\right) \cos(\omega_c t)$$

(2.27)

Multiplying the terms and using trigonometric identities gives

$$x(t) = \underbrace{A \cos(\omega_c t)}_{\text{Carrier}} + \underbrace{\frac{1}{2} A \sum_{i=1}^{N} a_i \cos[(\omega_c - \omega_i)t - \theta_i]}_{\text{Lower sidebands}}$$

(2.28)

$$+ \underbrace{\frac{1}{2} A \sum_{i=1}^{N} a_i \cos[(\omega_c + \omega_i)t + \theta_i]}_{\text{Upper sidebands}}$$

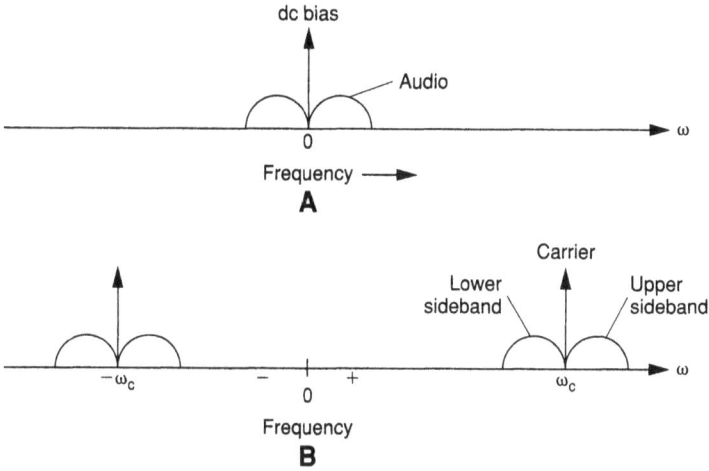

Figure 2.16 Spectra of AM waveforms. Spectrum of (A) modulating and (B) modulated waveforms.

Spectra of typical modulating and modulated waveforms are shown in Figure 2.16.

A corresponding rotating vector representation of a single tone-modulated waveform is shown in Figure 2.17. The upper and lower sideband rotating vectors combine vectorially to add or subtract in phase with the carrier vector so that the overall resultant is a vector in phase with the carrier vector but having a sine-wave fluctuation in amplitude. For more general modulating functions, the upper and lower sideband vectors still combine to form a resultant vector in phase with the carrier and having an amplitude fluctuation proportional to the modulating function.

From Equation 2.26, the carrier power is $P_c = A^2/2$. The average sideband power P_{sb} is

$$P_{sb} = \frac{A^2}{2} \frac{E[m^2t]}{e_p^2} = \frac{A^2}{2} \frac{1}{R_m} = \frac{P_c}{R_m} \tag{2.29}$$

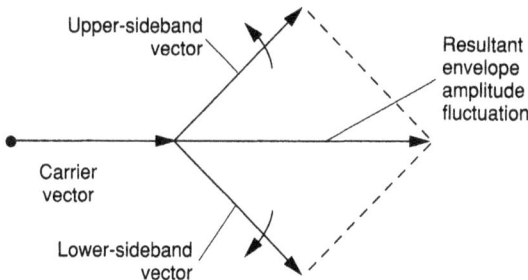

Figure 2.17 AM rotating vector diagram (single-tone sine-wave modulation).

where $E[\,]$ = average or expected value operator
R_m = peak-to-average power ratio of the modulating function

The total average power P_{avg} is

$$P_{avg} = P_c + P_{sb} = P_c\left(1 + \frac{1}{R_m}\right) = \frac{P_c(1 + R_m)}{R_m} \tag{2.30}$$

For 100% modulation where the peak value of $m(t)$ equals e_p, the peak voltage is $2A$ and the peak instantaneous output power P_p is

$$P_p = 4A^2 = 8P_c \tag{2.31}$$

The receiver audio output signal power is proportional to the received sideband power, so that the output signal-to-noise ratio is maximized by maximizing sideband power. Most transmitters are, however, peak-power-limited rather than average-power-limited, and it is therefore desirable to relate the sideband power to peak and average power. From Equations 2.29 through 2.31,

$$P_{sb} = \frac{P_{avg}}{1 + R_m} \tag{2.32}$$

and

$$P_{sb} = \frac{P_p}{8R_m} \tag{2.33}$$

Transmitter peak power ratings are usually specified in terms of peak envelope power P_{ep} which is half the true peak instantaneous power rating P_p. We shall henceforth give peak ratings in terms of peak envelope power. The relationships between peak envelope, average, carrier, and sideband power for the AM family are summarized in Table 2.1.

DSBSC modulation. Double-sideband suppressed-carrier modulation is identical to full-carrier AM except that the carrier frequency is partially or totally suppressed. For full-carrier suppression the wave is given by

$$x(t) = A\frac{m(t)}{e_p}\cos(2\pi f_c t) \tag{2.34}$$

which is equivalent to AM except that the modulating function is not provided with a dc bias term. When the modulating function changes sign, the modulated carrier makes sudden 180° phase shifts.

TABLE 2.1 Power Relationships for the AM Family

Type of modulation	Equation For	Equation In terms of	Equation
AM	P_{sb}	P_c	$P_{sb} = P_c/R_m$
		P_{avg}	$= P_{avg}/(1 + R_m)$
		P_{ep}	$= P_{ep}/(4R_m)$
	P_{avg}	P_c	$P_{avg} = P_c(1 + 1/R_m)$
	P_{ep}	P_c	$P_{ep} = 4P_c$
DSBSC	P_{sb}	P_{avg}	$P_{sb} = P_{avg}$
		P_{ep}	$P_{sb} = P_{ep}/R_m$
DSB with pilot carrier	P_{sb}	P_{avg}	$P_{sb} = P_{avg}(1 - P_c/P_{avg})$
		P_{ep}	$= P_{ep}(1 - \sqrt{P_c/P_{ep}})^2/R_m$
	P_{ep}	P_{sb} and P_c	$P_{ep} = (\sqrt{R_m P_{sb}} + \sqrt{P_c})^2$
SSB	P_{sb}	P_{avg}	$P_{sb} = P_{avg}$

Unlike AM, the absolute value of the envelope is anything but an undistorted version of the modulating waveform, and an envelope detector is not a suitable demodulator. The spectrum of the modulated signal is the same as for AM, as shown in Figure 2.16, except that the carrier frequency is absent.

Since all the power is now in the sidebands, the sideband power is equal to the average power:

$$P_{sb} = P_{avg} = \frac{A^2}{2} \frac{E[m^2(t)]}{e_p^2} = \frac{A^2}{2R_m} \qquad (2.35)$$

Since the maximum value of $m(t)/e_p$ is 1, the peak power is

$$P_p = A^2 = 2R_m P_{sb} \qquad (2.36)$$

where the last expression was obtained by solving Equation 2.35 for A^2 and substituting. The peak envelope power is therefore

$$P_{ep} = R_m P_{sb} \qquad (2.37)$$

The power relationships are summarized in Table 2.1. It should be noted from this table that the sideband power is always greater for a given peak envelope or average power for double sideband (DSB) than for AM because no power is wasted in transmission of the carrier. For synchronous detection, the output signal-to-noise ratio is proportional to the sideband power, and DSB is therefore more efficient than AM.

While DSB can be synchronously demodulated without a carrier, a pilot carrier is sometimes added as an aid to synchronous detection. The carrier has values between full AM and zero. In the general case, a DSB signal with pilot carrier is given by

$$x(t) = A\left(c + \frac{m(t)}{e_p}\right) \cos\left(2\pi f_c t\right) \tag{2.38}$$

The relationships between peak envelope, average, sideband, and carrier power can be solved by means similar to those used for AM and DSBSC, and the results are summarized in Table 2.1.

SSB modulation. Full-carrier AM consists of a carrier plus upper and lower sidebands. In SSB both the carrier and one sideband are removed so that only the upper or lower sideband remains. The three common methods of SSB generation are the filter method, the phase-shift method, and the Weaver method. Today the filter method is used almost exclusively in analog circuit implementations, while the phase-shift and Weaver methods are finding favor in digital circuit implementations.

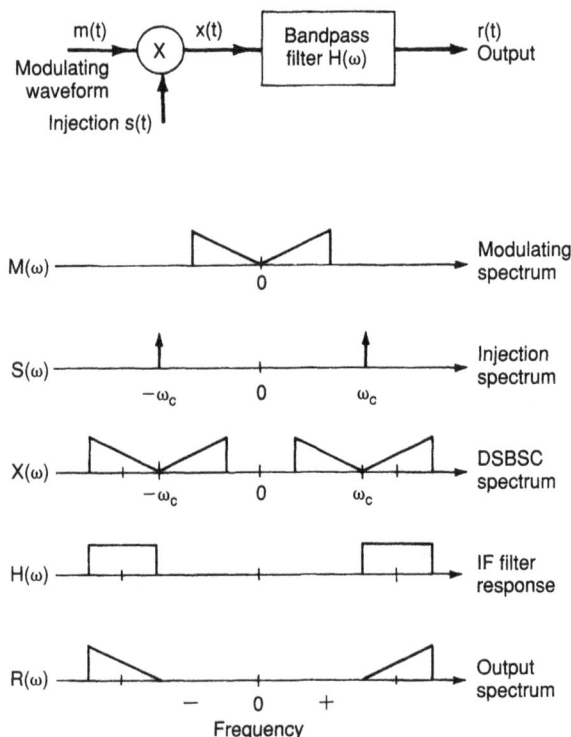

Figure 2.18 The filter method of SSB modulation.

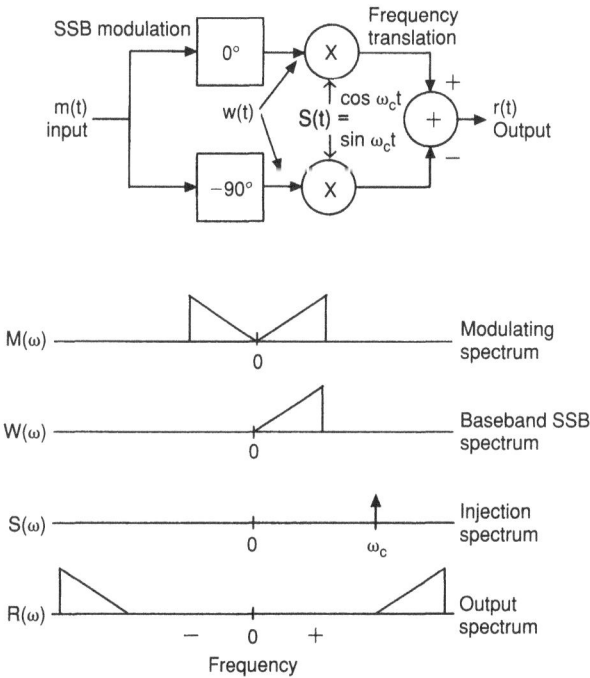

Figure 2.19 Phase-shift method of SSB modulation.

In the filter method an AM or DSBSC signal is generated and the result is bandpass filtered to remove one sideband and any carrier (see Figure 2.18 where the frequency spectrum of $x(t)$ is $X(\omega)$, etc.).

In the phase-shift method of Figure 2.19 a 2-phase baseband SSB signal is generated as discussed under **Hilbert transform** and the result is frequency translated to the desired RF frequency as discussed under **Frequency translation**. Since only a 1-wire, 1-phase signal is required at RF, only half of the full 2-phase frequency translator is implemented and the RF output spectrum is therefore two-sided.

In the Weaver method the center of the audio passband is mixed to zero frequency so that the 2-phase audio sideband extends over equal positive- and negative-frequency bandwidths. The lower sideband is removed by low-pass filtering both the I and Q wires. The resultant 2-phase baseband signal is translated to RF. Again, since only the I output of the 2-phase mixer is implemented the output spectrum is two sided.

It is readily apparent that the SSB frequency spectrum is the single-sided audio spectrum translated to an RF or IF. While the modification of the signal spectrum is quite simple, the modifications of the signal time waveform and envelope are quite complex. This is most easily seen by studying the phase-shift method shown in Figure 2.19, where the output is

$$r(t) = m(t) \cos \omega_c t - \hat{m}(t) \sin \omega_c t$$

$$= \underbrace{\sqrt{m^2(t) + \hat{m}^2(t)}}_{\text{SSB envelope}} \cos [\omega_c t + \phi(t)] \qquad (2.39)$$

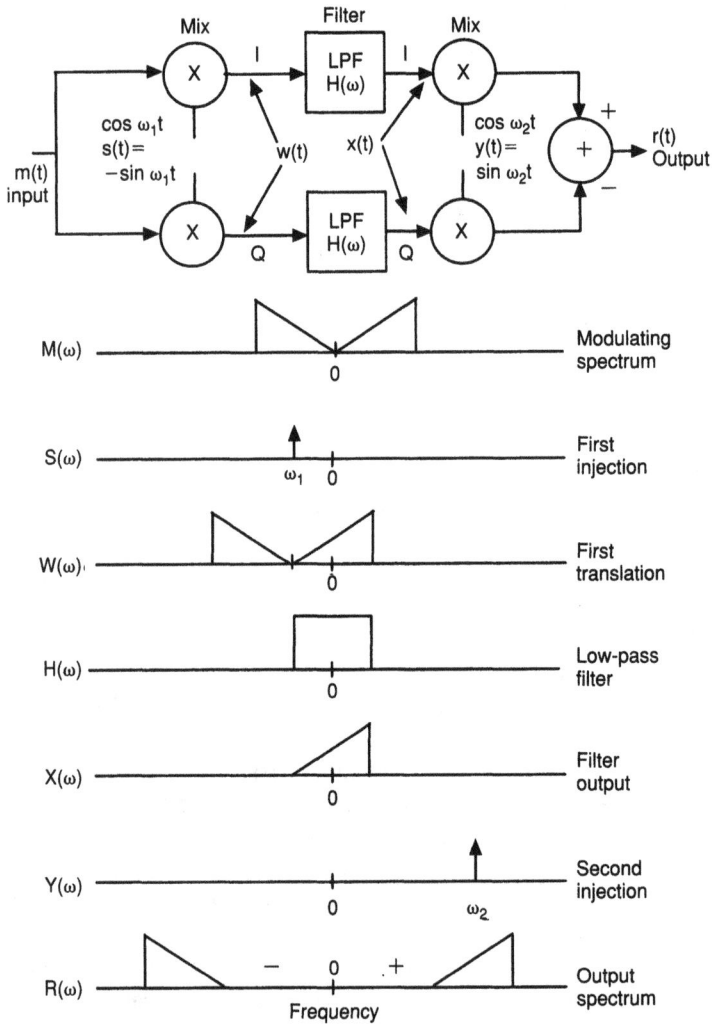

Figure 2.20 Weaver method of SSB modulation.

where $\hat{m}(t)$ is the Hilbert transform of $m(t)$ and $\phi(t)$ is the four-quadrant arctangent given by

$$\phi(t) = \arctan\left[-\hat{m}(t), m(t)\right] \tag{2.40}$$

Since all the power is in the SSB,

$$P_{sb} = P_{avg} \qquad P_c = 0 \tag{2.41}$$

As noted in Equation 2.39 the SSB waveform, or any other RF waveform for that matter, can be represented either in terms of the single time function $r(t)$ or in terms of an instantaneous envelope $e(t)$, an instantaneous phase $\phi(t)$, and a reference frequency ω_c. In some cases this type of representation is used in actual implementations. In digital implementations the signal is often represented in terms of the three components—envelope, phase, and reference frequency—while in analog implementations, such as for example in envelope elimination and reconstruction EER, the envelope is represented by one audio channel while the reference frequency and phase are represented by a constant amplitude, phase-modulated RF channel [10–13]. In the latter case the phase of the RF signal is retained without the envelope by limiting the original RF signal and amplifying it in a high-efficiency, constant-amplitude RF power amplifier. The envelope of the original RF signal is extracted and amplified in a high-efficiency AF power amplifier in the envelope path. The AF power amplifier is used as a high-level amplitude modulator for the RF power amplifier to restore the envelope on the RF output signal.

It is interesting to note that even though the original RF signal is bandwidth limited, the envelope and phase functions may not be bandwidth limited. For example, if the original RF signal consists of two equal amplitude sine waves, the envelope is a fullwave rectified sine wave of the difference frequency. The envelope function is no longer bandwidth limited due to the sharp cusps when the envelope goes to zero. At these same points in time the phase function makes instantaneous 180° phase flips. Of course it is not necessary to provide an infinite bandwidth in the envelope and phase paths, but it is necessary to provide much wider bandwidth paths than required for the original RF signal in order to avoid producing excessive out-of-band intermodulation distortion products. It is also necessary to closely match the time delays in the envelope and phase signal paths.

If $m(t)$ is a single sine-wave tone, $\cos \omega_a t$, the Hilbert transform is $+\sin \omega_a t$ and the SSB envelope is a constant. Moreover, the SSB waveform is a sine wave, so that the peak-to-average power ratio of the SSB signal is equal to that of the modulating waveform. While the peak-to-average power ratio of SSB modulation is never smaller than the modulating function, it can be much greater. An extreme case is a wideband square-wave modulating function that alternates periodically between plus and minus 1 and for which the peak-to-average power ratio R is unity (0 dB). However, the Hilbert transform of an ideal square wave has infinite peaks, and therefore the peak-to-average power ratio for ideal square-wave SSB modulation is infinite, as shown in Figure 2.21. Any practical SSB modulation system, however, is bandwidth limited, and R for these systems is finite. A square-wave modulating function is of some interest in that it approximates, to a degree, an infinitely clipped speech waveform.

Most SSB systems have a ratio of high-frequency to low-frequency cutoff of no more than 13 to 1, for which the peak-to-average ratio for square-wave modulation is 8 dB.

For waveforms of fluctuating amplitudes such as speech, however, audio clipping reduces the peak-to-average power ratio of the SSB waveform, as shown in Figure 2.22. The results will depend on the bandwidth and phase linearity of the postmodulation IF filter, and the results shown must be viewed as typical. Results for symmetrical peak clipping of the SSB IF waveform and refiltering are also shown in Figure 2.22. It should be noted that IF clipping is more effective for SSB modulation and can reduce the SSB peak-to-average power ratio by 6.5 dB, for 18 dB of clipping, as compared to 4.0 dB for audio clipping. The SSB results are much less dramatic than those shown in Figure 2.8, largely because of the narrow bandwidth and sharp cutoff of the postclipping IF filter.

In some cases a low-level pilot carrier is retained as a phase reference for synchronous SSB demodulation. The average power P_{avg} is the sum of the sideband power P_{sb} and the carrier power P_c.

$$P_{avg} = P_{sb} + P_c \qquad (2.42)$$

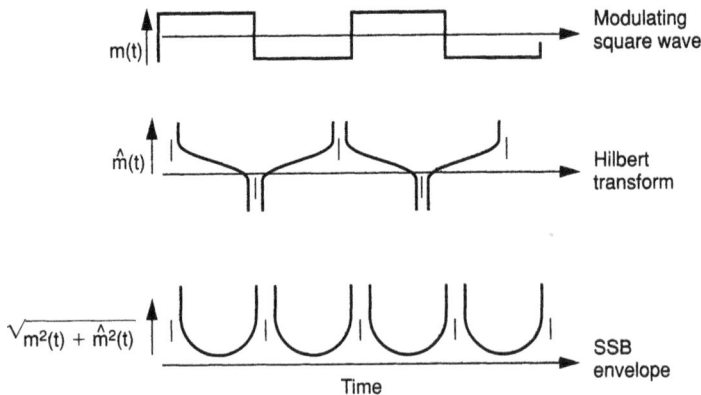

Figure 2.21 SSB square-wave modulation functions.

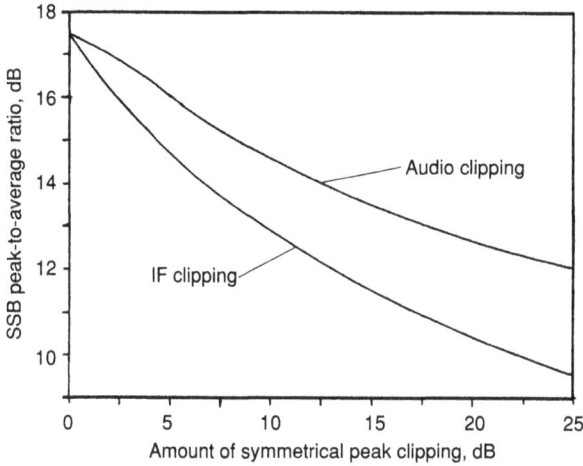

Figure 2.22 The effect of audio and IF peak clipping on the peak-to-average power ratio of SSB signals (no preemphasis; preclipping audio bandwidth is 70 to 12,000 Hz, postmodulation and post-IF clipping bandwidth is 300 to 3200 Hz).

while the peak envelope power P_{ep} is obtained from the sum of the peak sideband voltage and the peak carrier voltage or

$$P_{ep} = (P_{ps}^{1/2} + P_c^{1/2})^2 \tag{2.43}$$

where P_{ps} is the peak envelope power of the SSB signal without the pilot carrier.

AME modulation. Amplitude modulation equivalent consists of SSB with a carrier of substantial amplitude added so that the signal can be demodulated using a conventional envelope detector while occupying only half the bandwidth of conventional AM. The carrier level is typically set so that P_{ps} does not exceed the carrier level, as discussed in greater detail in Chapter 4. The average and peak envelope power are the same as for SSB with pilot carrier since only the relative magnitude of the carrier is different.

Demodulation

The demodulated audio output average signal-to-noise density ratio $(\overline{S}/N_0)_{AF}$ for all of the AM family is

$$\left(\frac{\overline{S}}{N_0}\right)_{AF} = \frac{P_{sb}}{N_{or}} \tag{2.44}$$

where P_{sb} is the total sideband power, and N_{or} is the RF noise power spectral density, both referred to the receiver input.

AM demodulation. Amplitude modulation or AME can be demodulated either by a synchronous detector or by an envelope detector. Synchronous detection gives maximum demodulation efficiency and gives the full performance indicated by Equation 2.44. In order to achieve this performance the recovered carrier must contain very little noise. The carrier can be recovered by using a phase-lock loop as discussed in the section titled "Pilot Carrier Recovery." Demodulation is achieved by multiplying the AM signal by the recovered carrier and lowpass filtering the result.

Amplitude modulation and AME can also be demodulated at IF using a simple envelope detector. At carrier-to-noise ratios below 0 dB there is appreciable loss in detection efficiency, and the performance of Equation 2.44 can no longer be achieved. However, for AM speech modulation the audio output signal-to-noise ratio will generally be too low to be useful before the carrier-to-noise ratio reaches 0 dB, and therefore a synchronous detector does not improve AM voice demodulator performance significantly unless the IF filter bandwidth is much greater than needed for the AM signal. Envelope demodulated AME will have some distortion since the envelope is not a perfect replica of the modulating waveform.

DSBSC demodulation. A DSBSC signal can be demodulated coherently by multiplying the IF signal by a re-created carrier of the proper frequency and phase, and then lowpass filtering the result. Under these conditions the demodulated signal will be a faithful reproduction of the modulating function. Multiplying the DSBSC signal of Equation 2.34 by a synthesized carrier, we have

$$x(t) \cos (2\pi f_c t) = A \frac{m(t)}{e_p} \left[\frac{1}{2} + \frac{1}{2} \cos (4\pi f_c t) \right] \qquad (2.45)$$

After lowpass filtering we obtain the demodulated output as

$$y(t) = \frac{A}{2 e_p} m(t) \qquad (2.46)$$

The required carrier for demodulating can be recovered either by using a frequency doubler or a Costas loop, described in the following paragraphs.

Frequency-doubler DSB demodulation. A frequency-doubler demodulator is shown in Figure 2.23. The received DSBSC signal is first squared in a frequency doubler. By squaring Equation 2.34, we obtain

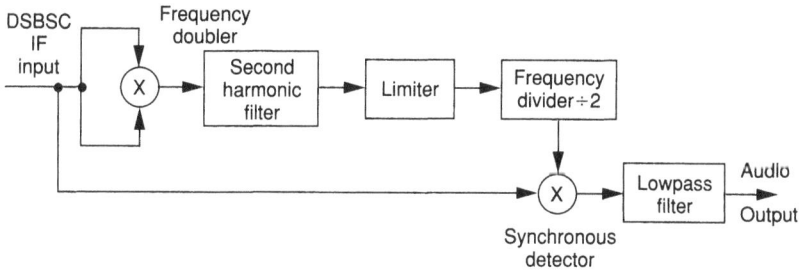

Figure 2.23 Frequency-doubler DSBSC demodulator.

$$y\,(t) = x^2\,(t) = A^2 \left(\frac{m\,(t)}{e_p} \right)^2 \cos^2\,(2\pi f_c t) \qquad (2.47)$$

$$= A^2 \left(\frac{m\,(t)}{e_p} \right)^2 \left[\frac{1}{2} + \frac{1}{2}\,\cos\,(2\pi 2 f_c t) \right]$$

While the original DSBSC waveform had abrupt 180° phase flips, there are no phase flips in the frequency doubler output because a 180° phase change of the input signal corresponds to 360° at the doubler output. The dc output component is removed by the second harmonic filter. This filter will also provide a flywheel or ringing action to provide a second harmonic component for periods when the signal may be zero. The filter output is limited or clipped to form a square wave which in turn drives a frequency divider to provide the synthesized carrier cos $(2\pi f_c t)$, which is used to synchronously demodulate the DSBSC signal. The filter action can also be provided by a phase-lock loop, as discussed below [14, 15].

Costas loop DSB demodulator. If the DSBSC signal includes a pilot carrier, the carrier can be recovered by using a very narrow phase-lock loop locked to the carrier. Since a phase-lock loop locks at 90° with respect to the input carrier phase, a 90° phase shifter is required in order to obtain an in-phase recovered carrier that can be used for synchronous demodulation, as shown in Figure 2.24. In the absence of a pilot carrier there are sudden 180° phase shifts and the phase-lock loop will continually break lock and relock. To avoid this, a phase deflipper is added in the loop so that the sense of the loop is reversed every time the signal phase changes 180°, thus avoiding loss of lock during phase flips. This Costas loop DSBSC demodulator is shown in Figure 2.24 [16].

SSB demodulation. Demodulation of an SSB signal is accomplished by mixing the received RF signal to audio. The average audio signal output power $\overline{S}$ is proportional to the RF input sideband power, while the audio

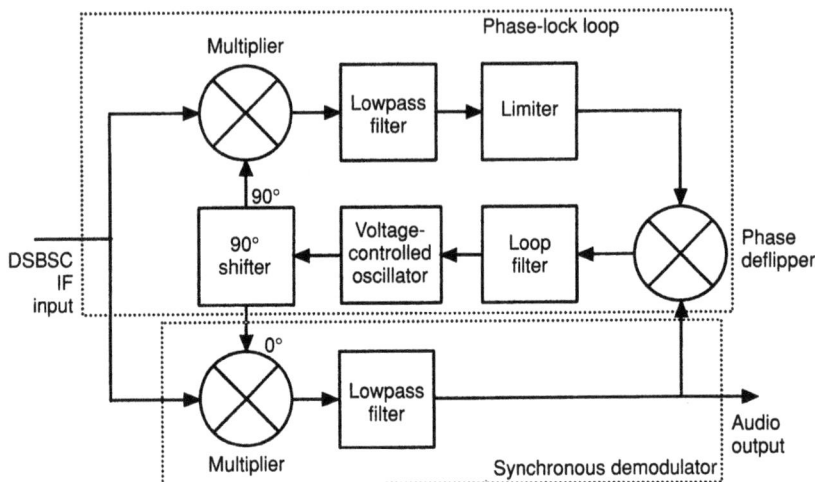

Figure 2.24 Costas DSBSC demodulator.

output noise density is proportional to the RF input noise density, thus giving Equation 2.44. For analog circuit implementations, the filter method of SSB demodulation is used almost exclusively. Here the IF signal is filtered to pass only the desired sideband and is then mixed to audio. If the mixer injection frequency and phase are exactly correct, the reconstructed audio waveshape will match the original modulating function. However, since there is no absolute phase reference in the case of SSB, the demodulated wave will not generally have the proper phase relationships between the frequency components, and the waveform may be severely distorted even though the power spectrum is unchanged. For voice communications these phase nonlinearities have no effect on intelligibility. In addition, there will be errors in the output frequencies because of injection frequency errors and Doppler shifts. Shifts of ±5 Hz have a small effect on speech subjective quality. For shifts between ±5 and ±50 Hz the quality suffers but intelligibility is not degraded significantly. The approximate AI reduction factor F due to a frequency offset is shown in Figure 2.25 [1, 17]. The AI for a frequency shift of ΔF is given by

$$\mathrm{AI}(\Delta f) = F(\Delta f)\,\mathrm{AI}(0) \qquad (2.48)$$

where AI (Δf) is the AI for a frequency shift of Δf.

Amplitude modulation and DSBSC can be detected by SSB demodulation of either the upper or lower sideband, but in so doing there is a 3-dB reduction in the audio output signal-to-noise ratio because of the loss of the coherent opposite sideband.

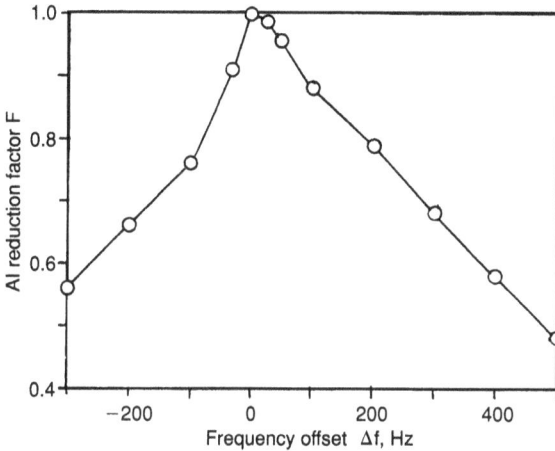

Figure 2.25 AI reduction factor versus frequency offset.

Pilot-carrier recovery

Manual frequency tuning of an SSB signal by "ear" is a tedious task requiring attention and skill. Moreover, if the tuning error is more than a few hertz then the subjective quality of the speech is degraded. In order to improve intelligibility and quality and to remove the operator work load, a pilot carrier is sometimes added to the SSB signal as a reference frequency so that the SSB signal can be tuned in automatically without frequency error. This is generally achieved by using a second-order phase-lock loop. This circuit and its operation are similar to the Costas loop of Figure 2.24 except that the phase deflipper is not needed because there are no phase flips on the pilot carrier.

A simplified block diagram of a pilot-carrier, phase-lock loop SSB demodulator is shown in Figure 2.26. The lowpass filtered output of the limiter and multiplier provide an output that is directly proportional to the phase difference over a range of ±90° and therefore act as a phase detector. If there is a phase error the phase-lock loop drives the phase error between the oscillator and the pilot carrier to zero. The SSB signal is demodulated by multiplying the SSB signal by the oscillator output, thus mixing the signal down to audio. The dc output of this product detector is proportional to the pilot-carrier strength, and therefore can provide AGC and can control the SSB link gain from transmitter input to receiver output to be 0 dB. It is thus possible to provide interfacing of full-duplex SSB links with standard two-wire land lines without regeneration or singing problems.

There are two phases in the operation of the phase-lock loop: acquisition, or capture, and tracking. During acquisition either the loop bandwidth must be wide enough to be able to acquire with the maximum frequency error or the oscillator frequency of a narrower loop must be slowly swept across the pilot-carrier frequency. In either case the loop bandwidth must be narrow enough to provide an adequate loop signal-to-noise ratio to allow capture. During tracking, the loop bandwidth must be narrow enough to maintain accurate tracking at poor input signal-to-noise ratios and during fades, but wide enough to accommodate the phase noise on transmitter and receiver frequency synthesizers and the maximum rate of change of frequency due to Doppler shifts without losing lock.

The loop may be operated in the wideband acquisition mode until phase lock is detected. At that time the loop is sometimes switched to a narrower bandwidth carrier-tracking mode. The bandwidth switching must be done without introducing a transient voltage into the voltage-controlled oscillator or the loop will lose lock.

Lock is usually determined by the ratio of the dc voltage of the output of the upper phase detector to that of the lower SSB demodulator. When lock is obtained the dc output of the SSB detector will be proportional to the magnitude of the pilot carrier, while the low-frequency output of the phase detector will be due only to noise close to the pilot carrier.

Phase-lock loop operation

The analysis and operation of phase-lock loops are well covered in the literature [14, 18]. The brief discussion to follow is based primarily on Gardner

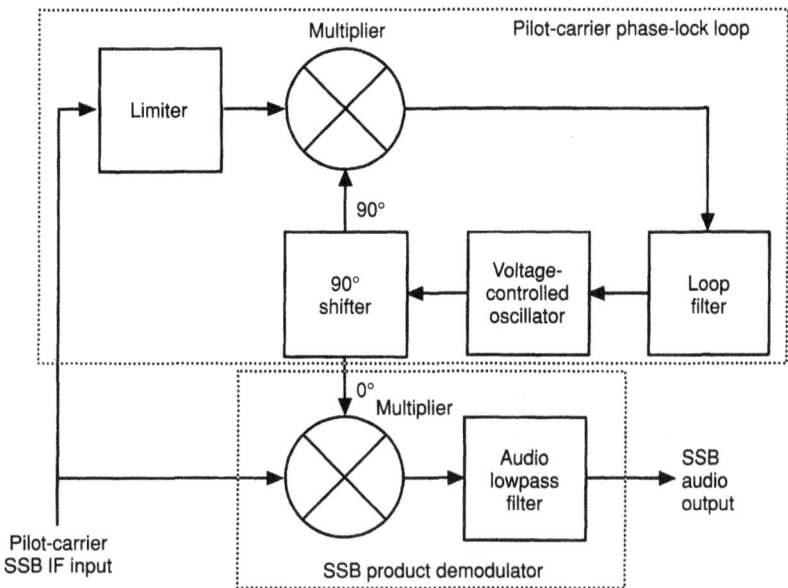

Figure 2.26 Pilot-carrier PLL SSB demodulator.

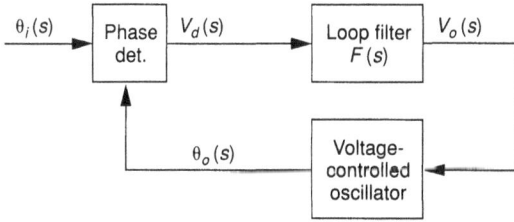

Figure 2.27 Simplified block diagram of a PLL.

[14]. Figure 2.27 is a simplified block diagram of a phase-lock loop. In Laplace transform notation the phase detector outputs a voltage $V_d(s)$ that is proportional to the phase difference between the input signal phase $\theta_i(s)$ and the oscillator signal phase $\theta_o(s)$ or

$$V_d(s) = K_d[\theta_i(s) - \theta_o(s)] \tag{2.49}$$

The loop filter output voltage $V_o(s)$ is

$$V_o(s) = F(s)V_d(s) \tag{2.50}$$

where $F(s)$ is the response of the loop lowpass filter. The voltage-controlled oscillator produces an output frequency ω_o that is proportional to the applied control voltage or

$$\omega_o(s) = K_o V_o(s) \tag{2.51}$$

but because the output phase is the time integral of the instantaneous frequency we have

$$\theta_o(s) = \frac{K_o V_o(s)}{s} \tag{2.52}$$

Combining the above equations to obtain the input-output transform $H(s)$ gives

$$H(s) = \frac{\theta_o(s)}{\theta_i(s)} = \frac{K_o K_d F(s)}{s + K_o K_d F(s)} \tag{2.53}$$

A typical active filter for a second-order loop along with the applicable transfer function are shown in Figure 2.28.

Gardner shows that the transfer function for a second-order loop with such an active filter is

$$H(s) = \frac{2\zeta\omega_n s + \omega_n^2}{s^2 + 2\zeta\omega_n s + \omega_n^2} \tag{2.54}$$

where

$$\omega_n = \sqrt{\frac{K_d K_o}{T_1}} \text{ and } \zeta = \omega_n \frac{T_2}{2} \tag{2.55}$$

and where, in servo terminology, ω_n is the natural frequency in radians per second and ζ is the damping factor of the loop. The one-sided noise bandwidth of the loop transfer function B_L is given by

$$B_L = \frac{\omega_n}{2}\left(\zeta + \frac{1}{4\zeta}\right)\text{Hz} \tag{2.56}$$

For carrier tracking, it is more instructive to specify loop performance in terms of B_L than in terms of ω_n since the noise bandwidth B_L must be chosen so as to give an adequate loop signal-to-noise ratio and is therefore a primary design parameter. If we solve for ω_n from Equation 2.56 we obtain

$$\omega_n = \frac{2B_L}{\zeta + \dfrac{1}{4\zeta}} \tag{2.57}$$

With this substitution the important loop performance parameters are as listed in Table 2.2. This table provides both the general formulas and the simplified equations for critical damping ($\zeta = 0.707$) since this is the most commonly used damping factor. The table gives the loop signal-to-noise ratio, rms phase-tracking jitter, and the mean time to loss of lock for a pilot-carrier carrier-to-noise density ratio C/N_0. For the noise-free case the table also gives the maximum frequency sweep rate without loss of lock, the maximum frequency step size without loss of lock, the maximum frequency capture range without cycle slip (snap-lock range), the approximate time for snap lock, and the maximum frequency pull-in range with the corresponding pull-in time.

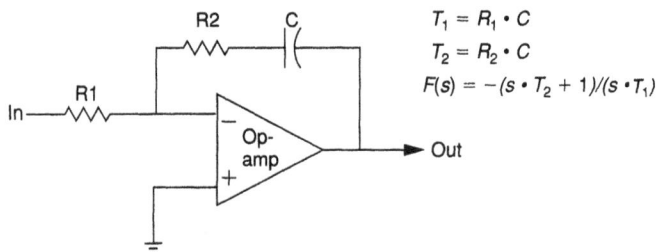

Figure 2.28 Active second-order loop filter block diagram.

TABLE 2.2 Performance Parameters for a Second-Order PLL with an Active Filter

Performance parameter	General equation	Equation for critical damping	Eq. num.
Loop signal-to-noise ratio, $(SNR)_L$	$(SNR)_L = \dfrac{C/N_0}{2B_L}$	Same	2.58
Rms phase jitter, $\Delta\theta_{rms}$ (rms radians)	$\Delta\theta_{rms} = \sqrt{\dfrac{1}{2(SNR)_L}}$	Same	2.59
Mean time to loss of lock, T_{av} (seconds)	Not available	$T_{av} = \dfrac{1.06}{B_L}\, e^{\pi\,(SNR)_L}$	2.60
Maximum tracking sweep rate, df/dt (Hz/sec)	$\dfrac{df}{dt} = \dfrac{2B_L^2}{\pi\left(\zeta + \dfrac{1}{4\zeta}\right)^2}$	$\dfrac{df}{dt} = 0.57\,B_L^2$	2.61
Maximum allowable frequency step size, Δf_{po} (Hz)	$\Delta f_{po} = \dfrac{1.8\,B_L}{\pi} \cdot \dfrac{1+\zeta}{\zeta + \dfrac{1}{4\zeta}}$	$\Delta f_{po} = 0.92\,B_L$	2.62
Maximum snap lock range, Δf_L (Hz)	$\Delta f_L = \dfrac{2\,B_L}{\pi} \cdot \dfrac{\zeta}{\zeta + \dfrac{1}{4\zeta}}$	$\Delta f_L = 0.42\,B_L$	2.63
Time for snap lock, T_L (seconds)	$T_L = \dfrac{\zeta + \dfrac{1}{4\zeta}}{2\,B_L}$	$T_L = \dfrac{0.53}{B_L}$	2.64
Maximum pull-in range, Δf_p (Hz)	$\Delta f_p = \dfrac{1}{\pi}\sqrt{\dfrac{2K_vB_L\zeta}{\zeta + \dfrac{1}{4\zeta}}}$	$\Delta f_p = 0.37\sqrt{K_vB_L}$	2.65
Pull-in time, T_p (seconds)	$T_p = \left(\dfrac{\pi}{2}\right)^2 \cdot \dfrac{\Delta f^2\left(\zeta + \dfrac{1}{4\zeta}\right)^3}{\zeta B_L^3}$	$T_p = 4.2\dfrac{\Delta f^2}{B_L^3}$	2.66

The pilot-carrier power should be kept as low as possible, commensurate with providing an adequate pilot-carrier signal-to-noise ratio and fast enough lock acquisition time. The signal-to-noise ratio can be improved by narrowing the loop bandwidth, but this limits the frequency error and speed with which acquisition can be obtained and the amount of frequency synthesizer phase jitter that can be tolerated during tracking. Thus there is a trade-off between capture range, pilot-carrier level, and loop bandwidth.

It is common practice to use a pilot-carrier level 10 dB below the total peak envelope power (PEP) rating of the transmitter or at an envelope voltage 32% of peak. From Equation 2.43 one finds that the peak envelope sideband power is 0.47 of PEP. Thus the pilot carrier has reduced the available SSB power by 3.3 dB.

In addition to the noise on the radio link, the speech sideband signal itself acts as a form of "noise" as far as the phase-lock loop is concerned. In order that this form of noise not prevent acquisition, it must be attenuated by the filter action of the loop so as to provide a loop signal-to-noise ratio $(SNR)_L$ of at least 6 dB. With a pilot carrier 10 dB below the transmitter PEP (6.7 dB below the sideband PEP) a loop bandwidth B_L of some 200 Hz is required. Assuming a critically damped loop bandwidth of 200 Hz and using Equations 2.63 and 2.64, we find that the maximum snap-lock frequency range is ±84 Hz and the snap-lock time is 2.65 ms. If frequency sweep is used to increase the acquisition range then we find, from Equation 2.62, that the maximum sweep rate is 184 Hz/sec.

The above calculations assume that the link noise does not significantly lower the loop signal-to-noise ratio compared with the noise due to the speech sidebands. We will now check this assumption. Table 2.3 shows that the minimum usable sideband PEP/N_0 for SSB is 53 dB-Hz. The pilot carrier C/N_0 is 10 − 3.3 = 6.7 dB less than this, or 46.3 dB. The signal-to-noise ratio in the loop is therefore given by Equation 2.58 as 46.3 − 26.0 = 20.3 dB. Thus the above assumption is justified.

While this example demonstrates the general method of analysis, each particular system must be analyzed based on its individual requirements and constraints.

Offset pilot carriers

In the case described above, the pilot carrier is inserted at the normal carrier frequency. However, since provisions are not generally made in the SSB transmitter to provide the carrier, nor in the receiver to pass such a carrier, an offset carrier is often used. Here the carrier can be provided by introducing an audio tone along with the speech either in the middle of the voice band (pilot tone-in-band) or above it (pilot tone-above-band). In either case this carrier must be filtered out of the audio output. While the details of implementation are slightly different, the performance and principles are exactly the same as with the normal pilot carrier.

Amplitude compandored SSB

In LINCOMPEX [19], SYNCOMPEX [20], and amplitude compandored SSB (ACSSB) [21] the carrier reference tone is placed above the speech band at around 3150 Hz and the tone is modulated to provide an amplitude control for compandoring such as shown in Figure 2.10 [9]. In LINCOMPEX the tone is FM modulated, for SYNCOMPEX the tone is modulated with digital data, and for ACSSB the tone is AM modulated. In all three cases the tone modulation bandwidth is less than 200 Hz. SYNCOMPEX and ACSSB provide both automatic fine tuning and volume compression and expansion in a nominal bandwidth of less than 5 kHz. Since the end-to-end audio gain of the radio link (channel) is controlled at 0 dB the link can be interfaced with full-duplex, two-wire land lines without singing due to regeneration. All are usable without operator attention and are therefore suitable for land mobile radio telephones.

Angle modulation family

Angle modulation is the second classical analog modulation system. Here the phase angle $\theta(t)$ of a reference carrier of frequency ω_c is varied under control of the modulating function while the amplitude A of the signal is constant. This modulated waveform can be expressed as

$$x(t) = A \cos [\omega_c t + \theta(t)] \tag{2.67}$$

In phase modulation (PM) the phase $\theta(t)$ is made to vary in direct proportion to the modulating waveform. Sine-wave phase modulation can be decomposed into an in-phase carrier that is DSB modulated by even cosine harmonics of the modulating tone plus quadrature DSBSC sidebands of odd sine-wave harmonics, as shown in Figure 2.29. For small peak deviations the primary sources of phase deviation are the fundamental quadrature sidebands, while the smaller in-phase sidebands, which are cosine-wave modulated by the second harmonics, remove most of the AM introduced by quadrature sideband modulation. For larger phase deviations there are higher-order sidebands of significant amplitudes. The modulating function is

$$\theta(t) = \beta \sin (\omega_m t) \tag{2.68}$$

where β is the peak phase deviation. Equation 2.68 can be expanded into

$$\begin{aligned} x(t) = A\,[&J_0(\beta) \cos (\omega_c t) + 2J_1(\beta) \sin (\omega_m t) \sin (\omega_c t) \tag{2.69}\\ &+ 2J_2(\beta) \cos (2\omega_m t) \cos (\omega_c t) + 2J_3(\beta) \sin (3\omega_m t) \sin (\omega_c t) \dots] \end{aligned}$$

where $J_n(\beta)$ is the Bessel function of the first kind of the nth order. For very small peak deviations the RF spectrum is largely contained within twice the audio bandwidth, as with AM. For peak phase deviations greater than 45° this approximation is grossly inadequate and second- and higher-order sidebands of significant amplitude are present.

Instantaneous frequency. If the total angle of the wave is $\phi(t)$ (the argument of the cosine) where

$$\phi(t) = \omega_c t + \theta(t) \tag{2.70}$$

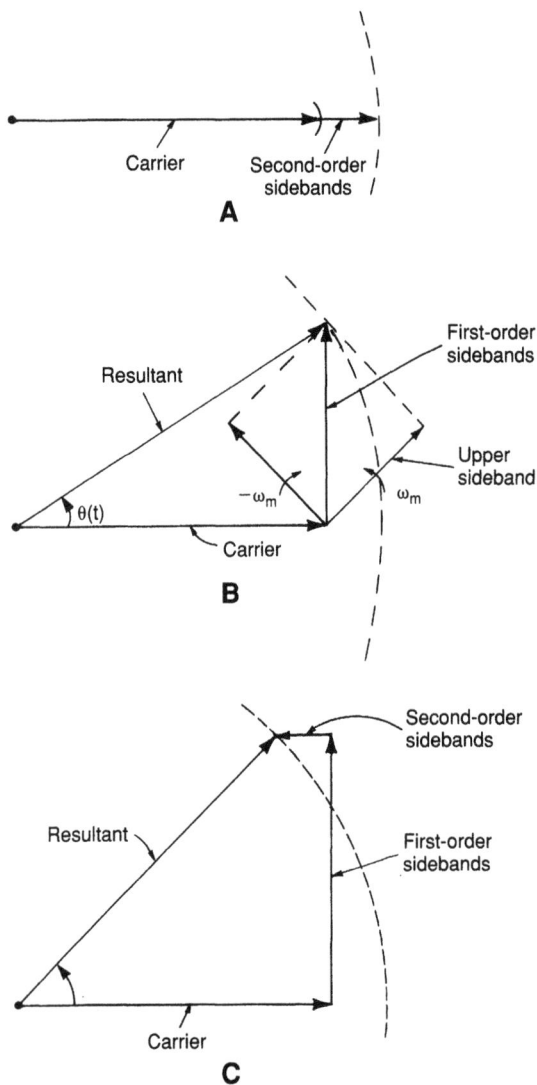

Figure 2.29 Rotating vector representation of a sine-wave-modulated FM signal. (A) $\theta(t) = 0$; (B) $\theta(t) = \frac{1}{2}$ radian; (C) $\theta(t) = 1$ radian. At $\theta(t) = 0$ the first-order USBs and LSBs are 180° out of phase and therefore cancel.

then the instantaneous radian frequency $\omega(t)$ is defined as the rate of change of phase, or

$$\omega(t) = \frac{d\phi(t)}{dt} = \omega_c + \frac{d\theta(t)}{dt} \tag{2.71}$$

For a constant carrier frequency the instantaneous frequency is equal to the carrier frequency. However, for many waveforms there is no simple relationship between the instantaneous frequency and the frequency spectrum. The former is time dependent and has only a single value at a given instant, while the true frequency spectrum is independent of time and can be composed of many frequency components. Instantaneous frequency is a useful concept but should not be confused with the frequency spectrum.

Frequency modulation. In FM the instantaneous frequency $\omega(t)$ is made to vary in direct proportion to the modulating function so that

$$\omega(t) = \omega_c + cm(t) \tag{2.72}$$

where c is a constant.

At high signal levels both FM and PM offer improvement in the receiver output signal-to-noise ratio when compared with AM. There is an IF carrier-to-noise ratio below which the full output signal-to-noise ratio improvement can no longer be maintained. This IF carrier-to-noise ratio is called the FM improvement threshold. The signal and noise characteristics of the FM system can therefore be divided into two primary regions: above and below the improvement threshold.

The receiver IF bandwidth B required to pass the significant spectrum components depends upon the amount of output signal distortion that can be tolerated [22]. One widely used rule of thumb is Carson's rule,

$$B = 2(\Delta f + b) \tag{2.73}$$

where Δf is the peak frequency deviation and b is the upper audio cutoff frequency. This rule allows rejection of all spectrum lines that contribute less than 1% of the total power of the FM spectrum. For high values of m, $(\Delta f > b)$, the value of B approaches $2\Delta f$. It has been empirically determined that for communications quality voice modulation, a suitable rule for the minimum bandwidth B is

$$B = \max[2\Delta f, 2b] \tag{2.74}$$

where max [ ] is the larger value of the two arguments. Additional bandwidths must of course be allotted for the frequency uncertainties in the system.

For minimum bandwidth systems the IF bandwidth can be set at $2b$ and the peak deviation Δf can be equal to b.

The threshold $(C/N)_{IF}$ value lies between 7 and 10 dB and is often considered to be a lower limit for acceptable FM system performance. However, for voice communications, usable intelligibility is reached at several decibels below threshold.

A completely rigorous expression describing the entire below-threshold region for the sine-wave modulation case is not available. Frutiger [23], however, has developed an equation that allows accurate determination of the audio frequency (AF) output signal-to-noise ratio $(S/N)_{AF}$ for sine-wave modulation at $(C/N)_{IF}$ values greater than 4 dB. Frutiger's equation is

$$\left(\frac{S}{N}\right)_{AF} = \frac{\dfrac{3}{2}m^2\,(B/b)\,(C/N)_{IF}}{1 + \dfrac{0.9\,(B/b)^2\,(C/N)_{IF}\,\exp\left[-(C/N)_{IF}\right]}{\left(1 - \exp\left[-(C/N)_{IF}\right]\right)^2}} \qquad (2.75)$$

where B is the IF bandwidth, b is the audio bandwidth, and m is the modulation index, as defined by

$$m = \frac{\Delta f}{b} \qquad (2.76)$$

where Δf is the peak instantaneous frequency deviation. Equation 2.75 has been employed to develop the set of $(S/N)_{AF}$ versus $(C/N)_{IF}$ curves illustrated in Figure 2.30, where the curve parameter is B/b and the deviation ratio m is equal to 1 for all curves. These curves can be adapted to calculate the sine-wave audio output signal-to-noise ratios for other values of the modulation index by adding $20\log_{10}(m)$ to the value of $(S/N)_{AF}$ read from the curves.

Voice modulation FM systems

The FM curves shown in Figure 2.30 are for sine-wave modulation. The output signal-to-noise ratio will be lower with voice modulation because the peak-to-average power ratio of speech is higher than that of a sine wave. The output signal-to-noise ratio as read from Figure 2.30 must be reduced for voice modulation by the factor

$$\frac{\text{Peak-to-average power ratio of speech}}{\text{Peak-to-average power ratio of sine wave}}$$

It is clear from this that peak clipping of the voice modulating waveform will improve the output signal-to-noise ratio for a fixed peak frequency deviation since it reduces the peak-to-average power ratio of the modulating waveform.

In calculating the intelligibility of speech, it is necessary to know the shape of the audio output noise spectrum. Above the FM improvement threshold the output noise is f^2 noise, while well below threshold the noise tends to be white. The transition from f^2 to white noise is rather sudden. Near the FM improvement threshold the output noise is made up of two ad-

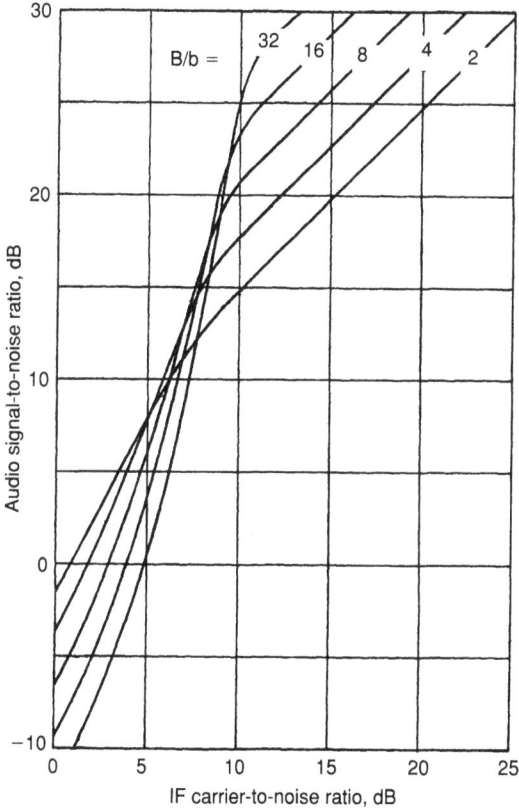

Figure 2.30 Output signal to noise at baseband as a function of input carrier-to-noise ratio for sine-wave modulation. The parameter is B/b. The deviation ratio is $m = 1.0$. For other deviation ratios, add 20 log (m) to the ordinate values. Note: these curves are normalized to facilitate calculation. The apparent improvement in performance with increasing IF bandwidth (B/b) is deceptive, because increasing the bandwidth lets in more noise, thus decreasing the IF carrier-to-noise ratio for the link.

ditive components. One is the f^2 Gaussian noise, which is inversely proportional to the carrier-to-noise ratio above the FM improvement threshold. The other is white impulse noise, which increases very rapidly with a reduction in the input carrier-to-noise ratio. These two components contribute equally to the output noise when the output signal-to-noise ratio is 3 dB below the straight-line extension of the above threshold curve. This point is referred to as the 3-dB FM improvement threshold.

In calculating intelligibility, we switch from f^2 to white output noise at the 3-dB FM improvement threshold in lieu of calculating the exact shape of the output noise spectrum.

We are now in a position to calculate the intelligibility of an FM voice system. The procedure is as follows:

1. Assume sine-wave modulation of the transmitter with the same peak frequency deviation as will be obtained with voice modulation.

2. Look up the output signal-to-noise ratio for the given input carrier-to-noise ratio using Figure 2.30 with corrections for the actual modulation index.

3. Subtract from the sine-wave output signal-to-noise ratio (in decibels) obtained in step 2 the difference between the peak-to-average power ratio of the speech waveform (in decibels) obtained from Figure 2.8 and the 3-dB peak-to-average power ratio of a sine wave to obtain the voice output signal-to-noise ratio.

4. Look up the AI on an appropriate intelligibility curve for the proper premodulation preemphasis and bandwidth, using a curve for f^2 noise when above the 3-dB FM improvement threshold or for white noise when below this threshold.

5. Repeat the above procedure for various carrier-to-noise ratios to obtain an intelligibility curve.

The intelligibility of two FM systems has been calculated by the above method and the results are shown in Figures 2.31 and 2.32. Both systems have +6 dB/octave preemphasis above 500 Hz, a lower audio cutoff frequency of 200 Hz, 21 dB of audio clipping, a predetection IF bandwidth of twice the peak frequency deviation, and a conventional FM detector. The first system has a peak frequency deviation equal to the upper audio cutoff frequency ($m = 1$), as shown in Figure 2.31. The second system has a peak frequency deviation equal to twice the upper audio cutoff frequency, as shown in Figure 2.32. The premodulation voice processing for these sample calculations was picked so as to optimize intelligibility under poor signal conditions.

The threshold of intelligibility (AI = 0.3) for the narrower band system ($m = 1$) requires 1 dB less carrier power than for the wider band systems ($m = 2$), and it is desirable to use an audio bandwidth of between 2 and 3 kHz for the former. All systems achieve the threshold of intelligibility well below the FM improvement threshold. Thus, if usable intelligibility under weak signal conditions is the primary criterion, a modulation index of 1 should be used. At this low modulation index there is little or no benefit from the use of FM threshold extension techniques.

Comparison of voice modulation systems

In the previous sections we developed the tools necessary for the calculation of the intelligibility of a wide range of voice communication systems. Modulation theory gives the audio signal-to-noise ratio, given the link sig-

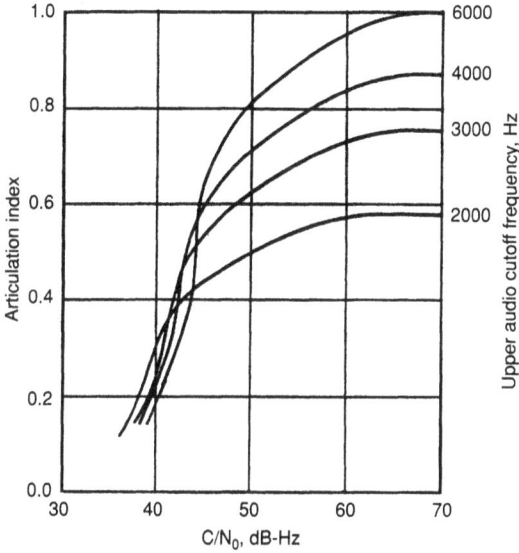

Figure 2.31 Articulation index versus carrier power-to-noise power density (C/N_0) for $m = 1.0$. Preemphasis: +6 dB/octave above 500 Hz; lower audio cutoff: 200 Hz; audio clipping: 21 dB; (R_m=7 dB); modulation index: $m = \Delta f/b = 1$; IF bandwidth: $2b$.

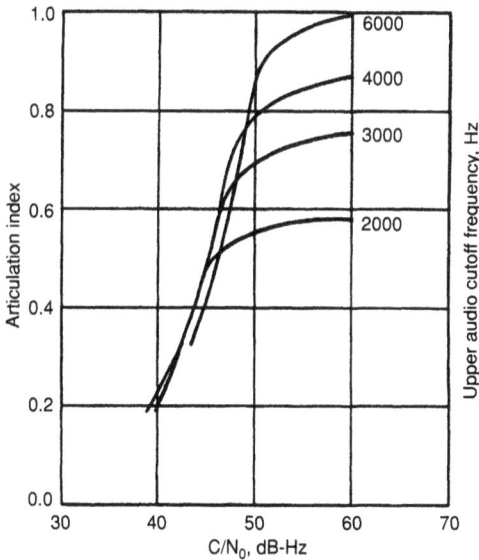

Figure 2.32 Articulation index versus carrier power-to-noise power density ratio (C/N_0) for $m = 2.0$ with preemphasis and audio clipping. Preemphasis: +6 dB/octave above 500 Hz; lower audio cutoff: 200 Hz; audio clipping: 21 dB ($R_m = 7$ dB); modulation index: $m = \Delta f/b = 2$; IF bandwidth: $4b$.

nal-to-noise ratio and amplitude statistics of speech, while the AI provides a method of calculating the intelligibility, given the audio signal-to-noise ratio. We have used these tools to compare the various modulation systems under typical conditions to exemplify the results that are obtainable.

The intelligibility for unprocessed voice modulation is shown in Figures 2.33 and 2.34. The audio bandwidth is 300 to 3000 kHz and there is no preemphasis or clipping of the audio modulating signal or the IF or RF signal. The curves show the performance of SSB, DSBSC, AM, and FM. Frequency modulation curves are shown for peak frequency deviations of 3 and 6 kHz. In each case, the receiver IF bandwidth is equal to twice the peak frequency deviation. Figure 2.33 shows the AI as a function of the RF average signal power-to-noise power density ratio P_{avg}/N_0, while Figure 2.34 shows AI versus the peak envelope signal power-to-noise density ratio, PEP/N_0. Single-sideband modulation is most efficient for average power-limited transmitters, while FM is most efficient for peak power-limited transmitters.

The intelligibility for processed voice modulation is shown in Figures 2.35 and 2.36. The audio bandwidth is again 300 to 3000 Hz, but here the speech waveform has been preemphasized by +6 dB/octave above 500 Hz and clipped to reduce the audio peak-to-average power ratio to 7 dB. In the case of SSB, the clipping, whether done at audio or IF, reduces the final RF peak

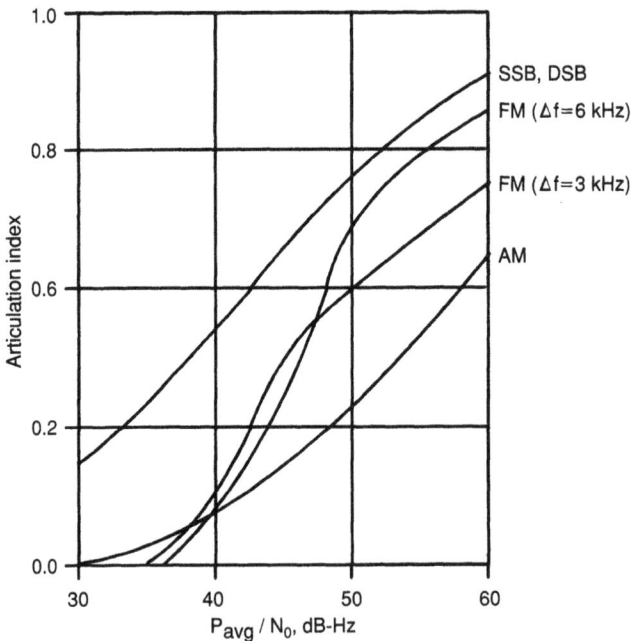

Figure 2.33 Articulation index versus P_{avg}/N_0 for unprocessed speech modulation (300- to 3000-Hz audio bandwidth).

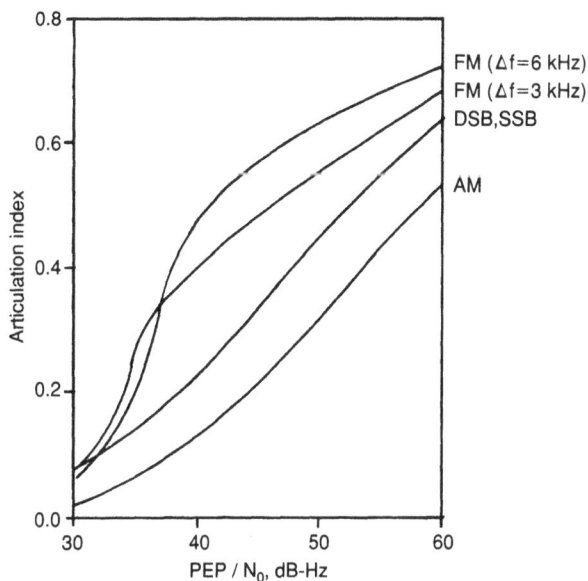

Figure 2.34 Articulation index versus PEP/N_0 for unprocessed speech modulation (300- to 3000-Hz audio bandwidth).

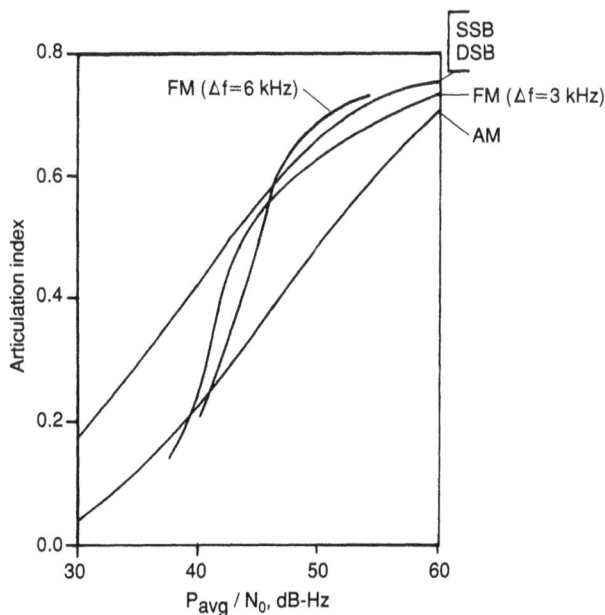

Figure 2.35 Articulation index versus P_{avg}/N_0 for preemphasized and clipped speech modulation (300 to 3000 Hz).

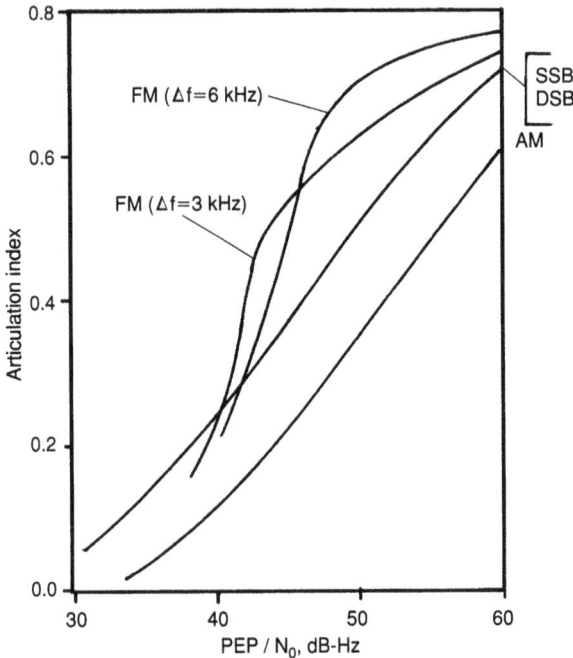

Figure 2.36 Articulation index versus PEP/N_0 for preemphasized and clipped speech modulation (300 to 3000 Hz).

envelope-to-average power ratio to 7 dB. This processing represents a reasonable compromise between quality and efficiency for links that must be operated frequently under marginal signal conditions. Again, SSB is generally best for average power-limited transmitters, while FM is generally most efficient for peak power-limited systems.

These results are summarized in Table 2.3. Most power amplifiers have both peak and average power limits, and which one actually limits performance is determined by the peak envelope-to-average power ratio of the RF signal waveform.

TABLE 2.3 Comparison of Voice Modulation Systems*

Modulation type	RF bandwidth, kHz	Signal-to-noise density ratio at AI = 0.3			
		Unprocessed speech		Processed speech	
		P_{avg}/N_0	PEP/N_0	P_{avg}/N_0	PEP/N_0
SSB	3	39	53	35	42
DSBSC	6	39	53	35	42
FM($\Delta f = 3$ kHz)	6	45	45	41	41
FM($\Delta f = 6$ kHz)	12	46	46	42	42
AM	6	54	59	43	48

* See text for conditions.

If the transmitter is average power-limited, then preemphasis and clipping reduce the required power by 4 dB for SSB, DSBSC, and FM and by 11 dB for AM. Single-sideband modulation has greater power and bandwidth efficiency than FM modulation regardless of the type of speech processing used.

If the transmitter is peak power-limited, preemphasis and clipping reduce the required power by 11 dB for SSB, DSBSC, and AM and by 4 dB for FM. For processed speech modulation the efficiency of SSB is approximately equal to FM. In addition, SSB voice occupies half the RF bandwidth and, as we shall see, is little disturbed by multipath propagation; therefore, SSB provides the best performance on HF links. However, for peak power-limited systems without speech processing, *very* narrowband FM has about 6 dB greater power efficiency than SSB.

Multipath propagation

Multipath propagation is the rule rather than the exception on long-haul HF links and mobile communications. In multipath propagation the signal is received by reflection from two or more paths. The distances traveled by the paths are different, resulting in time delays between the received signals. Delayed signals may tend to cancel or reinforce each other. For two-path propagation, a received sine wave $x(t)$ is of the form

$$x(t) = \sin 2\pi f t + \rho \sin \left[2\pi f(t - \tau)\right] \qquad (2.77)$$

where τ is the multipath differential time delay and ρ is the relative voltage of the delayed path. The delayed path may add or subtract from the direct path signal, depending on their relative phase, thus causing reinforcement or fading. The frequency response of the medium is periodic in frequency f with a period Δf given by

$$\Delta f = \frac{1}{\tau} \qquad (2.78)$$

At HF, multipath differential delays range between 1 and 5 ms, so that deep frequency-selective fade troughs are typically separated by 200 to 1000 Hz. For VHF and UHF mobile communication links the delays are much shorter. The fading troughs are much farther apart in frequency, causing a relatively flat fading over the signal bandwidth. The peak-to-trough power ratio in two-path selective fading is $(1 + \rho^2)/(1 - \rho^2)$, and multipath propagation can cause 6-dB reinforcement at peaks and complete cancellation in troughs. Articulation index calculation for two equal paths indicates that multipath selective fading has only a small effect on SSB intelligibility, as shown in Figure 2.37. Here the total average power is held constant, implying that path loss for each of the two paths is 3 dB more than on the single

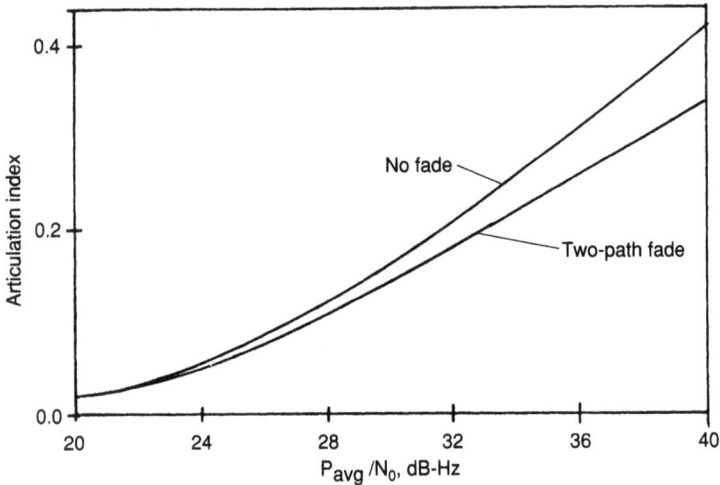

Figure 2.37 Articulation index versus average signal power-to-noise density ratio for SSB with and without two-path multipath (two equal paths with 1.66-ms differential delay).

nonfading path. While SSB is affected only slightly by multipath fading, the attenuation of the carrier in AM and FM introduces severe distortion in demodulation, causing significant degradation in intelligibility when the carrier falls in the fading trough. Thus, all things considered, SSB is superior to other types of modulation under adverse HF propagation conditions.

We have addressed intelligibility for the classical modulation systems under typical conditions. The general method of analysis presented, however, applies to a wide range of conditions. The examples presented are useful as a general guide but it is often desirable to derive results for the particular case under study.

2.2 Collocated System Design

When a transmitter and receiver are operating simultaneously in close proximity, there are several ways in which receiver operation may be degraded. Considering receiver limitations, the most severe degradation is receiver burnout, or more likely, opening of a protective antenna relay. A second is desensitization or blocking caused by overload of the receiver front-end circuits. Somewhat related is cross-modulation, which is a transferal, due to receiver nonlinearity, of transmitter modulation to that of the received signal. Reciprocal mixing may occur, a phenomenon that utilizes a strong out-of-band signal to heterodyne local oscillator noise sidebands into the receiver IF passband. Finally, higher-order mixing products, discussed in Chapter 4, "Receiver Design," may result in spurious responses to strong unwanted signals. Transmitter limitations that may degrade collocated receiver operation include harmonic and intermodulation distortion products and broadband noise components that fall within the receiver passband.

A special problem may arise when two transmitters are in operation at the same time in the vicinity of a receiver. First, coupling between the two transmitting antennas transfers some of the power from one transmitter to the other. This transferred power is reflected to the PA output where it intermodulates with its signal to create new frequencies, one of which may fall within the receiver passband. This phenomenon is known as *back-intermodulation* (back-IM). Second, these two transmitted signals, even if they do not create back-IM, may create IM distortion in the receiver front end. Since the same two frequencies are involved, these distortion products also have the same frequencies as those produced by back-IM.

All these degraded reception effects are made worse when there is a strong coupling between transmitter and receiver antennas; hence it is important to reduce the coupling between them as much as possible. Practically, this means maintaining as much physical separation as allowed and orienting the receiving antenna to place it as close to a null or minimum in the transmitting antenna pattern as permitted by the constraints of the site.

Figure 2.38 shows the elements of a communication system designed for simultaneous operation of the transmitter and receiver. The main differences between it and a simple half-duplex installation are the use of separate transmitting and receiving antennas, a preselector and/or adaptive canceler to reduce the level of the transmitter at the receiver input, and a postselector to reduce exciter noise and distortion products. The power amplifier may also be designed for lower noise and distortion output. Typically this might be accomplished by heavier filtering of the PA output.

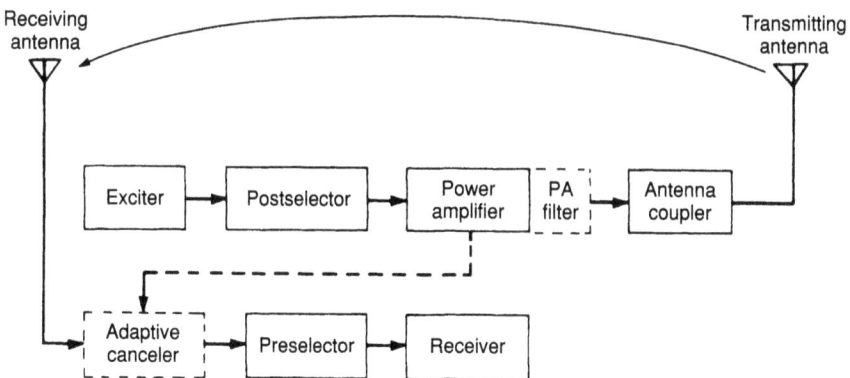

Figure 2.38 Elements of a communication system designed for simultaneous operation of the transmitter and receiver.

Quasi-minimum noise considerations

In determining the amount of acceptable interference to a receiver from a col-located transmitter, it is useful to consider to what extent naturally occurring noise levels limit the sensitivity of the receiver. In the absence of other noise sources, thermal noise sets a lower limit to receiver sensitivity. The magnitude of this noise density is kT_0–W/Hz bandwidth, where k is Boltzmann's constant and T_0 is the absolute temperature (in Kelvin). Any practical receiver adds noise beyond the thermal limit, and the ratio of its noise output to that which would result from thermal noise alone is the receiver noise figure. Beyond this, naturally occurring atmospheric and galactic sources, as well as artificially created radio interference, contribute additional noise. Although these levels can vary widely with location and time, it is useful to lump them together and define quasi-minimum noise (QMN) as a practical minimum noise design level. This level, expressed in decibels, varies approximately log linearly from about 53 dB above thermal at 2 MHz to about 21 dB above thermal at 30 MHz. Quasi-minimum noise serves as a floor to limit receiver sensitivity, for no matter how efficient the antenna, the signal-to-noise ratio is ultimately limited by the ratio of the field strength of the signal to that of the QMN. The antenna need only be efficient enough to ensure that the level of QMN delivered to the receiver is about equal to, or exceeds slightly, the receiver noise as determined by its noise figure. In fact, when a nearby transmitter is operating, it is desirable that the receiving antenna be as inefficient as permissible to limit the power coupled from the transmitter. Quasi-minimum noise also functions as a useful limit of allowable transmitter interference to the receiver, as will be shown.

Simultaneous operation of one transmitter with one or more receivers

This is perhaps the most common simultaneous operation scenario. Interference to the receiver is due to performance limitations in both the receiving and the transmitting system. Receiving system limitations are caused by its nonlinearities, finite dynamic range, and synthesizer broadband noise and spurious outputs. Performance degradation is manifested in a number of different ways such as

1. Overload and distortion in the adaptive canceler and/or the preselector
2. Receiver blocking, cross-modulation, and intermodulation
3. Spurious mixer products
4. Reciprocal mixing of receiver synthesizer noise and spurious outputs into the IF passband by the transmitter signal
5. Transmitter broadband noise and spurious outputs
6. Transmitter harmonics and IM products

Each of these factors will be considered in turn.

Overload and distortion in the adaptive canceler. An adaptive canceler operates by utilizing a sample of the transmitter output to cancel the transmitter signal induced in the receiving antenna. It does this by adjusting the phase and amplitude of the transmitter sample such that it is equal and opposite in phase to that picked up by the receiving antenna. Typically this may be done by deriving an in-phase and quadrature-phase signal from the transmitter sample and adjusting the amplitude and polarity of each, such that when recombined, the resulting signal has the same amplitude but opposite phase to that arriving from the receiving antenna. In producing the proper amplitude and phase of the canceling signal, care must be taken not to produce distortion products. Attenuators using positive intrinsic negative (PIN) diodes, although perhaps the lowest distortion solid-state amplitude-control devices, are not immune to distortion generation when signal levels are very high, as for example when the transmitting and receiving antennas are aboard an aircraft. For this reason, applications for adaptive cancelers are presently somewhat limited, although current development work may lead to wider usage in the future. They have the advantages of being capable of broadband operation and, when combined with automatic fast-nulling circuitry, being useful for ECCM operation.

Overload and distortion in the preselector. Modern HF receivers are generally designed with an untuned, broadband antenna input. The only protection against strong signals within the receiver's tuning range is an overload relay that typically operates at a level of around 0 dBW (1 W). Thus when a receiver must operate simultaneously with a nearby high-powered transmitter within the receiver's tuning range, a frequency-selective receiver preselector may be employed to prevent the receiver overload relay from opening up and preventing reception. The preselector must have sufficient out-of-band signal handling capability to prevent internal damage to itself, and it must have sufficient selectivity to protect the receiver from the effects of the transmitter signal. A purely passive preselector using no semiconductor components or magnetic core inductors would produce no distortion. Using varactor diodes for tuning, or PIN diodes for switching tuning elements, introduces a potential for distortion. Since varactor diodes introduce much more distortion than PIN diodes, they are seldom used as tuning elements of a preselector.

A preselector has a selective specification, a maximum voltage rating, and a distortion specification. A typical preselector might have the characteristics shown in Table 2.4. Specifications are given for an assumed 10% minimum spacing between the transmitter and receiver, typical for many systems with a collocated transmitter and receiver.

TABLE 2.4 Typical Specifications for a Receiver Preselector

Parameter	Specification at 10% away from resonance
Relative selectivity	–40 dB
Maximum input level	200 V or 200 W (+23 dBW) available power from a 50-ohm antenna
–10-dB cross-modulation level (Modulation transfer for a 30% modulated interfering signal)	50 V or 12.5 W (+11 dBW) available power from a 50-ohm antenna)
Center frequency loss	5 dB

An associated receiver might have a maximum input level specification of –20 dBW. Using a preselector with the specifications of Table 2.4, the maximum interfering signal level separated in frequency from the receiver by 10%, increases to +25 dBW. However, the preselector performance is limited by its cross-modulation specification to an available input of +11 dBW.

Receiver blocking, cross-modulation, and intermodulation. All three of these effects are due to nonlinearities in the preselector and receiver. Most modern HF receivers use no RF amplification before the first mixer. The output of the mixer feeds a narrowband IF filter, protecting the rest of the receiver from interference by off-tune signals. Hence the first receiver mixer and possibly the following crystal filter are primarily responsible for interference produced by the presence of the transmitting signal.

A typical good-quality HF receiver may have a third-order out-of-band intercept point (IP or ip) (see Chapter 4 for the definition of intercept point) of 0 dBW. The onset of blocking may be defined as the 1-dB compression point, and for a typical receiver with no RF amplification, the 1-dB compression point occurs at about 10 to 15 dB below the intercept point or around –10 to –15 dBW. When third-order nonlinearity predominates, the modulation transfer ratio is

$$\frac{m'}{m} = \frac{4P_c}{P_{ip}}$$

where m' = transferred modulation percentage
  m = modulation percentage of the interfering signal
  P_c = interfering power level that causes the specified modulation transfer ratio
  P_{ip} = out-of-band intercept point.

Setting $m'/m = 0.1$ as for the preselector, and $P_{ip} = 1$ W (0 dBW), Table 2.5 compares the preselector and receiver cross-modulation performance. In this example, the preselector is the limiting factor in cross-modulation performance. Nonetheless, it does provide for 27 dB greater interference handling capability over that of the receiver alone.

Spurious mixer products. A receiver mixer produces not only the desired sum or difference frequency between the signal (f_1) and the synthesizer (f_2), but a multitude of products having frequencies of $|mf_1 \pm nf_2|$. For a given m, n, and f_2, other values of f_1 can be found that produce the same IF as when m and n are unity, as discussed in Chapter 4. Generally, the lower m and n, the stronger is the spurious response. Ideally such spurious responses should be no greater than the noise level in the IF bandwidth. A typical well-designed mixer might meet this criterion with an unwanted signal up to −40 dBW. If a preselector having 40-dB attenuation to the unwanted signal precedes such a mixer, the unwanted signal level from the receiving antenna could have a level up to 0 dBW.

Reciprocal mixing of receiver synthesizer noise into the IF passband. When a strong out-of-band signal feeds the receiver mixer, it can heterodyne noise sidebands of the receiver synthesizer or local oscillator into the IF passband. As described in Chapter 10, in the section entitled "Loop Noise Sources," the noise sidebands of the synthesizer consist of a number of components from various parts of the synthesizer circuitry. It is convenient to lump all noise sources together for the purpose of analyzing the noise effect on a collocated receiver and transmitter. As a general trend, this noise decreases as frequency separation increases from the synthesizer frequency, but at a given frequency offset the noise level might vary up to 30 dB depending upon the quality of the synthesizer design.

In order to isolate the synthesizer oscillator (VCO) from the mixer it feeds, and to increase its amplitude where necessary, a broadband buffer amplifier is frequently used. Beyond a certain frequency offset, the VCO phase noise falls below the broadband buffer noise output and may be neglected. The noise then shelves out and remains more or less constant over the bandwidth of the buffer amplifier. Using a low-noise buffer, this noise shelf power density may be as low as 160 dB below the carrier level. Further details of phase noise in oscillators may be found in references 24 through 28.

TABLE 2.5 Comparision of Preselector and Receiver Cross-Modulation

Component	Interfering level for 10% cross-modulation
Preselector output	−29 dBW
Receiver input	−16 dBW

A process known as reciprocal mixing, shown in Figure 2.39, can heterodyne synthesizer noise into the IF passband and appreciably degrade the desired signal-to-noise ratio if the interfering signal is sufficiently strong, or if the
noise sidebands of the receiver synthesizer are not sufficiently attenuated.

Suppose we were to determine the level of an interfering signal that will
heterodyne synthesizer broadband noise into the IF passband at the same
level as that of the equivalent received noise consisting of that due to the
receiver noise figure plus QMN. Suppose that to minimize interference,
the receiving antenna system is such that the QMN noise level delivered to
the receiver equals the receiver noise level. This would result in an increase
in the effective receiver noise figure of 3 dB. A typical receiver noise figure
might be 12 dB, so that the sum of receiver noise and QMN is 15 dB above
thermal. The synthesizer noise sidebands vary with distance from their output frequency, but typically fall off to a shelf level that is more or less constant beyond a certain frequency separation. This separation may vary
depending upon the loop bandwidth of the synthesizer, which is a function
of the loop settling time (see Chapter 10).

For a noise shelf of –150 dBc/Hz, the frequency separation at which this
occurs may typically lie between 100 and 200 kHz for a good design. Since
we have been assuming a 10% minimum frequency separation between
transmitter and receiver, it is desirable that the phase noise of the receiver
synthesizer drop below the noise shelf at a frequency offset of 200 kHz
(10% of 2 MHz) or less. In any case, assume the separation is adequate to
ensure that a noise level of –150 dBc/Hz has been achieved.

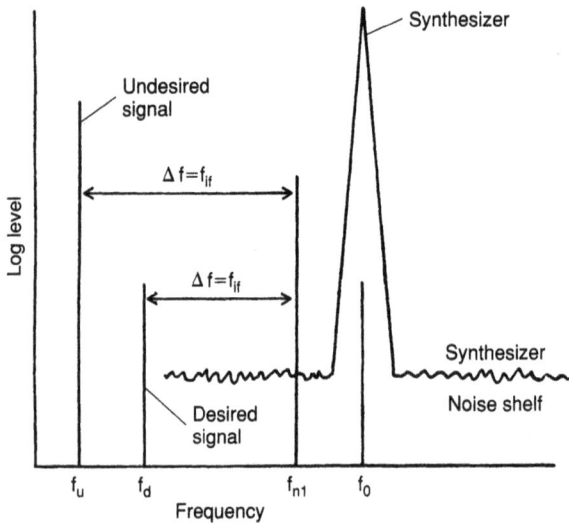

Figure 2.39 Reciprocal mixing mechanism. $f_{if} = f_0 - f_d = f_{n1} - f_u$.

Since the thermal noise level at room temperature is about –204 dBW/Hz, the equivalent receiver noise is 15 dB above this or –189 dBW/Hz. To heterodyne this amount of noise into the receiver IF requires an interfering signal 150 dB stronger, or – 39 dBW. If the receiver front end is broadband, this is the level of interference that will cause the equivalent receiver noise to increase 3 dB. Note that if the preselector of the previous example is used, there is a 5-dB loss at the receiver frequency. To make up for this loss, it will be assumed that the receiving antenna is improved to increase its output by 5 dB. This, of course, also restores the QMN at the preselector output to its original level. Then, preceding the receiver with such a preselector increases the allowable level of the interfering signal by the amount of the preselector selectivity, 40 dB. If the interfering transmitter has a power of 1000 W (30 dBW), 29-dB attenuation is required between the transmitter output and the receiver antenna output to reduce the receiver's interfering level to –39 dBW. If no preselector is used, this attenuation must be increased by 40 to 69 dB, a difficult task when the spacing between transmit and receiver antennas is limited, such as on a ship or aircraft.

It might also be noted that the allowable maximum signal level at the preselector input for reciprocal mixing performance (+6 dBW) is less than that for cross-modulation performance (+11 dBW) and sets the maximum interfering signal level allowed at the receiving antenna.

Transmitter broadband noise. Most HF transmitters consist of an exciter followed by a power amplifier. Modern HF solid-state amplifiers operate broadband, without tuned circuits that would progressively attenuate noise at frequencies away from the signal. Hence noise from the exciter and power amplifier can extend over a wide range, as shown in Figure 2.40, unless some selective filtering is provided. Noise from the exciter comes primarily from two sources: synthesizer noise sidebands that are translated to the transmitted output signal, and broadband amplifier noise following the last IF filter in the exciter. The noise contribution from each of these sources must be added to determine the overall noise level.

Figure 2.41 shows a simplified block diagram of a typical transmitter from the last mixer to the PA output.

In the example shown, the sum of the noise from the mixer output from the synthesizer noise and the preamp noise is –188.5 dBW. With no preselector, the noise contribution of the power amplifier is negligible, so that the transmitter noise output is increased by the 70-dB RF gain to –118.5 dBW. Since this noise power density is presumed to exist at the receiver frequency, no receiver selectivity can aid in reducing it. To reduce the transmitter noise level to the equivalent receiver noise level, including a QMN of –189 dBW/Hz, 70.5 dB of attenuation is required between the transmitter output and the receiver input. This attenuation is the sum of space loss, antenna gains, and antenna mismatch losses.

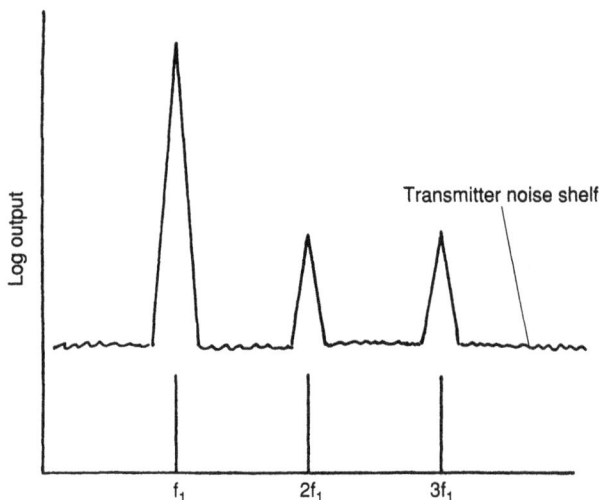

Figure 2.40 Transmitter output spectrum.

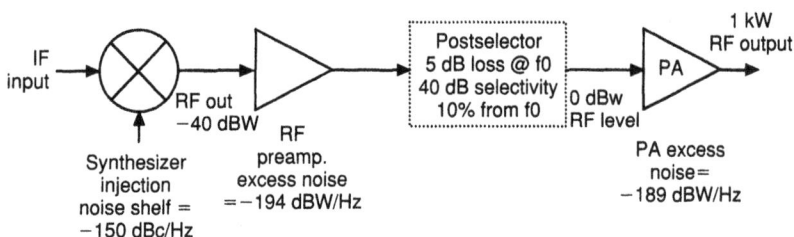

Figure 2.41 Last mixer and output stages of a transmitter. The postselector may or may not be used.

Next consider the effect of using a passive postselector. In the example, the noise at the postselector output at 10% from resonance equals the noise of the RF preamp input increased by the gain of the RF preamp and decreased by the postselector selectivity. Since the RF gain and selectivity exactly offset each other, the noise level at the output of the postselector is −188.5 dBW/Hz. Combining this with the PA equivalent excess input noise gives a noise level of −185.7 dBW/Hz. Adding the 30-dB PA gain gives a transmitter noise output level of −155.7 dBW/Hz at 10% from the PA output frequency. Since this is 37.2 dB less than if no postselector were used, the transmitter-to-receiver isolation can be decreased by the same amount, to 33.3 dB. Note that this is 4.3 dB more than the 29-dB isolation required for the same amount of reciprocal mixing noise. If the postselector were at the PA output, the required isolation for the same PA output noise could be decreased by another 2.8 dB. However, in light of the comparative cost and size of postselectors operating at a level of 0 dBW or at +30 dBW, a 0-dBW level postselector may be a good compromise between noise output and required antenna isolation.

Combined effect of reciprocal mixing and transmitter broadband noise

If no preselector were used in this example, and the transmit-to-receiver isolation is 69 dB (the level at which reciprocal mixing noise equals the sum of QMN and receiver noise), the sum of reciprocal mixing noise (–189 dBW/Hz) and transmitter broadband noise (–187.5 dBW/Hz) is 185.2 dBW/Hz. The combined effect of reciprocal mixing and transmitter broadband noise is to raise the effective receiver noise input by 5.3 dB.

To decrease the required spacing where antenna separation is limited both a preselector and postselector are required. If one or the other is omitted, either reciprocal mixing or broadband transmitter noise will greatly predominate and little will be gained.

When using both the preselector and postselector with a transmit-to-receive isolation of 29 dB, which makes the receiver reciprocal mixing noise from a 1-kW transmitter equal to the same level as in the case with no preselector (–189 dBW/Hz), the transmit broadband noise level at the receiver is (–155.7 – 29 = –184.7 dBW/Hz). The combined effect of reciprocal mixing and transmitter broadband noise is to raise the effective receiver noise input by 5.7 dB.

Increasing the transmit-to-receive isolation by 4 dB to 29.4 dB will make the added noise practically equal to that with no preselector or postselector. The required isolation using both preselector and postselector is then decreased by 39.6 dB, almost equal to the 40-dB selectivity of the preselector and postselector at 10% from resonance.

In general, to minimize the transmitter broadband noise output the following principles should be observed (refer to Figure 2.41).

- Minimize the noise figure of the exciter amplifier following the last mixer.

- If using a postselector between the exciter and the PA, minimize the PA noise figure.

- Design the postselector for the highest selectivity possible at the required transmit-receive frequency spacing, consistent with cost and size factors.

- Although, other conditions remaining the same, operating the postselector at the PA output minimizes the transmitter broadband noise, a good compromise between postselector size and cost and minimum broadband noise would operate the postselector at a level such that the noise at its output from previous stages is attenuated to the same level as that due to the PA noise figure at the desired transmit-receive frequency separation.

- Minimize the noise of the synthesizer to as low a level as practical, both by improving the noise characteristic of the VCO (see references 26–28), and by using a low-noise VCO buffer amplifier.

This latter principle is just as important in the receiver to reduce reciprocal mixing noise. In addition, if a postselector is used, a preselector should also be used. A nearly optimum system design results if the preselector has the same selectivity as the postselector, hence the same basic design can be used for both.

Transmitter harmonics and IM products. When the transmitter output signal contains more than one frequency, IM products between the various frequencies can be formed. With two frequencies f_1 and f_2, the IM products and harmonics are located at $|mf_1 \pm nf_2|$, where m and n are any integers, including zero. The amplitude of these products is determined by the amount and kind of nonlinearity. Class A amplifiers, for example, can have lower distortion levels than that of class AB amplifiers, and generally their higher-order products drop off in amplitude at a higher rate. Because of the lower signal levels in the exciter, it is seldom the limiting factor in determining the transmitter distortion. This is even more true if a postselector is used to attenuate wideband noise and harmonics. Since solid-state amplifiers are inherently broadband, solid-state power amplifiers seldom employ tuned circuits. To attenuate harmonic and sum IM products, lowpass or bandpass filters are used instead. The filters are often arranged to cover a frequency range of about $\sqrt{2}$. This ensures that at the lowest edge of the band in use there is enough spectral space to gain sufficient attenuation to the second harmonic, which then lies at a frequency ratio of $\sqrt{2}$ beyond the upper edge of the band in use. Levels of distortion in the transmitter are such that it is seldom considered feasible to attenuate harmonics and sum IM distortion products to the same level as transmitter broadband noise, say in a 3-kHz bandwidth. Considering the previous example, even without the postselector, the output noise density of –118.5 dBW/Hz corresponds to a noise power in a 3-kHz bandwidth of –83.7 dBW, or a level –113.7 dB below the 1-kW transmitter rated output. Harmonic levels are specified at a level considerably higher, with the expectation that a collocated receiver will not simultaneously operate at transmitter harmonic frequencies. Specified harmonic levels vary, depending upon application, but second and third harmonic requirements may typically range between 40 and 90 dB below the fundamental frequency.

Simultaneous operation of more than one transmitter with one or more receivers

When two nearby transmitters are operating simultaneously, in addition to the receiver problems caused by each transmitter operating independently, an additional source of potential interference is caused by back-IM distortion produced by two transmitters coupling to each other and/or by IM distortion produced in the preselector or receiver, as shown in Figure 2.42.

In the case of two transmitters, like the IM products produced by a single power amplifier, the distortion level depends on the degree of amplifier nonlinearity, and again, class A amplifiers show considerably less back-IM distortion than class AB amplifiers. Besides the problem of IM distortion generation, power coupled between transmitters appears as reflected power and may result in power turndown by standing wave ratio (SWR) protection circuits. The presence of the two strong transmitter signals at the receiver or preselector input can also produce IM products, the level of which depends upon the input signal levels and the intercept points (see Chapter 4) for the various orders of distortion.

At any rate, the location of transmitter harmonics and IM frequencies is predictable because they occur at integral multiples of the sums and differences of the two transmitter frequencies, for example,

$$\left| mf_1 \pm nf_2 \right|$$

for all integer values of m and n, including zero.

In principle, then, it is possible to determine sets of transmitter frequencies f_1 and f_2 and the receiver frequency f_3 such that f_3 avoids transmitter harmonic and IM frequencies. Intermodulation products do not fall between the transmitter frequencies. Hence it is possible to operate the receiver be-

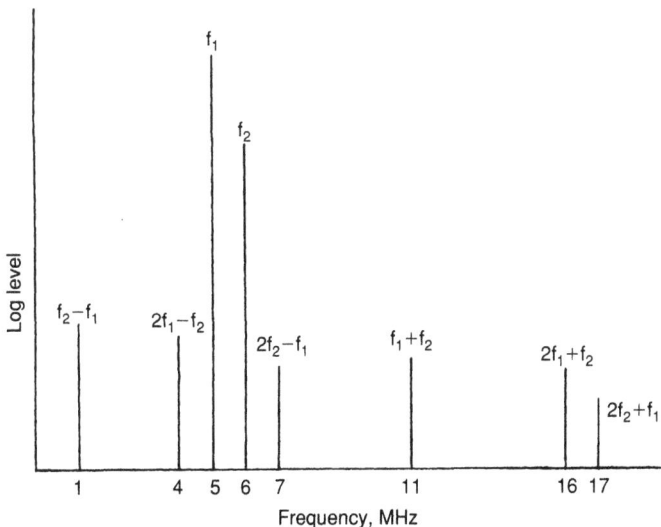

Figure 2.42 Low-order intermodulation products from two frequencies.

tween the transmitter frequencies without interference if transmitter harmonic regions are avoided, and as long as the separation from such transmitter frequencies is sufficient to avoid cross-modulation, IM products of each transmitter separately, transmitter broadband noise, and reciprocal mixing noise. Similarly, there are gaps between different orders of IM products that might also be used by the receiver.

Under conditions where it is difficult or impossible to avoid interference because of back-IM products, further filtering may be employed at the PA output to reduce those products. Reduction of back-IM products is helped both by the additional reduction in fundamental power coupled to the power amplifier from the other transmitter and by the further reduction of generated IM products by the selectivity of the filter.

2.3 ECCM Design Considerations

One of the reasons for renewed interest in HF communications, particularly by the military services, is their use as a backup system for satellite communications. To provide a measure of privacy, secrecy, and antijam capability, several schemes for spectrum spreading the signal have been proposed. Two basic types of spread-spectrum signals have gained wide acceptance. One of these involves using a pseudonoise (PN) code operating at high speed to switch the carrier phase at rates typically in the megahertz range. This produces a white-noise-like signal occupying several megahertz of bandwidth. Because of the wide bandwidth, the power density of this direct-spread signal is quite low, creating a minimum amount of interference to conventional narrowband signals in the band.

The second type of spread-spectrum signal in wide use utilizes frequency hopping over the band to spread the spectrum. The frequency hopping rate may vary from a few per second to thousands per second. Where the RF band occupancy is high, such as in the HF band, this is often the preferred ECCM method, because by proper choice of frequencies, strong fixed-channel stations in the band and interference to friendly signals can be avoided. A combination of the two methods, using a limited amount of direct-sequence band spreading along with frequency hopping, may also be employed.

Since the subject of this book is SSB systems and circuits, a comprehensive discussion of ECCM communication is beyond its scope, but those aspects that touch upon the design of SSB radio equipment often employed in ECCM communication will be covered. However, a discussion of modems and ECCM controllers, including selection of frequency hop sequences and synchronization, is beyond the scope of this book. Some useful introductions to spread-spectrum techniques are given in references 29 through 31.

Frequency hopping SSB

When frequency hopping is employed, there exists the possibility of direct transmission of the analog SSB voice signal without digital encoding. While not providing the voice privacy of which digital encoding is capable, there may be applications where hopping the analog signal alone is sufficient.

Several potential problems exist when an attempt is made to frequency hop an analog voice signal. To some extent these are common to all frequency hopped signals, but those most directly applicable to analog SSB voice transmission are synthesizer switching speed, phase discontinuities, and AGC.

During the time the synthesizer is switching from one frequency to another, the analog voice signal is lost. The effect this has on voice intelligibility depends both upon the percentage of time that the signal is lost and the length of time of the lost segment. Ideally the length of time of a lost segment should not exceed that of the shortest voice sound, or phoneme. Practically, this means that the longest gap should not exceed about 20 to 40 ms. The percentage of time that the signal is lost should not exceed about 20% or the signal will sound excessively chopped up. If, then, 20 ms represents the frequency transition period and this occurs 20% of the time, the hop rate is 10 per second, represented by 80-ms segments of the voice signal separated by 20-ms gaps.

It is, of course, possible to design synthesizers to switch frequencies much faster than 20 ms, and this then raises the possibility of faster SSB voice frequency hopping. In the laboratory, using the same synthesizer for the transmitter and receiver, it is possible to demonstrate excellent SSB voice transmissions using hop rates of thousands per second. In practice, however, when separate transmitter and receiver synthesizers are used, very close tracking of the phase of the two oscillators must occur or there will be corresponding phase discontinuities in the SSB voice signal when switching frequencies. These are not of great importance at low hop rates with considerable dead time between the hop periods because the correlation between adjacent speech signals on either side of the dead period is small. As the dead period becomes shorter, however, the correlation between adjacent transmitted speech segments increases. Phase discontinuities in the signal tend to destroy this correlation.

Even if the transmitter and receiver synthesizers track phase perfectly, phase discontinuities in the demodulated speech may result from differences in the propagation of the different hop frequencies. The problem is particularly severe with skywave propagation, where differences in propagation paths can completely destroy any phase coherence of the signal between hops.

The third problem with a hopping SSB voice signal is that of varying received signal strength at the different hop frequencies. Again, the problem is worse for skywave or for any multipath propagation. This problem is also more severe at high hop rates. It is very desirable to have the AGC stabilize in a short fraction of the on-time of the hop period. At the same time, it is desirable to have the AGC remain constant during the on-time so as not to further modulate the signal envelope and have it be capable of fast readjustment to the required AGC voltage for the next hop period. At low hop rates, the AGC design need not be much different than for conventional SSB voice receivers, but at high hop rates the attack time would have to be much faster, and the problem of rapid readjustment of the AGC voltage up or down, as required for the next hop period, is more severe.

Thus, although further experimentation with higher hop rates for specialized communication scenarios might lead to different conclusions, for general-purpose HF analog voice SSB communications, hop rates higher than around 10 per second are impractical.

Where the excision of the portion of the speech between frequency hops is objectionable, a method exists to eliminate this dead period at the expense of more circuit complexity, a total time delay of at least two hop periods, and a wider transmitted bandwidth, as follows. The speech wave is divided into segments equal to the hop period. Each segment in turn is filtered, sampled, digitized, and stored. The stored digitized speech segment is then read out of memory at a faster rate to correspond with the on-time of the hop period. It is converted to an analog signal, filtered, and used as the baseband signal to modulate the transmitter during the on-time of its hop period. Meanwhile the next speech segment is similarly processed. Since the stored speech is read out faster than it was stored, all its frequency components are increased by the ratio of the total hop period to the hop period on-time, so that if there is a 20% dead time, the bandwidth of the transmitted signal is increased by a ratio of 1 to 0.8, or an increase of 25%. At the receiver, the analog signal is again sampled, digitized, and stored. It is then read out of memory at a rate corresponding to the total hop period and reconverted to an analog signal. The reconstructed segment then occupies the same time span as the original, and the speech frequencies return to their correct pitches. In the meantime the next component speech signal is being received, digitized, and stored. This procedure can virtually eliminate the speech signals otherwise lost during the synthesizer switching time. The phase discontinuity problem still exists, but at low hop rates is not likely to be severe.

Equipment design considerations

When designing radio equipment for frequency hopping, factors that must be taken into account beyond that of conventional equipment design include

1. Synthesizer switching speed

2. Spectrum splatter

3. Receiver AGC

4. Bandwidth of receiver front end

5. Bandwidth of transmitter amplifier stages

6. Antenna and antenna coupler bandwidth

Synthesizer switching speed. The required synthesizer switching speed depends upon the maximum hop rate desired. In order to preserve the maximum amount of energy per hop, it is desirable to have the synthesizer settle quickly at the new frequency after the command to change frequency is given. If a 1-dB loss in energy is allowable, the maximum switching period is about 20% of the hop period. Synthesizer switching speed is discussed in Chapter 10, "Synthesizers for SSB."

Spectrum splatter. Spectrum splatter is a transient phenomenon that occurs when the frequency is changed. For a pulse of constant amplitude A and length T, the spectrum has the form

$$G(f) = AT \frac{\sin \pi T f}{\pi T f} \tag{2.80}$$

If such a pulse modulates a constant frequency f_1, the spectrum $G(f)$ is translated to f_1, with f in the expression for $G(f)$ replaced by Δf, where $\Delta f = f - f_1$. If f_1 changes each time period, as in a frequency hopping signal, the spectrum becomes the sum of the spectra of all the pulses. If the frequencies are produced by switching back and forth between a pair of oscillators, there will generally be no phase correlation between them, and the spectrum around each hopping frequency will drop off as $1/\Delta f$, as given above. This is typical of the spectrum of a discontinuous function. If the function itself is continuous, but has a discontinuous first derivative, the spectrum drops off as $1/(\Delta f)^2$, and in general, continuity of the function and its first m derivatives guarantees that the spectrum will eventually drop off as $1/(\Delta f)^{m+1}$. However, the frequency separation at which the ultimate drop-off first occurs may vary, so that for a given transition time between adjacent hopped frequencies and a given Δf, the spectrum may not always be lower for the transition having the highest number of continuous derivatives.

Phase continuity results if the same oscillator is used to generate a changing frequency without stopping and restarting each time the frequency is changed. However, the slewing of the oscillator adds to the generated spectrum. Also, if the frequency must be changed very frequently, it may be necessary to widen the frequency control-loop bandwidth to a point

where synthesizer wideband noise is excessive. It then may become necessary to switch alternately between a pair of synthesizers, each of which is
allowed an entire hop period to stabilize at the new frequency. As indicated
above, this leads to a $1/\Delta f$ spectrum rolloff. In order to produce a faster
spectrum rolloff, pulse shaping may be used. Its disadvantage, of course, is
that it reduces not only the spectral components, but also the total power
in the pulse. For example, a cosine squared (or raised cosine) pulse shape
as shown in Figure 2.43 has the spectrum

$$G_1(f) = \frac{AT}{2} \frac{\sin \pi Tf}{\pi Tf[1 - (Tf)^2]} \tag{2.81}$$

By comparison with the spectrum of a rectangular pulse (Equation 2.80),
the average amplitude of the raised cosine pulse is half that of the rectangular pulse of the same peak amplitude A and time duration T. A comparison of the spectra of the two pulses shows that the two numerators are the
same. Comparing the denominators, it is seen that the cosine squared pulse
spectrum eventually falls off as $1/f^3$ [or $1/(\Delta f)^3$ for a pulse modulated sine
wave]. This agrees with the conclusion drawn from the fact that the function and its first derivative are continuous for a single raised cosine pulse.
At $Tf = 0.5$, however, relative to their amplitudes at $f = 0$, the raised cosine
pulse spectrum is 4/3 times greater than that of the rectangular pulse, illustrating that for close-in spectra the modulating waveform with the higher
number of continuous higher-order derivatives may not have the highest attenuation. At $Tf = 5/2$, however, relative to their amplitudes at f_0, the relative amplitude of the raised cosine pulse is 5.25 times lower than that of the
rectangular pulse and its phase is inverted as well.

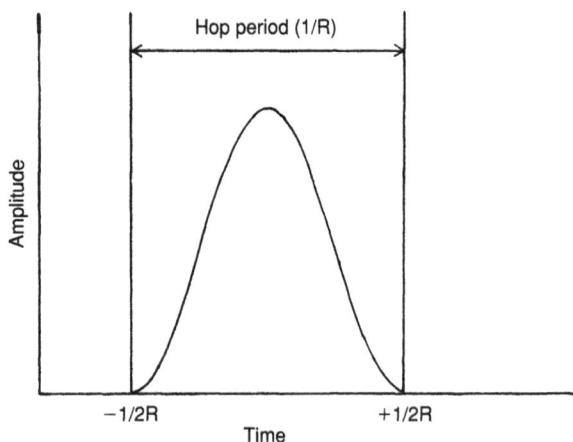

Figure 2.43 Raised cosine window.

Many other pulse modulation shapes may be used. A comprehensive analysis of the spectrum of a number of these is discussed in reference 32. Most of those in the cited reference, however, were evaluated on the basis of the spectral width of the main lobe and the strength of close-in spectral side lobes rather than on their ultimate rate of spectral drop-off.

Shaping the pulse may be done at any point in the transmitter where its shape will not be further distorted before transmission. Thus, in an exciter employing frequency translation, it should not be done before stages having selective circuits at their output, since these can stretch the shaped pulse so that its amplitude is not zero when the synthesizer changes frequency. Suitable places for the shaper would be just before or just after the last mixer that translates the IF to the RF band, as shown in Figure 2.44.

The shaper is basically a modulator. The desired modulating waveform may be stored in a read-only memory (ROM) that is read out to a digital-to-analog (D/A) converter synchronously with changes in frequency. Compensation can be made in the stored waveform for modulator nonlinearities and any D/A converter output filtering required.

Shaping is also desirable in the front end of a hopping receiver. A process similar to that in the exciter can create spectrum splatter at the desired signal frequency from undesired signals on nearby frequencies. In the presence of a strong off-tune signal at its input, when a rectangular pulse of oscillator voltage is applied to the mixer, a rectangular modulated pulse of the translated undesired signal results at the mixer output, displaced from the IF by the amount of its frequency offset from the desired frequency. The spectrum of such a pulse can extend to the IF passband, where it interferes with the desired signal. Preshaping the RF by modulating it before or just after the mixer, but before IF selectivity, as shown in Figure 2.45, with a pulse shape like that of the exciter has the same effect on reducing the spectrum produced by the interference as it does on the exciter spectrum.

Receiver AGC. Because of the differing atmospheric attenuation at different frequencies, a received frequency hopping signal is subject to change in amplitude from hop to hop. It is desirable to level out these changes in amplitude before detection. A simple way to do this for a constant amplitude

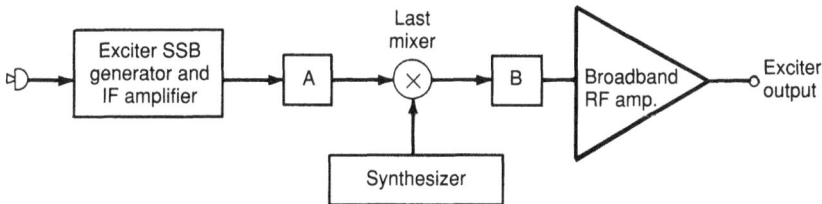

Figure 2.44 Exciter hopping pulse shaping modulator may be located at points A or B.

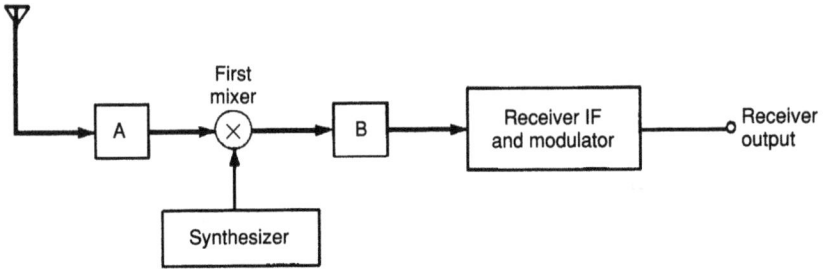

Figure 2.45   Receiver hopping pulse shaping modulator may be located at points A or B.

data signal is by means of a limiter. For a weak signal, limiting degrades the signal-to-noise ratio slightly, but the simplicity of the method commends its use where ultimate performance is not required.

If the best possible weak-signal performance is required, AGC must be used and adjusted to its proper level on a hop-by-hop basis. In the usual closed-loop feedback type of AGC, this requires that the AGC loop stabilize within a small fraction of the dwell period. Normally the RF pulse will be shaped to reduce the spectral splatter, and the detector and its matched filter will be designed to give the best output signal-to-noise ratio with that shaped pulse. A fast-acting AGC will try to undo this shaping. What is needed is a way to preset the AGC at each hop frequency so that the pulse shape is retained and the output is constant from hop to hop. The proper gain control voltage for each frequency can be approximated by averaging the required AGC voltages for past dwell periods on that frequency. This can be accomplished as follows. An AGC storage memory location is provided for each frequency. Initially it is zeroed. The AGC time constant is such that it allows the proper AGC voltage to be reached at the end of the dwell period. This value is digitized and stored. At the next frequency location the AGC voltage again starts from zero and its final value is stored, and so on for all frequencies. The next time the same frequency is encountered, the stored AGC voltage for that frequency is applied. However, the AGC voltage is allowed to readjust itself up or down. At the end of that dwell period the final AGC voltage is averaged with the initially applied value and stored as the new value for that frequency. This method gives the most weight to the most recent AGC value, and weighs preceding sample values less and less the further they are removed from the present sample. When the receiver gain is controlled in this manner, on a hop-by-hop basis, each controlled stage must have a rapid response time so that the gain can be stabilized quickly for each hop period.

Bandwidth of receiver front end.   High-frequency receivers that employ broadband front ends are favored for ECCM because there are no RF circuits that need to track the hopping signal. As will be seen in Chapter 4, this fits in well with modern receiver design unless a preselector is used. If a preselector is employed, either it must be a broadband type that covers all the hopping frequencies in one band or it must be capable of tuning rapidly between frequencies. For slow hopping speeds, relay-switched tuning elements can be used, but rapid frequency changes require the use of electronic rather than mechanical tuning. A further discussion of preselectors is found in Chapter 9.

Bandwidth of transmitter amplifier stages. Broadband amplifier stages are preferred for ECCM, again because there are no RF-tuned circuits that need to track the hopping signal. Fortunately, wideband amplifiers are easily designed using solid-state amplifiers. However, a problem that occurs with broadband amplifiers is that wideband noise is amplified together with the signal, as discussed in section 2.2. Since this noise can cause interference to nearby receivers, it is desirable to suppress it. Much of this noise may be suppressed by a postselector placed between the exciter and power amplifier. At this power level, often in the range 0.1 to 1 W, the power is low enough that a common design for the postselector and preselector may be used. As for the preselector, the design may be either broadband to cover all the hopping frequencies in one band or narrowband to tune rapidly between frequencies. For further attenuation of noise outside the hopping band, a wideband filter covering just the hopping band might be used at the transmitter output.

Antenna and antenna coupler bandwidth. A broadband antenna is desirable for ECCM transmission because this eases the burden on the antenna coupler. The antenna SWR can be further reduced by using wideband matching networks.

When the antenna bandwidth is narrow, it is desirable to tune it with the coupler to present the proper load to the power amplifier. If the hop rate is low, such a coupler may be digitally tuned using fast-acting relays. The limited lifetimes and response times of relays make it desirable to look for better switches. The solid-state switch offering the most promise is the PIN diode, because of its fast switching speed, relatively low distortion, and almost unlimited lifetime if operated within its ratings. Some of the biggest problems to date in employing PIN diodes for couplers are their limited power-handling capability, the high dc voltage required to back-bias "off" diodes, and the relatively high forward current for the "on" diodes. The switching of these high voltages and currents to the diodes at fast hop rates and the RF interference caused by the sudden application and removal of these high voltages and currents is also a potential problem. With currently (1995) available PIN diodes it is possible to construct frequency hopping couplers in the 100- to 1000-W range, depending upon the antenna impedance characteristics. A further discussion of antenna couplers is given in Chapter 15.

References

1. K. D. Kryter, "Methods for the Calculation and Use of the Articulation Index," *J. Acoust. Soc. Am.* 34 (1962): 1689–1697.

2. K. D. Kryter, "Validation of the Articulation Index," *J. Acoust. Soc. Am.* 34 (1962): 1698–1702.

3. L. L. Beranek, "The Design of Speech Communication Systems," *Proc. Inst. Radio Eng.* 35 (1974): 880–890.

4. H. R. Bertscher and J. C. Webster, "Intelligibility of UHF and VHF Transmissions at Fifteen Representative Air Traffic Control Towers," *J. Acoust Soc. Am.* 28 (1956): 561–564.

5. Harvey Fletcher, *Speech and Hearing in Communications* (Princeton, NJ: D. Van Nostrand, 1953).

6. N. R. French and J. C. Steinberg, "Factors Governing the Intelligibility of Speech Sounds," *J. Acoust. Soc. Am.* 19 (1947): 90–119.

7. *American National Standard Methods for the Calculation of the Articulation Index*, ANSI 53.5-1969 (New York: American National Standards Institute, 1969).

8. J. C. R. Liklider and I. P. Pollack, "Effects of Differentiation Integration and Infinite Peak Clipping upon the Intelligibility of Speech," *J. Acoust. Soc. Am.* 20 (1948): 42–51.

9. J. M. Frazer, H. H. Hass, and M. G. Schachtman, "An Improved High-Frequency Radiotelephone System Featuring Constant Net Loss Operation," *Bell System Tech. Jour.* (April 1967): 677–720.

10. L. R. Kahn, "Single Sideband Transmission by Envelope Elimination and Reconstruction," *Proc. of IRE* 40 (July 1952): 803–806.

11. F. H. Raab, "Envelope Elimination and Restoration System Concepts," *Proc. of RF Expo East* (Boston, MA) (Nov. 11–13, 1987): 167–177.

12. F. H. Raab, "Envelope Elimination and Restoration System Requirements," *Proc. of RF Technology Expo* (Anaheim, CA) (Feb. 10–12, 1988): 499–512.

13. F. H. Raab and D. J. Rupp, "Class S High-Efficiency Amplitude Modulator," *RF Design* (May 1994): 70–74.

14. F. M. Gardner, *Phase Lock Techniques* (New York: John Wiley & Sons, 1966).

15. W. C. Lindsey, *Synchronization Systems in Communication and Control* (Englewood Cliffs, NJ: Prentice-Hall, 1972).

16. J. P. Costas, "Synchronous Communications," *Proc. of IRE* 44 (1956): 1713–1718.

17. J. F. Nickerson and D. K. Weaver, Jr., "A Study of the Effect of Frequency Translation Error on the Intelligibility of Speech in the Presence of Noise," unpublished internal document, Electronics Research Laboratory Endowment and Research Foundation, Montana State College, Bozeman, MT, January 1959.

18. W. C. Lindsey and M. K. Simon, *Telecommunication System Engineering* (Englewood Cliffs, NJ: Prentice-Hall, 1973).

19. R. O. Carter and L. K. Wheeler, "LINCOMPEX—A System for High Frequency Radio-Telephone Circuits," *British Communications and Electronics* 12, 8 (August 1965).

20. S. M. Chow et al., *Syncompex Voice Processing* (Ottawa, Canada: Communications Research Center, 1980).

21. Lusignan, "Amplitude Compandored SSB in Mobile Radio Bands," Stanford Electronic Laboratory, Stanford University, July 1980.

22. S. C. Plotkins, "FM Bandwidth as a Function of Distortion and Modulation Index," *IEEE Trans. Commun. Tech.* 15-3 (June 1967): 467–470.

23. P. Frutiger, "Noise in FM Receivers with Negative Frequency Feedback," *Proc. of IEEE* 54 (November 1966): 1506–1520.

24. John Grebenkamper, "Phase Noise and its Effects on Amateur Communications," Part 1, *QST* (March 1988): 14–22.

25. John Grebenkamper, "Phase Noise and its Effects on Amateur Communications," Part 2, *QST* (April 1988): 22–25.

26. Ulrich L. Rohde, "All About Phase Noise in Oscillators," Part 1, *QEX* (December 1993): 2–6.

27. Ulrich L. Rohde, "All About Phase Noise in Oscillators," Part 2, *QEX* (January 1994): 9–16.

28. Ulrich L. Rohde, "All About Phase Noise in Oscillators," Part 3, *QEX* (February 1994): 15–24.

29. R. C. Dixon, *Spread Spectrum Systems* (New York: John Wiley & Sons, 1976).

30. R. C. Dixon (ed.), *Spread Spectrum Techniques* (Piscataway, NJ: IEEE Press, 1976).

31. W. E. Sabin, "Spread Spectrum Applications in Amateur Radio," *QST* 67 (July 1983): 14–19.

32. F. J. Harris, "On the Use of Windows for Harmonic Analysis with the Discrete Fourier Transform," *Proc. of IEEE* 66 (January 1978): 51–83.

3

High-Frequency (HF) Link Establishment

**Dr. David H. Bliss, James V. Harmon,
Daniel P. Roesler, Joseph C. Whited**

3.1 Overview of the Link Establishment Process

High-frequency communication link design and establishment is the subject of Chapter 3. Section 3.2 provides a review of system design trade-off considerations and an overview of propagation and system performance predictions. This is followed by a design example showing how a radio link may be designed. Next, a new tool is introduced that provides automated assistance for HF skywave frequency selection and applications planning.

The communications link design example demonstrates the need for link establishment and frequency management techniques, including the highly desirable feature of automatic connectivity. The latter subject is covered extensively in section 3.3, which gives general design principles and an example of a highly developed system in current use for global communications. Lastly, section 3.4 addresses HF network design issues based on automatic link establishment implementation and operations.

In general, a complete HF link management system includes the following functional elements to some degree [1, 2]:

1. Preparatory and/or on-line propagation and performance prediction algorithms and programs

2. A control mechanism, including an "engineering orderwire" function

3. Channel evaluation and/or sounding and noise monitoring

4. Selective calling and a frequency-scanning receiver, if automatic connectivity is required

5. On-line performance monitoring and adaptation

Figure 3.1 represents some of the relationships between these functions in the complete HF link call process, as discussed below. The optimum combination of the five functional elements and the implementation of each is dependent upon the overall system requirements and the available funding; that is, frequency management capability and hence system effectiveness must be traded off against capital and operational costs. When optimized data transmission is required, some form of adaptive data terminal, having variable modulation and coding types and rates in its library, is highly desirable.

Propagation/performance predictions

Considerations for propagation predictions are discussed in section 3.2. Computer, microprocessor, data base, propagation prediction, and supporting-sensor technologies are evolving very rapidly [2, compared with 3, 4, and 5]. For instance, it is now possible to receive continuous broadcasts of

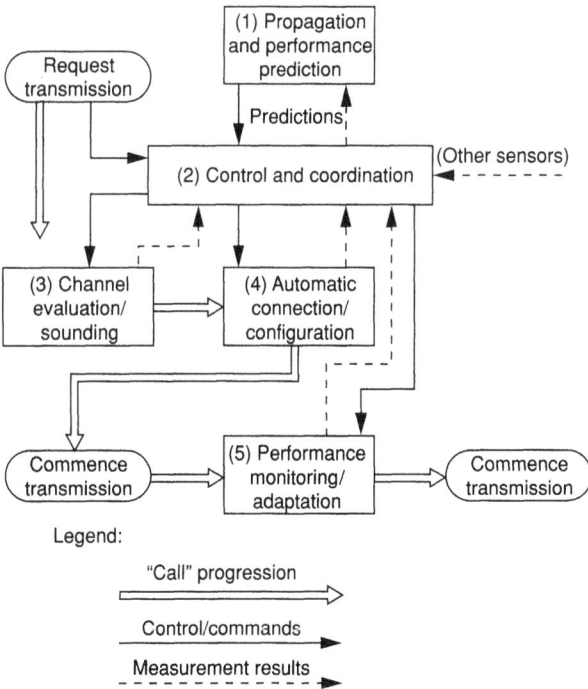

Figure 3.1 Functional representation of frequency management and link-establishment process for HF communications.

solar, ionospheric, and geomagnetic data from satellites using a simple earth-terminal receiver [6], which can be utilized in refining propagation predictions, as described in section 3.2. Clearly, near-real-time propagation predictions will become an essential part of HF frequency management in the near future, as they are already being employed in selecting scanned frequencies for some operational HF systems.

Control mechanism

The degree of sophistication and cost of the frequency management control mechanism will vary greatly from one system to another. For the simplest operations, voice calls (QSY) on one or two fixed frequencies are employed. In other, more advanced systems, the control mechanism is built into the automatic connectivity function implementation. Thus, once a connection is established, it is possible to automatically pass additional control information between stations using the same signaling equipment and formats as used in automatically establishing the link. Automatic link coordination is discussed extensively in section 3.3.

Channel evaluation and sounding

The core of transmission performance and throughput optimization, especially in a dynamic and stressed environment, is the channel evaluation process. The rationale for the development of real-time channel evaluation (RTCE) techniques is simple. Significant improvements in the use of the HF propagation medium can be achieved only if a communicator, or HF systems controller, using a specific path at a given time, has access to real-time data on the path parameters rather than having to rely on frequency predictions, which can be subject to appreciable errors [7]. But further development and guidance in RTCE subsystem design and selection is needed. MIL-STD-188-100 [8] and MIL-STD-188-317 [9], for example, are intended to be used in the design and installation of new military HF radio communications systems, subsystems, and equipment. However, they do not provide specifications for acceptable limits of certain parameters introduced by the ionospheric propagation channel or designations of methods and techniques for measuring those parameters. Most notably lacking from definition and specification are the channel parameters of multipath time-delay profile and rms spread, (Doppler) frequency profile and rms spread, and fading depth and rate [1, 10, 11]. These, coupled with the burst characteristics of atmospheric and other impulsive noise, determine the ultimate performance of higher-speed (>75 bits/sec) data transmission over the HF circuits. Indeed, for transauroral paths, it has been shown that all of these factors are important for transmission rates as low as 75 bits/sec, a standard teletype channel rate [12]. Although the types of radio equipment under consideration are not the cause of these channel-induced parameters, the effects of the parameters in defining an "irreducible" bit error rate (BER) for data transmission must be taken into account in the HF system design and operation (e.g., see references 13 and 14).

A wide variety of channel evaluation techniques have been designed and experimented with in the past. Vertical-incidence and oblique-incidence ionospheric sounders, while popular in the past, are expensive to purchase because they utilize separate RF equipment from that employed for communications (see references 15–17 for examples). Generally they require trained interpretation of the resultant ionograms to determine the optimum frequency. The current trend is to utilize the system's RF equipment (employed for communications) as the primary tool for probing or sounding the channel, with high-speed, digital signal processors added at the terminal to extract useful channel parameters from the received IF or baseband signal. One example is an advanced link quality analyzer (ALQA), designed by Rockwell Corp., which measures many HF parameters and is compatible with an existing automatic connectivity subsystem for conventional, processor-controlled HF radios [12, 18].

Automatic connectivity

Provision of automatic connectivity is perhaps the key building block in advanced HF systems design and implementation. The primary function of automatic connectivity is to quickly and reliably bring two or more stations into HF communications with each other, without the need for manual trial-and-error optimum frequency searching. Two examples of fielded automatic connectivity subsystems currently being utilized are automatic link establishment (ALE) and RACE [19]. The ALE system provides automatic connectivity within 10 seconds of call initiation; details of ALE design, operation, and performance are provided in sections 3.3 and 3.4.

Performance monitoring

The degree of complexity required for on-line performance monitoring varies greatly with the intended application of the HF system. Ordinary voice circuits are adequately monitored by the users themselves, who can readily detect when the signal quality deteriorates sufficiently to require a change in operating frequency or configuration. For high-speed data transmission, the situation is different; often the users are not directly monitoring the quality of the output or the channel conditions. Several schemes for automatic, on-line HF data transmission performance monitoring have been devised and tested. Some of these employ measures of analog signal distortion in the demodulator (similar to the "eye pattern" used in wireline communications). Others operate with a parallel "pseudoerror" generation device or algorithm at the demodulator to detect in advance when the channel is deteriorating. With a larger number of HF data circuits now employing forward error correction (FEC) coding, error detection and correction (EDAC) coding, or automatic repeat request (ARQ) operation, it is practical to monitor the performance by using the error detection circuitry. Especially with the advent of soft-decision decoding techniques, the trend will be toward using the BER indicator as the primary on-line performance monitoring method. Specifications and standards cited in section 3.3 include performance monitoring considerations.

It should be noted, however, that such performance monitors will not necessarily provide a meaningful indication of which channel parameters are contributing to the degraded BER performance. Hence, reversion to a channel evaluation technique must likely be employed to determine what form of adaptation is required (frequency or terminal configuration) to recover from the situation. For practical purposes, there is no universal on-line performance monitor applicable to the wide variety of HF communications waveforms currently employed. The selection of the technique(s) to use will be dependent on the form of modulation and coding chosen for data transmission in the overall system design.

3.2 System Planning Utilizing Propagation Predictions

There are many factors that can be traded off in the process of arriving at the most effective system design for HF communications. Some of the parameters in the system design that can be varied in the trade-off process are listed here:

- Signal-to-noise density ratio for required quality of service, that is, voice quality or data BER, SN_0R_R, dB-Hz
- Data transmission rate, R, bits/sec
- Transmit power amplifier output level, P_t, dBW
- Transmit antenna coupler loss, L_c
- Transmit antenna gain, "takeoff" angle, and efficiency, G_t, dB
- Receive antenna gain, G_r, dB
- Receiver noise figure
- Modulation type and parameters
- Demodulation and processing type and parameters
- Source and channel coding types and level
- Propagation loss (groundwave versus skywave mode), L_p, dB
- Miscellaneous cabling, distortion, and intermodulation losses, L_m, dB
- Received external (atmospheric and artificially created) noise density, N_0, dBW/Hz

Illustrations of specific trade-offs are given in several of the references (see references 2, 13, 14, and 20, for example).

The received external noise can generally be assumed to be independent of the receive-antenna gain, based on an approximately uniform distribution of noise in the space surrounding the receive antenna. Furthermore, good receiver design will provide a receiver noise figure such that the noise internal to the receiver can be neglected compared with the rather high level of external HF noise. Using the above-defined symbols for the terms, it is readily apparent that the *available* signal-to-noise density ratio at the input to the receiver SN_0R_A is (in decibel notation):

$$SN_0R_A = P_t - L_c + G_t - L_p + G_r - L_m - N_0 \text{ dB-Hz} \qquad (3.1)$$

Thus, the objective of HF system design is to choose the system parameters such that SN_0R_A equals or exceeds the required signal-to-noise density SN_0R_R a sufficient percentage of the time (over the distribution of environmental and temporal factors to be considered) such that the desired system reliability can be achieved with reasonable confidence. The choices of parameter values such as required quality of service, data transmission rate, modulation, demodulation, processing and error correction coding types, and configurations all impact the required SN_0R_R for the service to be provided.

Many of the system design factors (that is, equipment performance parameters) are described and analyzed in detail elsewhere in this text. Guide-lines for HF antenna design and/or characterization can be found in several of the references (see 13, 21–24, as examples).

Use of propagation and link performance predictions (forecasts)

The topic of HF radiowave propagation is fairly complex and is treated extensively in references 1, 2, 5, and 26. An overview of prediction techniques is provided by several of the references (see 1, 2, and 27, for instance). Methods of HF radiowave propagation and link performance predictions can be categorized for practical application purposes by the type of computational tools employed: manual methods, micro-/minicomputer-based methods, and large, mainframe computer-based methods. Examples of these categories are discussed here. The system design methodology presented here can now be conducted using modern microcomputers.

Manual methods were the first-used approach to radiowave propagation predictions. These methods generally employ charts, nomographs, and/or tables of propagation and noise statistical parameters. Use is made of overlays on a specially constructed map of the earth so that geographical (latitude and longitude) effects on propagation can be taken into account by laying out the great-circle propagation path. Examples of propagation prediction principles are given in references 20, 25, and 28 through 30.

Graphical presentations of expected atmospheric radio noise are available for use in manual link analysis [31]. These data are also employed in the form of a data base in automated methods using computers. Manual methods are no longer used for commercial design of HF communications systems because of the length of time and expense of manual effort required to consider the multitude of possible design combinations, especially for a system or network with a large number of possible point-to-point or mobile-terminus paths.

In recent years, simplified approaches to computer prediction of ionospheric-model(s) propagating frequencies, maximum usable frequency (MUF), and, to some extent, propagation loss have been developed. These tools are intended more for operational (day-by-day) use than for large-system design applications. In general, they were developed for military applications, although they are now widely applied to other uses as well. They can be employed on minicomputers or even on desktop or laptop microcomputer-based processors. Early examples of these small computer-based tools are the computer programs MINIMUF and PROPHET [3, 4]. (Because of their original government development and the wide proliferation of a number of versions of these programs, especially MINIMUF, applications support for their use may be difficult to obtain.)

The current U.S. standard large-computer propagation and system performance prediction program is IONCAP, developed by the U.S. Department of Commerce [13]. Due to growth in processing speed and memory capacity, IONCAP can now be run on many personal computers, as discussed at the end of this section. Other versions of large, mainframe computer programs for the same purposes are available elsewhere from international agencies (see reference 28, for example).

Example of system design

As an example of a typical HF SSB system design process, a circuit from Cedar Rapids, Iowa, to Hong Kong is considered. This circuit has a great-circle path length of nearly 12,500 km (about one-third the circumference of the earth), demonstrating the potential of HF SSB systems for long-haul communications. It illustrates the care and trade-offs required in selecting power levels, antenna types, frequency assignments, and operational procedures/tools for a long-haul circuit. As one might expect, reliable ionospheric skywave communications are not always available on a circuit of this type, and operating frequencies and contact times must be selected judiciously.

The propagation and performance prediction tool chosen in illustrating the design process is the IONCAP computer program. There are several methods of analysis available in using this program, which provide versatility in the types and forms of outputs available from the analysis, only a few of which are illustrated here. Also, some data reduction and presentation tools that make use of the IONCAP outputs are utilized here to facilitate rapid processing and condensation of the outputs into summarized form. (These latter tools have been developed within Rockwell for on-line and off-line application of a microcomputer terminal to the reduction process.)

The IONCAP program method chosen to begin the system design in this case is called Method 16. The first step is to run some sample month and sunspot number combinations for hypothetical ideal *isotropic* antennas (that is, antennas with uniform gain in all directions of spherical space). Thus, the antennas are first modeled as lossless radiators (collectors) having a constant gain versus elevation angle and azimuth angle of 0 dBi (gain relative to that of a reference isotropic antenna). Knowing that this path length is quite long (just over 12,000 km) and that a minimum of three maximum-length (4000-km) "hops" will normally be required, a relatively high power level of 10 kW is initially assumed.

Also, since long hops are involved, the trial operating frequencies are initially chosen in the 8- to 26-MHz portion of the band, rather than in the 2- to 10-MHz spectrum normally associated with short path lengths. If higher or lower frequencies are required, the initial predictions will indicate this by the occurrences of MUFs outside the chosen band for some times of day and/or month sunspot combinations.

Another input factor to be chosen in the performance analysis is the required signal-to-noise density ratio SN_0R_R. The value of SN_0R_R used in the analysis must be chosen very carefully, with due consideration for fading and atmospheric noise effects on the demodulation integrity/fidelity, as a few decibels variation in its value can cause a substantial variation in the predicted (and experienced) system performance. The required value is derived from section 2.1 results as follows. Assume that 90% intelligibility of words from a 256-word vocabulary is required, which is adequate for operator-to-operator ("order-wire") communications. From Figure 2.1, this requires an articulation index (AI) of approximately 0.36. From Figure 2.37 (extrapolated), SSB operations with speech preemphasis in the presence of two-path multipath fading requires a P_{avg}/N_0 of approximately 41 dB-Hz for an AI of 0.36. In systems analyses, transmitted power levels (and hence received signal power levels) are usually specified in terms of peak envelope power. In comparing Figures 2.34 and 2.35, it is noted that, for an AI of 0.36, the difference between the required PEP/N_0 and P_{avg}/N_0 is 7 dB. Thus, using the results from section 2.1, the required value to be used here is given by

$$SN_0R = \text{PEP}/N_0 = 41 + 7 = 48 \text{ dB-Hz} \qquad (3.2)$$

(This is also the value specified in the IONCAP user's manual [13] for operator-to-operator quality of service for 90% intelligibility of related words for nondiversity, SSB, suppressed-carrier communications.)

If higher-quality commercial-grade voice communications or high-speed digital data transmission is to be provided, a significantly higher value of SN_0R_R must be used in the analysis to provide a realistic system performance prediction. The specification of SN_0R_R for representative data systems is beyond the scope of this section; examples for typical data modulation schemes are given in references 9, 13, and 14.

Another input parameter to be chosen judiciously is the approximate level of artificially created electrical (RF) noise density at the receiving sites. A value of –148 dBW/Hz (specified at 3 MHz, with a known rolloff in level with increasing frequency) is used in this system design example. This is typical of levels experienced in rural locations [13]. Thus, the location assumed here is that of a typical station situated at the outskirts of a residential area, away from large industrial plants and heavily traveled highways. (Somewhat higher [typically +12 dB] values are experienced in residential urban areas and lower [typically –16 dB] values are found in remote, unpopulated areas.) Atmospheric noise tends to dominate at the lower frequencies and at nighttime, while artificial noise can dominate at the higher frequencies, especially in daytime. It is readily apparent that the predicted performance of the system is directly dependent on the *total* RF noise density (atmospheric and artificial) N_0 at the receiver site. The IONCAP program contains a data base of expected atmospheric noise levels as a function of location (latitude and longitude), season of year, and time of day (TOD). It sums the expected atmospheric noise level for a receiver location with the value of artificially made noise at the frequency of the calculation, based on the value specified at 3 MHz.

Typical values for the month and sunspot activity level must be chosen for the initial analysis, as conducting the analysis for all possible month/sunspot combinations is excessively lengthy, expensive, and unnecessary. Predictions for June and December (the solar solstice months) and for sunspot number (SSN) values of 10 and 130 essentially bracket the expected variations in propagation conditions and are adequate for system design purposes [9]. A slightly more conservative value for maximum SSN of 110 is used in this example. Also, predictions for an SSN of 60 are included to represent the long-term solar-cycle average conditions. The maximum practical time interval between times of day for which the analysis is performed is 4 hours. Because the circuit under consideration is a difficult one and careful, detailed selection of TOD for reliable operations is required, the example analysis is conducted here for all 24 hours of the day (at intervals of 1 hour).

As described above, initial system performance predictions are made using Method 16 for isotropic antennas at both locations. Figure 3.2 shows a sample tabulated output for this case for one month (June) and one SSN (10) for three selected hours of the day (universal time, or UT, which is the same as Greenwich mean time, or GMT). (In some of the figures that follow,

```
                        METHOD 16    IONCAP 78.03   PAGE   5
             JUN                    SSN =  10.
CEDAR RAPIDS,IA TO HONG KONG                 AZIMUTHS           N. MI.      KM
42.00 N    91.50 W - 22.00 N   114.10 E     334.45   20.23   6709.6   12425.3
                               MINIMUM ANGLE   1.0  DEGREES
ITS- 1 ANTENNA PACKAGE
XMTR   2.0   TO  30.0  CONST. GAIN  H   0.00 L    0.00 A    0.0  OFF AZ    0.0
RCVR   2.0   TO  30.0  CONST. GAIN  H   0.00 L    0.00 A    0.0  OFF AZ    0.0
POWER =  10.000 KW   3 MHZ NOISE = -148.0 DBW   REQ. REL = .90  REQ. SNR = 48.0


  UT  MUF

 13.0 13.7 11.9   8.0 10.0 12.0 14.0 16.0 18.0 20.0 22.0 24.0 26.0 FREQ
      F2F2 F1F2  EF2 F1F2 F1F2 F2F2 F2F2 F2F2 F2F2 F2F2 F2F2 F2F2 MODE
       5.4 10.0  1.0 10.0 10.0  4.3  8.1  8.1  8.1  8.1  8.1  8.1 ANGLE
      14.0 14.0 14.0 14.0 14.0 12.0  3.0  1.0  8.1  8.1  8.1  8.1 ANGLE
      44.2 43.5 42.6 43.3 43.6 44.2 44.0 43.8 43.8 44.3 44.3 44.3 DELAY
      415. 302. 177. 260. 311. 420. 376. 368. 384. 434. 434. 434. V HITE
       .50  .82 1.00  .97  .81  .43  .08  .00  .00  .00  .00  .00 F DAYS
      149. 152. 159. 156. 151. 151. 179. 221. 276. 334. 335. 335. LOSS
       21.  17.   6.  12.  18.  19.  -8. -49.  ****  ****  ****  **** DBU
     -109 -111 -119 -115 -111 -110 -139 -181 -235 -294 -294 -294 S DBW
     -162 -159 -152 -156 -160 -163 -166 -168 -170 -171 -172 -173 N DBW
       54.  48.  33.  41.  49.  53.  27. -65.  ****  ****  ****  **** SNR
       20.  26.  28.  27.  25.  21.  47.  87. 139. 183. 182. 181. RPWRG
       .61  .50  .02  .18  .52  .59  .15  .00  .00  .00  .00  .00 REL

 14.0 15.0 13.0   8.0 10.0 12.0 14.0 16.0 18.0 20.0 22.0 24.0 26.0 FREQ
      F2F2 F1F2  EF2 F1F2 F1F2 F2F2 F1F2 F1F2 F1F2 F1F1 F1F1 F1F1 MODE
       5.2 10.0  1.0 11.5 10.0  7.8  3.9  3.9  3.9  3.9  3.9  3.9 ANGLE
      12.0 14.0 14.0 14.0 14.0 10.0 10.0  3.0  1.0  3.9  3.9  3.9 ANGLE
      44.6 43.7 42.7 43.4 43.4 44.6 43.5 43.6 43.4 43.4 43.4 43.4 DELAY
      478. 325. 181. 275. 275. 462. 319. 319. 350. 308. 308. 308. V HITE
       .50  .90 1.00 1.00  .98  .74  .24  .02  .00  .00  .00  .00 F DAYS
      145. 147. 159. 154. 152. 142. 161. 199. 261. 357. 371. 371. LOSS
       25.  22.   6.  13.  17.  28.  10. -27. -88.  ****  ****  **** DBU
     -105 -107 -119 -113 -111 -101 -121 -159 -220 -317 -330 -330 S DBW
     -164 -161 -151 -156 -160 -163 -166 -168 -170 -171 -172 -173 N DBW
       60.  55.  32.  42.  49.  62.  45.   9. -50.  ****  ****  **** SNR
       14.  19.  28.  24.  25.  12.  29.  64. 124. 219. 218. 217. RPWRG
       .72  .63  .01  .22  .51  .75  .44  .03  .00  .00  .00  .00 REL

 15.0 16.0 13.9   8.0 10.0 12.0 14.0 16.0 18.0 20.0 22.0 24.0 26.0 FREQ
      F2F2 F1F2  EF2  EF2 F1F2 F1F2 F2F2 F2F2 F2F2 F2F2 F2F2 F2F2 MODE
       5.0  9.8  1.0  1.0  9.0  9.6  5.0  4.3  4.3  4.3  4.3  4.3 ANGLE
      10.0 14.0 14.0 14.0 14.0 14.0 10.0  3.0  4.3  4.3  4.3  4.3 ANGLE
      44.4 43.8 42.7 42.7 43.4 43.8 44.4 44.1 44.7 44.7 44.7 44.7 DELAY
      452. 338. 183. 190. 285. 339. 453. 437. 527. 527. 527. 527. V HITE
       .50  .82 1.00 1.00  .96  .81  .50  .14  .02  .00  .00  .00 F DAYS
      142. 145. 157. 149. 146. 145. 142. 161. 220. 265. 312. 342. LOSS
       30.  25.   8.  18.  23.  25.  20. -47. -91.  ****  ****  **** DBU
     -101 -104 -117 -109 -105 -104 -101 -120 -180 -225 -272 -302 S DBW
     -166 -163 -151 -156 -160 -163 -166 -168 -170 -171 -172 -173 N DBW
       65.  59.  34.  47.  55.  59.  65.  48. -10. -53. -99.  **** SNR
        9.  15.  25.  14.  15.  15.   9.  26.  84. 127. 173. 188. RPWRG
       .80  .70  .02  .45  .65  .71  .80  .50  .00  .00  .00  .00 REL
```

Figure 3.2 Sample IONCAP Method 16 system performance prediction tabulation.

the time is also indicated by LMT, the local mean time at the transmitter.) The results shown in the first full column of the figure are for the MUF, indicated on the top line of the column for each TOD. The results for the frequency of optimum transmission/traffic, or FOT (nominally about 85% of the MUF), are shown in the second column. The results for each even integer frequency from 8 to 26 MHz are shown in the remaining columns in each TOD block.

Several calculated results are tabulated for each frequency for each hour. MODE indicates the skywave propagation mode (layer) for the transmit-end hop and the receive-end hop, respectively. Similarly, ANGLE indicates the wavepath takeoff angle (elevation angle) for the two ends of the path. (There may be other hop modes in addition to those indicated at the transmitter and receiver, especially for a path of this great length.) DELAY is the calculated value of propagation time (in milliseconds) for the path of the modes shown. F DAYS is the probability that the specific frequency listed will be exceeded by the predicted MUF, indicating the fraction of days in the month that propagation by these modes is expected (in a statistical sense) at a given frequency for the tabulated TOD. LOSS is the median system loss (in decibels) for the most reliable mode. DBU is the median field strength expected at the receiver location (in decibels) above 1.0 μV/m. S dBW and N dBW are, respectively, the median signal and noise power density levels expected at the receiver input terminals (in decibels) above 1 W. The difference between these values is the available signal-to-noise density ratio, shown as SNR (in dB-Hz).

The results of primary interest are the communications reliability values (REL) listed in the last line in each TOD block. Each of these values is essentially the probability that the required SN_0R_R of 48 dB-Hz will be exceeded (provided) by the available SN_0R_A over the month for this TOD for the frequency listed. The calculated result, termed RPWRG, is the required decibel power gain in the system needed to obtain SN_0R_R for a specified percent of time (REQ REL), 90% in this figure. It is seen that only at the frequencies where the predicted reliability, REL, is already moderately high is the calculated value of RPWRG nominally 10 dB or less. Thus, proper choice of operating frequency is a necessity. The meanings of the remaining parameters tabulated in Figure 3.2 are explained subsequently, as needed for the system design process.

Figure 3.3 is an IONCAP Method 24 tabulation of the voice communications reliability summary for selected frequencies from 6 to 25 MHz for all 24 hours of the day for June with the SSN value of 60. This summary is for the case of 0-dBi transmit and receive antennas and for a nominal transmit power of 10 kW. It is apparent that the frequency of maximum reliability (FMR) varies significantly with the diurnal cycle. Also, the reliability is much higher for some times of day than for others. Figures 3.4A and 3.4B are graphical representations of the diurnal variation in maximum reliability and frequency of maximum reliability for this sunspot value for June and December, respectively. The maximum reliability plotted here is the largest value of reliability selected for each hour from tables such as that shown in Figure 3.3 for the FMR. (For this to be meaningful, it is assumed that some form of "adaptive" communications system is employed that utilizes automatic connectivity and optimum frequency selection.) It is seen that the reliability is higher for a longer interval of the day during June than during December. In general, these reliabilities are considered marginally or unacceptably low, so some improvement in the system design is required to provide an acceptable system.

```
                        METHOD 24    IONCAP 78.03    PAGE   1
          JUN                       SSN =  60.
CEDAR RAPIDS,IA TO HONG KONG              AZIMUTHS          N. MI.        KM
42.00 N    91.50 W - 22.00 N    114.10 E    334.45    20.23    6709.6    12425.3
                                MINIMUM ANGLE    1.0  DEGREES
ITS- 1 ANTENNA PACKAGE
XMTR   2.0  TO  30.0  CONST. GAIN  H    0.00 L    0.00 A    0.0  OFF AZ    0.0
RCVR   2.0  TO  30.0  CONST. GAIN  H    0.00 L    0.00 A    0.0  OFF AZ    0.0
POWER = 10.000 KW   3 MHZ NOISE = -148.0 DBW   REQ. REL = .50  REQ. SNR = 48.0

                        FREQUENCY / RELIABILITY
```

GMT	LMT	MUF	FOT	6	8	10	12	14	16	18	20	22	25	MUF
1.0	18.9	17.9	.02	.00	.00	.00	.04	.01	.09	.16	.07	.01	.00	.14
2.0	19.9	17.7	.01	.00	.00	.00	.00	.01	.00	.10	.03	.01	.00	.11
3.0	20.9	17.7	.00	.00	.00	.00	.00	.00	.00	.09	.03	.01	.00	.02
4.0	21.9	17.2	.00	.00	.00	.00	.00	.00	.00	.02	.01	.00	.00	.01
5.0	22.9	16.2	.00	.00	.00	.00	.00	.00	.01	.04	.00	.00	.00	.00
6.0	23.9	15.2	.00	.00	.00	.00	.00	.00	.02	.01	.00	.00	.00	.01
7.0	.9	14.4	.00	.00	.00	.00	.00	.01	.03	.00	.00	.00	.00	.01
8.0	1.9	13.5	.00	.00	.00	.00	.00	.04	.02	.00	.00	.00	.00	.02
9.0	2.9	12.8	.00	.00	.01	.00	.12	.10	.01	.00	.00	.00	.00	.14
10.0	3.9	12.4	.00	.00	.03	.00	.16	.08	.00	.00	.00	.00	.00	.22
11.0	4.9	12.5	.02	.00	.00	.05	.13	.11	.01	.00	.00	.00	.00	.20
12.0	5.9	14.3	.07	.00	.00	.00	.10	.38	.16	.02	.00	.00	.00	.35
13.0	6.9	15.7	.63	.00	.00	.00	.54	.80	.76	.53	.20	.01	.00	.76
14.0	7.9	16.2	.70	.00	.00	.06	.61	.70	.81	.62	.35	.05	.00	.81
15.0	8.9	17.3	.78	.00	.00	.00	.58	.72	.89	.66	.20	.00	.00	.77
16.0	9.9	18.2	.79	.00	.00	.00	.50	.71	.79	.80	.49	.03	.00	.79
17.0	10.9	18.8	.81	.00	.00	.00	.52	.67	.80	.83	.65	.00	.00	.82
18.0	11.9	19.1	.80	.00	.00	.00	.30	.59	.76	.83	.16	.01	.00	.76
19.0	12.9	19.1	.77	.00	.00	.00	.08	.52	.71	.77	.14	.01	.00	.19
20.0	13.9	16.9	.00	.00	.00	.00	.00	.00	.02	.07	.05	.00	.00	.04
21.0	14.9	18.2	.08	.00	.00	.00	.00	.00	.23	.50	.16	.00	.00	.49
22.0	15.9	18.4	.00	.00	.00	.00	.00	.00	.06	.18	.25	.05	.00	.38
23.0	16.9	18.7	.02	.00	.00	.00	.00	.00	.14	.12	.16	.03	.00	.16
24.0	17.9	17.3	.02	.00	.00	.00	.00	.00	.02	.10	.14	.02	.00	.12

Figure 3.3 Sample IONCAP Method 24 communication reliability summary tabulation.

Further system analysis and design

The most practical way to improve the performance of this hypothetical analog voice system is to increase the transmit and receive antenna gains for the frequencies and takeoff angles of interest. It is seen from Figure 3.2 and other similar tables (not shown) that the most-usable frequencies are in the range of 6 to 22 MHz, and that the required takeoff angles are typically 15 or less at these frequencies. To indicate a possible solution, Figure 3.5 shows a tabulation and graph of the gain versus takeoff angle and frequency for a typical horizontal log periodic array (LPA) antenna. This type of antenna has high gain at low takeoff angles and medium to high frequencies, as needed for this path.

(This pattern and several others are stored on computer magnetic tape and are available for use in IONCAP analysis. In addition, IONCAP has the capability of calculating and tabulating the gain versus frequency and takeoff angle for any antenna that can be modeled by one of several available configurations. In either case, IONCAP uses the specified antenna pattern results in the system performance predictions.)

Since fixed (ground) stations in rural areas are assumed, LPA antennas mounted at about 50 ft (15.2 m) in height appear to be a practical solution to the increased gain requirement. If cost, real estate requirements, or vul-

Maximum reliability

Link:42.00 N   91.50 W - 22.00 N 114.10 E   Month: Jun Tx pwr(kW): 10.00
Distance (km): 12425.3   SSN: 60.   Req'd SNR: 48.

A

Maximum reliability

Link: 42.00 N 91.50 W - 22.00 N 114.10 E   Month: Dec   Tx pwr (kW): 10.00
Distance (km): 124425.3   SSN: 60.   Req'd SNR: 48.

B

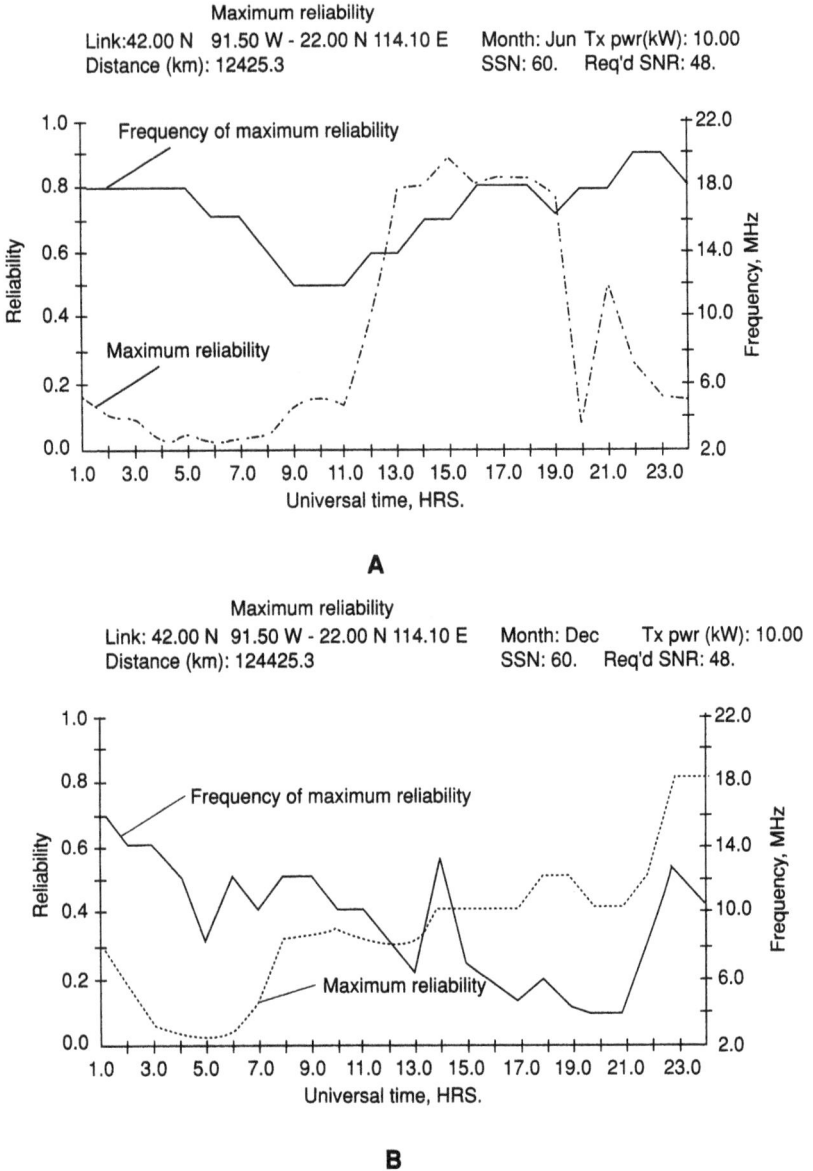

Figure 3.4 Sample computer-generated graph from IONCAP Method 24 data, showing maximum reliability and frequency of maximum reliability (FMR) for (A) June and (B) December; SSN = 60 for both.

nerability to icing and high winds rule out this type of antenna, some other choice must be made and a reduction in reliability and/or flexibility may result. This is typical of the trade-offs that must be made in HF systems design.

The remainder of the system analysis is for the assumed configuration of LPA antennas at both ends of the circuit. Figure 3.6 shows a sample reliability summary table for the same conditions present in Figure 3.3, but with LPA antennas employed. It is apparent that the additional 15 dB or more of system gain results in a significantly improved SN_0R_A for this circuit. Figures 3.7A and 3.7B plot the maximum reliability and FMR for the same conditions as in Figures 3.4A and 3.4B, but with the improved antennas. These improvements result in a circuit that is usable for operator-to-operator quality voice communications about 5 hours of the day for June, SSN = 10, and 8 hours of the day for June, SSN = 110. The diurnal availability for the month of December is somewhat less. Contacts between the two stations would have to be very carefully scheduled and would last for only 1 or 2 hours daily during the winter months for all SSN values. Some means of automatic selection would be essential to successful operations. If additional hours of contact and/or a higher SN_0R_A are required, the system designer must consider the cost and value of adding more transmitter power to the system. Because the present level is assumed to be 10 kW, a further increase in power output level would be quite expensive. The designer and the user representative must carefully consider the benefits to be gained by further power increases versus increased costs and make a decision judiciously before proceeding.

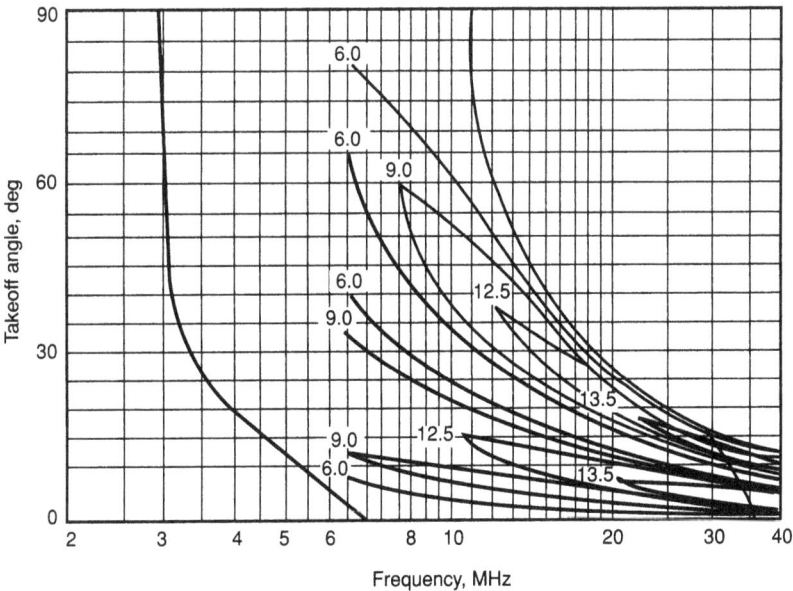

Figure 3.5 Gain versus takeoff angle for a typical LPA antenna from 6 to 30 MHz. (Above) Graph of gain contours (in decibels) above an isotropic radiator; (following) tabulation.

METHOD 15   IONCAP 78.03   PAGE   1

ITS- 1 ANTENNA PACKAGE   ANTENNA PATTERN
FREQUENCY RANGE   ANTENNA TYPE   HEIGHT   LENGTH   ANGLE
2.0 TO 30.0   237B-3 50 FT   0.000   0.000   0.000

	2	3	4	5	6	7	8	9	10	11
90	-45.3	-45.3	-45.3	-45.3	-20.0	-20.0	-20.0	-20.0	-20.0	-20.0
88	-45.3	-45.3	-45.3	-45.3	3.5	-.2	-4.6	-10.4	-10.8	-5.8
86	-45.3	-45.3	-45.3	-45.3	4.9	1.6	-2.5	-8.0	-8.5	-3.4
84	-45.3	-45.3	-45.3	-45.3	5.8	2.7	-1.2	-6.6	-7.2	-2.1
82	-45.3	-45.3	-45.3	-45.2	6.4	3.5	-.2	-5.4	-6.4	-1.3
80	-45.4	-45.3	-45.3	-45.2	7.0	4.2	.6	-4.4	-5.8	-.8
78	-45.4	-45.3	-45.3	-45.2	7.4	4.8	1.4	-3.5	-5.5	-.5
76	-45.5	-45.4	-45.3	-45.2	7.8	5.3	2.1	-2.6	-5.2	-.4
74	-45.5	-45.4	-45.3	-45.2	8.1	5.8	2.8	-1.8	-5.0	-.4
72	-45.6	-45.5	-45.4	-45.1	8.4	6.3	3.4	-.9	-4.7	-.6
E 70	-45.7	-45.5	-45.4	-45.1	8.7	6.8	4.0	0.0	-4.3	-.9
L 68	-45.8	-45.6	-45.4	-45.1	9.0	7.2	4.6	.9	-3.7	-1.4
E 66	-45.8	-45.6	-45.4	-45.1	9.3	7.6	5.2	1.8	-2.9	-2.0
V 64	-46.0	-45.7	-45.5	-45.1	9.5	8.0	5.8	2.7	-1.9	-2.6
A 62	-46.1	-45.8	-45.5	-45.0	9.8	8.4	6.4	3.5	-.8	-3.0
T 60	-46.2	-45.9	-45.6	-45.0	10.0	8.7	6.9	4.4	.5	-3.1
I 58	-46.3	-46.0	-45.6	-45.0	10.2	9.1	7.5	5.2	1.7	-2.6
O 56	-46.5	-46.1	-45.7	-45.1	10.4	9.4	8.0	6.0	2.9	-1.5
N 54	-46.7	-46.3	-45.8	-45.1	10.5	9.7	8.5	6.7	4.0	0.0
52	-46.8	-46.4	-45.9	-45.1	10.7	10.0	8.9	7.4	5.1	1.5
A 50	-47.0	-46.6	-46.0	-45.2	10.8	10.2	9.4	8.1	6.1	3.1
N 48	-47.3	-46.7	-46.1	-45.2	10.9	10.5	9.8	8.7	7.1	4.5
G 46	-47.5	-46.9	-46.3	-45.3	11.0	10.7	10.1	9.3	8.0	5.8
L 44	-47.7	-47.2	-46.4	-45.4	11.0	10.9	10.4	9.8	8.8	7.0
E 42	-48.0	-47.4	-46.6	-45.5	11.1	11.0	10.7	10.3	9.5	8.1
40	-48.3	-47.7	-46.8	-45.7	11.1	11.1	11.0	10.7	10.1	9.1
I 38	-48.6	-47.9	-47.1	-45.8	11.0	11.2	11.2	11.1	10.7	9.9
N 36	-49.0	-48.2	-47.3	-46.1	10.9	11.2	11.3	11.4	11.2	10.6
34	-49.4	-48.6	-47.6	-46.3	10.8	11.2	11.4	11.6	11.6	11.2
D 32	-49.8	-49.0	-48.0	-46.6	10.7	11.1	11.5	11.8	11.9	11.8
E 30	-50.2	-49.4	-48.3	-46.9	10.5	11.0	11.4	11.9	12.1	12.2
G 28	-50.7	-49.8	-48.8	-47.3	10.2	10.8	11.4	11.9	12.3	12.5
R 26	-51.3	-50.4	-49.2	-47.7	9.9	10.6	11.2	11.8	12.4	12.7
E 24	-51.9	-50.9	-49.8	-48.2	9.5	10.3	11.0	11.7	12.3	12.7
E 22	-52.5	-51.6	-50.4	-48.7	9.0	9.9	10.6	11.4	12.2	12.7
S 20	-53.3	-52.3	-51.0	-49.3	8.4	9.4	10.2	11.1	11.9	12.5
18	-54.1	-53.1	-51.8	-50.1	7.8	8.7	9.7	10.6	11.5	12.2
16	-55.0	-54.0	-52.7	-50.9	7.0	8.0	9.0	10.0	11.0	11.7
14	-56.1	-55.1	-53.7	-51.9	6.0	7.1	8.1	9.2	10.2	11.1
12	-57.4	-56.3	-55.0	-53.1	4.9	6.0	7.1	8.2	9.3	10.2
10	-58.9	-57.8	-56.4	-54.6	3.5	4.6	5.7	6.9	8.0	9.0
8	-60.8	-59.7	-58.3	-56.4	1.7	2.9	4.0	5.2	6.4	7.3
6	-63.2	-62.1	-60.7	-58.8	-.7	.5	1.7	2.9	4.1	5.1
4	-65.2	-64.7	-64.1	-62.2	-4.1	-2.9	-1.7	-.4	.8	1.8
2	-65.4	-64.7	-64.2	-63.7	-10.0	-8.8	-7.6	-6.3	-5.1	-4.1
0	-65.4	-64.7	-64.2	-63.7	-20.0	-20.0	-20.0	-20.0	-20.0	-20.0
	2	3	4	5	6	7	8	9	10	11

FREQUENCY IN MEGAHERTZ

ANTENNA EFFICIENCY

-50.0	-50.0	-50.0	-50.0	0.0	0.0	0.0	0.0	0.0	0.0
2	3	4	5	6	7	8	9	10	11

FREQUENCY IN MEGAHERTZ

Figure 3.5 Continued.

```
AZIMUTH    EX(1)      EX(2)      EX(3)      EX(4)   CONDUCT.  DIELECT.
 0.000     0.000      0.000      0.000      0.000      .010    15.000
  12    13    14    16    18    20    22    24    26    28    30
-20.0 -20.0 -20.0 -20.0 -20.0 -20.0 -20.0 -20.0 -20.0 -20.0 -20.0 9
 -4.1  -2.3  -1.5  -3.6  -8.6 -13.5  -6.4  -4.0  -5.8 -10.8 -13.7 8
 -1.3    .5   1.3   -.5  -5.5 -10.3  -3.2   -.6  -2.2  -7.1 -10.4 8
   .2   2.1   2.9   1.3  -3.5  -8.6  -1.3   1.3   0.0  -4.7  -8.6 8
  1.3   3.2   4.1   2.7  -1.9  -7.5   -.2   2.7   1.6  -2.8  -7.6 8
  2.0   4.0   5.0   3.8   -.6  -6.7    .6   3.7   2.9  -1.2  -6.8 8
  2.5   4.7   5.6   4.7    .6  -6.1   1.1   4.5   4.0    .4  -6.1 7
  2.9   5.1   6.2   5.5   1.8  -5.4   1.3   5.0   4.9   1.8  -5.1 7
  3.1   5.5   6.6   6.2   2.8  -4.6   1.2   5.5   5.8   3.1  -3.7 7
  3.2   5.7   7.0   6.8   3.8  -3.5   1.0   5.7   6.5   4.3  -1.9 7
  3.1   5.8   7.3   7.4   4.8  -2.1    .4   5.8   7.0   5.4    .1 7
  2.9   5.8   7.4   7.9   5.7   -.5   -.4   5.7   7.5   6.5   2.1 6
  2.5   5.7   7.5   8.3   6.6   1.1  -1.4   5.4   7.8   7.3   3.9 6
  2.0   5.5   7.5   8.7   7.4   2.7  -2.4   4.9   7.9   8.1   5.5 6
  1.3   5.1   7.4   9.0   8.1   4.2  -2.6   3.9   7.8   8.7   6.9 6
   .4   4.6   7.1   9.1   8.7   5.6  -1.4   2.6   7.5   9.0   8.1 6
  -.7   3.8   6.7   9.2   9.2   6.9    .8    .8   6.9   9.2   9.0 5
 -1.8   2.9   6.1   9.2   9.7   8.0   3.1  -1.2   5.9   9.0   9.6 5
 -2.6   1.6   5.4   9.0  10.0   8.9   5.1  -1.9   4.3   8.6  10.0 5
 -2.6    .1   4.3   8.7  10.2   9.7   6.9    .1   1.9   7.7  10.0 5
 -1.4  -1.5   2.9   8.2  10.2  10.3   8.4   3.1  -1.0   6.2   9.6 5
   .5  -2.6   1.2   7.5  10.1  10.8   9.6   5.8  -1.6   3.8   8.7 4
  2.5  -1.9   -.9   6.4   9.7  11.0  10.5   7.9   1.8    .2   7.0 4
  4.3    .1  -2.6   5.0   9.1  11.0  11.2   9.5   5.2  -1.9   4.2 4
  5.9   2.5  -2.0   3.0   8.2  10.8  11.6  10.7   7.9   1.8   -.1 4
  7.4   4.7    .6    .2   6.9  10.3  11.7  11.5   9.8   5.8  -1.4 4
  8.6   6.5   3.3  -2.7   4.9   9.4  11.5  12.0  11.2   8.7   3.7 3
  9.7   8.1   5.6  -2.3   2.1   8.0  10.9  12.1  12.0  10.6   7.7 3
 10.6   9.4   7.5   1.2  -1.9   5.8   9.8  11.7  12.4  11.8  10.2 3
 11.4  10.5   9.1   4.5  -3.4   2.5   8.0  10.9  12.2  12.4  11.8 3
 12.0  11.4  10.3   7.0    .9  -2.6   5.2   9.3  11.5  12.4  12.6 3
 12.5  12.1  11.3   9.0   4.9  -2.2    .3   6.6  10.0  11.8  12.7 2
 12.8  12.6  12.1  10.5   7.7   3.3  -4.1   1.9   7.3  10.2  11.9 2
 13.1  13.0  12.7  11.6   9.7   7.1   2.2  -4.7   2.3   7.2  10.1 2
 13.2  13.3  13.1  12.5  11.2   9.7   6.9   2.0  -5.2   1.4   6.6 2
 13.1  13.3  13.3  13.0  12.3  11.5   9.9   7.3   2.9  -4.6  -1.1 2
 12.9  13.2  13.3  13.4  13.0  12.7  11.8  10.4   8.2   4.8   -.8 1
 12.5  12.9  13.2  13.4  13.3  13.4  13.1  12.4  11.2   9.5   7.3 1
 11.9  12.4  12.7  13.2  13.3  13.7  13.8  13.5  13.0  12.2  11.1 1
 11.1  11.6  12.0  12.6  13.0  13.6  14.0  14.0  13.9  13.6  13.2 1
  9.9  10.5  11.0  11.7  12.2  13.1  13.6  13.9  14.1  14.1  14.1 1
  8.3   9.0   9.5  10.4  11.0  12.0  12.7  13.2  13.6  13.8  14.0
  6.1   6.8   7.4   8.3   9.0  10.1  11.0  11.6  12.1  12.5  12.9
  2.9   3.6   4.1   5.2   5.9   7.1   8.0   8.7   9.3   9.8  10.4
 -3.0  -2.2  -1.7   -.6    .2   1.4   2.4   3.1   3.8   4.3   5.0
-20.0 -20.0 -20.0 -20.0 -20.0 -20.0 -20.0 -20.0 -20.0 -20.0 -20.0
  12    13    14    16    18    20    22    24    26    28    30
```

FREQUENCY IN MEGAHERTZ

ANTENNA EFFICIENCY

```
 0.0   0.0   0.0   0.0   0.0   0.0   0.0   0.0   0.0   0.0   0.0
  12    13    14    16    18    20    22    24    26    28    30
```

FREQUENCY IN MEGAHERTZ

```
                        METHOD 24    IONCAP 78.03    PAGE   1

           JUN                    SSN =  60.
CEDAR RAPIDS,IA TO HONG KONG              AZIMUTHS          N. MI.       KM
42.00 N   91.50 W - 22.00 N  114.10 E    334.45   20.23   6709.6   12425.3
                             MINIMUM ANGLE   1.0  DEGREES
ITS- 1 ANTENNA PACKAGE
XMTR   2.0  TO  30.0  237B-3 50 FT H    0.00 L    0.00 A    0.0  OFF AZ    0.0
RCVR   2.0  TO  30.0  237B-3 50 FT H    0.00 L    0.00 A    0.0  OFF AZ    0.0
POWER =  10.000 KW   3 MHZ NOISE = -148.0 DBW    REQ. REL = .50  REQ. SNR = 48.0

                    FREQUENCY / RELIABILITY
```

GMT	LMT	MUF	FOT	6	8	10	12	14	16	18	20	22	25	MUF
1.0	18.9	17.9	.18	.00	.00	.00	.03	.13	.34	.48	.41	.14	.00	.47
2.0	19.9	17.7	.14	.00	.00	.00	.00	.10	.22	.42	.26	.10	.00	.48
3.0	20.9	17.7	.00	.00	.00	.00	.00	.00	.15	.40	.17	.10	.00	.33
4.0	21.9	17.2	.00	.00	.00	.00	.00	.00	.15	.40	.11	.04	.00	.21
5.0	22.9	16.2	.00	.00	.00	.00	.00	.00	.15	.21	.03	.01	.00	.14
6.0	23.9	15.2	.00	.00	.00	.00	.00	.04	.13	.10	.00	.00	.00	.15
7.0	.9	14.4	.00	.00	.00	.00	.00	.18	.17	.03	.00	.00	.00	.19
8.0	1.9	13.5	.01	.00	.00	.00	.18	.26	.15	.00	.00	.00	.00	.28
9.0	2.9	12.8	.08	.00	.00	.09	.45	.31	.04	.00	.00	.00	.00	.43
10.0	3.9	12.4	.12	.00	.01	.11	.46	.28	.02	.00	.00	.00	.00	.51
11.0	4.9	12.5	.16	.00	.02	.14	.51	.34	.04	.00	.00	.00	.00	.49
12.0	5.9	14.3	.38	.00	.00	.18	.46	.69	.35	.12	.02	.00	.00	.61
13.0	6.9	15.7	.93	.00	.00	.02	.88	.97	.95	.84	.53	.06	.00	.96
14.0	7.9	16.2	.96	.00	.00	.06	.91	.96	.97	.92	.72	.19	.00	.98
15.0	8.9	17.3	.98	.00	.00	.00	.92	.95	.90	.90	.50	.01	.00	.96
16.0	9.9	18.2	.98	.00	.00	.00	.90	.95	.98	.96	.77	.11	.00	.96
17.0	10.9	18.8	.98	.00	.00	.00	.68	.95	.98	.97	.83	.02	.00	.95
18.0	11.9	19.1	.98	.00	.00	.00	.58	.92	.97	.96	.20	.03	.00	.90
19.0	12.9	19.1	.96	.00	.00	.00	.15	.88	.95	.91	.18	.03	.00	.39
20.0	13.9	16.9	.01	.00	.00	.00	.00	.04	.37	.34	.30	.08	.00	.42
21.0	14.9	18.2	.64	.00	.00	.00	.00	.03	.77	.85	.36	.00	.00	.83
22.0	15.9	18.4	.20	.00	.00	.00	.00	.01	.58	.59	.64	.24	.00	.79
23.0	16.9	18.7	.44	.00	.00	.00	.00	.03	.61	.63	.49	.21	.00	.56
24.0	17.9	17.3	.27	.00	.00	.00	.00	.28	.63	.49	.23	.04	.00	.51

Figure 3.6 Sample reliability summary tabulation with LPA antennas added; for June, SSN = 60.

Maximum reliability

Link: 42.00 N   91.50 W—22.00 N   114.00 E   Month: Jun   Tx pwr (kW): 10.00
Distance (km): 12425.3   SSN: 60.   Req'd SNR: 40.

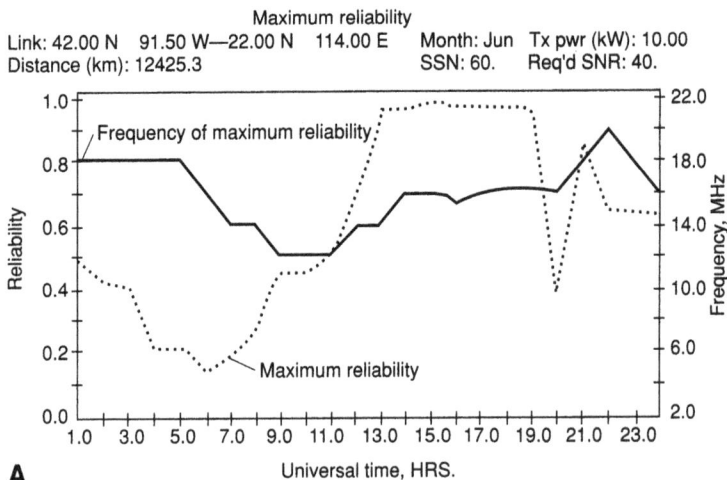

A

Figure 3.7 Sample computer-generated graphs of maximum reliability and FMR with LPA antennas added for (A) June and (B) December; SSN = 60 for both.

Maximum reliability
Link: 42.00 N   91.50 W—22.00 N   114.10 E   Month: Dec   Tx pwr (kW): 10.00
Distance (km): 12425.3   SSN: 60.   Req'd SNR: 48.

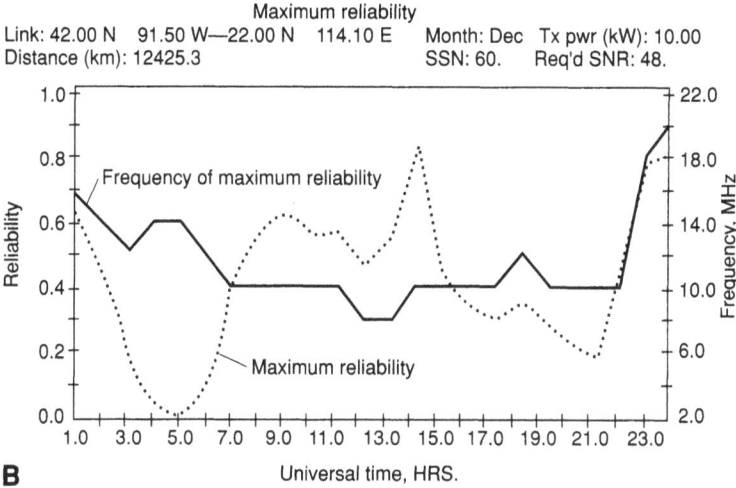

B   Universal time, HRS.

Figure 3.7 Continued.

Frequency management techniques

Tables such as those shown in Figures 3.2 and 3.3 are of value in manually
selecting the operating frequency to try first when contact is attempted or
in selecting frequency presets for automated operations, and they illustrate
the variability of the FMR with the many HF propagation factors. These results clearly demonstrate the need for some form of automated frequency
management technique for selecting the optimum frequency or scanned
frequencies and terminal configuration in HF communications operations.
One such automated planning and operational tool is described next.

PropMan™, an automated planning tool

PropMan™ (an acronym for PROPagation resource MANager) is a software
product developed by Rockwell [32, 33]. It is a tool that provides easy-to-use,
fast, automated assistance for HF skywave frequency/channel selection and
planning. It runs a well-known propagation prediction program and takes
into account those parameters that affect HF propagation, including station
locations, sunspot number, time of day and year, and equipment and antenna
parameters. PropMan generates and displays tabulated and graphical results
showing the current and 24-hour HF skywave propagating frequency band.
While continuously running, it automatically updates the displayed results as
time progresses. PropMan, as described here, includes some recently developed enhancements not available in the latest product version.

High-frequency skywave propagation prediction programs are useful because of the variability of skywave propagating frequencies versus station

locations, sunspot number, and time of day and year. Many well-known programs are available including IONCAP [13], AMBCOM [34], MINIMUF [3], and PROPHET [4]. PropMan runs the IONCAP program (an acronym for IONospheric Circuit Analysis Program). Comparisons of HF skywave field-strength prediction methods for seven models are presented in a recent government report [35].

The IONCAP propagation prediction program requires a correctly formatted input parameter file. When creating the file, a user can easily make time-consuming and sometimes undetected mistakes. IONCAP can be instructed to produce any one of a number of output files. These output files may contain a considerable amount of tabulated data to be analyzed by the user.

When running PropMan, the user is spared the task of directly interfacing with IONCAP. Based on user inputs, PropMan correctly creates and formats the IONCAP input parameter file. It automatically analyzes the contents of the IONCAP output parameter file, extracting and displaying the propagating frequency band.

The PropMan software runs on IBM-PC™ 286/386/486 or compatible computers with a math coprocessor and color VGA monitor. The program requires 2 MB of disk space when loaded on the computer and about 490K of RAM when running.

PropMan generates and displays HF skywave propagating frequency information for a user-selected pair of stations. It contains a data base of about 4000 stations with coordinates from around the world. Various user-selected equipment parameters are taken into account to include transmit power (in watts), transmit and receive antenna gains (in decibels above isotropic), and the minimum useful antenna takeoff angle (in degrees). The user enters the time (UT) and date in which to begin operations. While the program is running, the user has access to a menu whereby the stations, equipment parameters, time, and date may be changed.

PropMan runs IONCAP for the stations, equipment parameters, time, and date selected by the user. It then displays the current optimum frequency (FOT), propagating frequency band, and user-channel recommendations as shown in Figure 3.8.

The following data are displayed in the upper left window:

- Current-time best frequency (FOT, in MHz) and signal-to-noise density ratio (SN_0R, in dB-Hz) at the FOT
- Current-time propagating frequency windows that exceed four SN_0R thresholds

Note that PropMan displays SN_0R in dB-Hz (as used by IONCAP, where noise is referenced in a 1-Hz bandwidth). A 0-dB SNR in a 3-kHz bandwidth equals an SN_0R of 34.8 dB-Hz. For a user-selected list of up to 20 channel frequencies, the program displays the following data in the lower left window:

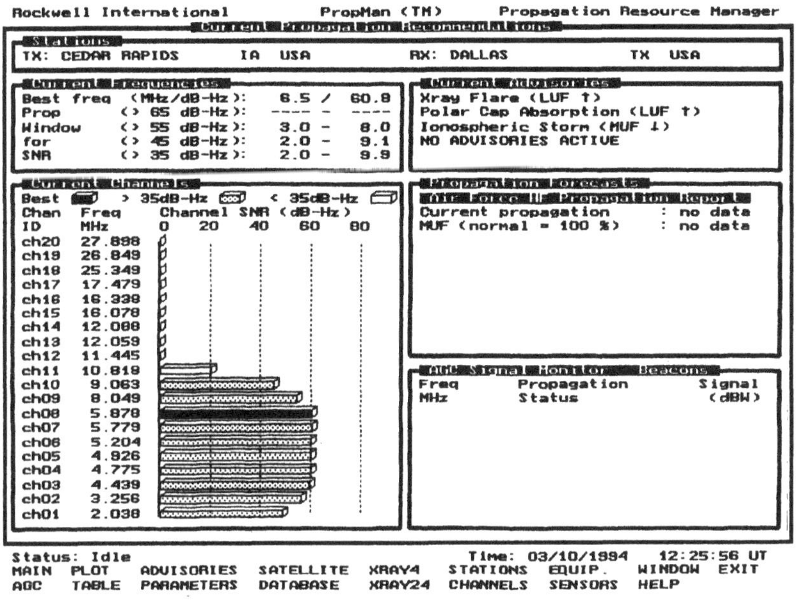

```
Rockwell International          PropMan (TM)        Propagation Resource Manager
█████████ Current Propagation Recommendations █████████
  ▀Stations▀
  TX: CEDAR RAPIDS        IA  USA        RX: DALLAS            TX  USA

  ▀Current Frequencies▀                  ▀Current Advisories▀
  Best freq  (MHz/dB-Hz):   6.5 /  60.8  Xray Flare (LUF ↑)
  Prop       (> 65 dB-Hz):  ---- - ----  Polar Cap Absorption (LUF ↑)
  Window     (> 55 dB-Hz):  3.0 -   8.0  Ionospheric Storm (MUF ↓)
  for        (> 45 dB-Hz):  2.0 -   9.1  NO ADVISORIES ACTIVE
  SNR        (> 35 dB-Hz):  2.0 -   9.9

  ▀Current Channels▀                     ▀Propagation Forecasts▀
  Best ▓▓ > 35dB-Hz ▓▓  < 35dB-Hz ▓▓     ▀Air Force HF Propagation Report▀
  Chan  Freq     Channel SNR (dB-Hz)      Current propagation    : no data
  ID    MHz      0   20  40  60  80       MUF (normal = 100 %)   : no data
  ch20  27.898  ▐
  ch19  26.849  ▐
  ch18  25.349  ▐
  ch17  17.479  ▐
  ch16  16.338  ▐
  ch15  16.078  ▐
  ch14  12.088  ▐
  ch13  12.059  ▐
  ch12  11.445  ▐
  ch11  10.818  ▐▬▬▬
  ch10   9.063  ▐▬▬▬▬▬▬▬▬▬▬        ▀HF Signal Monitor    Beacons▀
  ch09   8.049  ▐▬▬▬▬▬▬▬▬▬         Freq    Propagation        Signal
  ch08   5.878  ▐                  MHz     Status             (dBW)
  ch07   5.779  ▐▬▬▬▬▬▬▬▬▬▬▬▬
  ch06   5.204  ▐▬▬▬▬▬▬▬▬
  ch05   4.826  ▐▬▬▬▬▬▬▬▬
  ch04   4.775  ▐▬▬▬▬▬▬▬
  ch03   4.439  ▐▬▬▬▬▬▬▬▬
  ch02   3.256  ▐▬▬▬▬▬▬▬▬
  ch01   2.038  ▐▬▬▬▬▬▬▬▬

  Status: Idle                           Time: 03/10/1994    12:25:56 UT
  MAIN  PLOT    ADVISORIES  SATELLITE  XRAY4    STATIONS  EQUIP.   WINDOW  EXIT
  AGC   TABLE   PARAMETERS  DATABASE   XRAY24   CHANNELS  SENSORS  HELP
```

Figure 3.8 Current frequency/channel recommendations.

- Current-time expected received SN_0R for each of the channel frequencies.
- Channel with highest SN_0R near the current-time FOT is highlighted.
- Channels falling within the current-time propagating frequency window are highlighted.

When optional real-time sensor data monitoring hardware is available, PropMan will generate warnings of solar and ionospheric disturbances that affect current propagation on the selected link. These warnings are displayed in the upper right window. The advisories are listed below:

- Solar x-ray flares
- Polar-cap absorption
- Ionospheric storms

Additional sensor inputs are displayed in the middle right and lower right windows.

PropMan automatically updates the current-time frequency and channel results as time progresses, running IONCAP as needed. Once the current-time propagation results have been generated and displayed, PropMan continues to run IONCAP until 24 hours of propagating frequency results have been generated and plotted. An example 24-hour propagation plot is shown in Figure 3.9. The FOT and the propagating frequency windows exceeding four SN_0R thresholds are displayed. A user can observe the variability in the propagating band over a 1-day period, starting at the time and date entered. The example in Figure 3.9 shows that, as is typical of an HF link, higher frequencies generally propagate during the day and lower frequencies are preferred at night. PropMan automatically updates the 24-hour frequency plot as time progresses, running IONCAP as needed.

PropMan is capable of receiving satellite-broadcast Space Environmental Services Center (SESC) sensor data [6]. These data consist of solar and ionospheric measurements updated every minute. It then uses the data to detect solar and ionospheric disturbances that affect HF skywave propagation. Direct hookup to the SESC satellite data requires a satellite-receiving antenna and receiver. The remote hookup requires a remote source of the satellite data accessible via a phone modem.

PropMan monitors the SESC data x-ray levels (watts per square meter measured in a 1- to 8-Angstrom wavelength channel), updated every minute, and computes a long-term x-ray background level. It detects solar flares when the current x-ray level exceeds a threshold above the background level. Solar flares affect HF propagation by increasing the ionospheric D-layer absorption, which degrades the lower frequencies within the propagating frequency band (large flares can sometimes cause propagation blackout). Propagation degradation typically lasts tens of minutes.

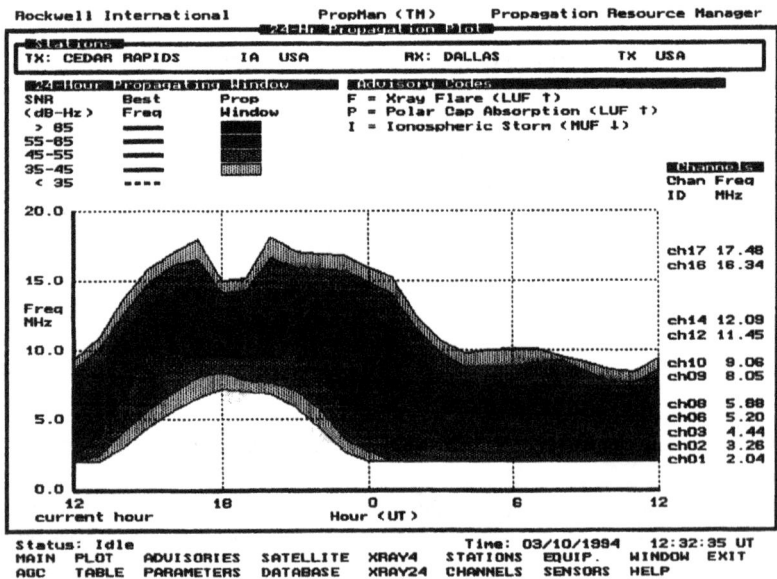

Figure 3.9 24-hour propagating frequency plot.

The program monitors the SESC data polar-cap absorption index (dB) and the SESC data 10 MEV proton level (number of protons with 10 MEV energy per second per square cm per steradian). By observing these values, it detects situations where high levels of signal absorption will degrade high-latitude skywave links. Polar-cap absorption indicates increased ionospheric D-layer absorption, which degrades the lower frequencies within the propagating frequency band.

PropMan monitors the SESC data 3-hour planetary K indices. The measured K index provides an indication of magnetic and ionospheric storms. Large K values (4 to 9) indicate potential storms expanding outwards from auroral latitudes (60 to 70 degrees) downwards toward lower latitudes. An ionospheric storm decreases F-layer ionization, which degrades the higher frequencies within the propagating frequency band. The storm can persist for a few days.

The program can optionally call (via a phone modem) the SESC bulletin board data base in Boulder, Colorado, to download and display a portion of an HF propagation report. That report contains current and near-term forecasts of propagation conditions covering the northern hemisphere.

PropMan can optionally monitor field-strength measurements of received on-the-air HF signals. It acquires this data through a serial connection to a receiver AGC monitor computer. The AGC monitor computer controls an HF receiver, scans through preset channels, collects AGC and field-strength measurements, generates long-term per-channel field-strength background levels, and sends the results to the PropMan computer. By observing the per-channel current signal field-strength measurements compared with long-term field-strength averages, periods of degraded propagation can be detected as they occur in real time.

PropMan can be used for systems design, mission control, and mission planning in the following ways. PropMan provides information for mission control by continuously generating the current optimum frequency and propagation frequency window between a selected pair of stations. It provides for mission frequency planning by generating a 24-hour propagating frequency window plot for a selected pair of stations. Channel selection for current use or mission planning is provided by entering the specific channel/ frequency list that will be displayed along with the propagating frequency data. During a mission, PropMan can provide real-time warnings of solar and ionospheric disturbances that can seriously degrade or black out HF communications (requires optional sensor input source). It can be used as a system design aid for estimating the required power amplifier output power and the transmit and receive antenna gains. This is accomplished by running the program for typical system links and modifying the equipment parameters until a propagating frequency band providing an acceptable receiver SN_0R is observed.

3.3 Automatic Link Establishment

Because of the complexity of the subjects to be discussed in the next two sections, the following approach will be used. A brief overview of the various topics will be given in this chapter. The software disk that comes with this book contains a word processor file that provides greater depth into those topics that cannot be adequately covered in this chapter. Chapter 17, "Software for SSB," describes the disk file that is involved.

Microprocessor technology has allowed electronic equipment designers to implement relatively complex logical control and data processing functions as an integral part of modern radios. A logical application of this powerful technology is the automation of HF SSB communications equipment to provide automatic connectivity, since establishing a link is by far the most difficult part of communicating at high frequency. Automating the labor-intensive tasks normally performed by a skilled HF radio operator can make an HF radio as easy to use as a telephone.

Principles of HF automatic link establishment

What does it do? High-frequency ALE systems greatly reduce operator workload and improve communications reliability by using preset channels and receiver scanning techniques to monitor multiple channels, thus eliminating the coordination problem and the need for frequency-time plans. No matter what channel is selected by the calling station, the receiving station's scanning receiver will hear the call if the selected channel is propagating at that time. In order to improve the probability of selecting a channel that is propagating at the time a call is placed, each HF ALE system uses sounding and link quality analysis (LQA) techniques to build and maintain its own local LQA data base. Automatic channel selection algorithms use the propagation information stored in the LQA data base to rank and select the calling channels in the order most likely to succeed. The HF ALE system listens on the selected channel for other traffic before transmitting a call. If a selected channel is busy, the system will not interfere but will choose an alternate calling channel and try again (see "Placing a scanning ALE call" in section A3.3-4 for a detailed discussion).

Operational advantages of the HF ALE system include a greatly simplified user interface. Since ALE calls are placed by station address, and the rest of the linking process, including channel selection, is automatic, the process is user friendly, with reduced opportunity for human error. The chance of missing an incoming call is virtually eliminated because the system processes the incoming call and completes the link before the operator is notified. The operator's receive audio is muted during scanning to reduce distractions and fatigue. With the automated features, communications can be handled by personnel with no special HF communications skills.

How does it do it? Automatic connectivity can be realized through the proper integration of the following capabilities into an HF radio system:

1. Selective calling

2. Preset channel scanning

3. Channel propagation evaluation

4. Automatic channel selection

5. Microprocessor radio control and interface logic

The block diagram in Figure 3.10 shows the major functional elements of an ALE processor capable of providing automatic connectivity when used with an appropriate remotely controllable HF radio.

Selective calling provides station identification via data transmission and reception of assigned station addresses. A data modulator/demodulator (modem) is required to facilitate the exchange of addresses and other control data between two automatic HF stations. The data modem design that is selected must be compatible with the bandpass characteristics of the equipment with which it is intended to operate. Also, the data format must quickly provide word and bit synchronization at any time during a transmission to be effective while scanning. Section A3.3-4 describes the addressing structure in more detail.

The incorporation of preset channels allows for rapid selection of any of several assigned operating frequencies. High-frequency ALE systems use an

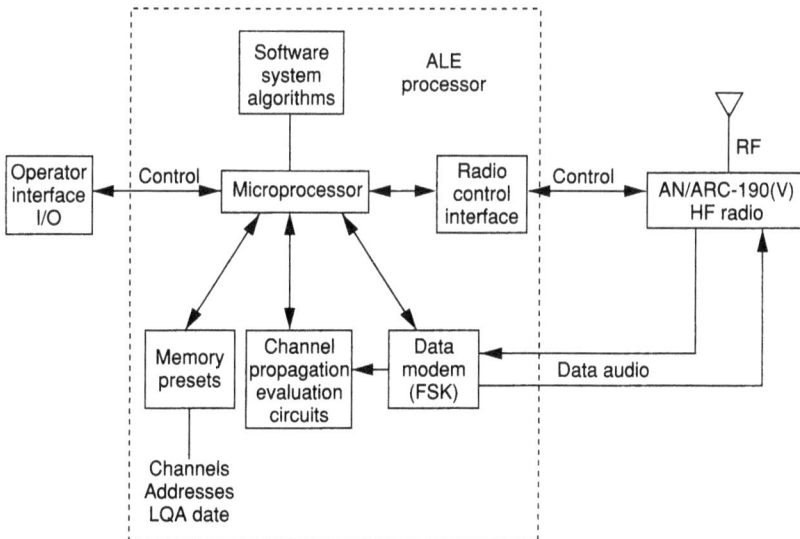

Figure 3.10 ALE system block diagram.

asynchronous receiver scanning approach that eliminates dependence on TOD clocks. This approach does not require the transmitter to scan, thus offering compatibility with fixed-station HF systems that use large slow-tuning transmitters as well as aircraft or vehicular HF systems with narrowband antennas and slow-tuning antenna couplers. Scanning dwell times must be selected to allow a minimum sampling period on each channel. The minimum sampling period is determined by the length of time required to decode sufficient data to make a stop-scan decision. It is therefore dependent on address length and the data format and coding redundancy factors chosen. High-frequency receiver tune time and synthesizer settling time must be added to the minimum sampling period to obtain the total dwell period.

Channel propagation evaluation data provides the basis for quick and accurate automatic channel selection, significantly improving the probability of success on the first call transmission. High-frequency ALE systems are capable of periodically transmitting "sounds" on each of the scanned channels. This allows other ALE scanning systems an opportunity to receive the sounds and evaluate the actual propagation conditions between the two stations. The sounding signal contains the ALE address of the station transmitting the sound. The process of evaluating received sounding signals and assigning relative path quality factors is referred to as link quality analysis (LQA). An LQA data base is maintained and continuously updated with current link quality information that allows the scanned channels to be automatically ranked and selected for the highest probability of successful link establishment. An automatic channel selection scheme of this type is essential to the establishment of a best-quality link in the shortest time.

A microprocessor or other data processing mechanism is required to provide the data manipulation, coordination, radio remote control, and operator input/output (I/O) interfaces necessary to make meaningful use of the other four basic functions. It is, of course, the various algorithms and protocols implemented in the software that coordinate the hardware functions and ultimately determine the operational characteristics of the system.

What are the important design considerations? The following is a representative list of some of the major system parameters that must be considered:

1. Scanning receiver dwell versus call transmission timing relationship
2. Scanning rate
3. Address data word format
4. Synchronization
5. Data modulation scheme
6. Data rate

7. Data error rate performance and error correction schemes

8. Address coding scheme and number of address code combinations

9. Effects of coded address length on address recognition reliability and false-alarm rate

10. Channel evaluation measurement techniques

Several of the relationships between these factors are discussed in section 3.3-2.

How well does it work?   If, at any time in a given network of ALE-equipped HF stations, an ALE call is placed from one station to any other station in the network, that link will be successfully completed if there is at least one scanned preset channel that is available and propagating well enough to support communications between the two HF stations. The better the selection of HF channels, the higher the probability of successful link establishment under all possible propagation conditions.

High-frequency ALE systems provide connectivity on demand, without prior coordination, and with a very high probability of success. A U.S. Air Force test performed with the Pacific fleet in the late 1970s, using Rockwell prototype ALE processors with AN/ARC-190(V) radios, claimed over 90% successful connectivity, up from an estimated 35% using traditional manual operating procedures with the same radios.

The U.S. Customs over-the-horizon enforcement network (COTHEN) is the world's largest operating HF ALE tactical network implemented to date. This network currently has over 250 mobile units and 13 ground stations. The ground stations are remotely controlled by a network of over 50 land-based control points at various geographical locations. According to the COTHEN network manager, the network is operating 24 hours a day with a connectivity success rate (the ratio of links established to calls placed) estimated at better than 80%.

HF ALE interoperability and performance standards

As different manufacturers developed and began producing their own proprietary HF ALE systems, it became obvious that the U.S. government would have to provide an interoperability standard in order to procure interoperable ALE systems from multiple sources. A family of military and federal standards were proposed and are being developed. See section A3.3-3 for a detailed discussion of these standards as they exist as of September 1994.

The development of HF ALE interoperability standards has provided an opportunity to improve on the first-generation ALE system designs. The new HF ALE standards incorporate the best ideas from the first-generation designs along with new capabilities and features to expand the system's ability to meet a broader spectrum of user requirements and provide improved system performance. All U.S. government procurements of HF ALE radio equipment or systems now require compliance with one or more of the military or federal ALE interoperability standards.

New interoperable HF ALE systems and capabilities

Since appendix A of MIL-STD-188-141A, Notice 2, is the most complete and most recently updated of the new ALE standards published to date (August 1994), the following discussions will be based primarily on that document, unless otherwise noted. Systems built to the new MS-141A standards will be referred to as second-generation ALE systems. We present here a brief summary and section A3.3-4 provides greater depth.

From the outside, the core ALE features that provide basic automatic connectivity capabilities in second-generation systems appear very similar to those found in some of the more advanced first-generation systems such as SELSCAN®. However, there are very significant differences between first- and second-generation ALE systems that make the newer systems more robust and capable of providing enhanced features not possible with first-generation systems. The technical features of the new MS-141A-based ALE systems are best described in terms of the waveform, signal structure, and protocols. Significantly more detail may be found in appendix A of MIL-STD-188-141A or in FED-STD-1045.

Waveform. The new interoperable ALE systems use an 8-Ary FSK waveform whose eight tones are evenly spaced at 250-Hz intervals between 750 Hz and 2500 Hz. Each of the eight tones represents one symbol or three bits of information. The transmitted data rate is 125 symbols per second (375 bits per second).

Signal structure. A 24-bit word format contains three 7-bit ASCII characters and a 3-bit preamble. Each preamble identifies one of the eight possible word types that can be used to convey the address, command, and message information needed to carry out the ALE sounding, linking, data message delivery, and protocol control processes.

It is often possible to establish a link under conditions that will not support voice communications. The sounding and LQA features of the system allow the automatic channel selection algorithms to select the best available channels for link establishment. When the best available channel is data-only quality the ALE system is capable of using the same FEC coding and triple-redundancy techniques used in the linking process to pass orderwire data messages.

The ability to achieve word synchronization is of paramount importance in an asynchronous scanning environment. It is necessary for a scanning system to be able to start listening to an ALE transmission at some unknown point and begin recovering error corrected data words as soon as possible.

A false alarm occurs when random noise received produces a bit pattern at the FSK modem's output that is decoded as that station's own address, its *self address*. The larger the number of unique bits being transmitted over the air, the lower the probability of false alarms. Address recognition reliability, on the other hand, decreases as the number of transmitted bits increases.

The MS-141A signal structure is unique in the way it combines Golay FEC coding, bit interleaving, and triple-redundant error reduction techniques to permit fast and reliable word/bit synchronization under scanning conditions, excellent address recognition reliability under poor signal conditions, and a false-alarm rate that is virtually nonexistent.

All ALE transmissions consist of a series of contiguous triple-redundant words known as a frame. Frames may contain up to three sections: the calling cycle section, the message section, and the conclusion section. The calling cycle section contains the address of the station, or stations if multiple addressees are involved, to which the transmission is directed. The message section contains orderwire data message and/or system control information and an optional means by which the transmitting station can be identified early in the transmission if a long data message is to follow. The conclusion section contains the address of the transmitting station and is always used to terminate a frame. Depending on the protocol, the message section (or both the calling cycle and message sections) may be omitted. However, when used, these sections will always be transmitted in the standard order shown in Figure 3.11.

Protocols.   Protocols determine the interaction between ALE stations during the sounding, linking, and orderwire message transmission processes. An almost overwhelming number of variations exist for each of the above categories, providing option upon option to allow a wide variety of users to tailor the powerful features of ALE to their own unique system requirements.

Sounding is the simplest of the ALE protocols. Sounds are one-way transmissions that identify the transmitting station and provide other listening stations an opportunity to update their LQA data bases with current information on actual propagation conditions between the transmitting station and the listening station.

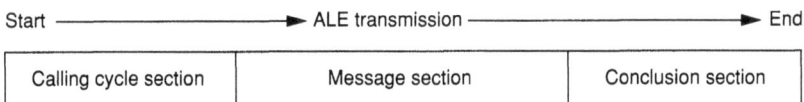

Start ────────────────▶ ALE transmission ────────────────▶ End

| Calling cycle section | Message section | Conclusion section |

Figure 3.11   Standard ALE transmission frame structure.

Five different automatic linking protocols are currently supported by MS-141A. They are associated with the *individual call*, the *star net call*, the *start group call*, the *allcall*, and the *anycall*. The standard *call* structure diagrammed in Figure 3.12 shows the basic sequence around which all the different call types are constructed.

The standard call involves a three-way handshake (data exchange) consisting of a call, a response, and an acknowledgment between the stations involved in the automatic linking process. Each segment of the ALE call structure shown in Figure 3.12 represents an ALE transmission frame as defined in Figure 3.11.

Placing a scanning ALE call. An ALE system's receive audio output is automatically muted while the receiver is scanning and unmuted when an ALE link has been successfully established. To place an ALE call, the operator selects/inputs the address of the station(s) to be called and presses the PTT key or enters an appropriate *initiate call* command via some other control means. There is no further operator interaction with the system until after the link has been successfully established. The ALE processor automatically selects the best available channel from the current list of scanned channels based on information available from the LQA data base. The receiver is tuned to the selected channel and a channel-busy determination is made before proceeding with the call.

The Individual Call. The Individual Call is the most commonly used of all the call types. It uses Individual Addresses to link a calling station with one other designated station in a two-way communications link. The calling station transmits the Call frame and returns to the receive mode to wait for the Response from the called station. If a response is received, the calling station transmits the Acknowledgment frame to confirm that a bidirectional link has been established.

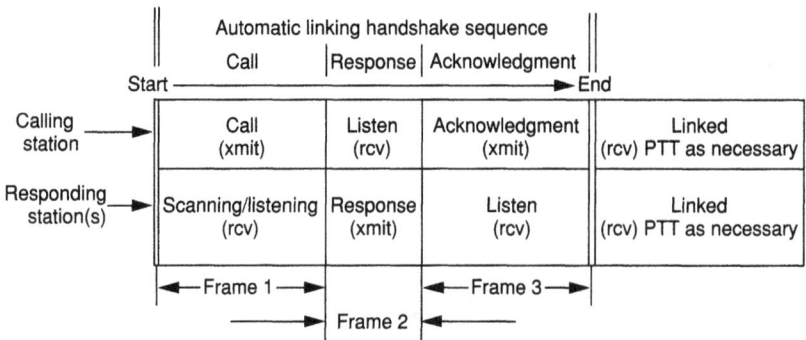

Figure 3.12 Standard ALE "call" structure.

Failure to receive the expected response within a predetermined period of time will end the call attempt on that channel and no acknowledgment frame will be transmitted. If a response is actually transmitted by the called station but not heard by the calling station, the called station will return to the scanning condition when an acknowledgment is not received within a predetermined period of time.

When a call is unsuccessful on one channel, due to a lack of response from the called station, the calling station will automatically continue the call attempt by repeating the process on the next-best alternate channel. This process is repeated until a link is established or until all the available channels have been tried. Periodically repeating the call on the best channel could improve the probability of a successful link if the called station was busy at the beginning of the call attempt.

The Star Net Call. This variation on the individual call allows a call to be placed to a single net address that is shared by multiple individual stations. Its purpose is to link an individual calling station with several other predesignated individual stations in a communications link where the caller can communicate bidirectionally with all the responding stations, even though the responding stations may not be able to communicate with each other on that channel.

The Star Group Call. The Star Group Call is similar to the star net call in that it allows the calling station to link with several other individual stations, but without the benefit of the predetermined response structure provided by the star net call. In other words, no single Net Address exists for the randomly selected group of individual stations to be called. The star group call requires the calling station to transmit each called station's Individual Address, sequenced in a modified calling cycle section of the call frame. Responding stations transmit their responses in slots similar to the star net slotted responses, except that the slots are not preassigned but derived by reversing the order in which they were transmitted in the calling cycle sequence.

The Allcall. The Allcall is used for general global broadcast-type calls. A response is not required from those that hear the call. All ALE stations will accept an allcall; they will stop scanning and listen but do not transmit a response frame. Both the called and calling stations proceed directly from the call frame to the linked condition. The calling station broadcasts its message to all listening stations.

The Anycall. The Anycall is a special-purpose type of call that solicits responses from any station that hears the call. It could be visualized as a cross between an Allcall and a Star Group Call. Like the Allcall, all stations who hear the Anycall will stop scanning and, similar to the Star Group Call, will transmit a slotted response, except that the slot position is randomly chosen. Response collisions are to be expected if a large number of responses are received.

Sounding. The sound is an identifying transmission that can be used for link quality measurements at the receiving station. It consists of the conclusion section of a standard ALE transmission frame repeated multiple times. Whenever a scanning receiver hears a conclusion section being transmitted, it pauses on the channel, measures and stores the LQA data relative to the transmitting station's address, and continues scanning. A sound repeats the conclusion section a sufficient number of times to allow scanning receivers an opportunity to hear it. In that respect, the sound has the timing characteristics of a calling cycle section.

Orderwire messages. There are three different orderwire message modes described in appendix A of MS-141A, but only one is considered mandatory. They are

1. Automatic Message Display (AMD) mode: This mandatory capability allows transmission and reception of simple text messages of no more than 90 characters using the 8-Ary FSK data modem, communications processing, and operator interface and control capabilities already in place to support the ALE automatic connectivity processes.

2. Data Text Message (DTM) mode: This optional orderwire message mode takes advantage of the 8-Ary FSK data modem, the triple redundancy, and the FEC coding offered by the ALE system's waveform and signal structure to support ASCII or unformatted binary data communications using an external data terminal device.

3. Data Block Message (DBM) mode: This optional orderwire message mode offers a higher-speed data transfer capability that takes advantage of the ALE system's waveform, but replaces the ALE triple redundancy and FEC coding with a more efficient and more robust, externally implemented block-encoding EDAC scheme appropriate for long or very long messages. An external data terminal device is also required.

Regardless of which orderwire message mode is used, the appropriate control and message information is transmitted in the message section of the standard ALE transmission frame.

Programmable presets and operating parameters. While the major features of the new second-generation ALE systems have been addressed in their most basic form in the preceding paragraphs, appendix A of MIL-STD-

188-141A contains many more variations and special features too numerous to mention. Because of the many options and possibilities, it is necessary to be able to program an ALE system to execute the desired options and thus determine the operational behavior of all the systems that must interoperate in a given network environment.

3.4 ALE Network Design

The primary benefits of ALE systems are reduced operator workload and improved connectivity. Once the ALE network is operating smoothly, simplicity and dependability will indeed be realized. This brief overview of ALE network design is supplemented in section A3.4 on the software disk.

The following paragraphs describe five design and planning steps that a network manager can use to put a high-quality network of HF ALE stations on the air. Steps 1 through 3 are not unique to ALE networks and are applicable to the design of conventional HF communications networks.

Step 1: Identify and physically locate each station in the network

This is not difficult if a finite number of fixed stations are involved. However, if mobile platforms are involved, it may be necessary to identify commonly used long-range routes of travel and/or short-range areas of operation instead of the individual platforms involved.

Step 2: Define the operational network requirements and structure

The reasons why communications are important, and the types of communications that are required, must be carefully analyzed. A purpose statement should be written and agreed upon by the network designer/manager and the users and operators to be served by the network. There are a number of possibilities ranging from a number of fixed stations providing point-to-point voice and/or low-speed data communications as a backup or substitute for normal long-distance telephone services to a worldwide network of ground entry stations providing two-way communications between aircrews and their bases of operation. Section A3.4-1 gives more details on this issue, but briefly a few questions are

1. Who needs to talk?
2. Who initiates calls?
3. What is the nature of the communication?
4. How are unmanned stations handled, if used?
5. How are network operations coordinated with time of day and frequencies?

Step 3: Consider the practical effects of HF propagation on the network design

A propagation analysis should be performed for all identified communications paths within the network. The purpose is to determine worst-case upper and lower frequency limits for the various paths over the 11-year sunspot cycle and the times of day that those paths need to be in service. A matrix showing the correct frequency range for each path will help to determine common frequency range requirements. Using operating frequencies that can be shared between various paths in the network helps to minimize the number of frequencies needed for satisfactory network operation. The PropMan™ automatic planning tool discussed in section 3.2 provides an accurate and convenient way to acquire propagation predictions.

If a particular path cannot reliably support the required communications with the application of reasonable assets, adjustments to the physical network configuration may have to be made to provide alternate routing. Adjustments to the output of step 2 should be made as necessary to finalize the physical network geometry.

The outputs of step 3 are frequency recommendations and an assignment table. The recommendation list identifies several ranges needed to support the network and the number of frequencies from each range. This list is used to apply for the needed frequency authorizations. The assignment table shows which network stations need which frequencies. Depending on the final network configuration, it is possible that the entire list of frequencies would be assigned to every station.

Step 4: Assign ALE addresses and determine ALE operating presets and parameters

Before a network designer/manager is ready to tackle step 4, he must have a good working knowledge of ALE addresses, scanning, and the scanning call process. Section A3.4-2 elaborates on these abbreviated points.

About ALE addresses. A basic principle of ALE is the notion of selective calling, where each ALE station is assigned one (or more) *addresses*, which can be thought of as digital *call signs*, used by the ALE system to identify the various stations in the network. Each user-programmable address is stored in an appropriate *address record*.

Individual Self Address. Each station must have a memory capacity for at least 20 different Individual Self Addresses (not shared with any other station in the net) to which it will respond when called. All Individual Self Addresses are treated equally when receiving and responding to an Individual Call, except if restricted through association with a particular group of channels (scan list). Individual Self Addresses are also used according to a Net Self Address association when responding to a Net Call. An ALE station always, without exception, identifies itself every time it transmits an ALE transmission frame by including its Individual Self Address in the conclusion section of that frame.

Net Self Address. Each station must have a memory capacity for at least 20 different Net Self Addresses (typically shared with a small number of other net members) to which it will respond when called. All Net Self Addresses are treated equally when receiving and responding to Net Calls, except when restricted through association with a particular group of channels (scan list). A Net Call normally requires each individual responding station to transmit its response in a preassigned time slot using a specific associated Individual Self Address.

Anycall Address. Each station is permanently programmed with the standard Anycall Address, as specified by MS-141A.

Allcall Address. Each station is permanently programmed with the standard Allcall Address, as specified by MS-141A.

Null Address. The Null Address is a permanently assigned three-character address used primarily for testing or as a placeholder. Automatic link establishment receiving systems ignore calls to the Null Address.

Other Addresses. Each station must have a memory capacity for at least 100 individual or net addresses of *other* stations in the network to which calls may be placed.

Individual Addresses. An Individual Address is simply some other station's Self Address and is used when placing a call to that other station.

Net Addresses. A Net Address is an address used to place a call to a preplanned group of stations that all share that Net Address as one of their Net Self Addresses. These stations will transmit responses in preplanned time slots when they receive a call to this Net Address.

Group Address. A Group Address is actually a sequence of other station's Individual Addresses. A group name or pseudoaddress may be assigned locally by an operator to facilitate temporary storage and recall of the group's list of Individual Addresses, depending on the particular manufacturer's design.

Floating Addresses. Floating Addresses are Other Addresses that were not programmed by the user. If implemented, they are programmed automatically by the ALE system as it hears other stations over the air.

Wild card characters in addresses. The MS-141A standard permits calls to be placed to Individual Addresses containing one or more question mark (?) characters known as *wild card characters.*

About scanning. There are a few key programmable parameters that determine how an ALE system handles the scanning process and calls to scanning stations. Note that receivers scan but transmitters do not. Transmitters are tuned to a selected calling channel and, once tuned, transmit a call or a sound for a period of time long enough for the other station's scanning receivers to have a chance to hear it. Tuning times for transmitters are not critical, but scanning receivers must be capable of operating at the two MS-141A required scan rates of 2 channels/sec and 5 channels/sec. The selection of the scan rate does not have as big an impact on connectivity time as might be expected because there are other factors that influence the actual time it takes a scanning receiver to complete one full scan cycle. A scan cycle has been completed when a scanning receiver has visited each of the scanned channels once.

The length of time the scanning receiver stays on a particular preset channel during the scanning process is called the *channel dwell time.* A programmable parameter specifies the minimum time a scanning receiver will spend on any one scanned channel, but that is truly a minimum. The maximum dwell time on a channel depends on what signals are heard and what decisions must be made before the scanning receiver can move on to the next channel in the scanning sequence. If ALE tones are detected, the dwell will be extended long enough to decode one word. If decoding one word is not sufficient to make a stop-scan decision, then the dwell must be extended ever further until such time as a firm decision can be made.

The length of time a call must be transmitted is determined independently of the scan rate. Therefore, connectivity time is influenced by the call duration time rather than by the scan rate.

Preset channels and scan lists. Each system must have a memory capacity for at least 100 preset channels. As a minimum, each preset channel record must be able to accommodate a pair of receive and transmit frequencies with their associated emission mode information and sounding control data. Preset channels may be programmed into what we will call scan lists. The number of scan lists provided and the maximum number of channels per scan list may be different from one manufacturer to another. Scan lists provide a convenient way for an operator to select the channels to be scanned. Section A3.4-3 gives more details.

LQA data base. The LQA data base is automatically maintained through sounding and LQA measurement activities. However, allowing the LQA data base to be programmed via datafill operations provides an opportunity to input an externally synthesized data base using tools such as PropMan™, discussed earlier in section 3.2. The benefit of using synthesized LQA data is to reduce the need for sounding. This might be a viable solution for large fixed stations operating in a network with many geographically dispersed mobile platforms in order to reduce the sounding requirements for the mobile platforms.

About programmable parameters. Programmable parameters fall into two major categories: those that affect interoperability and those that don't. An example of a programmable parameter that does not affect interoperability might be one that determines the length of time a call alert signal is sounded when a call is received. Parameters that modify protocol-related timing functions are certainly more critical, and in most cases will affect interoperability.

The network designer/manager must be concerned with any parameters that directly affect interoperability or network performance and efficiency. Although the following is not intended to be a complete list, a few of the more important parameters and their purpose are

1. Scan rate: Determines the minimum receiver dwell time on a scanned channel.

2. Call duration: Length of time the calling cycle section of an ALE transmission frame is transmitted. It depends on the number of channels being scanned and the probability of encountering ALE signals on those channels.

3. Call reject duration: Maximum time a scanning receiver will wait on a calling channel for the conclusion section of the ALE transmission frame.

4. LQA reject threshold: Precludes marginal quality links by ignoring incoming calls with LQA values below a chosen value.

5. Command LQA: Uses AMD orderwire message capability to request and receive reverse-path LQA values from another station.

6. Wait-for-tune and response time: Sets length of time a calling station will wait for a called station to tune its transmitter and transmit a response frame before repeating the call on the next best channel.

7. Net slot response variables: Number of null slots required to cover the slowest tuning station in the network, net member individual addresses and slot assignments, net wait time, scan list association, etc.

8. Call attempts limit: Number of call attempts after which an ALE system will stop and declare the call unsuccessful.

9. Return-to-scan time: Elapsed time after last PTT key activity when ALE system automatically terminates the link and returns to scan.

10. Keep-alive transmission interval: Elapsed time after the start of a continuous transmission when the ALE system automatically transmits an ALE keep-alive frame to keep the listening station from timing out and returning to scan.

11. Terminate link transmission: Command transmitted by a terminating station that causes the other linked stations to return to scan without waiting for individual station time-outs.

12. Sounding interval: The period of time between sounds. Sounding may be disabled entirely at selected stations in the network.

13. Sounding duration: Length of time a sound is transmitted on each channel. It is similar to call duration but may be separately programmable.

14. Inhibit anycalls: Prevents acceptance of anycalls.

15. Inhibit allcalls: Prevents acceptance of allcalls.

16. AMD in call or acknowledge frame: Determines if the AMD message section will be inserted in the call frame or the acknowledge frame.

Step 5: Build the Datafill files required to load each ALE system in the network

Once the network design details have been finalized, it is necessary to prepare the datafill files needed to electronically transfer the programming data into the various systems in the network. Systems built by different manufacturers will likely require slightly different information in a different format, which may make it more difficult for the network designer/manager to know when two different systems will behave in an interoperable manner. Intimate familiarity with each different type of system is the only answer. Lab testing in a multisystem simulated network environment is highly recommended. This approach speeds acquisition of the required system familiarity, confirms interoperable system behavior, and validates the datafill generation techniques for different manufacturer's systems prior to fielding of the datafill files. Except for self addresses all other programmable data should be identical for all systems built by the same manufacturer.

References

1. Kenneth Davies, *Ionospheric Radio* (London: Peter Peregrinus Press, 1989).

2. John M. Goodman, *HF Communications Science and Technology* (New York: Van Nostrand Reinhold, 1992).

3. Robert B. Rose and J. N. Martin, *MINIMUF-3.5: Improved Version of MINIMUF-3, A Simplified HF MUF Prediction Algorithm*, Tech. Doc. NOSC TD 201 (San Diego, CA: U.S. Naval Ocean Systems Center, October 26, 1978).

4. Robert B. Rose, "PROPHET—An Emerging HF Prediction Technology," in John M. Goodman (ed.), *Effect of the Ionosphere on Radiowave Systems* (Washington, DC: U.S. Government Printing Office, 1982). (Book based on IES '81 Ionospheric Effects Symposium, Alexandria, VA, April 14–16, 1981.)

5. M. Daehler, *An HF Communications Frequency-Management Procedure for Forecasting the Frequency of Optimum Transmission*, NRL Memorandum Rep. 5505 (Washington, DC: U.S. Naval Research Laboratory, December 31, 1984).

6. J. A. Joselyn and K. L. Curran, "The SESC (Space Environment Services Center) Satellite Broadcast System for Space Environment Services," paper 4-1 in *IES '84 Proceedings: The Effects of the Ionosphere on C^3I Systems*, document ADA 163622 (Springfield, VA: National Technical Information Service, 1985. (Book based on IES '84 Ionospheric Effects Symposium, Alexandria, VA, May 1–3, 1984.)

7. M. Darnell, "Channel Evaluation Techniques for Dispersive Communications Paths," in J. K. Skwirzynski (ed.), *Communication Systems and Random Process Theory*, pp. 425–460 (Alphen aan den Rijn, The Netherlands: Sijthoff and Noordhoff, 1978). (Book based on NATO Advanced Study Institute Series E, No. 25, Darlington, UK, August 8–20, 1977.)

8. *Common Long-Haul and Tactical Communications Systems Technical Standard*, MIL-STD-188-100 (Washington, DC: Dept. of Defense, November 17, 1976).

9. *Standards for Long-Haul Communications: Subsystems Design and Engineering Standards and Equipment Technical Design Standards for High Frequency Radio*, MIL-STD-188-317 (Washington, DC: Dept. of Defense, March 30, 1973).

10. David H. Bliss, "Channel Characteristics and Anomalies," sec. 5.1 in Charles A. Harper (ed.), *Handbook of Electronic Systems Design*, pp. 5-15–5-18 (New York: McGraw-Hill, 1980).

11. L. W. Pickering, "The Calculation of Ionospheric Doppler Spread on HF Communications Channels," *IEEE Trans. Commun.* COM-23 (May 1975): 526–537.

12. David H. Bliss, " Automated Channel Evaluation for Adaptive HF Communications," in *HF Communication Systems and Techniques '85*, pp. 37–41. IEEE Conference Pub. 245 (London: 1985).

13. L. R. Teters, J. L. Lloyd, G. W. Haydon, and D. L. Lucas, *Estimating the Performance of Telecommunications Systems Using the Ionospheric Transmission Channel: Ionospheric Communications Analysis and Prediction Program User's Manual (IONCAP)*, NTIA Rep. 83-127 (PB84-111210) (U.S. Dept. of Commerce, National Telecommunications and Information Administration, Institute for Telecommunications Sciences, Boulder, CO: July 1983).

14. C. C. Watterson, *Methods of Improving the Performance of HF Digital Radio Systems*, NTIA Rep. 79-29 (PB80-128606) (U.S. Dept. of Commerce, National Telecommunications and Information Administration, Institute for Telecommunications Sciences, Boulder, CO: October 1979).

15. R. B. Fenwick and T. J. Woodhouse, "Real-Time Adaptive HF Frequency Management," paper 5 in V. J. Coyne (ed.), *Special Topics in HF Propagation*, NATO-AGARD Conference Proceedings 263, document ADA080855 (London: Technical Editing and Reproduction, 1979). (Book based on Symposium of the Electromagnetic Wave Propagation Panel, Lisbon, Portugal, June 1, 1979.)

16. L. E. Hoff and R. L. Merk, *Performance Measures for an Automated Navy Tactical Sounder System (NTSS)*, NOSC Tech. Rep. 409 (ADA072198) (San Diego, CA: U.S. Naval Ocean Systems Center, July 1, 1979).

17. Bodo W. Reinisch and Klaus Bibl, *Ionospheric Research Using Digital Ionosondes*, U.S. Air Force Geophysics Laboratory Tech. Rep. 83-0184 (Lowell, MA: University of Lowell, July 1983).

18. David H. Bliss, "A New HF-Link Parameters/Quality Analysis Approach," in *MILCOM '83 Conference Proceedings*, pp. 530–534 (Washington, DC: IEEE, 1983).

19. S. M. Chow et al., *'RACE'—An Automatic High-Frequency Radio Telephone System for Communications in Remote Areas*, CRC Rep. 1338-E (Ottawa, Canada: Communication Research Centre, Canadian Dept. of Communications, December 1980).

20. Gerhard Braun, *Planning and Engineering Shortwave Links*, pp. 15–77, 124–252 (London: Heyden & Son, 1982).

21. Gerald L. Hall (ed.), *The ARRL Antenna Book*, 14th ed. (and subsequent editions) (Newington, CT: American Radio Relay League, 1982).

22. R. C. Johnson and H. Jasik, *Antenna Engineering Handbook*, 2nd ed. (New York: McGraw-Hill, 1984).

23. L. A. Moxon, *HF Antennas for All Locations* (Bath, UK: Pitman Press, 1982).

24. M. F. Radford, "High Frequency Antennas," Chapter 16 in A. W. Rudge et al. (eds.), *The Handbook of Antenna Design*, vol. 2, pp. 663–724 (London: Peter Peregrinus, 1983).

25. P. David and J. Voge, *Propagation of Waves* (New York: Pergamon Press, 1969).

26. Karl Rawer, *The Ionosphere: Its Significance for Geophysics and Radio Communications* (New York: Frederick Ungar, 1957).

27. J. A. Betts, *HF Communications*, pp. 8–16, 84–94 (New York: American Elsevier, 1967).

28. *CCIR Interim Method for Estimating Skywave Field Strength and Transmission Loss at Frequencies between the Approximate Limits of 2 and 30 MHz*, CCIR Report 252.2 in *New Delhi Assembly*, 1970/1974, ITU-CCIR, New Delhi, 1970/1974. See also *Second CCIR Computer-Based Interim Method for Estimating Skywave Field Strength and Transmission Loss at Frequencies between 2 and 30 MHz*, CCIR Doc. 6 1070-E, Draft Supplement to CCIR-Report 252-2, submitted to XIV Plenary Assembly, Kyoto, Japan, 1978.

29. George Jacobs and Theodore J. Cohen, *The Shortwave Propagation Handbook*, 2nd ed. (Hicksville, NY: CQ Publishing, 1982).

30. P. N. Saveskie, *Radio Propagation Handbook* (Blue Ridge Summit, PA: TAB, 1980).

31. *World Distribution and Characteristics of Atmospheric Radio Noise*, CCIR Report 322 in *1963 Geneva Assembly*, ITU-CCIR, Geneva, 1964. (Revised in 1974 as CCIR Report 322-1.)

32. Daniel P. Roesler, "HF/VHF Propagation Resource Management Using Expert Systems," *IES '90 Ionospheric Effects Symposium*, pp. 313–321 (Washington, DC: U.S. Government Printing Office, 1990).

33. Daniel P. Roesler, "HF Propagation Management Tools System," *MIL-COM '92 Conference Proceedings*, pp. 3.6.1–3.6.5 (IEEE, October 1992).

34. Georgellen Smith and V. Elaine Hatfield, *AMBCOM User's Guide for Engineers*, Contract F08606-85-C-0018 (Menlo Park, CA: Stanford Research Institute, January 1987).

35. David B. Sailors and Robert B. Rose, *HF Sky-Wave Field Strength Predictions*, Tech Report 1624 (ADA 275891) (San Diego, CA: U.S. Naval Command, Control and Ocean Surveillance Center, September 1993).

4

Receiver Design

David B. Hallock

This chapter will present the results of the evolution of receiver design with emphasis on the circuits, components, and techniques that are important for SSB reception.

4.1 Introduction

The block diagram and circuit of any receiver are strongly if not totally determined by the available components. Wave-filtering components having sufficient selectivity to select SSB signals are available only with fixed center frequency designs. Therefore, the standard receiver design for HF SSB has become the superheterodyne which uses frequency translation to convert the HF signal to a fixed intermediate frequency where selectivity is practically obtainable. The modern superheterodyne receiver will typically use several intermediate frequencies to obtain selectivity for image and spurious signal rejection in addition to sideband selection. An advantage of the superheterodyne is that the majority of the gain needed to amplify picowatt signal levels to the level needed for human audibility can be constructed and controlled in the fixed-IF circuits. Constant-frequency amplifiers are generally better behaved in the areas of gain stability, noise figure, and distortion.

4.2 HF Receiver Requirements

The technical, physical, and cost requirements of receivers are as varied and complex as there are applications and users. However, there are several basic requirements, plus service and special needs, that commonly enter into the design of a receiver, be it simple or complex. It is helpful and necessary to understand the basis for the technical needs, and to know the range of values to be expected in each parameter. First, the basic requirements will be examined.

Sensitivity

High-frequency receiver sensitivity is often expressed as the number of microvolts necessary to achieve some value of signal-to-noise ratio at the output. More specifically, the signal level is measured in open-circuit generator voltage together with a generator source resistance. Closed-circuit voltage, or the voltage across the receiver input terminals, is sometimes used to specify sensitivity. This is technically imprecise since the receiver input resistance is then an undefined variable, which makes the power flowing into the receiver input an unknown. Figure 4.1 illustrates the generator and receiver input circuit model. If R_{in} is exactly equal to R_g, the receiver is accepting the generator's available power ($V_{oc}^2/4R_g$) and V_{in} is one-half of V_{oc}. But, if V_{in} were to be specified for sensitivity, the power flowing into the receiver could theoretically approach infinity if R_{in} were allowed to approach zero.

Industry jargon often calls open-circuit voltage "hard" and closed-circuit voltage "soft." It is harder to achieve a certain output signal-to-noise ratio with one hard microvolt than it is with one soft microvolt. A source of confusion is that most signal generators are calibrated in terms of the closed-circuit voltage across a load resistance equal to the generator source resistance. The open-circuit voltage is then twice the indicated value. Often a 6-dB attenuator is added at the generator output so that the indicated voltage can be read directly as open-circuit voltage. Typical HF receiver

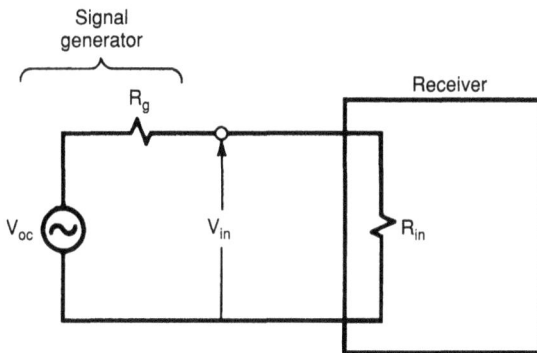

Figure 4.1 The closed-circuit input voltage V_{in} is a function of the open-circuit generator voltage V_{oc}, the generator resistance R_g, and the receiver's input resistance R_{in}.

sensitivity for a 10-dB output signal plus noise-to-noise ratio lies in the region of 1.0 μV, open circuit, for a nominal 50-ohm source resistance. Sensitivity is a function of bandwidth, since the signal (for test purposes) is a single frequency that occupies no bandwidth while the noise fills the receiver's bandwidth.

The open/closed voltage specification confusion is eliminated when sensitivity is stated in terms of available signal power; for example, dBm, decibels with respect to 1 mW. This is unambiguous and also appears directly on signal generator displays. (In this chapter we will use the more common abbreviation, dBm, for the precise designation, dBmW.) A typical sensitivity power level of –113 dBm is equivalent to 1.0 μV, open circuit, for a 50-ohm source resistance.

Noise figure

The ratio of the available signal-to-noise ratio at the input of a receiver (or any two-port network) to the available signal-to-noise ratio at the output is called *noise figure* when given as a decibel ratio and *noise factor* when expressed as a power ratio. (For a complete and basic foundation of noise factor, see Friis [1].) Noise figure is independent of bandwidth because the available noise powers are from the same bandwidth and cancel when the ratio is taken. High-frequency SSB receivers have noise figures in the 12- to 17-dB region, which is all that is usually necessary because of the generally high level of atmospheric noise prevalent at HF. An important exception occurs in special receiving systems with antennas that are very inefficient. This results in available signal and noise powers commensurate with the self-noise of the receiver. In these systems, RF gain is usually required to reduce the noise figure.

The interrelation of noise figure and sensitivity is readily done when the receiver's effective noise bandwidth is known because the overall system is linear. If, for example, a certain signal level produces a 10-dB output signal-to-noise ratio and the receiver has a 15-dB noise figure, the available input signal-to-noise ratio must be 25 dB, the sum of the output ratio and noise figure. The noise power available from a room temperature source (+290 K) is kTB, where k is Boltzmann's constant (1.38×10^{-23} W·s/K), T is the absolute temperature, and B is the bandwidth. An easily remembered fact is that in a 1-kHz bandwidth, the available noise power is –144 dBm. (A negative gross of dBms!) For an SSB receiver bandwidth of 3 kHz, the noise is about 5 dB greater ($10 \log [3 \text{ kHz}/1 \text{ kHz}]$) or –139 dBm. In our example the signal power is 25 dB stronger than the noise power, and so is –114 dBm. For a 50-ohm signal source, the corresponding open-circuit generator voltage is about 0.9 μV. Note that we did not know or mention the receiver's input resistance, it being sufficient only to know the source resistance and the resulting noise figure together with the noise bandwidth.

Intermodulation distortion

When two relatively large signals combine in a nonlinear stage to produce one or more new frequencies, the results are called *intermodulation distortion* (IMD). In a receiver, we are concerned with distortion from signals both within and outside the IF bandwidth, and so will refer to in-band and out-of-band IMD, respectively. For a first-cut explanation, the output of the receiver can be considered to be an ascending power series of the input such as

$$E_{\text{out}} = A_0 + A_1 E_{\text{in}} + A_2 E_{\text{in}}^2 + A_3 E_{\text{in}}^3 + \ldots \tag{4.1}$$

When E_{in} is replaced by the sum of two sine waves in this series, it is found that new frequencies are produced which occur at spacings equal to the difference between the original pair of signals. If it so happens that the IMD frequency is within the receiver's bandwidth, a signal is heard that is, in reality, not present. For example, if a receiver is tuned to 10 MHz and there are two large signals at 4 and 14 MHz also present, the difference frequency of 10 MHz may be generated somewhere in the receiver by nonlinearity. This example is termed *second-order* IMD because it arises predominantly from the A_2 term in the power series. It is readily eliminated by passive filtering at the receiver input because the large signal frequencies are much different from the desired signal frequency.

When the A_3 term in the series is included in the computation of distortion products, it is found that IM frequencies of the form $2f_2 - f_1$ and $2f_1 - f_2$ are generated, f_1 and f_2 being the large signal frequencies. These are called *third-order* IMD, and can be very difficult to reduce in level because the undesired signals can be very close to the passband frequencies and even within the passband. Third-order IMD is also generated by the higher odd-order terms of the series. In our 10-MHz receiver example, suppose there are two large signals at 11 and 12 MHz. The second harmonic of a 1-MHz signal minus the 12-MHz signal is exactly 10 MHz. Likewise, the second harmonic of a 9-MHz signal minus another 8-MHz signal will produce 10-MHz IMD. Higher odd-order IMD products are also sometimes specified and are of concern in receiver design. When three or more strong signals are present, the possible number of interference combinations grows astronomically (Chapter 14).

Fortunately, well-designed solid-state receivers have sufficient signal handling capability to be useful with normal on-the-air signal strengths. It is typical to have two signals 80 to 100 dB above the receiver's sensitivity level produce IM products less than the sensitivity level.

Intercept point

The concept of intercept point, developed by McVay [2], wraps the IMD performance of a receiver into one number. If we plot the two-signal output level and the distortion level (in dBm) versus the two-signal input level (in dBm) as shown in Figure 4.2, we find that the second-order distortion increases 2 dB for each 1-dB increase in both input signal levels. (This is also predictable from the series expansion of Equation 4.1.) Further, a 3-dB increase is found for third-order distortion when both signals are increased 1 dB. In general, there will be an n-dB increase in nth-order IM for each decibel of input signal increase. If the receiver did not limit, the distortion products would increase until they equaled the output signal level. This point is called the *second-* or *third-order intercept*, respectively. If output power is used as the dimension, the term *output intercept* is used. Receivers are usually specified in terms of input intercept, using the input signal level.

Useful relationships exist between IMD level, signal level, and the intercept point. For second-order intercept the input intercept is

$$_2I_{in} = \text{Input} + \Delta \tag{4.2}$$

where Input is the two-tone signal level per tone (in decibel units such as dBm), and Δ is the ratio between the two-tone level and the IMD level (in decibels).

The third-order input intercept is found from

$$_3I_{in} = \text{Input} + \frac{\Delta}{2} \tag{4.3}$$

Generally the nth-order input intercept is equal to

$$_NI_{in} = \text{Input} + \frac{\Delta}{N-1} \tag{4.4}$$

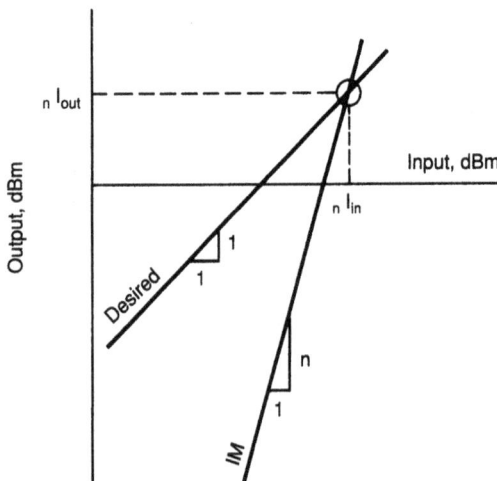

Figure 4.2 The intercept point is at the intersection of the desired output power and the IM output power lines.

The third-order input intercept for a high-grade commercial HF SSB receiver is in the +20- to +35-dBm region, with intercepts below 0 dBm being on the low-performance side. Second-order input intercept values are higher, +50 to +70 dBm being typical.

Dynamic range

The simplest definition of dynamic range is the ratio between the maximum signal that the receiver is capable of handling before overloading and distorting and the minimum signal producing a useful signal-to-noise ratio. This is the in-band dynamic range and would be 120 dB for a 1-V to 1-μV input ratio. It is a function of the receiver's bandwidth and the effectiveness of the AGC circuits.

A more useful measure of dynamic range is the ability of a receiver to receive weak desired signals in the presence of strong out-of-band signals. If a pair of strong signals has an IM product that lies in the receiver passband, sensitivity will be degraded by the interfering signal. Since the ear can recognize a desired signal at or somewhat below the receiver ambient noise level, one way to specify dynamic range is to state how far above the equivalent input noise level is each signal of a two-signal set whose IMD is equal to the input noise level. This definition of IM dynamic range is a function of the receiver's bandwidth because noise power is proportional to bandwidth as defined by the expression

$$N = kTBFG \qquad (4.5)$$

where F = noise factor
G = receiver gain

Referred to the input, the noise power is just $kTBF$. This product expression is more easily handled by working with power in decibel units. Suppose the receiver's third-order input intercept point is Δ dB above the input noise. Since the third-order IMD slope is 3 (see Figure 4.2), the per-signal input level of a two-signal set producing IM at the noise level is $\Delta/3$ below the intercept. The third-order IMD dynamic range (DR$_3$) is the difference between the per-signal level of the two-signal set and the input noise level. In equation form that is

$$\mathrm{DR}_3 = 2\,\frac{(_3 I_{\mathrm{in}} - kTBF)}{3}\ \mathrm{dB} \qquad (4.6)$$

where $kTBF$ is understood to be expressed in decibel power units. A receiver having a 14-dB noise figure, 3-kHz noise bandwidth, and third-order input intercept of +20 dBm will have a dynamic range of

$$DR = \frac{2[20 - (-144 + 5 + 14)]}{3} = 96.7 \text{ dB}$$

From a practical standpoint, an off-frequency pair of signals, each having a level of –28.3 dBm, will produce a third-order IM product that is equal to the receiver's input self-noise level of –125 dBm. With an antenna connected to the receiver, it is likely that the –125-dBm product will be further masked by external noise. However, the undesired signal level will not have to rise appreciably above –28 dBm before the IM interference is heard, since the absolute power level of the product rises 3 dB for each decibel of undesired level increase.

Dynamic range can also be limited by noise sidebands present on the injection power to the first mixer. This effect, called *reciprocal mixing*, was mentioned in Chapter 2 and is often noticed at collocated transmitter/receiver sites. Figure 4.3 illustrates the process that translates noise into the intermediate frequency when a strong off-frequency signal, which is ideally spectrally pure, is translated to the IF output together with the injection sideband noise. The translated signal frequency is then outside the IF passband while a portion of the noise spectrum lies within the passband. The mechanism in the mixer is simply preservation of instantaneous phase in the time domain as the injection vector is randomly advanced and retarded, causing similar perturbation of the IF vector. When the off-frequency signal is close to the receiver's tuned frequency, 100 kHz away for example, the amplitude of the noise at the IF center may exceed the receiver self-noise. When the injection spectrum is known from measurement, it is readily possible to calculate the magnitude of reciprocal mixing by remembering that the IF spectrum is a re-creation of the injection spectrum. If the injection noise is 100 dB below carrier at 100 kHz away from the injection frequency, then the IF noise due to reciprocal mixing will also be 100 dB below the IF carrier level at the mixer output.

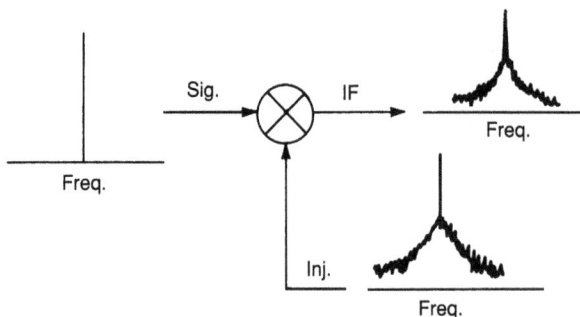

Figure 4.3 The mixer injection spectrum is heterodyned by a strong off-frequency pure signal into a spectrum offset from the IF and having noise sidebands identical in ratio to those of the injection.

4.3 Topology

A simple form of HF receiver is the homodyne or direct-conversion receiver. This receiver uses a conversion mixer preceded by some RF selectivity to heterodyne SSB or CW signals directly to audio frequency, using a carrier frequency local oscillator (LO) for mixer injection. A lowpass filter for SSB or a bandpass filter for CW directly follows the mixer to provide selectivity. This selectivity is less costly to implement than the bandpass filter used in a superheterodyne and is integrable into small size using either analog active filter or charge switching filter technology. Most or all of the receiver's gain is obtained in the AF amplifiers following the audio filter. This presents a challenge in audio amplifier design since noise and hum pickup must be kept very low. Another design and construction problem with the homodyne is leakage of the local oscillator into the RF filter and thence into the mixer. The mixer acts as a phase detector between the leakage and injection signals. If the oscillator has mechanical or electrical frequency instability, the resulting frequency modulation will cause audio output from the mixer. Also, since the RF selectivity is insufficient to reject the undesired sideband, this receiver has no unwanted sideband rejection. Despite these drawbacks, the direct-conversion receiver performs well, considering its simplicity, and is a practical design for hobby use.

Sideband selection can be obtained in a homodyne receiver by separately mixing the incoming signal with a local oscillator split into two paths having a 90° relative phase difference [3]. The resulting in-phase and quadrature audio channels are then phased-shifted a relative 90° with wideband networks and combined vectorially to output either the upper or lower sideband (USB or LSB). Lowpass selectivity then provides the receiver's major selectivity element. The ratio of desired to undesired sideband response (in decibels) may be found from

$$R = 10 \log \frac{1 + 2G \cos \phi + G^2}{1 - 2G \cos \phi + G^2} \text{ dB} \tag{4.7}$$

where G is the ratio of the I and Q channel voltage gains at the summing point and ϕ is the total phase error from 90° in the RF and audio phase shifters. With practical circuits, ratios of 30 dB are possible, which is low performance when compared to the 60 dB or more available with good bandpass filters in the superheterodyne.

The superheterodyne receiver, whose block diagram is shown in Figure 4.4, is capable of providing excellent selectivity by using fixed-frequency filters after conversion of the signal to IF. The conversion is done with mixers whose injections are generally obtained from synthesizers having a stable reference source. In simpler designs, crystal or LC oscillators can provide

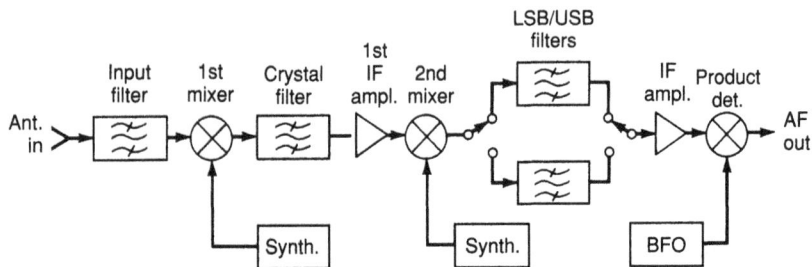

Figure 4.4 The superheterodyne uses multiple heterodyning and filtering to select the desired sideband.

adequate frequency stability. The various signal path blocks shown in Figure 4.4 perform the following functions:

- Input filter: Provides bandpass or lowpass selectivity to prevent first-mixer overload and spurious signal response.

- First mixer: Converts the signal frequency to the first intermediate frequency, which may be either lower or higher than the signal frequency.

- Crystal filter: Provides the first narrowband selectivity element in the receiver. Protects the following amplifier from overload by off-frequency large signals and attenuates the second-mixer image response. Also called a "roofing filter."

- First IF amplifier: Provides gain and low noise figure as the first stage in the IF section. An IF amplifier may directly follow the first mixer when sensitivity is paramount.

- Second mixer: Heterodynes the first IF to a new frequency where the major amount of the receiver's selectivity can be obtained with minimum cost.

- LSB/USB filters: High-performance bandpass filters with 2- to 3-kHz passbands used for lower or upper sideband selection. A single filter may be used in low-cost designs.

- IF amplifier: An integrated circuit (IC) or distributed component amplifier that provides a major amount of the receiver's gain. Includes gain control elements and selectivity to reduce wideband noise.

- Product detector: Heterodynes the IF signal to audio using injection from the following beat frequency oscillator (BFO).

- BFO: Provides injection power for the product detector. May be frequency-coordinated with the second-mixer injection to allow use of a single IF filter for LSB or USB selection.

The audio circuits following the product detector may contain functions in addition to gain and power output generation. Audio selectivity may be incorporated, especially as the IF selectivity is narrowed by filter selection for CW. Syllabic rate-detection squelch circuits will also be found in the audio section. Squelch and other audio processing circuits are discussed in Chapter 7. Audio compression may be necessary in some designs, although that function is better accomplished with AGC ahead of the product detector.

Historically, the frequency plan of the superheterodyne was first implemented with downconversion to successively lower frequencies where low-cost and high-performance selectivity could be obtained. This resulted in the requirement for high-performance and costly input filter selectivity to reject the image frequency. For example, a first conversion from 20 to 3 MHz using a 23-MHz injection would have an image response at 26 MHz, which must be reduced with selectivity. The more recent availability of very high-frequency (VHF) crystal filters using overtone-mode quartz crystal resonators has changed the frequency topology of modern HF receivers very often to upconversion schemes [4]. For example, the first IF might be 50 MHz, with the second IF as low as 455 kHz. The first injection for this case might range from 52 to 80 MHz for a signal range of 2 to 30 MHz, the first conversion being termed a *high-side difference mixer*. Sum mixing would require an injection frequency span from 48 to 20 MHz. In either case, the image frequency band lies well above 30 MHz and so is readily filtered by fixed tuned bandpass or lowpass filters. This upconversion and broadband approach has great advantages in component cost, accuracy, and performance. It will be examined in detail next.

4.4 Frequency Schemes

The optimum frequency scheme or plan for the HF superheterodyne involves trade-offs between performance, circuit complexity, and cost. In this section we will discuss topics that influence the designer's choice of IF and injection frequencies, with emphasis on the spurious responses that are inherent in any scheme.

IF selection

The first important choice in selecting a frequency scheme is the determination of the IF. As we have just seen, the upconverting superheterodyne is the primary topology for HF SSB use, so the IF must be above the highest signal frequency, or generally above 30 MHz. Crystal filters in production quantity are available up to about 150 MHz, with the cost increasing with frequency. There are several discrete frequency choices that may offer cost advantages because of high-volume usage by equipment in other services. These include

- 45 MHz, used in cellular radiotelephones
- 70 MHz, a standard military IF for ultrahigh-frequency (UHF) equipment
- 75 MHz, the marker beacon frequency in civil aviation

These and other "round number" IFs may not be optimum when the overall frequency scheme, including that in the synthesizer, is considered. One long-used IF at 109.35 MHz was based on the references used in a two-loop, sliding divisor synthesizer, the standard frequency, and the ease of multiplying the standard to get injection frequency for second-IF conversion.

The effects of IF upon receiver performance can be conflicting. As the IF is increased, the susceptibility of the receiver to spurious responses will diminish. These responses will be described in detail in the next section. An upconverting first mixer with a 45-MHz IF can be used to illustrate one class of spurious responses. If a relatively large signal at 22.5 MHz produces second-harmonic distortion in the mixer, the resulting 45-MHz signal will directly enter the IF. For an upper signal limit of 35 MHz, as set by the stopband of a lowpass filter before the mixer, a 75-MHz IF would reject the second-harmonic spurious but would be susceptible to third-harmonic interference. To reject third harmonics, an IF above 105 MHz would be required. As the IF is increased, however, the first injection carrier-to-noise ratio will generally degrade because the synthesizer must generate higher frequencies with constant step size. This means that reciprocal noise mixing (refer to section 2.2) with strong off-frequency signals will become more noticeable as the IF rises. The apparent stopband selectivity of the overall IF will also be degraded as synthesizer noise enters the IF passband. Also, if receiver operation down to near-zero frequency is required, the injection noise sidebands will directly enter the IF through mixer imbalance.

Intermediate frequencies above the readily available crystal filter range are not normally used in HF SSB receivers. Lumped or distributed constant LC filters, surface acoustic wave (SAW) transversal filters, SAW resonator filters, and chemically etched thin resonator filters all have practical problems with low Q, high insertion loss, and power-handling capability.

Spurious responses

The desired output frequency of an ideal mixer consists of the sum or difference of the signal and injection frequencies. The real mixer may also produce an identical output frequency when the input signal is at any one of numerous frequencies. This reality may be examined in detail by considering a general expression for the output of a mixer:

$$mSIG + nLO = IF \qquad (4.8)$$

In this equation m and n are signed integers defining which harmonics of the signal (SIG) and local oscillator (LO or injection) can add or subtract from one another to produce an IF output. Depending upon the frequency scheme, m and n may be both positive, alternately positive or negative, but never both negative. In a receiver, the mixer is followed by narrowband selectivity (the crystal filter), so that the only mixing products of concern are those equal to the desired IF.

An example of a particular spurious response is helpful in seeing how Equation 4.8 applies to a mixer. Illustrated in Figure 4.5 is a mixer that up-converts the 2- to 30-MHz HF range to 100 MHz by adding a 98- to 70-MHz local oscillator to the signal. The plus signs at the signal and LO ports indicate the sign of that addition. Thus, the desired mixing process has $m = +1$ and $n = +1$. Suppose that the desired signal is at 24 MHz, so that the local oscillator is set at 76 MHz to produce the desired IF of 100 MHz by summation. Further suppose that there is a large undesired signal at 26 MHz that can easily reach the signal port through wideband filters to create second-harmonic distortion in the mixer. The local oscillator also creates the second harmonic of 76 MHz in the mixer. For these harmonics, the arithmetic of Equation 4.8 becomes $2 \times 76 - 2 \times 26 = 100$; that is, the second harmonic of the 76-MHz injection mixes subtractively with the second harmonic of the undesired 26-MHz signal to produce the 100-MHz IF. This IF signal passes through the receiver's IF, detection, and audio circuits and appears as an output even though the receiver indicates it is tuned to 24 MHz rather than the undesired 26 MHz. This example response is sometimes referred to as a *2-by-2 response*, corresponding to the magnitudes of m and n in the order of signal and local oscillator, respectively. It is also called a *fourth-order spurious response* because the sum of the harmonic magnitudes is 4.

Another example of spurious responses, the crossover, may be found in the foregoing example when the receiver is tuned to 25 MHz. The local oscillator is at 75 MHz, and its second harmonic mixes with the second harmonic of 25 MHz to again generate the 100-MHz IF. This case is called a *crossover response* because the desired and undesired signal frequencies are equal. If the frequencies are exactly 25 and 75 MHz, the resultant IF vector voltage is

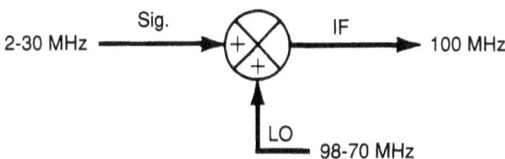

Figure 4.5 An example mixing scheme in which the HF signal is added to a backward-tuning LO to produce a 100-MHz IF.

only perturbed by a constant amplitude and phase change due to the spurious response. When the 25-MHz signal is an SSB signal with varying instantaneous frequency, the crossover response causes in-band distortion in the audio output if the mixer is poorly designed. Fortunately this example response can be very well suppressed with well-designed balanced mixers.

Determination of spurious response locations can be an intuitively difficult task without some kind of aid. Equation 4.8 can be expanded into a simultaneous set by using the subscripts d for desired responses and u for undesired responses:

$$m_d\text{SIG} + n_d\text{LO} = \text{IF} \tag{4.9a}$$

$$m_u\text{UND} + n_u\text{LO} = \text{IF} \tag{4.9b}$$

In the second equation of this set UND represents the undesired spurious response, which may or may not cross the desired frequency SIG. If UND = SIG, that is, if there is a crossover response, a graphical aid may be constructed by equating the set to give

$$\text{SIG} = \text{LO}\,\frac{n_u - n_d}{m_d - m_u} \tag{4.10}$$

This is the equation of a straight line on axes of SIG and LO passing through the 0,0 origin and having a slope of $(n_u - n_d)\,/\,(m_d - m_u)$. Using the prior upconversion example, it can be seen that $m_d = +1$, $n_d = +1$, $m_u = -2$, and $n_u = +2$, so that SIG = LO/3. This slope of 1/3 line is plotted in Figure 4.6 to-

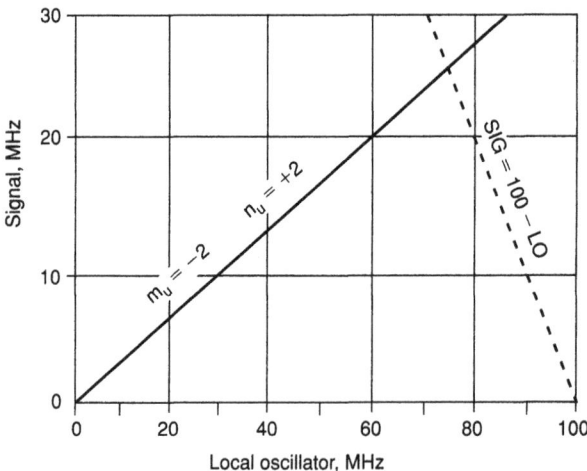

Figure 4.6 A spurious crossover response graph showing the loci of the fourth-order 2 × 2 responses for a sum mixer (solid line) and the intersecting dotted line of an upconverter with 100-MHz IF.

gether with a dotted line representing the relations between the signal and local oscillator, which produces a 100-MHz IF. The intersection at SIG = 25 MHz is, as shown before, the frequency at which the IF produced by the desired SIG + LO response is indistinguishable from the undesired –2SIG + 2LO response.

This type of crossover graph may be further generalized by considering both sum and difference mixing and finding the line slopes for higher-order responses.

The use of the crossover chart is not restricted to the fixed-IF example. It sometimes happens in receivers and often in complex frequency synthesizer schemes that all three mixer port frequencies are varying and are related by some independent variable. If the relationship is linear, the desired mixing line is straight; otherwise it is curved. The intersections still give the crossover points at which multiple responses occur.

The general simultaneous equation set may be manipulated in almost endless ways to create graphs, pocket calculator routines, and computer programs fulfilling special needs. For example, the simultaneous equations may be normalized by one of the input frequencies to form a spurious graph with frequency ratio coordinates (see reference 5). Another variation is to express the undesired response frequency separation from the desired in percent [5]. This is useful for finding how far away a noncrossing response is from the signal frequency. Spurious response separation can also be plotted versus signal frequency by defining the ordinate to be Δ_{freq}, the algebraic difference between undesired and desired response frequencies. This presentation succinctly displays the entire response picture, both crossing and noncrossing. Chapter 17, "Software for SSB," contains a description of a program that performs the analysis of the harmonic IM performance of a variety of mixer circuits. Please refer to that chapter.

Internal signals

Another source of undesired response interference is that emanating from sources within the receiver. These responses do not depend upon external signals and are often called *birdies* because of their characteristic sound as the receiver is tuned. In an HF receiver it is practically impossible to avoid having oscillators within the tuning range of the receiver. The internal frequency standard, operating in the 3- to 10-MHz region for best stability, is an example of such an internal source. The only way to avoid in-band internal signals is to shield and filter to the point of inaudibility; a difficult task if low-cost, open, printed circuit board construction is used. High-performance receivers require shielding of signal generator quality to avoid unusable frequencies.

The frequency scheme itself should be analyzed to avoid building in birdies that can be avoided without other penalty. In Figure 4.7, for exam-

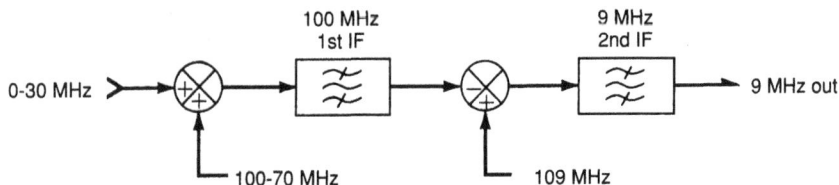

Figure 4.7 A frequency scheme having a 100-MHz first IF followed by a 9-MHz second IF has the second injection at 109 MHz to avoid an internal spurious beat with leakage from the first injection.

ple, a second mixer and IF are added to the 100-MHz upconverter scheme to allow adding selectivity at the 9-MHz second IF. There are two choices for the injection frequency into the second mixer, 91 or 109 MHz. When the 91-MHz frequency is used, leakage from the first injection at 82 MHz (corresponding to a signal frequency of 18 MHz) could pass around or through the 100-MHz crystal filter, enter the second mixer, and produce the 9-MHz IF. This situation is avoided by picking a 109-MHz second injection so that the first injection can never be 9 MHz away from the second injection. The frequency inaccuracy of the second injection is partially canceled by the other injection inaccuracies such that the overall stability of the receiver is equal to the frequency standard.

When harmonics of the injection sources are considered, it becomes difficult to foresee the many internal spurious signals that may be generated. This analysis problem is similar to that for spurious responses and is readily attacked with computer programs, which methodically plod through all the combinations and present only those that can pass through the IF selectivity. Knowing the oscillators, harmonics, and circuit paths responsible for the birdies, the designer can make appropriate choices of frequency scheme, shielding, and filtering to reduce or eliminate them.

RF selectivity

We have seen that the noncrossing spurious responses can be attenuated by providing selectivity ahead of the first mixer. The questions of how much selectivity and how it is to be implemented involve the usual trade-offs between cost, complexity, and performance. One response that must be attenuated to a great degree is the image. In our 100-MHz IF sum upconverter example, the image moves from 198 to 170 MHz as the signal tunes from 2 to 30 MHz. Throughout the world this range is well occupied with large signal sources from television transmitters and other emitters. Typically a 2- to 30-MHz bandpass filter will be used between the antenna and the first mixer to eliminate the image response, this architecture being termed *wideband*. This requires no switching in the receiver input and so is low in cost.

In frequency hopping receivers the fixed-input filter eliminates hopping noise generation that could occur if switched filters interrupted strong CW signals.

The penalty for the wideband architecture is increased spurious responses due to harmonics of HF signals being generated in the mixer. If a receiver tuned to 10 MHz has no selectivity at 5 MHz, a strong signal at 5 MHz with no harmonic energy of its own will produce 10 MHz through mixer distortion. An effective topology is to use *half-octave* filters which are switched as required across the 2- to 30-MHz range. A half-octave filter is a bandpass filter whose passband is about one-half octave wide, such as 2 to 3 MHz. These filters and narrower preselection filters are discussed in Chapter 9. Also see Chapter 17 for some software.

4.5 Block Diagram Design

In this section we will examine some of the techniques for designing an SSB receiver at the block diagram level. This will include the aspects of gain and noise figure distribution, optimization for IMD, synthesizer injection noise effects, and AGC.

Level diagrams

A level diagram is a pictorial way of describing any quantity or property in the receiver's signal path so that individual blocks are defined or optimization can be made. For example, we might wish to make a gain distribution diagram to show how a 1-μV signal is amplified throughout the receiver stages to produce a 1-V audio output signal. On the same diagram, the actual voltage, or power level, at each point can also be shown. Knowing the input and output power levels, the cumulative IMD or intercept points can be plotted. Cumulative noise figure is also a useful line item on a level diagram.

Figure 4.8 illustrates the use of a level diagram to describe the gain and noise figure of stages up to the output of the second mixer. The diagram is prepared by filling in the individual stage gain and stage noise figure rows with values that experience or manufacturer's data indicate can be obtained. The total gain and cumulative noise figure rows can then be calculated, the latter using the Friis equation in which F_{in} is

$$F_{in} = F_A + \frac{F_{out} - 1}{G_A} \qquad (4.11)$$

the input noise factor (a power ratio) to the stage, F_A is the amplifier or other block intrinsic noise factor (measured as if there were no subsequent noise sources), G_A is the amplifier available power gain (again, a numeric

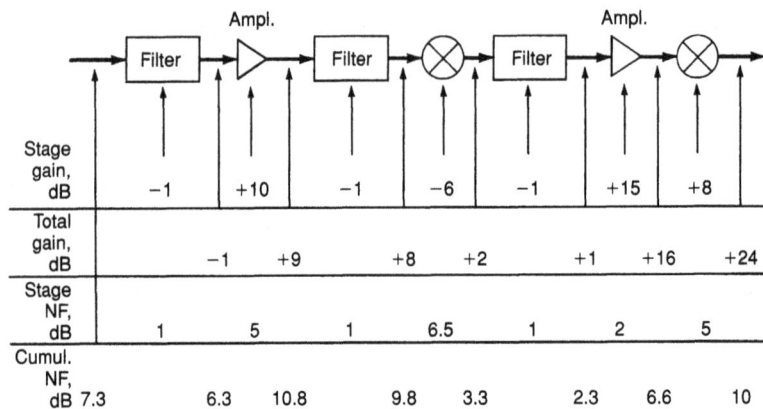

Figure 4.8 A level diagram that describes the gain and noise figure distribution up through the second mixer.

ratio), and F_{out} is the output noise factor terminating the stage. The noise figure (NF) (in decibels) is found from the noise factor by

$$NF = 10 \log_{10} (F) \text{ dB} \tag{4.12}$$

These calculations are quickly made with a hand calculator having a 10^x or Y^x function in addition to logarithms. An important fact that facilitates computation is that the noise factor of a passive network is the reciprocal of the available gain (see Friis [1]). In decibels, the passive network noise figure is equal to the loss.

If a great amount of component selection trade-off studies and comparisons are to be made, a very efficient analysis tool is one of the spreadsheet programs for personal computers. The spreadsheet columns can be related by any mathematical expression and become a powerful way of quickly seeing the influence of changes anywhere in the block diagram. Chapter 17 contains a PC program that performs cascaded noise figure and intercept point calculations using a Windows environment.

There is one other detail that should be understood before leaving the subject of gain and noise figure diagrams. When calculating noise figure, the precisely correct form of gain is the available power gain as defined by Friis [1]. In equation form that is

$$G_{av} = \frac{P_{out\ avail}}{P_{gen\ avail}} \tag{4.13}$$

where $P_{out\ avail}$ is the power available at the output terminals of the stage, that is, a conjugate match. $P_{gen\ avail}$ is the available generator power. The more commonly measured value of gain is transducer gain, which is defined by the ratio

$$G_T = \frac{P_{\text{deliv to load}}}{P_{\text{gen avail}}} \tag{4.14}$$

where $P_{\text{deliv to load}}$ is the actual power delivered to a specific test load on the stage. These two definitions of gain are equivalent when the output reflection coefficient is zero, a condition that may not always be true in a receiver cascade. For practical purposes, G_T is often used interchangeably with G_{av}. The error so caused is not large when the second term of Equation 4.11 is small compared with the first term. Also note that the resistance level at each point in the block diagram need not be known when transducer gain is used. If voltage levels are to be shown, the stage interface resistances need to be accounted for so that the power levels are consistent with the transducer gains.

The level diagram can also be used to compute the input intercept point of the receiver by using a stage-by-stage calculation similar to that used for noise figure. The third-order intercept cascading relation can be developed from the intercept defining Equation 4.3 and the cascaded stages shown in Figure 4.9, where a stage having a gain G_1 and an input intercept $_3I_1$ is followed by the *rest* of the system, which has an input intercept $_3I_2$.

In Figure 4.9, the two-tone signal power (per tone) at each point is S_i and the distortion power is D_i. All quantities are expressed in power units rather than dBm. D_{in} is the total distortion power, referred to the input. The distortion voltage of stage one, referred to its input, and the distortion voltage of the rest of the system, divided by the voltage gain of stage one, are added exactly in-phase as a worst case condition to get the input distortion voltage. The resulting input distortion voltage is

$$\underbrace{\sqrt{D_{\text{in}}}}_{\substack{\text{total}\\\text{distortion}}} = \underbrace{\sqrt{D_1}}_{\substack{\text{stage one}\\\text{contribution}}} + \underbrace{\sqrt{\frac{D_2}{G_1}}}_{\substack{\text{stage two}\\\text{contribution}}} \text{ volts} \tag{4.15}$$

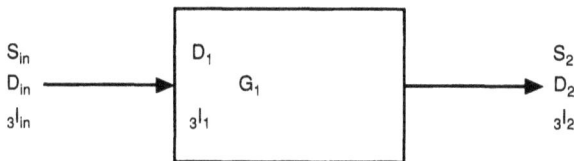

Figure 4.9 Block diagram for deriving the expression for the third-order input intercept of two cascaded stages.

where the square root converts the distortion powers to voltages with an implied resistance of 1.0 ohm.

Equation 4.3 provides the intercept point as

$$_3I_i\,(\text{dBm}) = S_i\,(\text{dBm}) + 0.5\,(S_i - D_i)\,(\text{dB}) \tag{4.16}$$

This can be rewritten, using watts rather than decibels and dBm, as follows

$$\frac{S_i}{_3I_i} = \sqrt{\frac{D_i}{S_i}} \tag{4.17}$$

Modify Equation 4.15 as follows:

$$\sqrt{\frac{D_{\text{in}}}{S_{\text{in}}}} = \sqrt{\frac{D_1}{S_{\text{in}}}} + \sqrt{\frac{D_2}{G_1 S_{\text{in}}}} \tag{4.18}$$

Substitute Equation 4.17 into each term of Equation 4.18 to get the result

$$\frac{S_{\text{in}}}{_3I_{\text{in}}} = \frac{S_{\text{in}}}{_3I_1} + \frac{G_1 S_{\text{in}}}{_3I_2} \tag{4.19}$$

$$\frac{1}{_3I_{\text{in}}} = \frac{1}{_3I_1} + \frac{G_1}{_3I_2} \tag{4.20}$$

This can be reciprocated to get the cascading equation for the third-order intercept point

$$_3I_{\text{in}} = \cfrac{1}{\cfrac{1}{_3I_1} + \cfrac{G_1}{_3I_2}}\ \text{watts} \tag{4.21}$$

Starting with the output termination of a receiver having infinite intercept watts, this relation can be used to find the cumulative input intercepts at each point in the block diagram. For example, in the cascade of a mixer, amplifier, and crystal filter shown in Figure 4.10, the individual stage intercepts are shown both in dBm and in milliwatts for clarity. The infinite intercept shown looking into the IF system is assumed to be valid in the stopband of the crystal filter, which precludes the entry of a strong off-frequency signal into the IF. Therefore, the intercept into the filter is +50 dBm or 100,000 mW. Looking now at the input to the amplifier, the cascaded intercept is found to be 99 mW, which is 19.96 dBm. The mixer gain is 6 dB, which is 0.25 in power ratio. Using that value for gain and the 25-dBm mixer input intercept together with the amplifier input intercept just determined, the total input intercept of 176 mW or 22.45 dBm is found.

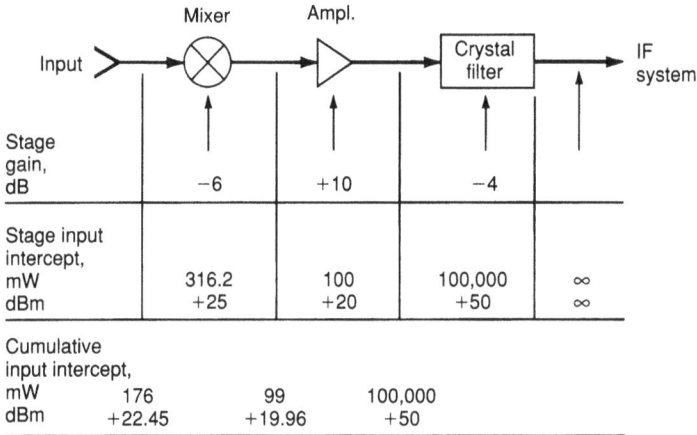

Figure 4.10   Example of a receiver front end to show the calculation of intercept points.

This example and Figure 4.9 illustrate some shortcuts to aid in quickly estimating the intercept point of a cascade. A stage terminated in an intercept much larger than its own intercept will have a net input intercept nearly equal to the stage value. Loss in the signal path increases the intercept and, if distortionless, increases the intercept by the amount of the attenuation. The receiver dynamic range as defined in this chapter is not increased by attenuation since the desired signal is reduced by the same amount. The input intercept of a stage will be 3 dB below that of the stage alone when the terminating intercept minus the stage gain is equal to the stage intercept.

Second-order intercept calculations may be accomplished with similar methods except that the cascading equation becomes

$$_2I_{in} = \frac{1}{\left(\sqrt{\dfrac{1}{_2I_1}} + \sqrt{\dfrac{G_1}{_2I_2}} \right)^2}$$
(4.22)

Formulas exist additionally for handling higher-order intercept calculations. Wilson [6] gives a concise introduction to this area. In IMD calculations it should also be clearly understood that all of the defining equations stem from the straight-line plots of IM versus signal level as shown in Figure 4.2. Not all active or passive elements are so well behaved. Doubly balanced diode mixers and bulk quartz crystal filters in particular can be difficult to model with straight-line equations, and departure from the theoretical performance can be expected in practical circuit designs.

Optimization

The receiver attributes of sensitivity and dynamic range are conflicting and cannot be simultaneously increased by manipulation of front-end topology and gain distribution. Noise figure can be improved by adding a low-noise RF amplifier before the first mixer, but this reduces the input intercept point by at least the amount of amplifier gain. When amplifier distortion is accounted for, the IM dynamic range of a receiver is actually decreased by adding an RF amplifier in front of the mixer. Sensitivity specifications for some unique applications may require an input noise figure lower than the 12 to 15 dB possible with a mixer-only front end, so that RF amplification is necessary. In some cases it is also necessary to use a low-noise IF amplifier between the first-mixer output and the following filter. When amplifier-mixer cascades are involved, trade-offs can be calculated to optimize the receiver's performance versus cost and complexity.

Before examining trade-off techniques, the reciprocal mixing phenomenon discussed in section 4.1 should be compared to the IM performance of the front end. It is inefficient to suppress IM products below the level of reciprocal mixing noise and, conversely, to produce a synthesized injection source that is purer than needed from an IM standpoint. To make this comparison, some idea or specification of the maximum undesired off-frequency signal strength is needed. Letting the undesired signal power be P_{in}, measured in units of power, and rearranging Equation 4.3 in terms of power, the following relation is obtained:

$$_3I_{in} = \frac{P_{in}^{3/2}}{P_{in}^{1/2}}$$
(4.23)

The equivalent input noise to the receiver comes from the input noise floor $kTBF$ and the reciprocal mixing noise referred to the input. Letting the total noise N_T be equal to the equivalent input IM power gives an expression for the maximum usable input intercept point:

$$_3I_{in} = \frac{P_{in}^{3/2}}{N_T^{1/2}}$$
(4.24)

A numerical example illustrates the calculation of the maximum usable input intercept for a particular case. Suppose we have a receiver with the following characteristics:

- Noise figure = 14 dB
- Bandwidth = 3 kHz
- Injection carrier to noise ratio = 100 dB (100 kHz away and measured in a 3-kHz bandwidth)

With a –20-dBm undesired input signal 100 kHz away from the desired frequency, the noise power due to reciprocal mixing is –120 dBm, or 1×10^{-12} mW. The equivalent $kTBF$ noise is found to be 3×10^{-13} mW, so that the total input noise is 1.3×10^{-12} mW. The input intercept is therefore

$$_3I_{in} = \frac{(1 \times 10^{-2})^{3/2}}{(1.3 \times 10^{-12})^{1/2}} \tag{4.25}$$

$$= 877 \text{ mW or } +29.4 \text{ dBm}$$

This analysis has ignored the practical fact that the interfering signal may itself have noise sidebands that lie within the passband of the receiver. This noise power available from the antenna acts to further reduce the intercept point needed in a practical sense.

Proceeding now to the problem of optimizing a practical receiver front end in terms of cost and complexity trade-offs for constant input noise figure and intercept point performance, consider the block diagram shown in Figure 4.11. The objective is to find the trade-off relationship between $_3I_1$ and $_3I_2$ while holding $_3I_{in}$ and F_{in} constant. The noise factor F_3 looking into the filter includes the filter loss plus the cumulative noise factor of the receiver's IF system. The filter's input intercept $_3I_3$ is an out-of-band intercept for the filter, the remaining IF stages being protected by the filter stopband attenuation. Application of the Friis [1] noise factor cascade equation to this block diagram gives

$$F_{in} = \frac{1}{G_1}\left(F_2 + \frac{F_3 - 1}{G_2}\right) \tag{4.26}$$

Also, the intercept cascading equation can be used to relate $_3I_1$ and $_3I_2$:

$$\frac{1}{_3I_{in}} = \frac{1}{_3I_1} + \frac{G_1}{_3I_2} + \frac{G_1G_2}{_3I_3} \tag{4.27}$$

For illustration, let this front end use a doubly balanced diode mixer with 6-dB loss and, neglecting any additive noise, a 6-dB noise figure. For calcu-

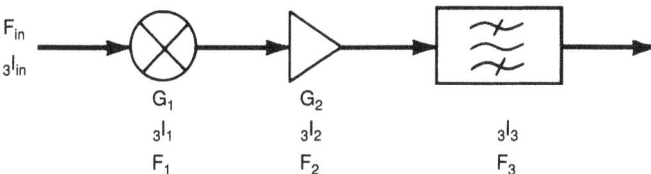

Figure 4.11 A front-end cascade having general values of gain, intercept point, and noise figure for optimization analysis.

lation purposes, let the IF amplifier gain G_2 be a variable that does not affect the amplifier noise figure. This is reasonably true for high-performance lossless feedback amplifier circuits. The amplifier noise figure will be 4.77 dB (power ratio of 3), which is representative of a high dynamic range VHF feedback amplifier. And let the IF system noise figure be 9 dB, which includes the combined termination loss and insertion loss of the filter plus the cascaded noise of the rest of the IF. Converting all numbers to power ratios and substituting into Equation 4.26 gives

$$F_{in} = 4 \left(3 + \frac{7.94 - 1}{G_2} \right)$$

Finally, let us ask for a receiver input noise figure of 12 dB, which results in an IF amplifier gain G_2 of 7.22 or 8.58 dB.

Now turning to the intercept cascade equation, let the filter input intercept be 50 dBm (10^5 mW). Equation 4.27 therefore contains input, mixer, and amplifier intercepts as variables. Figure 4.12 shows the results of solving the equation with the input intercept as a running parameter. As can be seen, mixer intercept can be traded for amplifier intercept within a range bounded by the input intercept asymptote on the mixer axis and the input intercept minus the mixer gain (–6 dB) asymptote on the amplifier axis. For a receiver input intercept of 20 dBm, the mixer intercept might be about 21.3 dBm when the amplifier intercept is 20 dB.

A poor design decision would put either the mixer or amplifier input intercepts near asymptotes since the cost of achieving large intercepts is high. The optimum decision can be a complex judgment based on the above

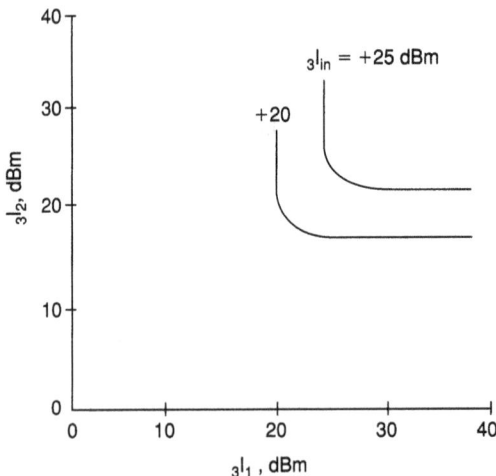

Figure 4.12 Contours of a constant-input intercept for a mixer-amplifier-filter cascade having a 12-dB input noise figure together with gain, intercept, and noise figure constraints.

analysis applied to the cost, circuit complexity, reliability, and power consumption of the actual circuits involved.

AGC design

The means for providing AGC for the overall receiver are important aspects of the block diagram design for an IIF SSB receiver. The receiver is operationally required to have nearly constant audio output as the signal input ranges from the 1-μV region to perhaps 1 V or more and to do so without having loop oscillation or distorting the received signal. A receiver primarily designed to be used by an operator for signal search or monitoring will typically allow an audio output rise of 6 dB or so over the input signal range. A receiver for data signals, on the other hand, can have a very flat output requirement so that the following data detection circuits have a constant-amplitude input.

Some insight into AGC loop design can be obtained by studying the receiver signal path shown in Figure 4.13, which includes voltage-controlled attenuators in the RF, first-IF, and second-IF sections. A detector at the output develops a dc output voltage proportional to the IF or AF output level. Intermediate frequency detection is normally used because it gives less time delay (more charging cycles per second) and has output even when the product detector audio output frequency is zero. After lowpass filtering, the voltage is compared with a reference such that gain control action commences when the signal input has reached some threshold level, such as that causing a 6-dB signal-to-noise ratio. As the input signal rises, attenuators at the end of the IF path are used first, preventing overload of the output stages as they reach their signal amplitude limit. Since attenuators at the input of the receiver are not yet being used, the gain into the first and second mixers does not change and the input noise figure is not materially increased. This results in a very nearly linear increase in output signal-to-noise ratio versus input signal increase. If all of the attenuation were placed at the receiver's input, the output signal-to-noise ratio would rise to the threshold value and stay there (or rise only slightly) as the signal level is increased. The delay blocks represent voltage delay circuits that enable the forward-placed IF and RF attenuators at signal levels which are ideally just in advance of succeeding stage overload or IM level specification limits. As the input attenuator begins to operate with rising signal strength, the output signal-to-noise ratio becomes constant with a value ranging from 30 to 50 dB in typical designs. The ultimate signal-to-noise ratio is usually limited by the synthesizer phase noise when the IF delays are properly placed.

The amount of dc gain needed in the AGC loop is dependent upon the output rise characteristics previously described. If the audio output as represented by the detected dc voltage is 1 V at threshold, and a 6-dB output rise is acceptable for input signals at some maximum level, the dc voltage

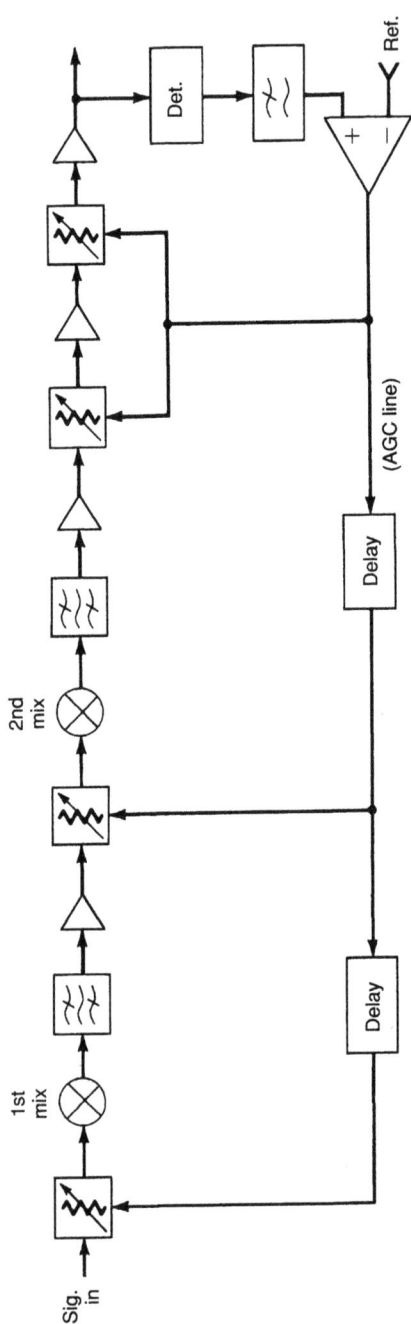

Figure 4.13 A double-conversion receiver has distributed AGC attenuators to prevent stage overload while also causing the output signal-to-noise ratio to rise with signal increases.

will rise to 2 V and the differential output will be 1 V. An AGC amplifier gain of only 10 would suffice to cause a 10-V AGC line increase to operate the RF and IF attenuators. On the other extreme, if virtually no output rise is desired, the AGC amplifier is replaced by an operational amplifier integrator whose huge low-frequency gain makes the output as constant as the nonlinearities of the final IF stage and detector will allow. When an integrator is used, it forms part of the lowpass filter which is required to remove envelope audio from the gain control voltage. This can result in unacceptable signal attack time delays unless nonlinear circuits are used.

The dynamics of the AGC loop constitute the most difficult part of the gain control design. An AGC loop is a feedback system having gain that is dependent upon the carrier strength in the forward path. Stability analysis is further complicated by the fact that in a sideband receiver, the AGC detector time constant holds the peak value for a period of time. Thus, when the signal level decreases, the AGC system becomes open loop and the receiver gain increases at a rate determined by the detector time constant. When the signal level is constant or increasing, the AGC loop can be described and analyzed as a linear servo-mechanism if the signal level and the attenuation are expressed in logarithmic units. Porter [7] gives a complete analysis of the closed-loop condition, including the case where one stage of delayed gain control is added to the simple one-loop system. The general result is that the attenuators should have logarithmic slope, with attenuation increasing by 1 dB per unit of control voltage throughout the control range. High attenuation slope, such as is found near the cutoff point of dual-gate field-effect transistor amplifiers, will decrease the loop phase margin, leading to overshoot or outright oscillation at a specific signal level.

Practical design of a multidelayed AGC system is done experimentally by setting up a stepped value signal source and observing the audio output and AGC voltage with an oscilloscope. The signal source steps can be obtained from a bus-controlled signal generator or an external attenuator driven by a low-frequency square-wave oscillator to produce steps in the 3- to 10-dB range at about a 1-Hz rate. The mean value of the signal should be moved from near zero to the maximum rated input signal level. Output overshoot, ringing, or oscillation problems are readily observed by this method.

The impulse-loading characteristic of the AGC system must be considered in the receiver block design so that overshoot in gain control is not excessive when a strong signal or noise impulse is received. A dual time constant approach has been used to solve the impulse-loading problem. The loop filter is designed to have some transmission at frequencies in the 100- to 300-Hz range, such that some gain control can be rapidly applied without causing a large amount of charge in the integrator or lowpass filter. The receiver gain will then rapidly recover after the noise burst or signal impulse has ceased so that a weak signal can be immediately recognized. This technique is a compromise, with envelope distortion caused by envelope frequency appearing on the control voltage to the attenuators.

Another problem associated with the narrow-bandwidth IF filters of the receiver is the absolute time delay incurred by the signal as it passes through the IF. When a rapidly rising signal is at the input, the AGC detector has no output until the signal propagates through the filters. The input stages then may be in overload during the delay time interval. When the signal does reach the detector, the system may overcompensate by reducing the gain more than necessary because of the energy stored in the IF filter. This effect can be practically reduced by using lead compensation on the voltage driving the attenuators ahead of the IF selectivity. The dual time constant filter mentioned above also helps reduce the effects of filter time delay.

A state variable type of AGC loop called *hang AGC* is very often used to obtain rapid recovery of receiver gain after signal cessation, while also having very low envelope distortion due to audio on the gain control line. The loop filter is designed to charge rapidly to follow the rising signal input and then to remain at that level after the signal drops. In effect, the receiver gain "hangs" for a preset amount of time. For voice systems the attack time should be in the 2-ms region, with a hang time of about 0.3 sec, followed by a gradual recovery time of up to 1 sec. Adaptive circuits can be designed that give shorter hang and recovery time as the signal on-time decreases, thus minimizing impulse loading of the AGC system.

Since a microprocessor is often used elsewhere in an otherwise analog receiver, it is an attractive possibility to consider a digitally closed gain control loop. The output of the AGC detector or even the detected audio can be sampled and processed for average, rms, or peak amplitude determination. After adaptive processing, the gain control is applied with D/A converters to the various attenuators in the receiver. Alternatively, step-value attenuators can be used as opposed to continuous-function attenuators. The latter generate IMD when the control current or voltage is low, which occurs near the minimum attenuation end of the range. Computed AGC can achieve infinite hang time when needed. An example of this would be to return the receiver to a set gain point after a companion transmitter completes its transmission, thereby readying the receiver for an expected signal strength.

4.6 Components

The design of a solid-state SSB receiver is heavily influenced by the available components together with their performance and cost. A thorough knowledge of components is as valuable to the design engineer as are the theoretical design details of receivers. This section will highlight some of the current state-of-the-art components that are applicable to HF receiver design and will show some of the associated circuit design techniques.

Front-end components

It is in the front end that the greatest amount of work and progress is made in the effort to improve IM and sensitivity performance. The input signal to the IF mixer, RF, or IF amplifiers, if used, and the first band-limiting filter are the critical components in front-end design.

Mixers

There are two broad classes of frequency mixers used in SSB receivers: passive and active. A passive mixer is generally one which has insertion loss and is often bidirectional, that is, one in which signal conversion can flow in two directions. The prime example of a passive mixer is the diode mixer, and more specifically, the doubly balanced diode mixer. The latter is the most widely available mixing component in the HF through microwave region and is obtainable with a variety of performance specifications. Active mixers contain elements having transconductance, and therefore have conversion gain. They are almost always unidirectional. Transistors, both bipolar and field effect, form the basis for active mixer designs, which are usually custom-designed for a particular receiving requirement. There can be exceptions to the passive and active classes, the parametric mixer being one example. This type of mixer contains varactor diodes as the active component. Although a varactor does not have transconductance, the parametric mixer can have gain, the added power being obtained by conversion of injection power to IF power [8].

In Figure 4.14 the schematic diagram of a doubly balanced mixer is drawn in two ways. In Figure 4.14A, the conventionally used schematic illustrates the cathode-to-anode connection around the ring of four diodes, often referred to as a *diode quad*. The diodes are usually silicon Schottky contact types having low capacitance and charge storage time that are matched for forward voltage drop and reverse capacitance over a wide dynamic range. The transformers use ferrite cores with transmission line winding construction when possible to maximize bandwidth. The combination of diode matching and high transformer coupling coefficients results in typical port-to-port isolation of 30 dB or more. Thus, the injection signal at the LO port is attenuated by 30 dB as it arrives at the RF or IF ports.

The redrawn schematic in Figure 4.14B perhaps more clearly illustrates the switching action of the mixer. At some instant the injection polarity might be such that forward current is flowing downward in the left-hand pair of series diodes. Since the diode forward voltage drops are matched, the potential at the cathode-anode node is zero with respect to ground by virtue of the center tap on the LO transformer. The RF signal voltage appearing across half the RF input transformer is switched to the IF output port. When the injection polarity reverses and the right-hand pair of series diodes is forward biased, the other end of the RF signal transformer is

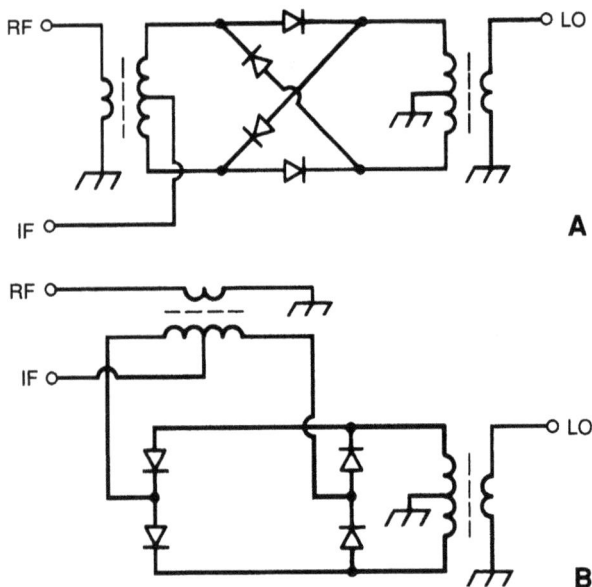

Figure 4.14 Schematic diagrams of the doubly balanced diode mixer. (A) Conventional circuit of a doubly balanced diode mixer. (B) Redrawn circuit emphasizes the commutation of the RF signal to the IF port by the LO.

switched to ground. The IF port then receives the RF signal with opposite polarity, a 180° phase shift. When the switching action is rectangular, the resulting IF spectrum consists only of the sum and difference of the RF and LO frequencies, with the RF and LO being suppressed in the IF output.

When one pair of diodes is forward current biased, the other pair is reverse voltage biased by the voltage drop of the forward pair. If the RF signal becomes comparable in power to the injection, the switching times become a function of the RF as well as LO signals. This results in single-signal compression and multiple-signal IMD. The signal handling capability can be increased in many ways, usually involving more diodes and injection power. Two or more diodes in series or parallel can be used in each diode position. Resistors can be placed in series with each diode to increase the voltage drop with an attendant increase in conversion loss.

Circuit modifications can increase the diode mixer's signal handling capability. In Figure 4.15 each single pair of diodes from Figure 4.14B has been replaced by a sampling quad of four diodes. Note that the diode quads are connected in a series-parallel arrangement rather than in the ring arrangement used in the simple doubly balanced mixer. The RF signal phase-reversing commutation principle is still the same, but the resistors in series with each diode quad cause large reverse bias voltage to appear across the nonconducting set. Greater than 30-dBm third-order input intercept points can be obtained with this mixer compared to 20 dBm for the four-diode circuit using 7-dBm injection power. The cost of increased mixer dynamic range is injection power, with levels of up to 1 W being used in commercially available mixers.

Another diode mixer improvement is to use square-wave as opposed to sine-wave injection. This minimizes the diode transition time from conducting to nonconducting states, making the switching more independent of signal level. Upconverting HF mixers with third-order input intercept points of 43 dBm or more have been constructed using square-wave drive from transistor-limiting amplifiers having transition times in the 1-ns region.

When a high dynamic range mixer is used as the first active stage in a receiver, there can be a very real problem with the noise of the injection entering the mixer, at both RF and IF. This desensitizes the receiver by decreasing the signal-to-noise ratio at the mixer output and is especially noticeable in low-front-end gain receivers with no RF stage. The extent of this problem is illustrated by considering a –113-dBm input signal to the mixer which is producing a 10-dB output signal-to-noise ratio. With 6-dB conversion loss, the IF signal power out of the mixer is only –119 dBm, and the equivalent noise level is about –129 dBm, neglecting additive sources within the mixer itself. For a mixer injection-to-IF port balance of 30 dB, a –99-dBm noise level at the IF would degrade the signal-to-noise ratio by 3 dB. If the injection power is +30 dBm for a large mixer, the injection carrier-to-noise ratio must be 129 dB. This large ratio is especially difficult to maintain when the injection frequency approaches the IF at the low-frequency end of the tuning range.

The injection noise problem can be minimized by a combination of approaches. The obvious improvement is to increase the mixer balance. The noise balance point may not be coincident with balance of second-order IMD, however. The bandwidth of the injection amplifier should just cover the necessary band. Low-frequency resonances in the amplifier dc feed circuits should be suppressed to prevent noise generation by high-impedance

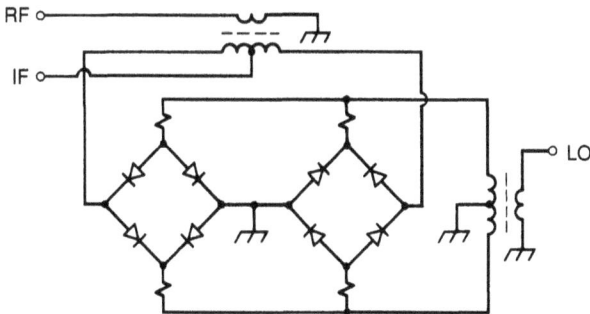

Figure 4.15 A high-level diode mixer using sampling diode quads for signal commutation.

collector loading. When linear injection amplifiers are used, linearity improvement to reduce second-order distortion will minimize translation of HF noise to VHF within the amplifier. Finally, the voltage-controlled synthesizer oscillator (VCO) should be operated at as high a level as is consistent with tuning range and tuning voltage limitations. This minimizes the amount of injection amplifier gain. It is not advisable to inject the mixer directly from a power VCO because large signal levels entering the mixer would then pass through the RF to injection port imbalance and perturb the oscillator.

The doubly balanced mixer is sensitive to reflection of conversion sidebands from its three ports. Intermodulation is especially degraded when reactive sources and loads are present at the RF and IF ports and is a strong function of frequency as the mixer is used over the HF band. The circuit blocks illustrated in Figure 4.16 cause matched or "50-ohm" terminations to be presented to the mixer at each port [9]. At the RF port, a complementary lowpass-highpass filter is used to get a wideband match. The filters are synthesized from Butterworth single-side-loaded prototypes [10] and are scaled to have identical cutoff frequencies above the HF range. The open-circuit ends of the filters are paralleled, the result being a constant-resistance node whose bandwidth is limited only by practical consideration of lumped component accuracy versus frequency. The injection amplifier output is attenuated by about 3 dB for sine-wave injection to improve the return loss seen by the mixer. The attenuator would not be used if square-wave injection were being applied. The circuits in the IF path are necessary to allow the use of a crystal filter directly after the mixer for best IM performance, while eliminating the poor termination on the mixer in the stopband of the filter where power is reflected back into the IF port. The desired IF is passed through an LC bandpass filter to the following 90° hybrid splitter.

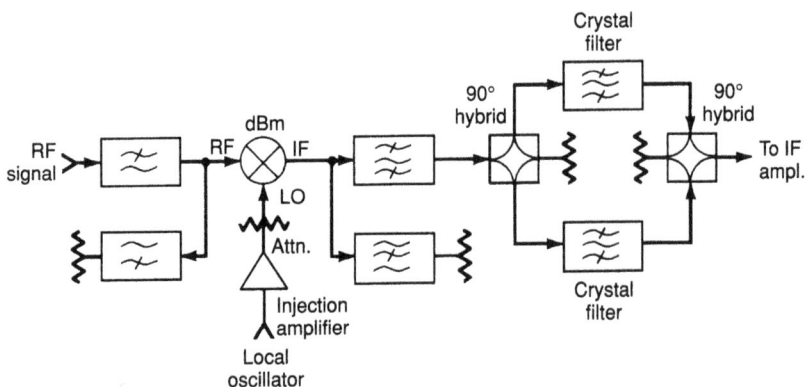

Figure 4.16 Improved IM performance is obtainable when low SWR terminations are used on the mixer ports.

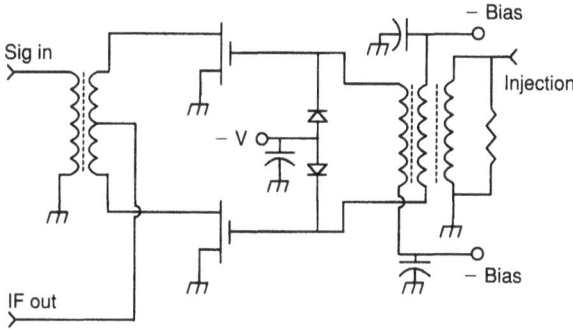

Figure 4.17 Low gate-capacitance microwave power FETs may be used as switches in a passive mixer.

Frequencies other than IF are passed to a 50-ohm load via an LC band-reject filter. The 90° hybrid has the valuable property that when its output ports are terminated in equal complex reflection coefficients, the input resistance remains 50 ohms. With two reasonably identical crystal filters in the hybrid outputs, the equal reflection coefficient condition is satisfied both in the passband and well into the stopband [11]. The 90° hybrid need not have extremely large bandwidth since the bandpass and band-reject filters can take care of the mixer loading well into the UHF region for an HF-to-VHF upconverting mixer. The crystal filter outputs must be combined in quadrature. A second hybrid may be used for this purpose or, alternatively, simple +45° and –45° L-section matching networks may be used to combine the outputs since the operating bandwidth is set by the crystal filters.

It is possible to construct *termination-insensitive* passive mixers by combining two or more mixers with hybrid transformers such that mixer products are returned to the mixer rather than to the external terminations. Since additional components are present, this type of mixer has greater insertion loss than the simpler doubly balanced type and has not been popular in HF receiver designs.

Passive mixers can be built with transistor switches substituted for the diodes of balanced mixers. For example, in Figure 4.17 the field-effect transistors (FETs) are gate controlled by the local oscillator to cause alternating polarity of the signal-to-IF path similar to that shown in Figure 4.14B. When fast microwave FETs such as the NEC NE 868299 are used in this circuit input, intercepts of 30 dBm can be obtained with injection power of well under 1 W. The balance of any mixer using discrete transistors will be poorer than the balance of a diode mixer because of the difficulty of matching the rather complex transistor parameters over the operating range. Matching can be greatly improved by integrating the transistors onto one custom chip when it is economically feasible to do so. Passive mixers using four complementary metal-oxide semiconductor/silicon-on-sapphire (CMOS/SOS) transistors in an integrated quad ring can be built having reliable input third-order intercept points of 35 dBm over the HF band in an upconversion (109.35-MHz) receiver [4].

A doubly balanced mixer illustrating the use of bipolar medium-power transistors is shown in Figure 4.18. The circuit is that of a pair of transconductance mixers with emitter resistors added for IM improvement. Resistors in the base and collector leads add loss of UHF to suppress parasitic oscillations caused by resonances formed by circuit and transistor capacitances together with the leakage reactance of the associated transformers. Injection is applied through a balancing transformer to the bases, which are overdriven, resulting in signal switching action. Collector supply voltage is applied to the output transformer centertap through a parallel resistor-inductor to further suppress oscillation. Although the transistors are operating as injection-controlled signal polarity switches, the impedance ratio between the signal emitters and the IF output collectors results in a modest gain of a few decibels. This can be very desirable when noise figure requirements make the loss of the diode mixer unattractive. Using transistors such as the Motorola MRF 517, this active mixer will give a 3-dB gain, 9-dB noise figure, and +25-dBm input intercept over the 2- to 30-MHz band as an upconverter to the 100-MHz range. This type of mixer is also available in IC form, the Motorola MC1596 being a low-level device and the Plessey SL6440 being a high-level receiving mixer [12].

Spurious signal responses in mixers are of great concern, especially in the first mixer of a receiver. The magnitude of these responses, whose frequency determination was discussed in section 4.4, can be estimated in the planning stage of a design when mixer data are available giving the ratio of desired output to spurious output for a 0-dBm or other signal input. At an-

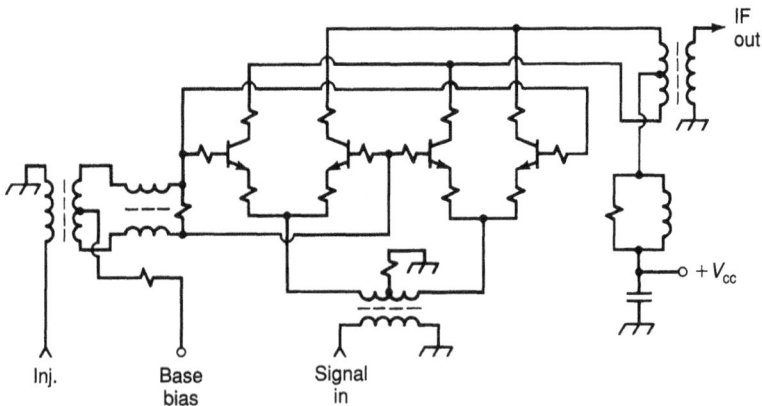

Figure 4.18 A doubly balanced active mixer uses bipolar transistors in a degenerated version of the balanced transconductance mixer.

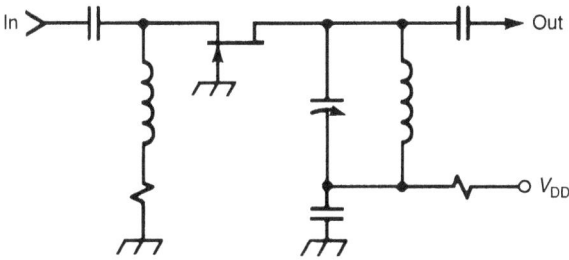

Figure 4.19   A simple grounded-gate FET amplifier can provide a 2-dB noise figure in the 100-MHz region.

other level, the absolute level of the spurious output (in decibel units) changes by the difference in input level times the harmonic number of the signal.

When experimental data are not available, the spurious signal suppression of a doubly balanced diode mixer can be calculated from the diode mismatch, forward diode voltage, and transformer imbalances as described by Henderson [13].

Amplifiers

The most critical amplifier in an HF receiver is the IF amplifier that follows the mixer, either directly or after a crystal filter. This amplifier must have a low noise figure because it is generally preceded by mixing and filtering components that have loss which add to the IF noise figure. It often must have a high-input intercept point, because even when selectively protected by a crystal filter, it will experience large off-frequency signals that are not attenuated greatly by a low-loss filter having limited stopband selectivity. Both bipolar junction transistors (BJTs) and FETs are used in circuits ranging from simple tuned amplifiers to push-pull designs with transformer-coupled RF negative feedback. As with mixers, a great variety of packaged amplifiers are available in the component market, ranging from units with a 2-dB noise figure to those with +40-dBm output third-order intercept points. Simultaneous low noise and high intercepts are a rare combination, however.

The simplest amplifier for the first IF stage is the FET common-base stage. A representative circuit diagram of this type of amplifier is shown in Figure 4.19. When a 2N5432 FET is used in this circuit, a 2-dB noise figure is obtainable with a 50-ohm system gain of 9 dB and an output third-order intercept point of +30 dBm when biased at V_{DD} of 12 to 15 V and 50 mA. The input coupling is low reactance, the optimum source resistance being about 50 ohms. The tuned output network is designed for a 450-ohm load on the drain. There are two drawbacks to this circuit: the input impedance is about

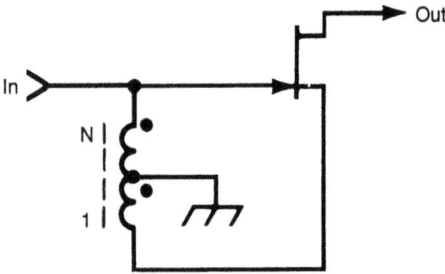

Figure 4.20 An ac circuit diagram of a transformer input technique for obtaining simultaneous noise and input match.

18 ohms, and it has some gain instability with temperature variation. The latter can be lessened by replacing the source bias resistor with an active constant-current sink. The former mismatch problem is a direct result of the inherent difference in amplifiers between the optimum source resistance for minimum noise and the input resistance. It is difficult to optimally terminate a crystal filter or mixer directly with the common-base or common-gate stage, although it is still a practical circuit.

Figure 4.20 illustrates a transformer feedback technique that can be applied to the input of a BJT or FET stage to obtain a simultaneous noise and input impedance match. That is, the input terminal resistance is equal to the optimum source resistance at the input terminal for the minimum noise figure. This is accomplished by presenting the optimum source resistance value to the transistor base-emitter or gate-source junction via a tapped autotransformer whose turns ratio N is determined from

$$N = g_m R_{S,\text{opt}} - 1 \tag{4.28}$$

In this equation, $R_{S,\text{opt}}$ is the optimum source resistance and g_m is the transconductance. The autotransformer is usually made with a large-area ferrite core to minimize flux density for negligible IMD. Using the 2N5432 as an example, g_m is $0.05S$ and $R_{S,\text{opt}}$ is 50 ohms, resulting in $N = 1.5$. The input resistance is

$$R_{\text{in}} = \frac{(1/g_m)N^2}{N + 1} \tag{4.29}$$

which is found to be 18 ohms for the example. If an 18-ohm source is presented to the input, the resistance looking into the transformer from the transistor junction may be computed to be 50 ohms, as desired. The 18-ohm level may be converted to any desired value by additional turns on the transformer or by a reactive matching network. One additional aspect of this circuit is that it may be neutralized for input/output isolation by adding capacitance from drain to source.

When an IF amplifier must provide a low SWR at both the input and output for mixer and filter termination, more complicated transformer feedback circuits may be used. One such feedback amplifier, shown in Figure 4.21, uses wideband transformers to sense drain current and voltage. Only the alternating current (ac) circuit details are shown using a metal-oxide semiconductor field-effect transistor (MOSFET) as the active component. The design equations for this amplifier, which force the input and output SWR to be equal and minimum, start with an independent choice of source (R_S) and load (R_L) resistances and the turns ratio M. Then

$$K = \frac{M^2 R_S}{R_L (M + 1)[(1/g_m) + R_L (1 + 1/M)/M]} \quad (4.30)$$

$$B = \frac{K}{g_m}$$

$$C = K R_S$$

$$N = \frac{B}{2} + \left[\left(\frac{B}{2} \right)^2 + C \right]^{1/2}$$

$$R_{\text{in}} = N \left(\frac{1}{g_m} + R_L \frac{1 + 1/M}{M} \right)$$

$$R_{\text{out}} = \frac{M^2}{M + 1} \left(\frac{1}{g_m} + \frac{R_S}{N} \right)$$

$$\text{Transducer gain} = \frac{4 R_S R_L N^2 (1 + 1/M)^2}{(R_S + R_{\text{in}})^2}$$

With large-power MOSFETs this circuit can have impressive performance. A push-pull version which consumed 0.9 A of drain bias current had a

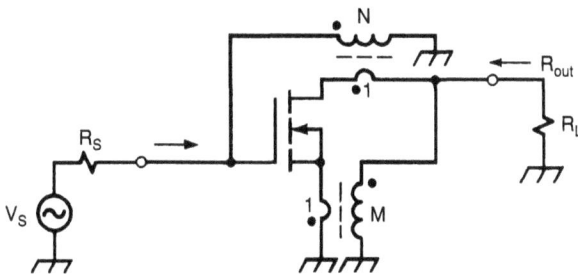

Figure 4.21 A transistor amplifier using both current and voltage feedback can have simultaneously low input/output SWR together with a moderate noise figure and high intercept point.

3.6-dB noise figure, a 9.6-dB gain, an input/output SWR of 1.48:1, and an output third-order intercept of +54 dBm.

Other amplifiers are used in the SSB receiver for IF and audio gain. These are discussed in the IF and baseband system sections of this chapter.

Filters

Filter components are used in all sections of the SSB receiver for RF selectivity, IF selectivity, sideband selection, and audio response shaping. The types of filters used are discussed in detail in Chapters 6 and 9. Intermediate frequency selectivity is detailed in Chapter 6, while RF preselection filters are discussed in Chapter 9.

IF system components

Filters for sideband selection, low-IF gain-stage components, attenuators, and product detectors are employed in the IF system following the front end. High-frequency IF to low-frequency IF mixers are needed as well, and can be similar to the passive and active types used in the front end.

IF gain

Depending upon in-band IMD requirements and thermal stability, IF gain is obtained either with linear ICs or with discrete-component circuits. The choice of ICs is quite narrow if most of the receiver's gain is to be developed in one IC. The Motorola MC1590 has a 50-dB gain at frequencies up to at least 30 MHz and is gain controlled by voltage applied to one of its pins. Its gain stability as a function of temperature is not outstanding, however, with a 10-dB gain change possible over a 150°C temperature range. The Plessey SL600 series of ICs may also be applied to SSB IF systems. Several chips are necessary to get sufficient gain and gain control. The Analog Devices AD600 and AD602 are more recently announced IF gain blocks that have 41 and 31 dB gain and feature accurate attenuation that is linear in decibels per volt. As mentioned in the section on AGC design, this characteristic is desirable for loop stability. There is a definite trend away from military temperature-range linear IF IC components caused by the economics of low military quantity compared with high consumer product quantity.

When high performance is needed, discrete-component amplifiers are often employed. Typical circuits use BJTs with emitter degeneration and low gain per stage to achieve low IM and thermal stability. Gain control is never done by changing the stage bias when low distortion is needed. The PIN diode is nearly always employed for gain control between BJT stages. Often several PIN diodes are used in series to get increased linearity at the expense of control range.

Figure 4.22 is a schematic diagram of a discrete-component IF amplifier stage that uses PIN diode attenuation for gain control. Transistors Q_1 and Q_2 form a common-collector, common-emitter pair having high input resistance and degeneration by the partially bypassed emitter resistance in Q_2. The gain is set by the ratio of collector-to-emitter resistance and is 26 dB for a representative circuit using 2N2222 medium-level transistors. The IF input series resistance and CR_1, a PIN diode with at least a 1.5-µs charge storage time, form the AGC attenuator. The series resistance in the AGC line is shunted by a capacitor for phase lead compensation. At 450 kHz, this gain block will accept an input level of up to 5 to 10 mV, rms per tone, for 65-dB down relative-output third-order IMD.

Product detectors

A mixer is required at the output of the IF system to convert signals to AF. The injection to the product detector is often called the *beat frequency oscillator* (BFO). A conventional diode balanced mixer can certainly be used for this function, with audio taken from the IF port that has response to zero frequency. Because of the normally large difference between IF and AF, however, a balanced mixer is not needed. For example, a dual-gate MOSFET with IF signal applied to the first gate and injection voltage applied to the second gate will perform well and has gain. Another type of balanced mixer with gain is the 1496/1596 family of linear IC doubly balanced modulator/demodulators. This circuit is a transconductance mixer similar to the mixer shown in Figure 4.18 without degeneration resistors. When operated with a 0.5-V (rms per tone) signal level, the IM ratio is below 50 dB.

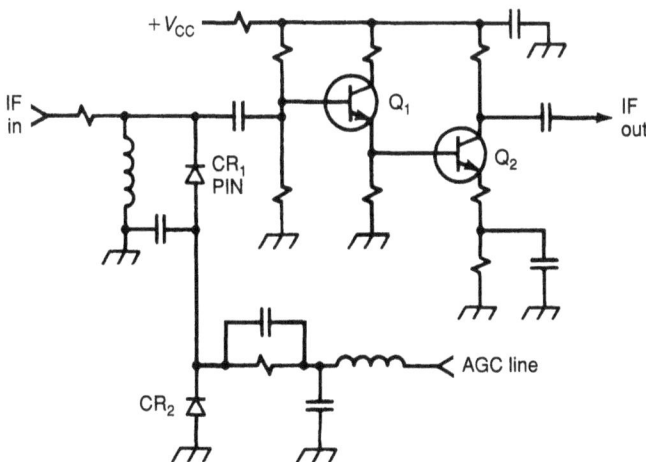

Figure 4.22 Schematic diagram of a discrete component 450-kHz IF amplifier stage that includes shunt PIN diode attenuation.

IF selectivity

In the IF section, the sideband selection and CW narrow bandwidths require filters with high Q resonators. Quartz crystal, mechanical, and, for low performance, piezoelectric ceramic resonator filters for IF use are nearly exclusively procured as purchased items from a large base of manufacturers. When time-delay equalization is required, allpass delay networks may be placed in the audio section and require coordination with the IF filter characteristics. Chapter 6 contains details of the filters usable in the receiver IF.

The distribution of IF selectivity and gain can degrade the output signal-to-noise ratio if care is not taken to limit the bandwidth immediately ahead of the product detector. In a receiver having the IF bandwidth restricted by the sideband selection filter immediately following the front-end mixer, and having the majority of the receiver's gain developed in an amplifier with unrestricted bandwidth, the noise developed by the wideband gain lies in both upper and lower sideband channels. It is translated to AF by the product detector. Although audio selectivity can be added to restrict the bandwidth to equal that from the sideband filter, the excess noise on the unused sideband remains in the audio passband and can degrade the signal-to-noise ratio. The addition of an IF filter directly ahead of the product detector will attenuate the undesired channel noise. The filter need not be especially high performance, needing only to drop the undesired sideband by 10 to 20 dB. An alternative to filtering is an image-canceling product detector using the circuits previously described for the single-sideband homodyne receiver. The relatively narrow IF bandwidth allows accurate formation of the I and Q channels. This can be followed by an "active" audio filter.

A recent trend in HF receiver design is to use digital computation to synthesize the receiver's selectivity, the digital output being converted back to analog with a digital-to-analog converter (see Chapter 8).

Baseband components

The audio or baseband (data) section of the receiver must provide the gain and power output to drive output transducers for voice and data lines. Filtering components are also needed in the baseband circuits.

Filtering is normally done with distributed component active filters to attenuate high-frequency hiss from wideband amplification after the IF filters. Analog filter circuits include the Sallen-Key operational amplifier (op-amp) circuits [14], as well as the more modern state-variable circuits [15, 16]. Switched capacitor IC components, intended for the telephone industry, are becoming available and offer extremely low shape factors and constant time delay [17, 18]. The operating bandwidth of these components is a function of the switch clocking frequency, and so is readily varied for operating flexibility.

There are a large variety of audio power output amplifiers available in plastic packages from the consumer industry. If the receiver is to be used in acoustically noisy environments, such as vehicles, at least 2 W of audio power is required. Output levels for driving headphones or telephone lines are nominally 0 dBm, with peak power capability of +20 dBm. Ordinary operational amplifiers have sufficient voltage and current swing capability to supply these types of outputs. With feedback, the audio circuit harmonic distortion of the output amplifiers is negligible compared with IF and product detector distortion contributions.

4.7 Design Example

In this section we will study a commercial SSB receiver design by examining the frequency scheme and block diagrams of each section. We will use level diagrams to explain the gain, noise figure, and IM distribution in the RF and IF sections of the receiver. Finally, the AGC system will be described using a level diagram showing signal levels throughout the receiver for several values of input signal.

Frequency plan

Figure 4.23 presents a block diagram showing the triple-conversion process from signal input to audio output. The first conversion produces the sum of the signal and first-injection frequencies to generate a 109.35-MHz first IF. The spurious responses of this conversion are chiefly caused by signal harmonics. The lowest-order spurious response that crosses the desired signal is fourth order (4×27.3375 MHz = 109.35 MHz). In this design, sum mixing has been selected to minimize noise on the first injection at the low-frequency end of the signal range where the synthesizer VCO is at its maximum frequency. The trade-off is that the VCO tuning ratio is greater for low-side compared with high-side injection.

The second injection at 118.8 MHz and the resulting 9.45-MHz second IF is selected for three reasons. The rather high second IF makes it simpler to reject the 128.25-MHz image frequency in the first-IF selectivity. Because

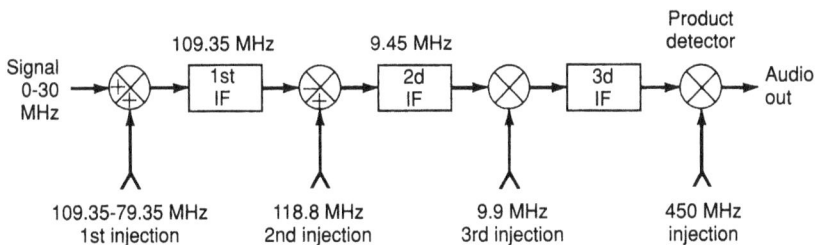

Figure 4.23 Example frequency plan is a triple conversion with sum mixing in the first frequency translation.

118.8 MHz is on the high side of the first-injection frequency it will not cause an internal birdie by mixing with the variable first injection to generate an IF. Finally, 118.8 MHz is the twelfth harmonic of the synthesizer frequency standard at 9.9 MHz and so is simple to generate.

The second IF uses a crystal filter whose bandwidth, like the first 109.35-MHz filter, is wide enough to simultaneously pass both sidebands. In some receivers, these filters are 12 kHz wide so that a total of four independent sideband channels may be processed by a multiple-channel IF system. The third injection at 9.9 MHz is conveniently obtained from the synthesizer standard. It should be obvious that excellent shielding and circuit bypassing are necessary to prevent the 9.9-MHz signal from reaching the front end of the receiver. Ideally, this internal signal should be inaudible; practically, reduction to a 0.5-μV equivalent level is possible.

Sideband selection is obtained in the 450-kHz output of the third mixer. The indicated filter may actually be several switched filters for differing bandwidths and center frequencies. Although the block diagram in Figure 4.23 shows fixed frequencies for the IF at each conversion, it is also possible to shift any one of the injections by an amount equal to the 450-kHz filter bandwidth so as to make a single filter select either sideband. This minimizes filter cost, but also means that the filter stopband must be fairly symmetrical to give equal-carrier and unwanted sideband rejection.

Finally, the IF signal is converted to audio by heterodyning with the 450-kHz injection signal in the product detector. When multiple-sideband selection filters are used, this final injection remains at 450 kHz. When a single filter is used, the 450-kHz injection must move from one stopband edge of the filter to the opposite stopband edge. In simpler receivers, the injection comes from switched crystal oscillators, or even an LC variable oscillator that is then often called a *beat frequency oscillator*. For CW (continuous-wave telegraphy) reception, a variable injection is desired by the operator to lessen listening fatigue by allowing variation of the resulting audio output frequency. The range of the BFO should result in an AF from zero to above 1 kHz. A direct digital synthesizer may be used when standard referenced stability is required, as for radio teletype reception.

With multiple mixers it is always possible that a combination of injection frequencies will occur which produces one of the IFs or the signal frequency. These internal birdies were mentioned in section 4.4 and are predictable. Table 4.1 shows the results of a simulation in which a line is printed whenever an oscillator combination produces a mixing product lying within the IF passband. Harmonic combinations up through the fifth order are allowed. For example, the first line predictably shows that the first injection enters the IF when the signal frequency is zero. Reception of the 9.9-MHz internal standard is indicated by the second line. Shielding and filtering will reduce these signals (except for the line 1 feedthrough) to the 1-μV (open-circuit) or less equivalent input level.

**TABLE 4.1 Internal Spurious Signal Analysis
for the Design Example Receiver**

```
INTERNAL TWEETS UP TO ORDER 5
LO1 =   79.350 to 109.350
LO2 = 118.800
LO3 =    9.900
IF1 = 109.350
IF2 =    9.450
IF BW = 0.012
```

HARMONICS			LO1	SIGNAL
LO1	LO2	LO3	FREQUENCY	FREQUENCY
1	0	0	109.350	0.000
1	0	1	99.450	9.900
1	0	2	89.550	19.800
1	0	3	79.650	29.700
-1	1	0	109.350	0.000
3	-2	0	82.350	27.000
3	-2	0	79.350	30.000

Hardware

A block diagram of the hardware making up the front end of the HF receiver is shown in Figure 4.24. Overload protection is required in the antenna input circuit to prevent physical and electrical damage when large signals from collocated transmitters are present. When the overvoltage is sustained, a detector-operated mechanical relay in the overload protector block is opened. Fast limiting for less than 1 ms is obtained with reverse-biased shunt diodes following the input lowpass filter.

Several lowpass and bandpass filters are needed to restrict the range of signals passing through to the first mixer. The lowpass filters primarily reject the first-IF image response, which lies in the band 188.7 to 218.7 MHz, a region with large RF field intensity from broadcast television transmitters. These filters also attenuate first-mixer injection energy that would cause possible interference to VHF receivers. The 0- to 530-kHz lowpass filter is used for low-frequency reception, which is low-frequency limited to about 100 kHz by synthesizer noise on the first injection. The commercial broadcast band is bandpassed by the 0.53- to 1.6-MHz filter to prevent strong HF signals from desensitizing the first mixer. The 1.6- to 30-MHz bandpass filter gives adequate selectivity to meet 80-dB spurious response rejection throughout the HF band with a well-designed first mixer. Optionally, half-octave filters could be fitted into the translator assembly to give additional selectivity. These filters are electrically switched with PIN diodes having 1-μs or greater storage time. A second lowpass filter ahead of the mixer further attenuates VHF signals.

The first mixer uses junction field-effect transistors (JFETs) in a singly balanced active mixer circuit having an input intercept of +25 dBm for

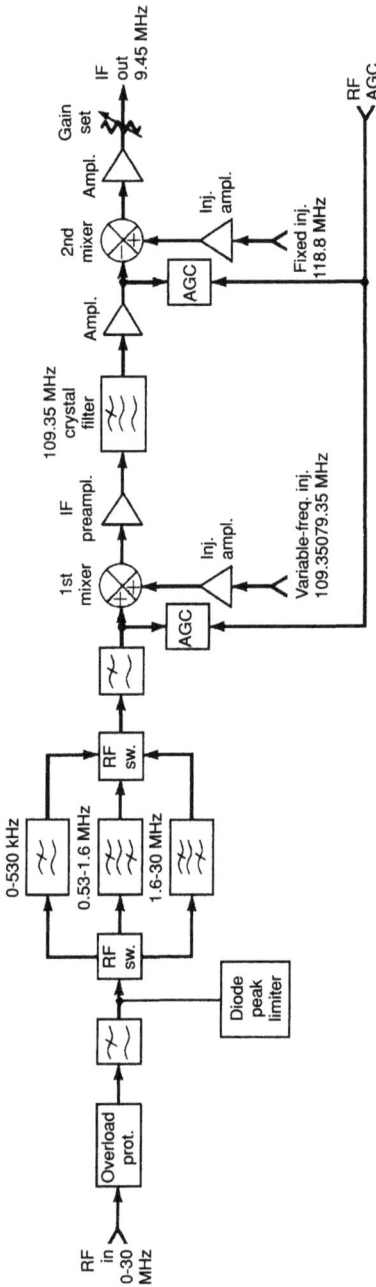

Figure 4.24 Front-end translator is a double-conversion design using 109.35- and 9.45-MHz IFs to give improved image reduction in the first-IF crystal filter.

third-order IM. The injection amplifier associated with this mixer includes bandpass elements to restrict the frequency range to 79 to 109.35 MHz to eliminate low-level signal frequency and VHF noise from the synthesizer. For special very-high-performance systems, a +35-dBm intercept doubly balanced switching MOSFET mixer is employed with added expense and injection power. This mixer must have improved balance plus a quiet injection amplifier to not be desensitized by synthesizer noise (refer to section 4.7). When high-level mixers are used, the power-handling capability and intercept point of the following 109.35-MHz overtone crystal filter becomes important and in some cases a limiting IM factor. In the lower-level first-mixer case, an IF preamplifier is used to improve the noise figure. At the low-signal frequency end, this preamplifier receives both sidebands from the first mixer since there is only matching circuit selectivity between the mixer and amplifier. For example, when receiving 2 MHz, the synthesized injection is at 107.35 MHz and both 109.35- and 105.35-MHz sidebands exit from the mixer. The signal power level is therefore double the expected amount and requires 3-dB additional intercept in the amplifier. When the +35-dBm intercept mixer is employed, a low-loss crystal filter is inserted between the mixer and preamplifier.

Delayed AGC is applied at the RF input to the first mixer and before the second mixer. The IF amplifiers at 109.35 MHz also isolate the crystal filter from termination impedance change due to AGC that would distort the passband shape.

The second mixer is a conventional +7-dBm doubly balanced diode mixer. At this point in the front end, out-of-band strong signal rejection has been obtained with the first crystal filter and the noise figure has been set by the first-mixer and IF preamplifier characteristics. After amplification and a gain adjustment, the resulting 9.45-MHz signal is passed to the IF system. The circuits comprising Figure 4.24 are packaged in a cast aluminum module with extensive cast-in shielded compartments to eliminate or reduce spurious responses and birdies.

Figure 4.25 shows the blocks of the IF system which include an active FET mixer third conversion to 450 kHz, the switched sideband selection filters, the IF gain, and the product detector. In multiple-channel receivers, this section becomes complex, with the output of the third mixer being split into as many as four independent channel paths. Mechanical filters are used in this example because their size and performance are excellent at 450 kHz. Crystal filters could be used, but would be larger in cost and volume at this frequency. The IF amplifier uses emitter degenerated bipolar transistors to obtain temperature-stable gain and low IM generation. Positive intrinsic negative diode attenuators are used for the gain control elements. For cost reduction, a narrowband IF image noise reduction filter is not used before the product detector. The gain distribution before and after the mechanical filters is such that the output signal-to-noise degradation is small.

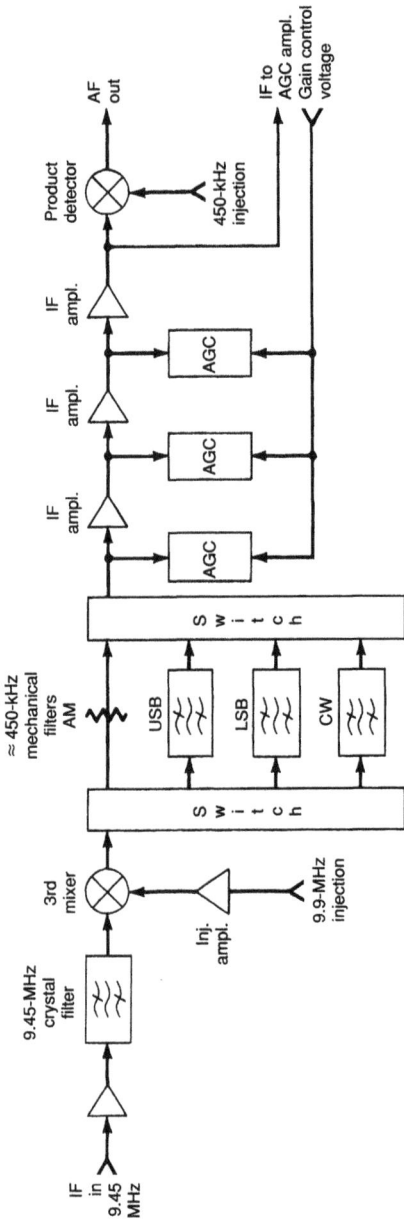

Figure 4.25 Third-conversion and IF amplifier/detector sections use mechanical filters at 450 kHz for sideband selection.

A doubly balanced transconductance mixer IC is used for the product detector. Printed circuit board construction is used for these circuits.

The remainder of the receiver's audio signal path is shown in Figure 4.26 along with the AGC detection, hold, and dumping circuits. This receiver has three independent audio outputs for a small built-in speaker, for 600-ohm balanced line driving, and for headphone output. The indicated audio amplifiers contain filtering components to restrict the high-frequency range. A transformer is frequently required for balanced line driving to isolate the receiver from common-mode hum and lightning surge voltages. The AGC circuits provide the features of selectable hang time and time-sequenced dump rates. A portion of the IF signal is amplified and average value rectified, then amplified and used to charge an RC lowpass filter whose following amplifier input impedance is high. Thus, the filter charges quickly up to the average signal value, with an attack time constant of about 2 ms for SSB and CW. The filter output is compared to a reference voltage, the resulting voltage operating the gain control elements, and the signal strength meter (S meter). The gain control versus voltage must be a stable function if the S meter is to be calibrated in a meaningful way.

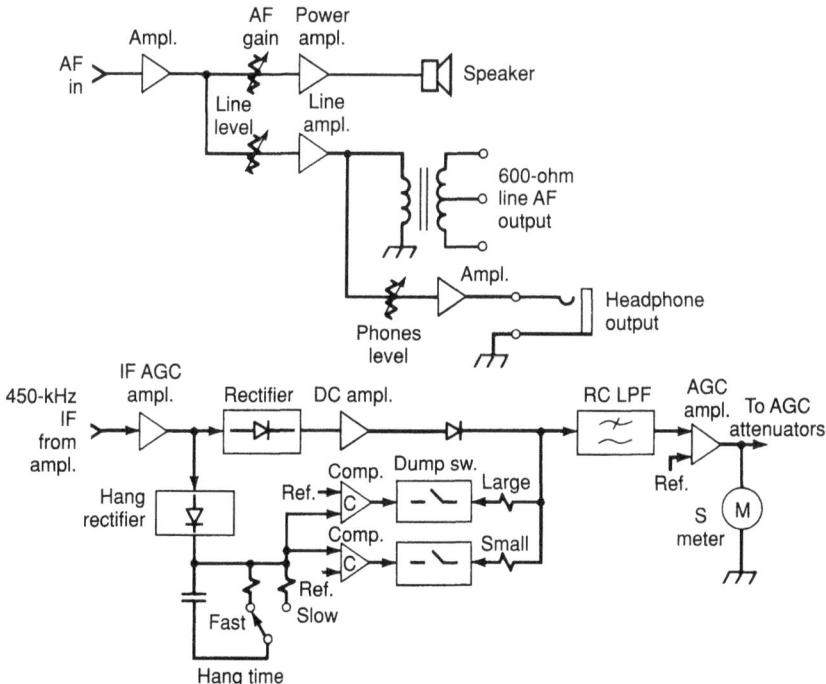

Figure 4.26 Audio and AGC system blocks provide three audio outputs and selectable AGC time constants.

The amplified IF signal is also rectified and charges a capacitor having an adjustable decay time. When the signal decreases, the capacitor voltage decreases to a reference value at which a long-time constant-charge dump is switched to begin discharge of the RC lowpass AGC filter. If the signal decrease persists, a second comparator reference is reached and a faster dump is turned on which quickly increases the receiver gain. The time constants are determined by operator subjective testing and typically range from 30-ms to 0.3-sec hang time and 0.1- to 1-sec dump time.

Level diagrams

The level diagram shown in Figure 4.27 shows the gain, noise figure, and third-order IM intercept points at each point in the front end of the receiver up to the second IF. In designing a receiver, conservative values should be used for the parameters of each stage so that a producible overall design is obtained. The noise figure of passive elements is equal to their loss. The noise figure of the second mixer (doubly balanced diode) is set 0.5 dB higher than its loss to account for small noise sources in the diodes. An adjustable-output attenuator (nominal 3 dB) is used to trim the overall gain to 20 dB to make front-end assemblies interchangeable. Intermodulation distortion calculations are not carried past the first-IF filter because selectivity prevents overload in subsequent stages of the IF. The input intercept point for the crystal filter is set at +50 dBm with the caveat that crystal filter IM is very ill behaved, does not follow a

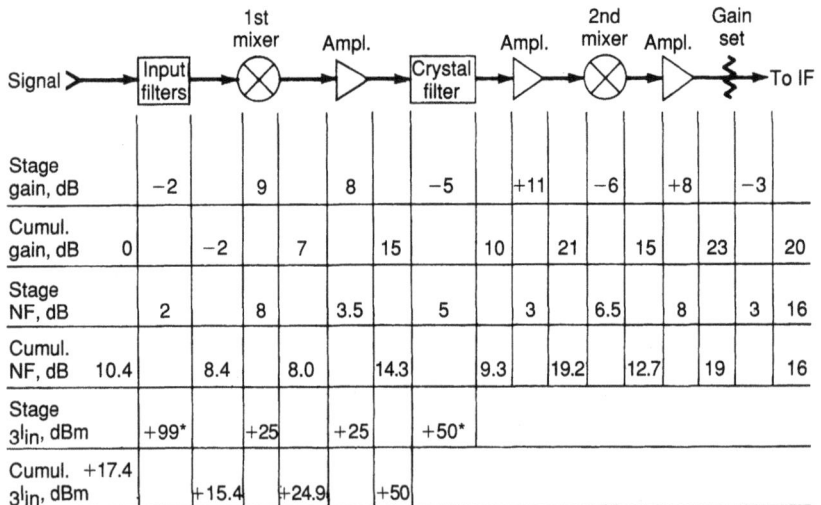

	Input filters	1st mixer	Ampl.	Crystal filter	Ampl.	2nd mixer	Ampl.	Gain set
Stage gain, dB	−2	9	8	−5	+11	−6	+8	−3
Cumul. gain, dB (0)	−2	7	15	10	21	15	23	20
Stage NF, dB	2	8	3.5	5	3	6.5	8	3	16
Cumul. NF, dB (10.4)	8.4	8.0	14.3	9.3	19.2	12.7	19	16
Stage $3I_{in}$, dBm	+99*	+25	+25	+50*				
Cumul. $3I_{in}$, dBm (+17.4)	+15.4	+24.9	+50					

Figure 4.27 A level diagram showing the gain, noise figure, and input intercept distribution from the signal input to the second-IF output. Note: For values with asterisks, see text.

3-dB output to 1-dB input slope, and can exhibit hysteresis. The intercept for the input filter group is estimated to be well above the intercept looking into the mixer and is therefore set at +99 dBm to have little effect on the cascade calculations. All lines of cumulative parameters are calculated using the previously discussed equations for cascaded gain, noise figure, and intercept point.

This front-end design is guaranteed to have a 14-dB noise figure and a +15-dBm third-order input intercept. In production, a typical input noise figure of 12 dB and an intercept of +20 to +25 dBm would be routinely expected.

The gain distribution of the rest of the IF system and the amount of signal attenuation due to AGC is described by a different type of level diagram, as shown in Figure 4.28. In this diagram, the IF output to the product detector is allowed to rise by 6 dB as the input signal level varies from 1 μV to 2 V. For convenience, the impedance level is kept at the 50-ohm input value throughout the analysis so that true stage power gain can be read off the curves. The lower curve establishes the receiver's net gain by showing how the input equivalent noise power $kTBF$ (–130 dBm) is amplified up to –30 dBm at the product detector input. The upper curve illustrates a maximum gain reduction condition, with a +20-dBm input causing both IF and RF attenuators to be fully used. The middle trace indicates an input signal condition where the IF gain has just started to decrease, but no RF attenuation has occurred, because of voltage delay in the RF AGC line to the front end. As the signal is further increased, the output signal-to-noise ratio will increase in proportion to signal level because the input noise figure is relatively unaffected by gain change deep in the IF chain.

This type of level diagram is useful for quick planning of gain distribution prior to detailed noise figure and IM calculations. Adequate sensitivity will be obtained if the signal level at any point is at least 5 dB above the input signal. High gain prior to selectivity can be avoided so that out-of-band IM is minimized. The attack points of AGC throughout the receiver can be distributed to maximize the ultimate signal-to-noise ratio without sacrificing distortion.

4.8 Direct Conversion Receivers

In section 4.3 of this chapter the direct conversion (DC) or homodyne receiver was discussed briefly and compared with the superheterodyne. In the simple form with only one mixer, the DC receiver has no sideband selectivity and has sensitivity degradation from the noise in the unwanted sideband. If two mixers with quadrature injection are used to generate in-phase and quadrature (I/Q) channels, then unwanted sideband and noise rejection can be obtained.

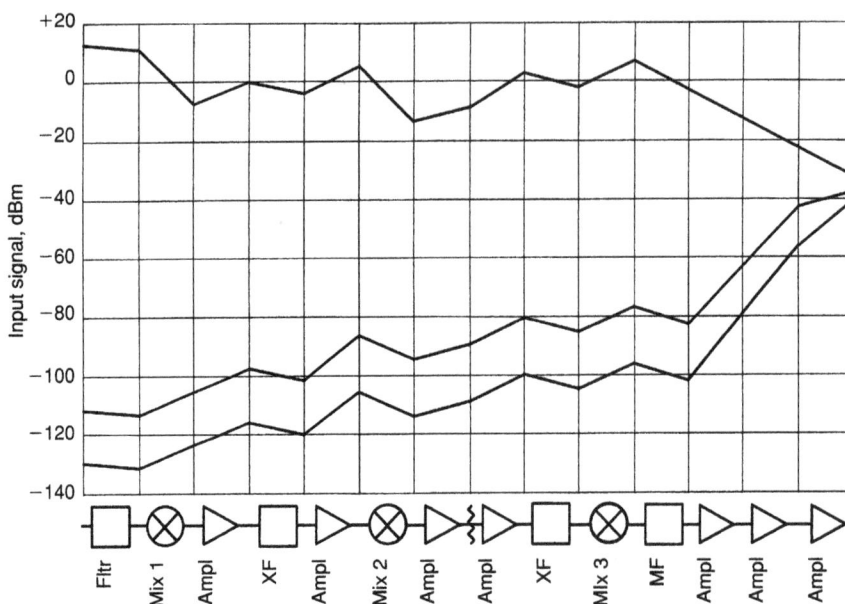

Figure 4.28 The power level for three conditions of signal input is plotted for each stage to plan the gain and AGC distribution.

There are several advantages to direct conversion that motivate refinement of the method. These advantages include

1. Simplicity of RF-to-AF conversion without an IF
2. Baseband filtering done at audio with active filters
3. Potential for miniaturization with IC and digital techniques
4. No internal birdies
5. Spurious responses are even order and cross only at 0 Hz (but they can be heard away from zero beat)
6. Only one synthesizer for injection frequency
7. Can readily be turned around for transmitting

There are some disadvantages also:

1. High AF gain with low noise figure is required
2. Simplicity becomes complexity when I/Q channel sideband rejection requirements exceed 30 dB
3. Rejection of large off-frequency AM signals is an inherent mixer balance problem

4. Vibration susceptibility

5. Local oscillator radiation

As in any receiver design, there are many trade-offs to be made when maximizing the advantages and minimizing the disadvantages. This section will expand on the techniques needed to get better sideband selection performance and will discuss other problems and strengths of the I/Q DC receiver concept.

Figure 4.29 illustrates the essential elements of a DC receiver, including a method of obtaining sideband selection by using quadrature networks in the mixer injection and IF output ports. The RF input blocks include a bandpass filter at the signal frequency and an RF amplifier. These are needed for second-order IM rejection, large off-frequency AM signal rejection, and attenuation of LO radiation. The I/Q channel technique has been widely used to suppress the image and its noise in microwave receivers and for SSB detection and generation before high-performance mechanical and crystal filters became available. Note that the 90° shift could be placed in the RF signal path but would then require a wideband 90° phase shifter that is signal linear. In the LO path quadrature mixer injection is readily obtained with digital circuits.

The +45° and –45° blocks represent networks whose relative output phase stays at 90° with constrained error over the desired audio bandwidth. Summation or differencing of the I and Q channels produces either upper or lower sideband. The I/Q DC receiver can have simultaneous USB/LSB output by using separate summing amplifiers. This feature requires separate IF filters, amplifiers, and product detectors in the superheterodyne receiver.

The LO in-phase and quadrature injection is most simply obtained with the digital divider shown in Figure 4.30. This circuit is called a Johnson or twisted-ring divider and divides the input signal by four while providing a 90° difference between the two Q outputs.

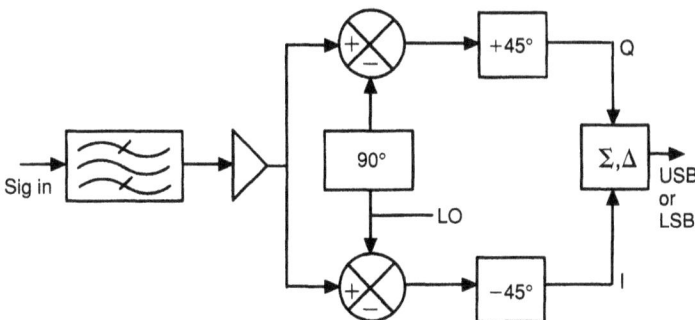

Figure 4.29 The essential elements of a direct conversion (DC) SSB receiver.

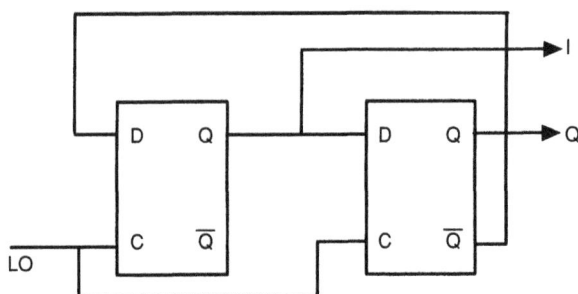

Figure 4.30 A divide-by-four digital circuit to generate quadrature LO drives.

Advanced CMOS (AC) and emitter coupled logic (ECL) dividers are good hardware choices. The fact that the driving synthesizer is at four times the signal frequency is not a large problem because the step size is also four times the desired signal step size. The accuracy of the 90° difference is dependent upon the differential clock to Q output propagation time. This is an unspecified parameter of dual flip-flops, but tends to be very close to zero when the flip-flops share the same die. For example, the clock pulse to Q output delay of a 54AC74 dual flip-flop is typically 6 ns. If the differential delay is one-hundredth of that delay then, at 30 MHz, the phase angle error is 0.648°. Using Equation 4.7 with $G = 1$ and $\phi = 0.648°$ gives 45 dB unwanted sideband rejection. This can be improved by adding a small capacitive delay on one of the Q outputs. It is also advisable to closely equalize the physical path distance between the divider and the two mixers.

Another method for quadrature injection is a dual-output direct digital synthesizer (DDS). If a DDS is used for the synthesizer for other reasons, then this is an economical choice, requiring an additional D/A converter. The DDS would operate at the signal frequency.

Yet another way to get quadrature injection is outlined in Figure 4.31, which shows $\omega = 1$ prototype values for an RL lowpass and an RC highpass filter combination whose parallel input resistance is 1 ohm at all frequencies. The filter output levels vary with frequency, but the relative phase is a constant 90°. The amplitude variation is removed with limiting amplifiers. This circuit requires high-speed limiting amplifiers to suppress amplitude-to-phase error conversion.

There are allpass LC lattice networks that can provide wideband 90° phase difference outputs. They are not capable of miniaturization and become complex when high accuracy is required. For example, a network having six inductors and six capacitors theoretically has 0.165° maximum phase error over the 2- to 30-MHz range. This type of network was first reported by Darlington and later refined by Bedrosian [19].

The mixers in Figure 4.29 require special consideration. Of first-order importance, both the signal and LO ports must be balanced. The signal port balance is required to minimize direct transfer of strong AM signals to audio by rectification, regardless of the injection frequency. This does not occur in a superheterodyne so the mixers in a DC receiver would have to have infinite second-order intercepts to eliminate this effect. The LO port balance is required to reduce the amount of conducted power out of the signal port, which is harmful in two ways. First, antenna radiation can interfere with other nearby receivers and is regulated in the United States by FCC regulations, Part 15. Second, LO energy that is reflected back into the mixer can cause severe microphonic response to physical vibration when any element in the path varies the phase of the reflection at an audio rate. The use of an RF amplifier, while not popular in superheterodyne receivers, is nearly mandatory in a DC receiver to minimize radiation. The amplifier should preferably be neutralized or *unilateralized*.

A design example illustrates the serious nature of oscillator radiation. Consider the blocks shown in Figure 4.32 for the input stages of a receiver including the antenna. The antenna SWR is 1.5:1, which is equivalent to a return loss of 14 dB, and the reverse transmission of the RF stage is estimated to be –30 dB. The RF stage forward gain is 8 dB and the mixer LO port balance is –30 dB to the +17 dBm injection power. The reflected signal into the mixer is then +17 – 30 – 30 – 14 + 8 = –49 dBm, which is 1.12 mV peak in a 50-ohm system. The mixer operates as a phase detector and has an output defined by

$$V_{dc} = 0.637 \, V_{pk} \cos \theta \qquad (4.31)$$

where transformer ratios are ignored and the mixer operates as a signal in-

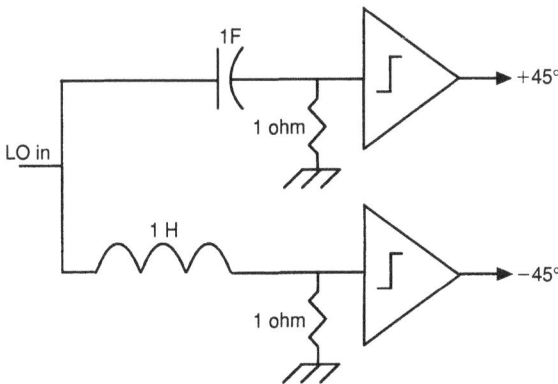

Figure 4.31 An RLC quadrature phase-shift network.

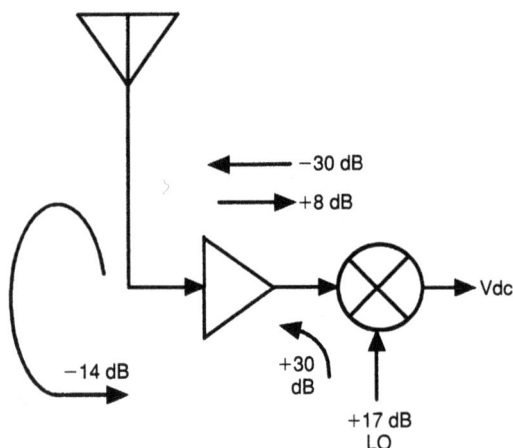

Figure 4.32 Local oscillator feedback problems in the direct conversion receiver.

verting switch with square-wave injection. Taking the differential of Equation 4.31 shows that

$$dV_{dc} \approx 0.637\, V_{pk}\, d\theta \qquad (4.32)$$

where θ is in radians and the returned signal is at 90° to the injection. A dV_{dc} value of 0.5-μV peak at the mixer output would cause substantial interference to a weak desired signal. With $V_{pk} = 1.12$-mV peak, $d\theta$ is 0.7 mrad or 0.04°. This is divided by two to get 0.02° phase error going one way up or down the antenna cable. Now suppose the antenna coaxial cable is 50 ft long with a velocity factor of 0.66 and the frequency is 30 MHz. The electrical cable length is then 831°. By proportionality

$$\frac{0.02}{831} = \frac{dL}{50}$$

which gives the change in cable length dL, a value of 1.45 mft or 17 mils. If the cable or its connectors are stretched 17 mils at an audio rate by vibration, substantial receiver output will be heard. Variation of the antenna return loss angle has the same effect and can be caused by any near-field periodic disturbances. In aircraft, rotor blades and turboprops can cause angle modulation at audio frequency.

When the balance requirements are considered, the doubly balanced diode mixer becomes a practical candidate for the mixers in an I/Q channel DC receiver. The multipoint diode matching in these mixers gives excellent balance versus frequency and temperature. A component refinement in this area would be a dual-channel mixer in which all the diodes and transformers have been matched for improved amplitude and phase balance.

The I/Q audio channels are presented in more detail in Figure 4.33. The DC port of the mixer is terminated in highpass/lowpass filters that absorb the sum product output and attenuate injection power present at the DC port. These filters are two- or three-element LC filters whose bandwidth is large compared to the final audio bandwidth and therefore contribute little gain or phase error. The following amplifier must be a low-noise device with feedback to reduce distortion and gain/phase errors. The gain should be in the 10- to 20-dB region and must be trimmable with stable components to equalize I/Q channel gains. Although somewhat protected by the LC lowpass filter, this amplifier must withstand multiple off-channel signals without distortion.

The next block is a bandpass filter that establishes the required passband of the receiver. This can be implemented with cascaded lowpass and highpass RC filters using operational amplifiers. The lowpass sections should be first, thereby immediately restricting the high-frequency audio bandwidth. Switched-capacitor bandpass filters may also be used and offer the advantage of variable bandwidth control by firmware. They are also available in dual units that facilitate channel matching. In any case, each channel must track the other in gain and phase over the passband. This is a difficult goal to attain when sideband rejection of 60 dB or more is desired. Component selection is a key issue, with ceramic capacitors being undesirable due to drift with temperature and generation of voltage when mechanically stressed with vibration.

The next block is an attenuator which is typically a multiplying D/A converter having the audio signal applied to the reference input. Thus, the output is the product of the signal and the digital word. This technique has large signal-handling capability, little distortion, and excellent gain/phase matching when dual-channel D/A converters are used. Fine step or continuous gain control is then placed in the USB/LSB audio path to fill in the digital steps. Additional attenuation control is also usually needed in the front-end RF stage to handle large signals.

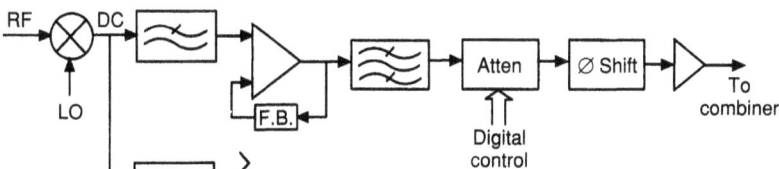

Figure 4.33 One audio channel of an *I/Q* direct conversion receiver.

The next phase network block is one-half of the 90° phase difference all-pass filter. One simple example of such a network is shown in Figure 4.34, which has the transfer function

$$\frac{V_{\text{out}}}{V_{\text{in}}} = \frac{1 - j\omega RC}{1 + j\omega RC} \qquad (4.33)$$

This function has unity magnitude for all frequencies and moves from 0° phase at zero frequency to −180° at infinite frequency. The shift is −90° when $\omega = 1/RC$. A cascade of these shifters will approximate a straight line of decreasing phase shift over a wide band when the individual RC values are correctly chosen. Another cascade will also have nearly straight-line decreasing phase, but with nearly a constant 90° difference with respect to the first cascade when the RC values are correctly interleaved between those of the first cascade. Finding the RC products is mathematically tedious, but computer routines have made the problem simple. The program in reference 20 is quick, user friendly, and gives the frequencies for each network, where $RC = 1/(2\pi f)$. If all the phase error in the DC receiver is allocated to the phase difference networks, 60-dB sideband rejection requires a total of six allpass sections, three in each channel.

The remainder of each channel consists of enough gain to satisfy the receiver overall gain requirement. A part of this gain can occur after the summing point, particularly when audio-derived AGC is used. At low audio frequency the AGC attack time is lengthened by the time it takes the audio to rise to a peak. This can be improved upon by realizing that if the I channel has an instantaneous signal $V_{\text{pk}} \sin \theta$, then the Q channel has $V_{\text{pk}} \cos \theta$. Applying the trigonometric identity $\sin^2 \theta + \cos^2 \theta = 1$, the peak value V_{pk} can immedi-

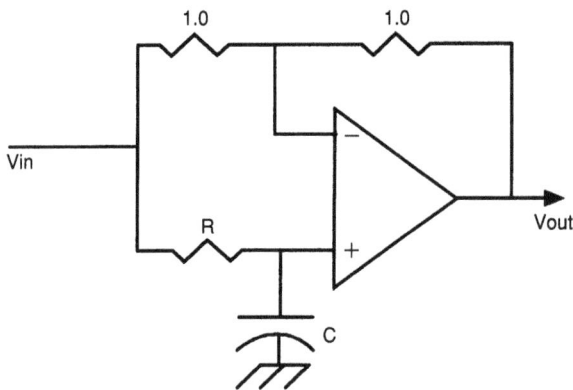

Figure 4.34 An allpass active op-amp circuit for phase-shift networks.

ately be determined. Thus, summing the output of two analog multipliers having I and Q channel inputs produces the squared peak envelope value, which may be used as is for AGC or passed through an analog square-root circuit to keep loop gain more constant. This level detection method is analogous to peak rectification of the IF signal in a superheterodyne receiver.

An optimistic view of the DC receiver is that the gain/phase balance requirements can be obtained with precise components, feedback, and perhaps self-calibration techniques. The pessimistic view is that a mixer with infinite second-order intercept will never be found, and that the superheterodyne will always be needed when performance requirements are high. Reality lies somewhere between these two viewpoints. One technique that is becoming increasingly powerful is to apply digital signal processing to the DC receiver. Each channel would be digitized just after the digital attenuator block in Figure 4.33, perhaps with additional gain to cause one- or two-bit quantization of input noise. The analog bandpass filter ahead of the attenuator would become a switched capacitor design with Gaussian amplitude response to minimize phase error due to component tolerance and drift. The final audio selectivity and the 90° differential phase shifting (the Hilbert transform, see Chapter 2) would all be done digitally to a precision determined by the computing processor, and with no drift versus time or components. It is possible that algorithms could be developed that would correct dynamically the amplitude and phase errors occurring in the analog portions of such a DC receiver.

4.9 IF Variable-Passband and Variable-Bandwidth Tuning

The preceding examples of IF selectivity have centered on highly selective fixed-frequency filters to reject undesired signals close to the desired signal frequency. Several filter bandwidth choices might be made available to the receiver user. When skilled HF communication operators and crowded frequency assignment conditions are involved, the simple fixed-IF filter may not be sufficient. For example, in voice SSB it is common to have a fixed-frequency interferer producing an annoying constant tone in the audio output. In an analog receiver it would be useful to shift the IF passband just enough to put the interferer in the filter stopband, while still allowing enough desired signal bandwidth to produce acceptable audio output. This concept, *passband tuning*, cannot be directly implemented with mechanical or quartz crystal filters, but can be made equivalent by synchronously moving mixer injection frequencies.

Another operational requirement is to reject interference by virtually changing the bandwidth of the IF filter, by moving either the upper or lower cutoff frequencies, or both. This effect, *variable bandwidth tuning*, can be obtained by cascading two fixed filters between mixers having variable injection.

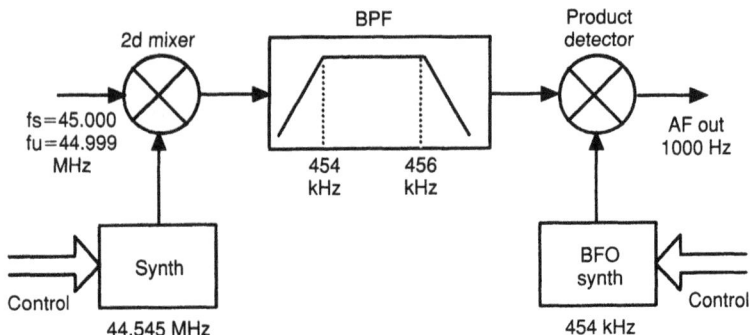

Figure 4.35 Passband tuning in a dual-conversion SSB receiver.

Passband tuning

In early vacuum tube HF receivers, passband tuning was accomplished by gang-tuning a multiresonator LC filter whose nominal center frequency was low enough to obtain reasonable insertion loss and shape factor. This meant that several frequency conversions were needed to get down to a low IF in the 100-kHz region. This is contrary to the modern requirement to keep the selectivity as close to the antenna as possible for undesired signal overload rejection. A way to avoid this is shown in Figure 4.35, which starts with a 45-MHz first IF that is then converted to a 455-kHz second IF, then to audio. The 455-kHz filter is centered on 455 kHz and is 2 kHz wide. First, consider a desired signal entering the IF at 45.000 MHz. With the second mixer synthesizer at 44.545 MHz, the signal is centered at 455 kHz in the second IF. With the synthesized BFO at 454 kHz, the desired audio output tone is 1000 Hz.

Next, the 45.000-MHz signal remains, but the second mixer injection is moved to 44.546 MHz and the BFO is moved to 453 kHz. The second-IF signal moves to the lower edge of the filter passband, but the audio output frequency stays constant because of the simultaneous injection shifts. Equivalently, the "fixed" filter has been moved up 1 kHz in frequency. If there had been an undesired signal present at 44.999 MHz it would have been moved to 453 kHz in the second IF and would be attenuated by the filter stopband. Moving the second injection down and the BFO up would, equivalently, move the filter down. It can be seen that if the desired signal is a band of frequencies, as in voice SSB, then part of the desired signal energy will be attenuated by the filter stopband. This will change the timbre or quality of the speech, but not the pitch, and is an operator's choice. Unfortunately, passband tuning may move a second interferer into the passband as the first interferer is rejected. In a crowded frequency band this can be a common occurrence.

There are several variations that may be made in Figure 4.35. First, the second injection need not be a variable synthesizer. The first injection in the receiver may be moved by adding or subtracting the passband shift to the synthesizer's frequency command generated in the controller. The 3-kHz total signal shift will easily be passed by the first-IF crystal filter, whose bandwidth is 12 kHz or so. A second modification might change the BFO to an adjustable LC oscillator in an effort to reduce cost. To cause an equal shift in either the first or second injections, a portion of the BFO output would be mixed into the first injection synthesizer loop. This increases the cost and often causes internal birdie problems as mixer products cross through the signal or IF ranges. With modern phase-lock loop (PLL) components such mixing hazards are avoided. Figure 4.36 illustrates this point by showing how a PLL with a voltage-controlled oscillator (VCO) operating at VHF, followed by a fixed-frequency divider, can generate the 454-kHz injection for the product detector in 100-Hz steps. Passband movement in finer increments than 100 Hz is not necessary because the filter attenuation versus frequency change usually does not warrant the increased resolution. With current components, the PLL division and phase comparison circuits are one 16-pin IC and the divide by 256 is an 8-pin IC. A DDS can also be used for the BFO, but currently requires two larger IC packages and probably costs more.

The strong undesired signal rejection characteristics of the receiver are not adversely affected by passband tuning because additional stages are not added. As in the fixed-tuned IF receiver, the relatively wide passband of the first-IF crystal filter allows strong close-in signals to arrive at the second mixer. Therefore, second mixer dynamic range and injection purity are important design problems. The total gain up to the second-IF filter should be kept as low as sensitivity requirements will allow.

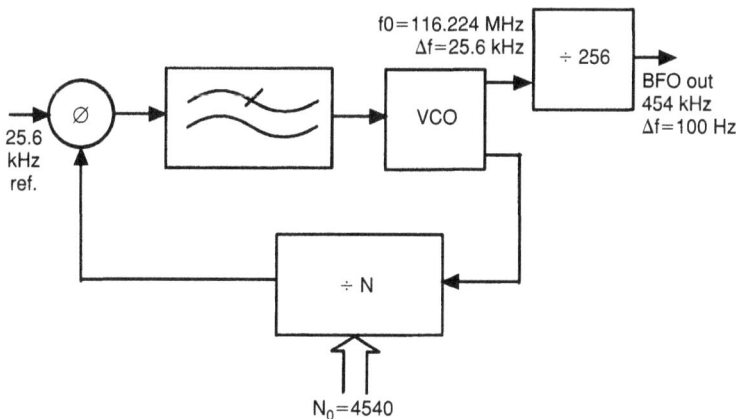

Figure 4.36 A digitally controlled BFO is generated by a VHF PLL and a fixed-output frequency divider.

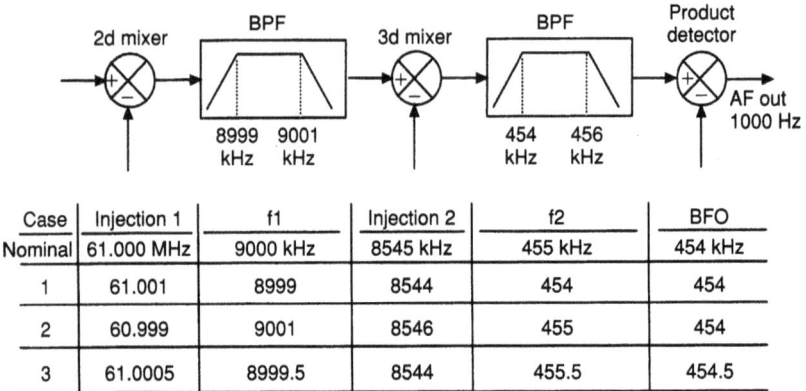

Case	Injection 1	f1	Injection 2	f2	BFO
Nominal	61.000 MHz	9000 kHz	8545 kHz	455 kHz	454 kHz
1	61.001	8999	8544	454	454
2	60.999	9001	8546	455	454
3	61.0005	8999.5	8544	455.5	454.5

Figure 4.37 Variable bandwidth options in a triple-conversion receiver.

Variable bandwidth tuning

It was noted in connection with passband tuning that a second interferer might come into the passband as the first one is rejected. The only way to avoid this is to reduce the IF filter passband by moving either the low- or high-frequency transition bands. Again, SSB voice quality will degrade, but may be acceptable as compared to the interfering signal degradation. The basic way to implement variable IF bandwidth in an analog receiver is to cascade two high-performance filters operating at different frequencies and separated by mixers whose injection frequencies may be varied. This may be thought of as two passband tuning systems whose combined linear response is the product of the two filter transmissions.

Consider an example block diagram in Figure 4.37, which shows the IF path of a triple-conversion receiver having a 70-MHz first IF, 9000-kHz second IF, and a third IF at 455 kHz. For analysis, a 70.000-MHz CW signal is entering the input mixer, and the nominal injections shown result in the signal being centered in the passbands of filters at 9000 kHz and 455 kHz. The product detector output is a 1000-Hz tone. The total IF bandwidth of 2 kHz is centered on the 70-MHz input.

Next consider the list of injection and intermediate frequencies for Case 1. Injection 1 has been moved up by 1 kHz and injection 2 has been dropped 1 kHz. The signal in the 9000-kHz filter has moved to the 8999-kHz lower edge, the 455-kHz filter signal is centered, and the audio output is still 1000 Hz. If the 70-MHz input frequency is lowered, the signal will start to be attenuated by the lower transition band of the 9000-kHz filter. As the input frequency is raised by 1 kHz and then more, the signal will go to the upper edge of the 455-kHz filter and then into the upper transition band. The overall bandwidth is now 1 kHz and, referred to the 70-MHz input, the narrowing has occurred on the low-frequency side.

Now consider Case 2, where injection 1 has been moved down by 1 kHz, injection 2 up by 1 kHz, and the audio output is still 1000 Hz. The equivalent bandwidth has again been narrowed to 1 kHz and the narrowing has occurred on the high-frequency side. Case 3 is a combination of cases 1 and 2, where injection 1 has been moved up 500 Hz, injection 2 down by 1 kHz, and to keep the audio out at 1000 Hz, the BFO injection has moved up by 500 Hz. The net result is a passband centered on 70 MHz that is 1 kHz wide.

Case 3, carried to the extreme of +1-, –2-, and +1-kHz injection shifts, respectively, results in a theoretically zero passband and immediate upper and lower transitions to the stopband. This suggests that the shape factor of the narrowed filter becomes greater than the individual filter shape factors. For example, if each filter has a 2:1 shape factor and the passband is 2 kHz, then the stopband is 4 kHz. When the injection shifts cause a 1-kHz passband, the stopband is now 3 kHz wide, which results in a 3:1 shape factor. Stated another way, the transition band slope of the narrowed filter is that of just one filter, not the linear product (decibel sum) of the two filters. It also happens that a very small shape factor in a narrow filter often causes poor transient response.

Although flexible bandwidth and position control are greatly appreciated by an experienced HF receiver operator, there are compromises that must be recognized. The shape factor increase is one compromise that limits the ability to reject very strong off-channel signals. This means that bandwidth reduction for the CW mode should preferably be done by switching to a narrow filter designed for that mode. More serious is the reduction in undesired signal dynamic range caused by the extra mixer and amplification associated with the added IF filter. Strong signals that are very close to the narrowed passband now move one more stage further down the IF chain. Undesired signal susceptibility and internal birdie suppression are degraded by the IF filter and injection frequencies placed in the HF band. This requires good shielding, power and control lead filtering, and amplifier reverse isolation to prevent direct pickup of 9000 kHz and an internal birdie at 8545 kHz in the example IF system.

Generation of the three injection frequencies may be done similarly to the passband tuning method by keeping injection 1 constant and varying the input signal mixer injection. The third injection is again best done with a PLL and divider. However, the 8545-kHz injection is too high in frequency to be done with small step size with a PLL and divider. A DDS can be used, but should be carefully analyzed for close-in spurious outputs that can degrade filter response or pass directly through the 9000-kHz filter passband. Another way is to translate a readily available fixed frequency to 8545 kHz with a low-frequency PLL/divider similar to the third injection. This arrangement is illustrated in Figure 4.38 where the low-frequency synthesizer has been placed at 345 kHz to avoid birdie generation at 455 kHz. The fixed 8200 kHz is used for mixer injection and is summed with 345 kHz to get

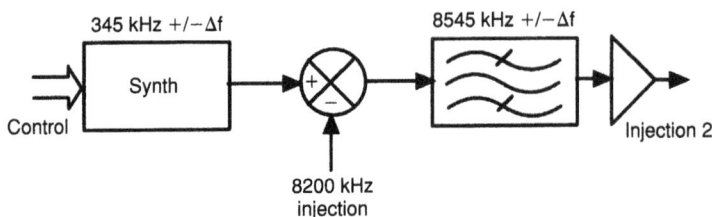

Figure 4.38 Generating a variable 8545 kHz using a synthesizer and a mixer.

8545 kHz. The injection frequency and other sideband outputs of the mixer are attenuated by a bandpass filter in the mixer output.

The use of firmware to control the oscillators needed for both passband and variable bandwidth tuning is encouraged. Analog HF receivers have been produced with complex mixing and drift canceling schemes for accomplishing the desired selectivity control. However, the circuitry tends to become complex, and it is virtually impossible to eliminate internally generated birdies.

References

1. H. T. Friis, "Noise Figures of Radio Receivers," *Proc. of IRE* 32 (July 1944): 419–422. (Note: Equation 15 in Friis, p. 421, is garbled and should read: $F_{ab} = F_a + (F_b - 1)/G_a$.)

2. Franz C. McVay, "Don't Guess the Spurious Level," *Electron. Des. 3* 15 (February 1, 1967): 70–73.

3. D. Norgaard, "The Phase-Shift Method of Single-Sideband Signal Reception," *Proc. of IRE* 44 (December 1956): 1735–1743.

4. William E. Sabin, "Use of Mixers in HF Upconversion Receivers/Exciters," *Wescon/81 Conf. Rec.*, Session Rec. 24/4 (San Francisco, September 15–17, 1981).

5. W. R. Olson and R. V. Salcedo, "Mixer Frequency Charts," *Frequency* 4 (March-April 1966): 24, 25.

6. Stuart E. Wilson, "Evaluate the Distortion of Modular Cascades," *Microwaves* 20 (March 1981): 67, 68, 70.

7. Jack Porter, "AGC Loop Design Using Control System Theory," *RF Design* (June 1980): 27–32.

8. Paul Penfield, Jr., and Robert P. Rafuse, *Varactor Applications* (Cambridge, MA: M.I.T. Press, 1962).

9. Michael Martin, "Stock Components Produce Improved Receiver Design," *Microwaves* 21 (August 1982): 59–61, 103.

10. *Reference Data for Radio Engineers*, 5th ed., Chapter 8 (Indianapolis: Howard W. Sams, 1968).

11. Reed Fisher, "Twisted-Wire Quadrature Hybrid Directional Couplers," *QST* 63 (January 1978): 21–23.

12. Peter E. Chadwick, "The SL6440 High Performance Integrated Circuit Mixer," *Wescon/81 Conf. Rec.*, Session Rec. 24/2 (San Francisco, September 15–17, 1981).

13. Bert C. Henderson, "Predicting Intermodulation Suppression in Double-Balanced Mixers," *Watkins Johnson Tech-Notes* 10, no. 4 (July/August 1983).

14. R. P. Sallen and E. L. Key, "A Practical Method of Designing RC Active Filters," *IRE Trans. Circuit Theory* CT-2 (March 1955): 74–85.

15. Arthur B. Williams, *Electronic Filter Design Handbook* (New York: McGraw-Hill, 1981).

16. L. P. Huelsman and P. E. Allen, *Introduction to the Theory and Design of Active Filters* (New York: McGraw-Hill, 1980).

17. *1984 Linear Supplement Databook*, pp. S9-1–S9-53 (Santa Clara, CA: National Semiconductor Corp., 1984).

18. *Analog and Telecommunications Data Book* (Melbourne, FL: Harris Corp., 1984).

19. S. D. Bedrosian, "Normalized Design of 90-Degree Phase Difference Networks," *IRE Transactions on Circuit Theory* (June 1960): 128–136.

20. Allan B. Lloyd, "BASIC Program for 90-Degree Allpass Networks," *RF Design* (April 1989): 47–53.

5

Exciter and Transceiver Design

Sylvan L. Dawson (Sections 5.1–5.5)
Robert H. Sternowski (Section 5.6)

An SSB transmitter system may contain several functional elements, as shown in Figure 5.1. These include a data terminal, remote control unit, exciter, power amplifier, antenna coupler, and antenna. More elaborate systems may also include switching units, RF output bandpass filters, or control processors. The essential elements, though, are the exciter, power amplifier, and antenna. They may be separate or combined into a single unit, as in a packset, for example.

The exciter is a key component of this system and performs several important functions. It translates one or more baseband input signals to the desired output frequency and drives the power amplifier at the proper level. It provides different emission modes such as CW or FSK. Usually it coordinates gain setting and tune cycle operations of the transmitter. It also controls such parameters as bandwidth, distortion, frequency stability, and spurious emissions.

This chapter discusses the performance requirements of an exciter, how its component parts are assembled in an architecture that meets the requirements, and how it performs the necessary functions.

A transceiver may be defined as an exciter and receiver combined in one functional unit. In applications where size is at a premium, such as in airborne or other mobile equipment, a transceiver configuration provides the most efficient use of available space. Transceiver design will also be discussed.

Antenna

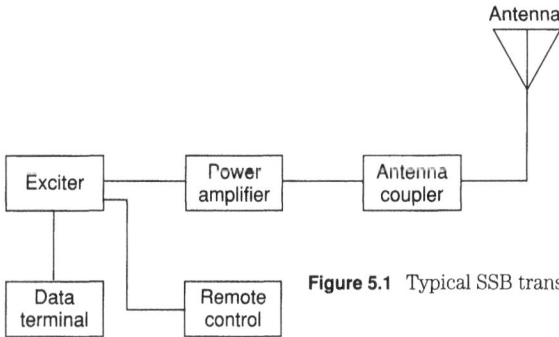

Exciter		Power amplifier		Antenna coupler

Data terminal		Remote control

Figure 5.1 Typical SSB transmitter system.

5.1 SSB Exciter Output Signal Requirements

Ideally the RF output from an SSB exciter would be an exact linear translation of the input baseband signal, with no amplitude or phase distortion or added noise, and at exactly the level required by the following power amplifier. Of course, this ideal is not achieved in practice, and the degree to which imperfections in performance can be tolerated determines to a great extent the complexity and cost of the exciter design. It is important, therefore, to determine the allowable levels of the various kinds of distortion in the exciter RF output.

The limits of distortion and noise pollution of the exciter HF signal are generally set by the signal fidelity needs of other communication system components and by government regulatory agencies. System components at the receiving end of a radio link generally place requirements on in-band characteristics of the signal. These are characteristics such as amplitude distortion, phase distortion, in-band IMD, and signal-to-noise ratio.

Out-of-band emissions from a transmitter can be in the form of either discrete spurious signals or broadband noise. Discrete spurious signals can be caused by IMD, leakage of mixer injections, and harmonics of the transmitted signal. Broadband noise is most commonly caused by amplifier noise and oscillator phase noise on mixer injections. The source of many of the spurious emissions of the transmitter system is the exciter.

Figure 5.2 is a spectral representation of the RF output of an exciter, showing the desired USB signal above f_c, the carrier frequency, and several of the possible unwanted outputs. The LSB signal is not perfectly suppressed and appears at the output. Harmonics of the USB appear around the carrier frequency harmonics, $2f_c$ and $3f_c$. Odd-order IMD products appear around the desired USB signal and its harmonics. Various other spurious emissions may appear because of combining of internal injection signals in the output stages. The power amplifier following the exciter will also generate IMD products. It may provide some selectivity, such as in the output circuit, to help suppress harmonics and similar products. A broadband amplifier, however, will not reduce these unwanted outputs. A principal function of the exciter design is to control out-of-band emissions.

Spurious emissions limitations

Governmental regulatory agencies prescribe certain limits to the levels of spurious emissions from radio transmitters. The International Telecommunications Union (ITU) sets regulations, observed by its member nations, regarding the levels of spurious emissions, frequency assignments, allowable signal bandwidth, and other significant parameters of radio transmission. Regulatory agencies of the member nations, such as the Federal Communication Commission (FCC), generate their own regulations for internal use within the ITU limitations.

Table 5.1 provides a few of the limitations on radio spurious emissions and frequency tolerances in the HF (generally 2- to 30-MHz) range. The limitations given may not be current and are provided only as an illustration.

In addition to the emissions limitations imposed by regulatory agencies, operational considerations may impose additional restraints. For example, when transmitters and receivers are located in close proximity, it is easy to imagine the interference that could be caused in the receivers by a transmitter that emits the broad spectrum of signals alluded to in Figure 5.2.

A situation in which a transmitter and a receiver are collocated is shown in Figure 5.3. The transmitter signal is at frequency f_t at a 1-kW level, +60 dBm. A receiver is simultaneously receiving a signal of frequency f_r of −116 dBm. The transmit and receive antenna separation provides isolation of 30 dB. If the acceptable interference level at the receiver input is −126 dBm,

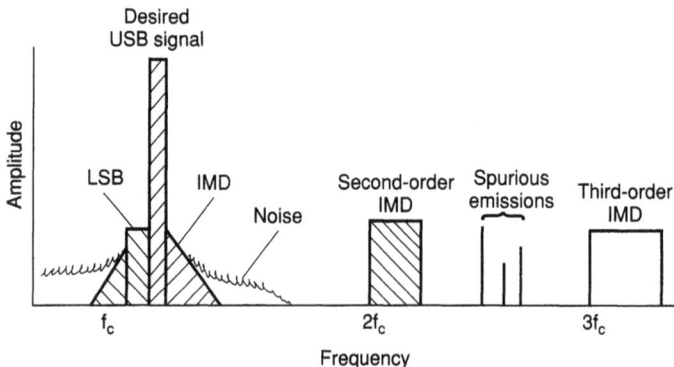

Figure 5.2 Possible spectral components of exciter output.

TABLE 5.1 Sample of Regulatory Agency Requirements

Frequency tolerance	Requirement
ITU, App. 7: 1.6 to 29.7 MHz	20 Hz
FCC, Part 81: 1.6 to 27.5 MHz	20 Hz
Spurious emissions	**Requirement**
ITU, App. 8: 9 kHz to 30 MHz	40 dB below mean power, 50 mW maximum
FCC, Part 81	50 to 150% of bandwidth from assigned frequency,* 25 dB below mean power
	150 to 250% of bandwidth from assigned frequency, 35 dB below mean power
	250% or more of bandwidth from assigned frequency, 43 dB + 10 log (mean power)

*Assigned frequency is center frequency of assigned bandwidth

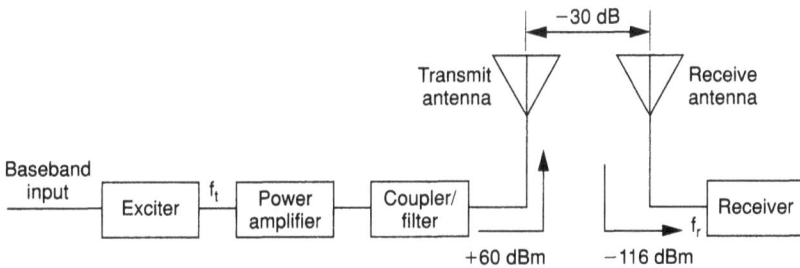

Figure 5.3 Collocated transmitter and receiver.

the allowable level of transmitter spurious emissions that fall on the receive frequency is –96 dBm, or 156 dB below the transmitter output signal. This is a very stringent requirement on the transmitter system. If filtering in the power amplifier and transmitter output filter provide 50 dB of selectivity, the required relative level of spurious signals at the exciter output is –106 dB. Note that this limit applies not only to discrete spurious signals, but to broadband noise as well.

It is apparent that the range of allowable spurious emissions levels can be quite wide, depending on end-user requirements.

SSB channel characteristics

The exciter, as it processes the baseband input to the final RF output, will add distortion and noise to the signal. A description of these effects and typical allowable levels are discussed below.

In-band IMD. Figure 5.4 shows a typical transmitter output within the passband. The in-band IMD products may interfere with signal components near the same frequency. Multitone data signals are particularly susceptible to this type of interference because the IMD products can fall into data tone slots.

The in-band IMD can be generated in the baseband AF processing circuits or in any of the IF or RF stages. Even-order and odd-order IMD can both be generated at baseband. In Figure 5.4, where the 800-Hz and 900-Hz tones are the only two input signals, the baseband circuits may generate second-order IMD products at 100, 1600, 1700, and 1800 Hz. Third-order IMD products are 700, 1000, 2400, 2500, 2600, and 2700 Hz. The 700-Hz and 1000-Hz products can also be generated in IF and RF circuits. Higher-order products may also be generated.

Radio frequency and IF stages generate IMD products, but only the odd-order products fall in-band; the even-order products fall around harmonics of the RF and IF frequencies, and are outside the passband. These are the stages with the highest signal levels, however, and can be the primary source of in-band odd-order IMD in the exciter.

In a transmitter, the highest levels of in-band IMD are usually generated by the power amplifier, with transmitter IMD limits close to actual power amplifier performance. This requires the exciter in-band IMD to be well below the transmitter requirement, so that the exciter and power amplifier combined will not exceed the limit. If the transmitter IMD limit is 35 or 40 dB below either of two equal output in-band signals, the exciter IMD is typically required to be no greater than –50 dB.

Noise. In-band noise appearing in the output of the exciter, if excessive, will degrade performance at the receiving end of the system. The noise may consist of broadband amplifier noise, discrete hum and power supply ripple components, and phase noise from injection signals. The hum components and phase noise modulate the signal, and are measured when the exciter has an output signal. Typical values of in-band noise, measured in a 3-kHz bandwidth, are 40 to 50 dB below the exciter output.

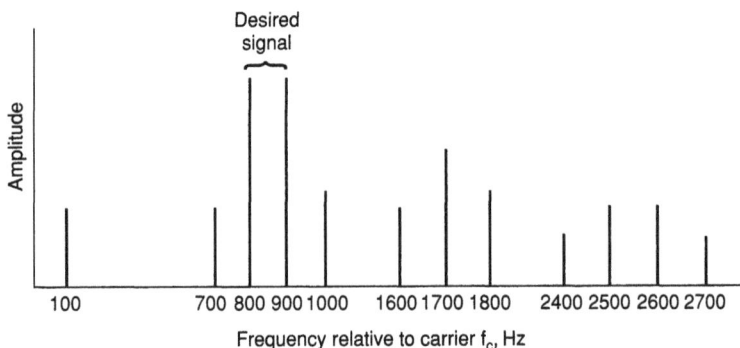

Figure 5.4 Possible in-band IMD products.

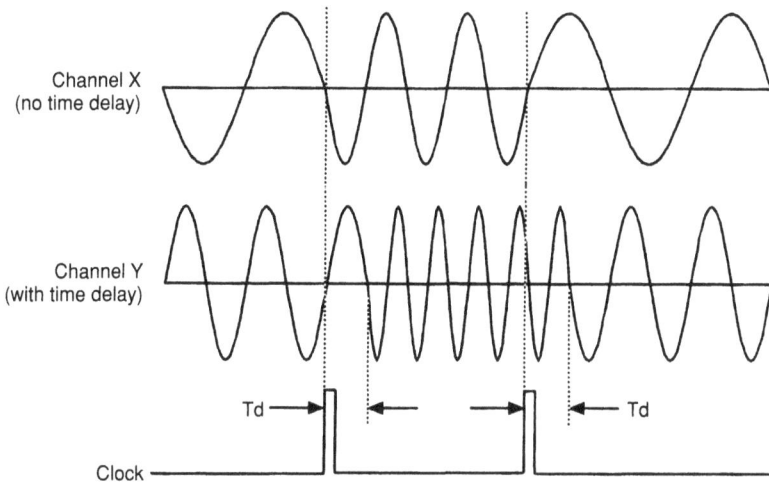

Figure 5.5 Effect of time-delay distortion.

Amplitude variation. This is the variation in RF output signal level as a function of baseband input frequency. The passband must be wide enough to accommodate the baseband bandwidth. For voice, a bandwidth of 2.5 to 3 kHz is commonly used, such as 300 Hz to 3 kHz. This bandwidth and frequency range also accommodate most single- and multitone data terminals producing FSK and differential phase-shift keyed (DPSK) tones. The 3-kHz bandwidth is also consistent with allowances of regulatory agencies for SSB transmission [1, 2].

The variation of amplitude response across the passband is also an important factor. For voice signals, variations of 2 to 4 dB are allowable. For data signals, data errors may increase at the receiver if a data tone is reduced in amplitude because of excessive amplitude variation in the exciter. A reasonable limit on passband variation for data signals is 2 to 3 dB, with some users requiring tighter limits.

Time-delay distortion. The effect of time-delay distortion is important to multitone data signals, as shown in Figure 5.5. The signals in two FSK channels are shown in the exciter output with modulation transitions separated by a time difference t_d. The time difference is caused by variations in group delay across the passband. If the time-delay difference (or time-delay distortion) is a significant part of the time 1/baud (per tone), modulation transitions will occur between clock intervals in the demodulator at the receiver, causing data errors to increase.

Typical data terminals operate at rates to 75 baud/channel, or modulation transitions 13 ms apart. Differential time delays of 500 to 700 μs maximum are normally allowed across that part of the passband occupied by the data signals.

Frequency stability. The frequency of the exciter output is determined by the various mixer injection frequencies along the exciter signal path. In modern exciters the injections are derived directly from a single reference frequency or by oscillators that are frequency locked to a reference signal. In either case the long-term frequency stability (in ppm) of the exciter output is the same as the reference.

The frequency reference is usually generated by a stable crystal oscillator, when the reference is part of the exciter. When better stability is required, reference oscillators using atomic devices are used. See Chapter 11 for more information on frequency standards.

Voice signals can tolerate frequency errors of about 20 Hz before the frequency shift becomes evident to the listener (see Chapter 2, section 2.1). Data signals vary in tolerance of frequency error, but some, utilizing narrow effective bandwidths per channel, may require tighter tolerances. As seen in Table 5.1, regulatory agencies impose limits apart from operational requirements. Typical frequency stability requirements are 0.5×10^{-6} or 1×10^{-8} for higher-performance equipment.

Phase stability. Voice signals are not affected by phase instability, but certain data signals are affected. An obvious example is a DPSK signal, where data is transmitted by phase shifting successive frames of the DPSK signal. If one or more of the mixer injections in the exciter introduces spurious phase changes from frame to frame, data errors may result. A typical allowable amount of short-term phase instability is 6° average per 13 ms, at 75 baud/channel, two channels per data tone (13 ms/frame, 45° phase-shift increments).

Exciter output signal level

The transmitter output power level is controlled by the exciter, which controls the RF drive level to the power amplifier, which is a fixed-gain amplifier. It changes the drive level in response to information fed back from power amplifier output level sensors.

The limitation on transmitter power output is normally the peak power output capability of the power amplifier. For single-frequency signals, the PEP and average power are the same, but for more complex signals the envelope peaks must be controlled so that the power amplifier output is maintained at an optimum level. The transmitter gain control (TGC) and automatic level control (ALC) functions are two ways to meet these requirements.

The TGC is a static gain control, while the ALC is dynamic. The TGC is set, when the transmitter is first tuned to a new frequency, by inserting a calibrated signal into the exciter output and adjusting the exciter output level for a desired power amplifier output. To do this, exciter gain is varied in response to sense signals fed back from the power amplifier. The calibration signal has a fixed relationship to the PEP during normal operation. However, the TGC alone is not sufficient to control the power amplifier PEP output.

The ALC controls the PEP output of the power amplifier. An envelope detector in the power amplifier senses the peak signal level and provides a gain control voltage back to the exciter. Its operation is similar to receiver AGC, responding quickly to signal peaks but decaying more slowly after the peaks have passed. This latter characteristic is needed to prevent distortion of the transmitter output caused by modulation of the desired signal by rapid variations of the ALC. This is discussed in more detail in section 5.3.

Ideally, ALC will "kick in" only when the baseband input has a high peak-to-average ratio. If ALC action is excessive, transmitter noise output will be high when no baseband input is present, such as during pauses in speech.

Audio compression amplifiers on microphone inputs can be used to reduce the wide dynamic range and high peak-to-average ratio of speech inputs. The compression amplifier introduces minor distortion on the voice signal because of the time constants and gain variations needed to be effective, but intelligibility is not reduced. If ALC in an RF or IF amplifier is used to perform this function, the resulting gain "pumping" will distort the data signals, degrading the data communications. In independent sideband (ISB) operation, where data and voice baseband inputs exist simultaneously, the data signal will be distorted by the speech signal acting on the ALC (see section 5.3).

Carrier levels. The carrier is suppressed in normal SSB operation, but it must be reinserted for certain emission modes or for tuning and setting up the transmitter output level. When the reinserted carrier is used to set gain its level has a fixed relationship to the PEP of the exciter output in normal operation. For example, if the power amplifier (and perhaps antenna coupler) is to tune up at half-power, the exciter carrier output during the setup or tune cycle will be 3 dB below the rated PEP output. A tune or gain setting level is chosen somewhat below maximum output in order to avoid power amplifier limiting or operation of protective circuits in the power amplifier or antenna coupler.

A particular emission mode may require transmitting a pilot carrier at a level, for example, 17 dB below transmitter PEP output. This level must be referenced to the tune power carrier level used during gain setup. In this example, if that is 3 dB below PEP, the pilot carrier level is set 14 dB lower.

Another commonly used mode that requires a reinserted carrier is amplitude modulation equivalent (AME). This carrier, in combination with an SSB signal, produces an envelope that can be received using an envelope detector. Maximum envelope amplitude occurs when the carrier is 6 dB below rated transmitter PEP. When the SSB signal and carrier are equal in amplitude, the PEP of the combined signals is at the rated PEP level. This carrier level is 3 dB below the tune power carrier used above.

An emission mode sometimes requiring a reinserted carrier is single-channel frequency shift keyed radio teletype (FSK RTTY). The carrier is shifted in frequency around the assigned carrier frequency. A common example is ±85 Hz. There is no baseband input; the only signal transmitted is the FSK carrier. Its level will be set to produce full PEP output. If CW operation is required using a keyed carrier, this carrier level must also produce full PEP output.

5.2 Exciter Architecture

A basic exciter block diagram is shown in Figure 5.6. The baseband input enters the baseband processing circuits, then goes to the modulator, where the first frequency translation takes place. The output of the modulator is the first IF. After filtering and amplification, the IF is upconverted to a second IF that is well above the HF range. After filtering in the second IF, the signal is downconverted by the output mixer to the desired HF.

This frequency translation scheme is the one most commonly used in modern exciters because of its simplicity and relative freedom from spurious emissions. The availability of crystal IF filters up to 100 MHz or more allows placing the second IF well above the output HF range, eliminating the presence of spurious second-IF signals in the exciter output. The filters also provide excellent rejection of the second mixer product $2 f_{01}$ removed from the desired product. Mixers and amplifiers for these frequencies are plenti-

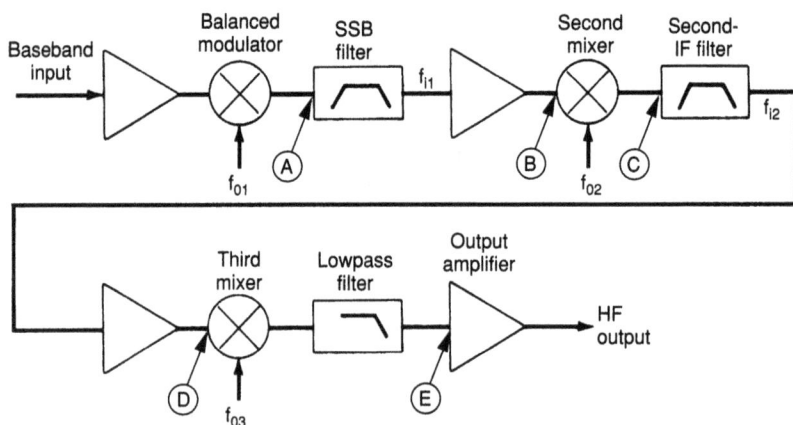

Figure 5.6 Basic exciter block diagram.

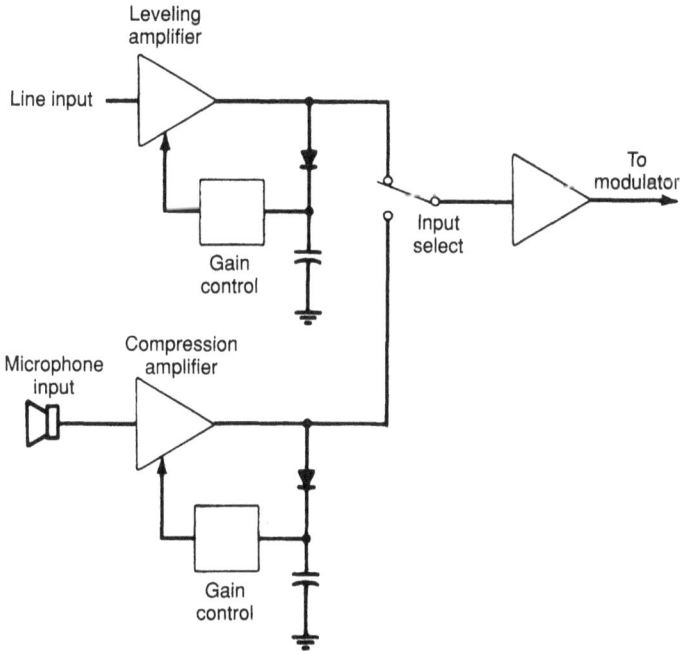

Figure 5.7 Baseband amplifier.

ful and have excellent performance. The first IF choice will usually be to accommodate an SSB filter design. For example, 500 kHz is a practical upper limit for mechanical filters.

The exciter architecture discussed in this chapter will use the upconversion approach. The basic blocks of the exciter will now be discussed in more detail.

Baseband amplifier and signal processing

The baseband signal processing section receives the baseband input and performs the necessary amplification, impedance changing, filtering, and other processing before it is modulated up to the first IF.

A block diagram of a typical baseband amplifier is shown in Figure 5.7. It has a 600-ohm line audio input that may range in level from –30 to +10 dBm, and a microphone input as low as –55 dBm. Each of these inputs passes through a gain-controlled amplifier. The purpose of the microphone amplifier is to compress the wide amplitude range of normal voice signals into a narrower range in order to increase the average power output of the peak-limited power amplifier. The desired effect is to maintain the average transmitter power output within 6 dB of the PEP output in speech transmissions.

Compression amplifiers are discussed further in Chapter 7. Preemphasis techniques are discussed in Chapter 2, section 2.1.

The function of the line-leveling amplifier is not to compress the range of the input signal, but to maintain a long-term relatively constant output level so that occasional peaks are at the desired PEP level. The input range for which leveling is desired will be 10 to 15 dB.

The outputs of the line-leveler and microphone amplifiers are connected one at a time by a switch into the output amplifier. This amplifier drives the output at the level and impedance needed by the modulator. Noise and IMD generated in these circuits should not be significant (see section 5.3), with IMD well below −60 dB if the amplifiers and other components are properly applied.

Filtering may be needed at this point to remove noise components below approximately 200 Hz, or noise at twice the first IF, which could be mixed into the modulator. These possibilities should be considered in the circuit design to determine if such filtering is needed. If additional speech processing is desired, Chapter 7 describes several other techniques.

Balanced modulator

The balanced modulator combines an LO injection with the input from the baseband amplifier and generates a DSBSC IF. The diagram in Figure 5.8 illustrates the process. The baseband signal into the modulator is f_b and the output IF signal has an upper sideband signal $f_{01} + f_b$ and a lower sideband signal $f_{01} - f_b$. The LO injection, or carrier at f_{01}, is suppressed below the

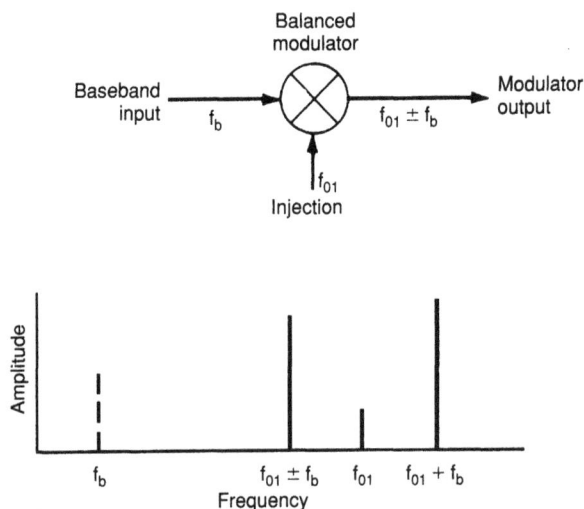

Figure 5.8 Balanced modulator output.

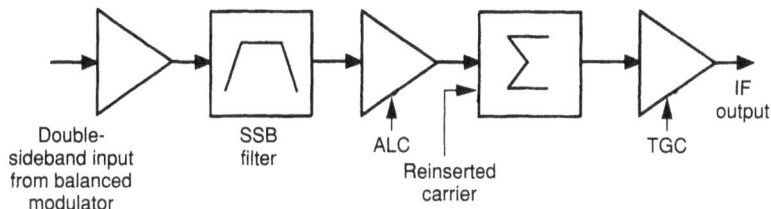

Figure 5.9 Intermediate frequency filter and amplifier.

two sideband levels by the balanced modulator. This is the conventional and most widely used method of producing the two SSB (USB and LSB) signals. An IF filter following the modulator will then select the desired sideband and also provide additional carrier rejection.

Suppression of the carrier in the output of the balanced modulator should be in the 35- to 45-dB range, and is achieved in the modulator by combining equal amplitude and opposite polarity injection signals. Careful design of printed circuit board layouts is needed to maintain isolation between modulator output circuits and LO injections to avoid compromising the inherent modulator balance.

The filter following the modulator selects the desired sideband. An alternative method uses phase shifting and combining techniques to produce an SSB output without filtering. It is used primarily when selection of the desired sideband by filtering is not desirable because of impractical filter bandwidth requirements, excessive time-delay distortion, etc. This modulator is discussed in section 5.3.

SSB IF filter and amplifier

The SSB filter and IF amplifier section of the exciter are shown in Figure 5.9. The double-sideband signal from the balanced modulator passes through an amplifier and then the SSB filter. The filter output contains only the selected USB or LSB signal, which then passes through additional linear amplifier stages to achieve the desired output level.

The SSB filter, because it is the narrowest selectivity in the exciter signal path, controls the amplitude and delay characteristics of the exciter passband. It also controls the level of the unwanted sideband and aids in carrier suppression. For a complete discussion of SSB filters, refer to Chapter 6.

The stages following the filter provide amplification, gain control of the exciter, and reinsertion of carrier, when needed. The gain control functions, as discussed previously, are the TGC and ALC. Referring again to Figure 5.9, the important thing to note is the relative position of TGC, ALC, and carrier reinsertion functions. Transmitter gain control is set up using a reinserted carrier of known level relative to PEP. The AME and pilot carriers, when enabled by the operating mode, are preset relative to the tune power level and are fixed, relative to rated PEP, once TGC is set. The gain from the point of carrier reinsertion should not be disturbed, once set, until the next tune cycle. For this reason the TGC function follows the point of carrier reinsertion while ALC precedes it.

Automatic level control controls the level of the SSB signal. Since ALC is a dynamic control, acting only after TGC is set, it should not affect the AME and pilot carrier levels.

Independent-sideband operation is achieved by adding a complete signal-processing path for a second independent baseband signal, as shown in Figure 5.10. In this case the added signal produces an LSB IF that is combined with the USB IF. Note that the IFs are combined before the ALC- and TGC-controlled stages, so the ALC acts on the combined signal. Another point is that the level of each IF should now be 6 dB lower than for just one sideband. This will keep the PEP of the combined signal within limits. The use of peak level control circuits in the baseband processing section (line levelers, microphone compression) is advantageous in ISB operation because it prevents one of the sideband signals from depressing the other by ALC action. For example, a strong voice signal in the USB channel will not depress or modulate a data signal in the LSB channel. The 6-dB reduction in IF level occurs before the point of carrier reinsertion, to avoid affecting carrier levels.

Four-channel ISB operation is sometimes required for very high capacity communications links. The arrangement of the four IF channels is shown in Figure 5.11. The inboard channels are arranged as in two-channel ISB operation, but the outboard channels have separate subcarriers, f_A and f_B. The injection signal into the modulator for the upper outboard channel is at frequency $f_B = f_{01} + 6.29$ kHz, and the injection into the lower outboard channel is $f_A = f_{01} - 6.29$ kHz. Thus, the lower outboard signal is a USB signal with respect to its subcarrier f_A, and the upper outboard signal is an LSB signal with

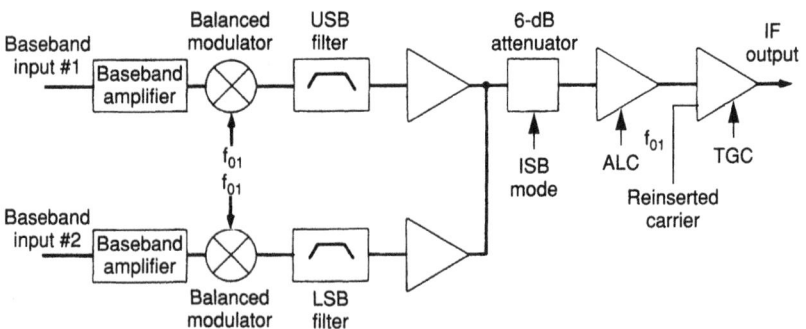

Figure 5.10 Independent sideband generation.

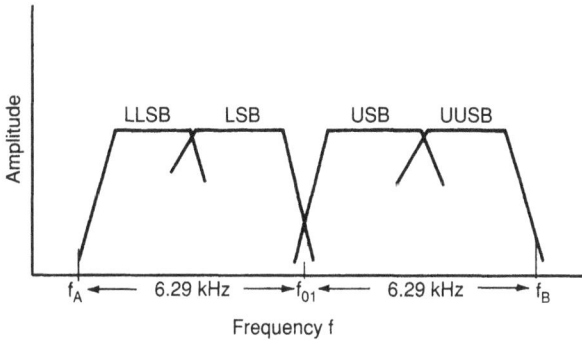

Figure 5.11 Intermediate frequency filter passband arrangement for four ISBs.

respect to its subcarrier f_B. The frequency spacing of the subcarriers from f_{01} used here is from current U.S. Department of Defense requirements. Note that other subcarrier frequencies may be required in a particular case.

The selectivity requirements are more severe for four-channel ISB operation than for SSB or two-channel ISB operation. For example, a filter used in this application has –2-dB passband edges at 250 and 3100 Hz from the carrier, and a –60-dB response at 3550 and –250 Hz. This places the passband edges of adjacent inboard and outboard channels a maximum of 90 Hz apart, requiring the very rapid reduction in response provided by the –60-dB specification. These filters required are discussed in Chapter 6.

Output frequency translator and amplifier

The purpose of this section of the exciter is to translate the output of the IF amplifier to the final output frequency and amplify it to the required level. Figure 5.12 illustrates the process. The signal from the IF amplifier output is translated up to a second, high IF in the second mixer. It is then filtered to remove the injection signal and unwanted mixer products, then mixed down to the desired IIF output frequency in a third mixer with a variable frequency injection input. The output of this mixer is passed through a low-pass filter with cutoff frequency above 30 MHz, to remove the injection signal and sum mix the product, and is then amplified to the required level.

The result of passing the first-IF signal through the second mixer is illustrated in Figure 5.13. The input from the first IF at frequency f_{i1} mixes with the LO injection f_{02} to produce the outputs shown. Not only are the two sidebands produced ($f_{02} + f_{i1}, f_{02} - f_{i1}$), but also sidebands around harmonics of the injection f_{02}. The LO injection frequency and its harmonics are also present in the output, even though it may be a high-quality double balanced mixer. The second-IF filter attenuates the unwanted products enough that they will not result in excessive spurious emissions in the exciter output.

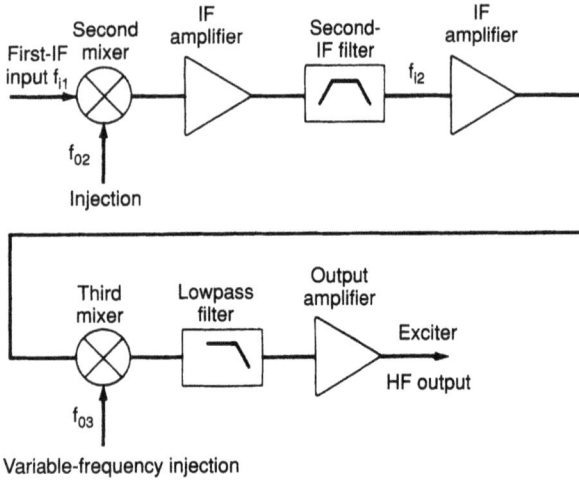

Figure 5.12 Output frequency translator and amplifier.

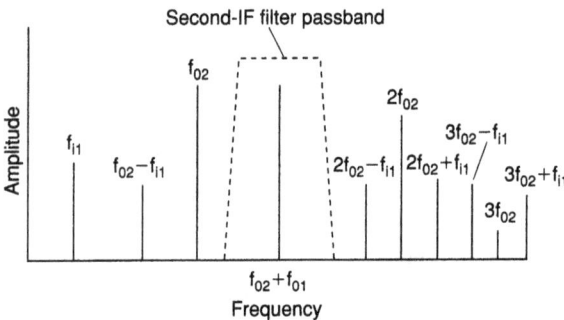

Figure 5.13 Second-mixer output signals relative to second-IF filter passband.

One example could be a double balanced diode mixer with injection f_{02} at 90 MHz at a level of +13 dBm. The first-IF input f_{i1} of 501 kHz is at −11 dBm PEP. The mixer conversion loss (per sideband) is 6 dB, and the rejection of the injection as it appears at the mixer output is 30 dB. The levels of the more important mixer output products can then be tabulated as shown in Table 5.2. The desired signal in Table 5.2 is the sum product, $f_{i1} + f_{02}$. The others may cause spurious signals in the exciter output and must be reduced by the second-IF filter.

The second-IF filter has a wider bandwidth that does not add significant selectivity or time-delay distortion to the desired signal. If the signals will be both USB and LSB, for example, the filter must be wide enough to pass both. It must also have enough selectivity to eliminate the unwanted mixer output signals listed in Table 5.2. A typical filter used for the second IF is a four-pole crystal filter with a center frequency of approximately 90 MHz. The –2-dB passband is 20 kHz, and the –60-dB stopband is 100 kHz wide. At 500 kHz below the passband, the selectivity is –70 dB.

To continue the example, we assume that the mixer outputs are amplified 7 dB in a broadband amplifier, then passed through a filter with selectivity characteristics as described above. The outputs of the filter, assuming a 4-dB passband loss, can then be tabulated as shown in Table 5.2. Note that the LO injection, which is only 500 kHz from the desired signal, is now –70 dB in relative level, and the unwanted difference product 1 MHz away is 70 dB below the desired signal. These unwanted products will now appear as spurious exciter outputs at these relative levels because they are too close in frequency to be filtered out after the second-IF filter output is downconverted to the final HF.

The filter output is amplified and fed into the output mixer at a level of –1 dBm. This mixer is also a double balanced diode mixer with an injection level of +23 dBm, at a frequency below 90 MHz. The mixer output level of –7 dBm passes through a 30-MHz lowpass filter to remove the injection frequency f_{03} and is then amplified to the proper exciter output level. There is further discussion of the IMD, spurious signal, and noise characteristics of the output translator in the next section.

Gain distribution, IMD, and noise

This section discusses methods of analyzing IMD and noise and for determining the proper gain distribution.

In-band IMD. As a multitone signal propagates through the exciter from baseband input to RF output, each stage along the path contributes to the level of in-band IMD. The problem is to maintain the signal level at critical points in the path at optimum levels so that excessive amounts of IMD are not generated by overdriving particular circuit elements. Also, signal levels

TABLE 5.2 Second-Mixer Outputs and Effect of Second-IF Filter

Second-mixer output	Frequency, MHz	Second-mixer output, dBm	Second-IF filter output, dBm
Injection, f_{02}	90	–17	–84
Sum product, $f_{02} + f_{if}$	90.501	–17 (PEP)	–14
Difference product, $f_{02} - f_{if}$	84.499	–17 (PEP)	–84
Injection second harmonic, $2 f_{02}$	180	–22	–89
Injection third harmonic, $3 f_{02}$	270	+3	–84
Injection fourth harmonic, $4 f_{02}$	360	–23	–90
Injection fifth harmonic, $5 f_{02}$	450	–1	–68

must not be too low, raising the output noise level and causing inefficient use of amplifiers. A method will be described that establishes signal levels at components that are critical (because of their IMD characteristics), then accounts for the contribution of each of these critical components to the total exciter output IMD.

Refer again to Figure 5.6, the exciter signal path. Certain critical points in the path are identified. The first critical point, A, is the input to the first-IF filter. The filter is critical because, as signal level is increased, it will be the first component to produce significant IMD. Preceding components (amplifiers, modulators) will be insignificant contributors if they are properly applied. It will be assumed that the IMD characteristics at A are due not only to the input signal to the filter, but also to the cumulative effect of preceding circuits. The critical point B is the input to the second mixer, whose IMD characteristic is related to its injection level. Point C is the input to the second-IF filter, which has definite upper limits on the signal level that can be applied. Point D is the output mixer and point E is the output amplifier, which will have IMD limitations at its output power level. At each of the critical points the level is determined by the IMD characteristics of that component, including the contribution of all circuitry back to the preceding critical point. The levels used will be the level per tone of a two-tone signal that produces the rated exciter PEP output.

An example will illustrate the method. The output requirement for the exciter will be assumed to be +20 dBm PEP, with in-band third- and higher-order IMD products each at most –50 dB. Considering the IF filter at point A, the allowable IMD level will be set at –65 dB. For the IF filter to be used, an input limit of –23 dBm/tone will be assumed.

The mixer at point B is a double balanced diode mixer with an injection level of +13 dBm, so an input third-order intercept of +18 dBm will be assumed. (See the material on mixers in section 5.3.) Input signal levels 35 dB lower, –17 dBm/tone, will produce third-order products (in the mixer and preceding circuits back to point A) of –70 dB. Point C, the second-IF filter, has a maximum input level of –10 dBm/tone, which is expected to produce third-order products suppressed to –65 dB. Note that neither of the IF filters can be characterized in terms of an input intercept. The IMD characteristics of mechanical or crystal filters are not predictable by an intercept concept, which requires empirically determining the allowed signal level at the desired IMD level. This is discussed further in Chapter 6.

Next is the output mixer, point D, which is a double balanced diode mixer with an injection level of +17 dBm, with a third-order input intercept of 25 dBm. The input signal level will be set 35 dB lower at –10 dBm/tone, producing third-order products suppressed to –70 dB. Finally, the output amplifier, point E, must provide the output power of +20 dBm PEP, or +14 dBm/tone. If the output intercept of the amplifier is +44 dBm, it will contribute third-order IMD products of –60 dB.

TABLE 5.3 Exciter Gain and IMD Performance

Critical point	Signal level, dBm/tone	Net gain, dB	Input 3d-order intercept, dBm	IMD-generated, dB	Sum of IMD products, dB
A	−23	6	—	−65	−65
B	−17	7	+18	−70	−61.1
C	−10	0	—	−65	−56.9
D	−10	−6	+25	−70	−55.3
E	−16		+14	−60	−51.3
Output	14 dBm	30			

The operating conditions that have been established at each of the critical points can now be tabulated and the gain distribution and total IMD performance determined. In Table 5.3, the signal levels and third-order input intercepts are entered. The net gain between critical points is the difference in signal levels. The IMD generated at a critical point is entered and the last column is a running total of the IMD level at each point. This is obtained by adding the IMD generated at a point to IMD input levels from preceding circuits. Figure 5.14 shows a curve that facilitates adding the IMD components. The larger of the two components being added is normalized to 0 dB. The relative level of the other, lower level, intersects the curve at a point representing the amount (in decibels) by which the sum exceeds the referenced level. This curve adds the two components in-phase, such as is most likely to occur in a broadband circuit.

The designer may not wish to use this worst-case approach when adding IMD components in filters. Their phase-shift characteristic makes in-phase IMD addition unlikely.

The bottom entry in the last column of Table 5.3 shows that the −50-dB level requirement will be met. These entries would also show where large in-

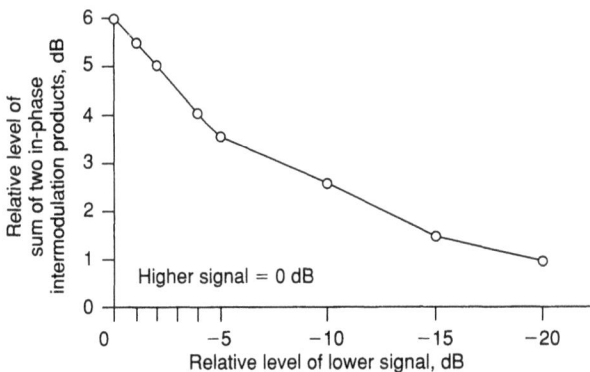

Figure 5.14 In-phase sum of two identical-frequency IMD products.

creases in IMD occur, indicating circuits that may need optimization, or perhaps where the design is too conservative. The process can, of course, be refined to add as many other significant points in the signal path as desired.

The gain values used should correspond to maximum gain. Transmitter gain control (TGC) or ALC action should only reduce the above levels. Intermodulation distortion caused by PIN diode attenuators commonly increases just at turn-on and should be checked against the allowable levels established in Table 5.4.

Out-of-band spurious signals. Spurious signals that are outside the exciter passband, that is, 20 kHz and farther off the output carrier frequency, will be generated between the output of the last-IF filter and the exciter output. The allowable level of these spurious signals at the exciter output will depend on how much additional selectivity is available in the power amplifier and the transmitter output spurious signal limitations. The power amplifier may have suboctave filters providing selectivity primarily at harmonics, or more selective bandpass circuits, or may be broadband with no selectivity. The transmitter spurious output limitations may be those of a regulatory agency, or they may be more severe requirements imposed by interference considerations with collocated receivers.

Referring again to Figure 5.6, the IF amplifier following the filter will generate second-, third-, and higher-order products, but they will be clustered around harmonics of the IF (neglecting in-band products). A simple low-pass filter in the output of the amplifier should reduce these products to the point that they can safely be ignored.

The output mixer generates spurious signals that are calculated as sum products, such as second and third order, of a two-tone in-band signal, and as harmonics of these signals. If the two equal-level in-band signals into the mixer are at frequencies f_1 and f_2, the so-called second-order sum product in the mixer output is at frequency

$$(f_1 - f_{03}) + (f_2 - f_{03}) = f_1 + f_2 - 2f_{03} \qquad (5.1)$$

TABLE 5.4 Third-Mixer Outputs

Mixer output signal	Level, dBm	Frequency, MHz
Desired signals	−16	2.001, 2.002
Second-order sum product	−78	4.003
Second harmonics	−84	4.002
	−84	4.004
Third-order sum products	−89	6.004
	−89	6.005
Third harmonics	−98	6.003
	−98	6.006

TABLE 5.5 Third-Mixer and Output Amplifier Contributions to Exciter Output

Exciter output signals	Third-mixer contribution, dBm	Output amplifier contribution, dBm	Exciter output levels, dBm
Desired signals	—	14.00	14.00
Second-order sum product	–48.00	–37.00	–34.84
Second harmonics	–54.00	–31.00	–30.41
Third-order sum products	–59.00	–46.00	–44.25
Third harmonics	–68.00	–42.00	–41.58

which is actually a fourth-order mixer product. For example, if $f_1 = 90.501$ MHz and $f_2 = 90.502$ MHz, and the desired output carrier frequency is 2 MHz, then $f_{03} = 88.500$ MHz, and the second-order sum product frequency at the mixer output is 4.003 MHz. The desired output frequencies due to f_1 and f_2 are 2.001 and 2.002 MHz, respectively. Similarly, the third-order sum products at the mixer output are actually sixth-order mixer products whose frequencies are given by

$$2\left(f_1 - f_{03}\right) + \left(f_2 - f_{03}\right) = 2f_1 + f_2 - 3f_{03} \qquad (5.2a)$$

and

$$\left(f_1 - f_{03}\right) + 2\left(f_2 - f_{03}\right) = f_1 + 2f_2 - 3f_{03} \qquad (5.2b)$$

Some of the more troublesome spurious signals generated in the mixer are tabulated in Table 5.4 using levels that may occur in a double balanced diode mixer. The injection level is +17 dBm and the IF input is –10 dBm, with frequencies f_1, f_2, and f_{03}. Output signals outside the 2- to 30-MHz range are not considered.

The output amplifier is also a source of out-of-band IMD. The amplifier will be assumed to have a third-order input intercept of +14 dBm and a second-order input intercept of +35 dBm. Table 5.5 lists the contribution of the mixer products to the exciter output spurious signals, the amplifier contribution, and the total. The two contributing sources are assumed to add in-phase. The levels shown in Table 5.5 for mixer contribution are for the mixer output referred to the amplifier output (add the amplifier gain to the mixer output levels shown in Table 5.4).

In the above example, the mixer is a significant contributor to the third-order products, while the amplifier controls the second-order IMD levels.

There are other spurious signals that can be generated, caused by the second-mixer injection frequency f_{02} leaking into the output mixer. These spurious signals are at frequencies

$$f_s = m\,f_{02} + n\,f_{03}$$

where f_s is the spurious frequency, and m and n are positive or negative integers. Figure 5.15 is a plot of output frequencies versus desired output frequency for the above example. Shown are the fourth-, fifth-, sixth-, and seventh-order products of the two injection frequencies used in the above example. The results will be different for various frequency schemes so the designer should check each case. The possible leakage paths for the injection signal at f_{02} and its harmonics include the second-IF filter and may also involve inadequate shielding and filtering. It is also found that using high-side injection, where both f_{02} and f_{03} are above the second IF, will produce fewer spurious signals.

Noise. The in-band noise at the exciter output comes from amplifiers and other active circuits in low-level stages, hum and power supply ripple components, and phase noise transferred to the signal from mixer injections. The noise from low-level stages should not be significant if the baseband and low-level IF stages are properly designed (see the material on baseband components in section 5.3). Hum components may be induced on the signal by the magnetic fields of power transformers or blower motors. A common problem is ac induced in the control line of a voltage-controlled oscillator in the frequency synthesizer. Physical separation and the use of steel or other magnetic shielding may reduce the effect. Equipment in an adjacent cabinet can also be a source of interfering fields.

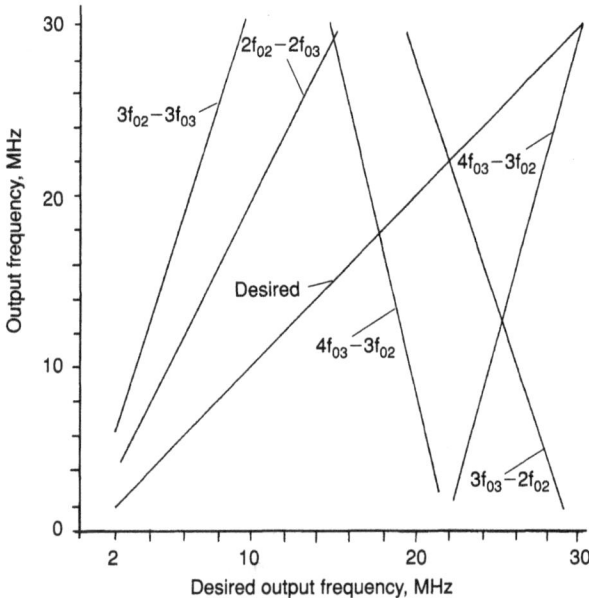

Figure 5.15 Spurious output signals caused by combining of two injections f_{02} and f_{03}.

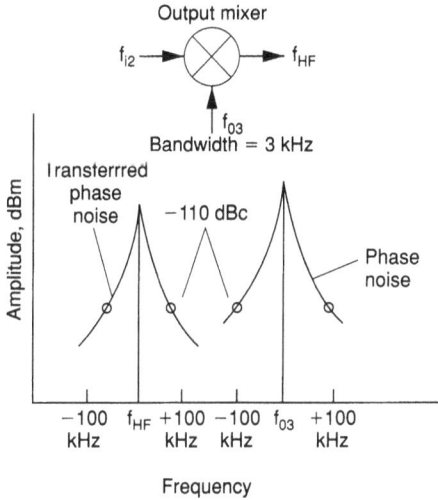

Figure 5.16 Transferal of phase noise on variable-frequency injection f_{03} to mixer output signal.

Phase noise on mixer injections affects both in-band and out-of-band noise because it is primarily associated with the variable-frequency oscillator (VFO) injection of the output mixer. The phase noise on the VFO is transferred directly to the signal passing through the mixer and is not reduced or affected by mixer balance. Figure 5.16 shows an injection f_{03} with phase noise superimposed. A mixer input at frequency f_{i2} produces an output at f_{HF} which has the phase noise transferred to it. Typical levels of this noise, relative to the VFO level, may be –50 dB in a 3-kHz band around f_{03}, –90 dB at 10 kHZ removed either side, and –110 dB removed 100 kHz, all measured in a 3-kHz bandwidth. If the in-band noise from low-level stages is at least –60 dB, phase noise will be the predominant source of in-band noise, which is usually the case. If the exciter output is at +20 dBm, the out-of-band noise output due to phase noise will be at –90 dBm at 100-kHz off-frequency.

Another source of out-of-band noise in exciters is the thermal and excess noise of the circuits between the output of the second-IF filter and exciter output. Figure 5.17 shows their effect. The noise factor of the cascade can be found by lumping together the two filters and the mixer into a single block with a noise figure of 8 dB (a gain of –8 dB). We then apply Friis's formula for a three-stage system

$$F_T = F_1 + \frac{F_2 - 1}{G_1} + \frac{F_3 - 1}{G_1 G_2} \tag{5.3}$$

but

$$G_2 = \frac{1}{F_2}$$

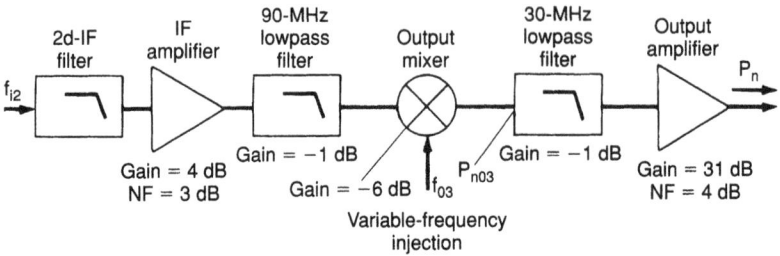

Figure 5.17 Noise sources in exciter output stages.

therefore

$$F_T = F_1 + \frac{F_2 - 1}{G_1} + \frac{F_2\,(F_3 - 1)}{G_1} = F_1 + \frac{F_2 F_3 - 1}{G_1} \tag{5.4}$$

Using the numbers in Figure 5.17, the total noise factor is 7.9 (9 dB). The output noise in a 3000-Hz bandwidth is then

$$N_{out}\ (\text{dBm}) = k\,T\ (\text{dBm}) + \text{NF}\ (\text{dB}) + 10\,\log_{10}\ (\text{BW}) + \text{Gain}\ (\text{dB})$$
$$N_{out}\ (\text{dBm}) = -174 + 9 + 10\,\log_{10}\ (3000) + 27 = -103.2\ \text{dBm} \tag{5.5}$$

The program in Chapter 17, NFIIP.EXE, for cascaded noise figures and intercept points can also be applied to problems of this kind. This noise is 36.0 dB greater than thermal noise and will be broadband in character, assuming a flat frequency response from 2 to 30 MHz. This compares with the –90 dBm due to phase noise, making phase noise predominant at 100-kHz frequency separation. Phase noise continues to decrease at greater frequency separations, but is dominant to wide frequency separations. Wideband amplifier noise often eventually overtakes phase noise.

Excess noise P_{n03} on the VFO in the 2- to 30-MHz range may require high-pass filtering in the VFO path. If this noise level is at –120 dBc (3-kHz bandwidth), for example, and mixer balance reduces it another 25 dB, the level at the mixer output would be –145 dBc, or –128 dBm for a VFO level of +17 dBm. It will be assumed that a small amount of filtering is used and that the contribution of P_{n03} in the exciter noise output is insignificant.

If the exciter is to be used in a transmitter that is collocated with receivers, the noise output of –110 dBc may be excessive. Obviously the VCO in the frequency synthesizer should be as clean as possible. Amplifier noise should be made insignificant by using a low-noise first stage and keeping the overall gain as low as possible. It may be necessary to add selectivity to the output amplifier to reduce off-frequency noise, and this can be complex and costly.

5.3 Circuit Components

Circuit components in this context are functional entities such as modulators, amplifiers, and gain control elements. Proper design and application of these components is critical to achieving the desired exciter performance.

Baseband components

The signal-to-noise ratio established at the baseband signal input must exceed the desired exciter output signal-to-noise ratio. The input signal-to-noise ratio is set by the level applied to the input of an amplifier and the internal noise level of that amplifier.

Figure 5.18A shows a typical balanced, transformerless line input amplifier. The potentiometer R1 is a termination for the line as well as a gain control. The amplifier is a JFET input operational amplifier with low input noise voltage and currents and excellent common-mode rejection. It will be

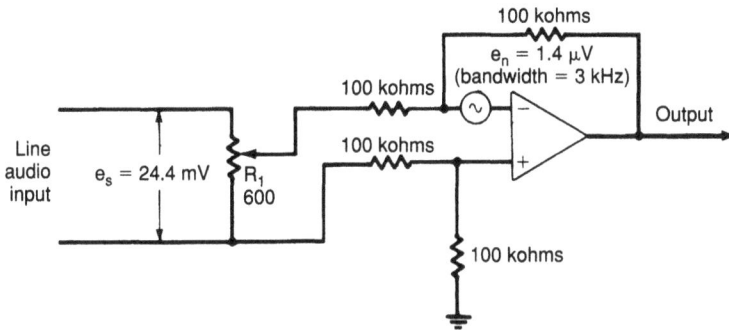

A

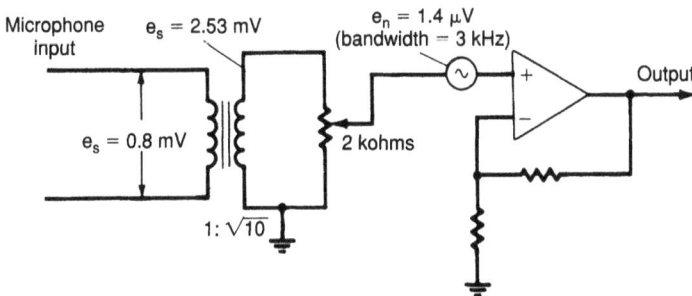

B

Figure 5.18 (A) Establishment of signal-to-noise ratio at the baseband line input amplifier. (B) Use of input transformer to improve signal-to-noise ratio.

assumed in this example that the equivalent input noise current of the operational amplifier is 0.01 pA/(Hz)$^{1/2}$, and the equivalent input noise voltage e_n is 25 nV/(Hz)$^{1/2}$. In a 3-kHz bandwidth, the noise current induces insignificant noise voltage in the source impedance of 50 kohms (< 0.1 µV), while the amplifier noise voltage is 1.4 µV.

If the line input in this example is a minimum of –30 dBm, $e_s = 24.5$ mV. The output signal-to-noise ratio will then be

$$20 \log \left(\frac{24.5 \times 10^{-3}}{1.4 \times 10^{-6}} \right) = 85 \text{ dB}$$

which is certainly adequate if the required exciter output signal-to-noise ratio is, say, 50 dB.

This amplifier could also be used as a preamplifier for low-level microphone inputs. A typical dynamic microphone input level is –55 dBm at 200 ohms, or 0.8 mV. This signal, applied directly to the above amplifier, would produce a signal-to-noise ratio of 55 dB. If a higher signal-to-noise ratio is needed an input transformer could be used that steps up the input voltage. Figure 5.18B shows a transformer with a 2-kohm to 200-ohm impedance ratio that provides this step up. The secondary signal voltage is applied to the amplifier input at full gain, giving a signal-to-noise ratio of 65 dB.

Distortion caused by this amplifier will be very low. Total harmonic distortion, with a voltage gain of 10 or less, should be less than 0.03% (< -70 dB). Second- and higher-order IMD products will be similarly suppressed.

Compression amplifiers and line-leveling amplifiers. Compression amplifiers and line-leveling amplifiers are similar in that they are AGC amplifiers and maintain constant output. The difference between the two is in the range of gain change and the time constants of the gain control action.

Figure 5.19 illustrates an AGC amplifier. The transistor Q1 is a JFET which acts as a variable resistor. The output of the amplifier is envelope rectified and the dc voltage stored in capacitor C1. The gain control amplifier varies the gate voltage of Q1, maintaining a constant output, because R1 and Q1 operate as a variable voltage divider. The voltage across C1 builds up quickly, follows signal peaks, and decays through R2.

The purpose of a line-leveling amplifier is to make minor adjustments in baseband amplifier gain to compensate for small changes in input line level. This is desirable in order to keep the transmitter operating at full-rated PEP. The range of gain adjustment will be 10 to 15 dB. The attack time (charge time of C1) must be fast enough to respond to signal peaks, on the order of 5 ms. But the decay time (discharge of C1 through R2) must be slow enough that the gain does not follow the signal envelope. To do so would introduce distortion, as discussed later for ALC operation. Since the line input signals may be data as well as voice, it is important that distortion

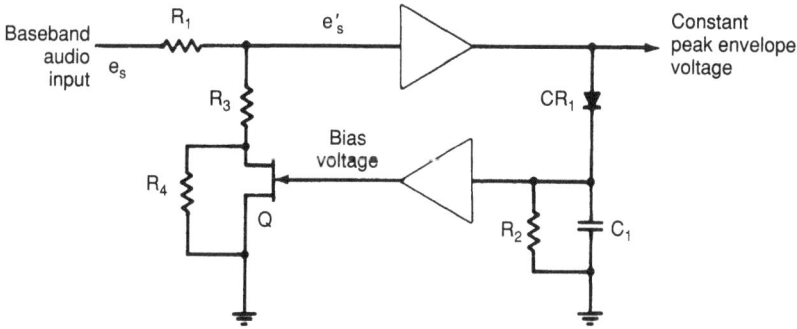

Figure 5.19 Automatic gain control audio amplifier.

not be introduced by the line leveler. The decay time should be relatively long, 5 to 20 sec.

The purpose of the microphone compression amplifier is to maintain a high average level of transmitter power output when wide swings of input voice level occur. This means that the attack time and decay time are both relatively fast, to follow the dynamics of normal speech. Compression amplifiers are discussed in Chapter 7.

Voice-operated transmit keying. Voice-operated keying (VOX) is a method of keying the exciter without operating a separate push-to-talk or other keying switch. It is especially useful when the user is patched in from a telephone line and does not have access to a key switch.

A VOX keying circuit is shown in Figure 5.20. The transmit audio is passed through a variable-gain amplifier, then detected. The dc output voltage is applied to the positive (+) input of a voltage comparator. When this input is more positive than the dc voltage applied to the negative (−) input, the comparator output turns on Q1, pulling the key line down to the (in this case) transmit condition.

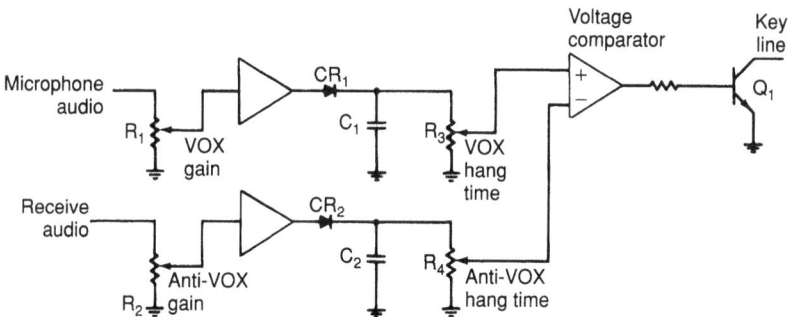

Figure 5.20 Circuit for voice-operated transmitter keying.

The dc voltage applied to the negative input of the comparator is derived from the receive audio output, which is most readily available in a transceiver. This is an anti-VOX function, to prevent false keying due to microphone pickup of a speaker output. Both the VOX and anti-VOX amplifier/detector have gain controls R1 and R2 to allow for adjustment for proper keying operation.

Associated with the VOX and anti-VOX detectors are decay-time adjustments R3 and R4. The purpose of these is to vary the hang time of the keying operation. When the transceiver is keyed and the voice input ceases, the adjustment of R3 determines how long the exciter will remain keyed with no voice input. It should be long enough that the key line is not released on short pauses or spaces between words. The anti-VOX hang time is also adjustable and should be long enough that the circuit sensitivity does not quickly increase, causing keying on background noise. The adjustment range of the hang time should cover 0.2 to 5 sec.

SSB modulators

Balanced modulator. By far the most popular form of SSB modulator is the balanced modulator, used in conjunction with a USB or LSB IF filter to remove the unwanted sideband. Two popular types of balanced modulators are the double balanced diode mixer and an integrated circuit implementation (an industry standard, type 1496) of a switching-type Gilbert cell modulator. This chip combines excellent carrier suppression, linearity, and moderate gain. Numerous descriptions of both types are available in manufacturers' data sheets.

Phasing modulator. Occasionally it is desirable to use a phasing-type modulator because of inadequate filter frequency response, excessive group delay distortion, or some other shortcoming of the balanced modulator and filter technique.

A simplified block diagram of the phasing modulator is shown in Figure 5.21. The baseband signal is applied to two balanced modulators, with the signal applied to one shifted from the other by 90°. The LO injection to one of them is also shifted from the other by 90° using digital circuits. The outputs of the two balanced modulators are then combined, resulting in the cancellation of one of the sidebands, because the components of that sideband in the two modulator outputs are out of phase.

The output signal of modulator A can be represented as

$$e_{0A} = K_A V_1 [\cos \ (\omega_c + \omega_1) t + \cos \ (\omega_c - \omega_1) t] \qquad (5.6)$$

and of modulator B as

$$e_{0B} = K_B V_1 \left\{ \cos \left[\left(\omega_c t + \frac{\pi}{2} \right) + \left(\omega_1 t + \frac{\pi}{2} \right) \right] + \cos \left[\left(\omega_c t + \frac{\pi}{2} \right) - \left(\omega_1 t + \frac{\pi}{2} \right) \right] \right\}$$

$$= K_B V_1 [\cos (\omega_c - \omega_1) t - \cos (\omega_c + \omega_1) t] \qquad (5.7)$$

where K_A, K_B = constants of proportionality between modulator output and
baseband input

V_1 = peak baseband voltage

ω_c = modulation injection angular frequency

ω_1 = baseband signal angular frequency

It is obvious that if $K_A = K_B$, adding together the above two modulator output signals, Equations 5.6 and 5.7 will result in cancellation of the upper sideband $\omega_c + \omega_1$. If we want to remove the lower-sideband signal, take the difference of e_{0A} and e_{0B} by shifting one of them 180°.

The effect of imperfect 90° phase shift in the baseband signal path to modulator B or the injection signal will be found by adding phase errors δ and Δ, respectively, to Equation 5.7.

$$e_{0B} = K_B V_1 \left\{ \cos \left[\left(\omega_c t + \frac{\pi}{2} + \Delta \right) + \left(\omega_1 t + \frac{\pi}{2} + \delta \right) \right] \right.$$

$$\left. + \cos \left[\left(\omega_c t + \frac{\pi}{2} + \Delta \right) - \left(\omega_1 t + \frac{\pi}{2} + \delta \right) \right] \right\} \qquad (5.8)$$

$$= K_B V_1 \{ -\cos [(\omega_c + \omega_1) t + \Delta + \delta] + \cos [(\omega_c - \omega_1) t + \Delta - \delta] \}$$

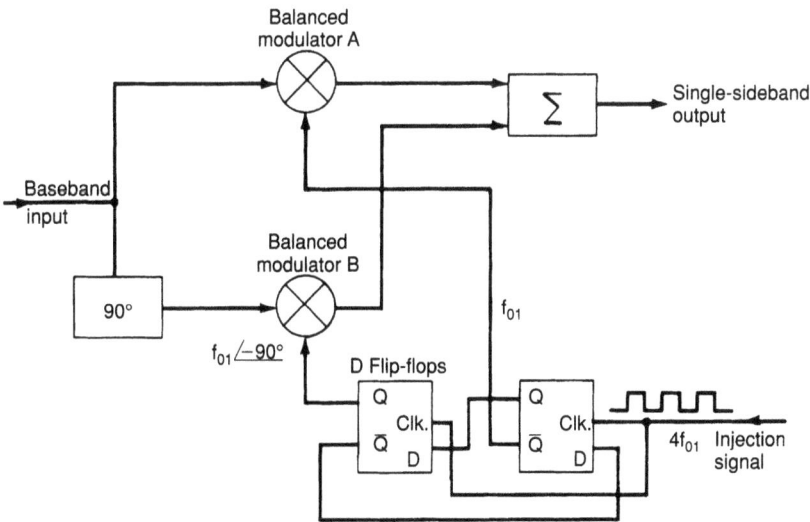

Figure 5.21 Phase-shift-type SSB modulator.

Combining the outputs of the two modulators gives

$$e_0 = e_{0A} + e_{0B} =$$

$$V_1 \sqrt{K_A^2 + 2K_A K_B \cos{(\Delta - \delta)} + K_B^2} \quad \cdot$$

$$\sin\left[(\omega_c - \omega_1)t + \tan^{-1}\left(\frac{K_A + K_B \cos{(\Delta - \delta)}}{K_B \sin{(\Delta - \delta)}}\right)\right]$$

$$+ \tag{5.9}$$

$$V_1 \sqrt{K_A^2 - 2K_A K_B \cos{(\Delta + \delta)} + K_B^2} \quad \cdot$$

$$\sin\left[(\omega_c + \omega_1)t - \tan^{-1}\left(\frac{K_A + K_B \cos{(\Delta + \delta)}}{K_B \sin{(\Delta + \delta)}}\right)\right]$$

This is the general expression for the combined modulator outputs. The amount of suppression a of the unwanted sideband ($\omega_c + \omega_1$) is given by

$$a = 10 \log\left(\frac{K_A^2 + K_B^2 - 2K_A K_B \cos{(\Delta + \delta)}}{K_A^2 + K_B^2 + 2K_A K_B \cos{(\Delta - \delta)}}\right) \tag{5.10}$$

This expression can be used to evaluate the effects of amplitude imbalance and phase-shift errors.

Figure 5.22 shows the degree of suppression of the unwanted sideband for various combinations of phase and amplitude error. For simplicity, it is assumed that only the baseband signal has phase-shift error, and amplitude imbalance is given as the ratio K_A/K_B or K_B/K_A, whichever makes the ratio less than 1.0.

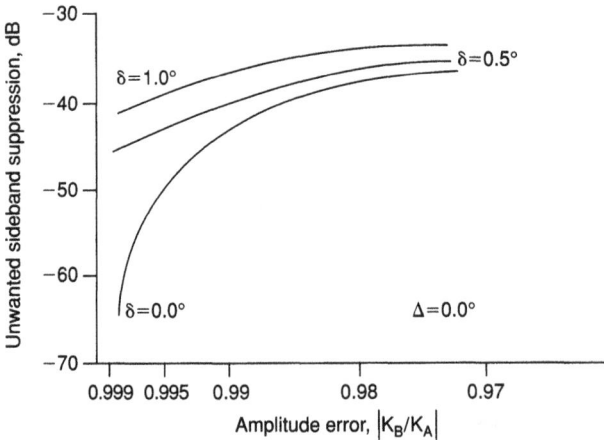

Figure 5.22 Suppression of unwanted sideband versus amplitude error.

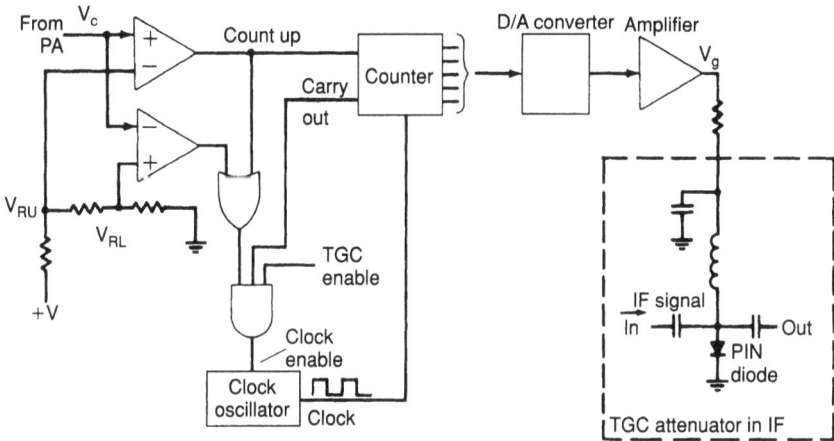

Figure 5.23 Transmit gain control circuit.

The biggest problem in designing phase-shift modulators is to provide an accurate 90° phase difference between the baseband inputs to the two modulators. This will not be further explored here, but there are some excellent techniques available that are described in references 3 and 4. The 90° phase shift of the injection signal can be quite accurately controlled by using flip-flops, as illustrated in Figure 5.22. At injections below 1 MHz, it should be possible to maintain a 90° shift to less than 1°.

Transmitter gain control methods

Transmitter gain control. The function of the TGC is to adjust the overall gain of the exciter and power amplifier during a tune-up or gain-setting cycle so that a reference carrier level produces a predetermined transmitter power output level.

Figure 5.23 illustrates a circuit for achieving this function. In general, the power amplifier generates a control voltage V_c that is positive and proportional to output power when driven by a single tone. A gain control circuit then adjusts gain in small steps at a fixed rate until the output power is properly set. In the diagram, V_c is determined by a dual-voltage comparator to be above or below the reference voltages V_{RU} and V_{RL}, which correspond to the upper and lower limits, respectively, of V_c, at which gain-setting action stops. The width of the window set by V_{RU} and V_{RL} may, for example, correspond to 0.5 dB in power output.

If V_c is above V_{RU}, a count-up command is applied to a counter and a clock oscillator is activated. The counter counts up at the clock rate and the counter outputs, which drive a D/A converter, causing the analog voltage V_g to increase, increasing attenuation in a PIN diode attenuator located in the

IF amplifier. If V_c is below V_{RL}, the counter will count down, decreasing attenuation in the attenuator. When the gain-change action causes V_c to fall between V_{RU} and V_{RL} the clock oscillator is shut off, stopping the counter. The enable signal is removed (by a separate timing process) and the counter maintains its output, holding exciter gain constant, until reappearance of the enable signal.

The range of TGC needed will depend on the expected worst-case gain variation of the power amplifier and exciter over the frequency range, as well as the variation in interconnect losses between the exciter and power amplifier. If the TGC range is, say, 15 dB, then an 8-bit counter (256 steps) will provide adequate resolution.

Automatic level control. Automatic level control is a dynamic function that operates while the transmitter is in normal operation. The primary function of ALC is to prevent overdriving the power amplifier on signal peaks. To achieve this, a peak detector in the power amplifier feeds an analog voltage back to the exciter that adjusts the gain of the IF amplifier. This voltage V_{ALC} is generated only when the power amplifier output exceeds a predetermined level which usually corresponds to its maximum allowable PEP.

As seen in Figure 5.24, the circuit concept is very simple: V_{ALC} is applied to an amplifier, causing the voltage V_a to increase, increasing attenuation in the PIN diode ALC attenuator in the IF amplifier. The capacitor C1 discharges or decays slowly through resistor R1 and the PIN diode when V_{ALC} drops, letting the gain increase.

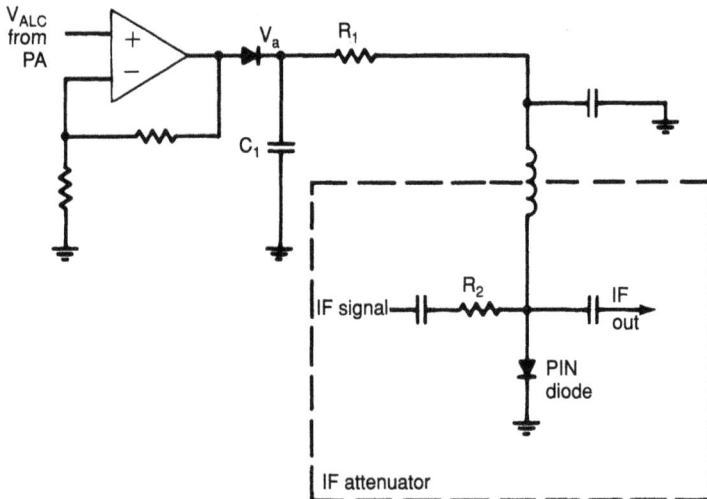

Figure 5.24 Automatic level control circuit.

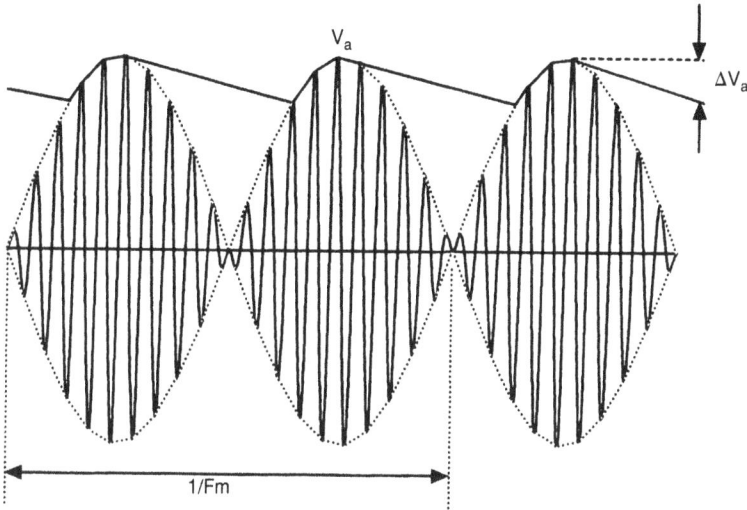

A. Two-tone PA output

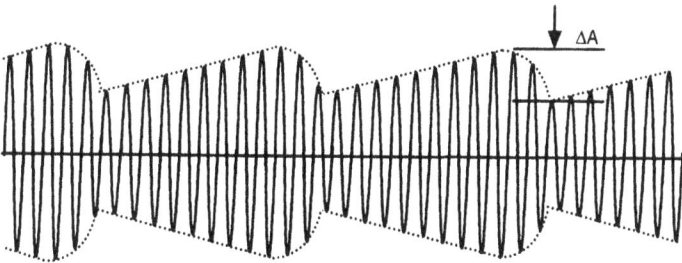

B. One-tone IF signal

Figure 5.25 (A) How ALC gain control voltage V_a tends to follow a two-tone signal envelope. (B) Modulation of each tone of the IF signal by ALC action.

An important consideration is the effect of the ALC action in producing distortion in the transmitter signal. The V_{ALC} follows the signal envelope, assuming fast attack and decay times of the power amplifier detector circuit. If the voltage V_a is allowed to follow the signal envelope, as for a multitone signal, AM distortion results. The time constant R1C1 must be long enough to reduce this distortion to acceptable levels.

Figure 5.25A illustrates a two-tone signal in the power amplifier output and the action of V_a in following the envelope. A variation $\Delta V_a(t)$ occurs, causing a change in IF gain ΔG

$$\Delta G = -K\Delta V_a(t) \tag{5.11}$$

where ΔG is the change in voltage gain in the attenuator, K is a positive constant for a narrow part of the gain control range, and the negative sign results because an increase in $V_a(t)$ causes a decrease in G.

Figure 5.25B shows how the resulting change in gain of the PIN diode attenuator would amplitude modulate each tone of the IF signal as it passes through the attenuator.

If the peak amplitude of the IF input signal is A, then

$$\Delta A = -A \, K \Delta V_a(t) \tag{5.12}$$

If $\Delta V_a(t)$ is approximated by a sawtooth waveform with a falling ramp as shown in Figure 5.25A, and calling the frequency difference between the two signal tones f_m, then

$$\Delta V_a(t) = \Delta V_a \left[\frac{1}{2} + \frac{1}{\pi} \sin \omega_m t + \frac{1}{2\pi} \sin 2\, \omega_m t + \dots \right] \tag{5.13}$$

The modulation index of an amplitude modulated signal is the envelope voltage divided by the unmodulated signal or carrier voltage:

$$m = \frac{(A_{\mathrm{pk}} - A_{\min})/2}{(A_{\mathrm{pk}} + A_{\min})/2} = \frac{\Delta A}{2A} = \frac{\Delta G}{2} = \frac{-K \, \Delta V_a}{2} \tag{5.14}$$

The Fourier expression for ΔV_a gives the frequencies of the modulation voltage. Each modulation frequency creates a pair of sidebands. If a modulation index m_n corresponds to the nth harmonic of the sideband, the relative amplitude of each nth harmonic sideband will be $m_n/2$. Calling the nth harmonic sideband level S_n, from Equations 5.13 and 5.14, then

$$S_n = \frac{m_n}{2} = \frac{K \, \Delta V_a}{4\pi n} \tag{5.15}$$

Modulation frequencies fall symmetrically on either side of the two tones at a spacing of $f_m = f_2 - f_1$. The S_2 of each of the two tones falls on S_1 of the other, and so on for higher orders of n.

The relative level of the sidebands at $\pm \omega_n$, produced by amplitude modulation of the signal passing through the attenuator, can now be calculated as

$$S_{\omega_m} = S_1 + S_2 = \frac{K \, \Delta V_a}{4\pi} \left(1 + \frac{1}{2} \right) = \frac{3 \, K \, \Delta V_a}{8\pi} \tag{5.16}$$

One example might be $K = 0.1$/volt and $\Delta V_a = 0.1$ V. The level of first-order upper- and lower-frequency sidebands is then $0.03/(8\pi) = 0.0011937$ or 58.5 dB *below* each of the two signal tones.

This will be the approximate amplitude of the first-order sidebands, depending on the accuracy of the approximations of the waveform of $\Delta V_a(t)$ as a sawtooth. It should also be taken into account that K will change over the range of V_a. It is also assumed in this discussion that the modulation process is inefficient enough that it is not regenerative.

Mixers

Mixers can be either active or passive. Active mixers use the nonlinear characteristics of active elements such as transistors or varactors to produce frequency mixing. Active mixers must be individually designed, are temperamental in behavior, and have no apparent advantage except for some power gain over passive double balanced diode mixers. Diode mixers have evolved into a highly developed art, provide excellent and predictable performance, and are available from many manufacturers in a wide range of characteristics and sizes. Therefore, the discussion here will be limited to diode mixers.

Of the several important characteristics of the mixer, one is the balance or isolation between the various ports. Figure 5.26 shows the balance parameters usually specified. These and other mixer characteristics are controlled and specified using standard impedance terminations on each port, usually 50 ohms. In HF exciters, the LO-to-IF and LO-to-RF port isolations are important because of the possibility of coupling noise or spurious signals appearing on the injection to these ports.

The signal level at which the mixer can operate is determined primarily by the injection or LO level. The RF input compression point depends on the LO level, but for HF applications, the IMD characteristics are usually the most important consideration.

In general, the harmonic IM levels of a double balanced diode mixer are related to the LO and RF input levels and their harmonic numbers, assuming proper terminations (50 ohms on all ports). Under these conditions, it

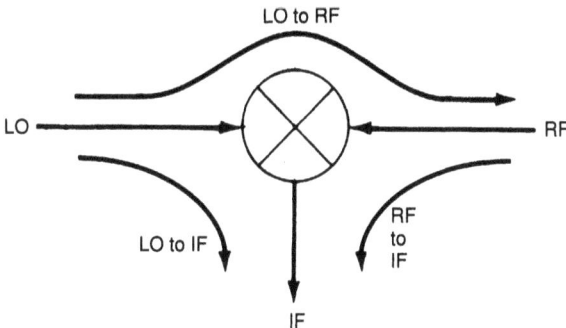

Figure 5.26   Mixer balance parameters.

has been found that mixer spurious levels can be estimated to a fair degree of accuracy [5, 6].

Table 5.6 lists the spurious frequencies and their predicted levels, based on the difference ΔP between the LO and RF input power levels. The spurious frequencies f_S are found from

$$f_S = m f_{LO} \pm n f_{RF} \qquad (5.17)$$

where m and n are the integers in the LO and RF columns, respectively. The levels stated are relative to the desired IF port output signal level. Note that the spurious levels associated with higher harmonics of the LO are still relatively high. An example is the calculation of relative spurious levels of a mixer having an RF input of -10 dBm at 9 MHz and an LO level of $+13$ dBm at 100 MHz. The products for $m = 3$, $n = 1$ are at 291 and 309 MHz. From Table 5.6, the estimated suppression of these products is only -10 dB from the desired IF output.

Table 5.6 does not predict two-tone IMD levels such as in-band third-order IMD. These can be predicted by estimating the input third-order in-

TABLE 5.6 Predicted Mixer Spurious Signal Levels

| Order of harmonic | | Spurious |
LO	RF	level, dB
1	1	0 (desired)
1	2	$\Delta P - 41$
1	3	$2\Delta P - 28$
2	1	-35
2	2	$\Delta P - 39$
2	3	$2\Delta P - 44$
3	1	-10
3	2	$\Delta P - 32$
3	3	$2\Delta P - 18$
4	1	-35
4	2	$\Delta P - 39$
5	1	-14
5	3	$\Delta P - 14$
6	1	-35
6	3	$\Delta P - 39$
7	1	-17
7	3	$2\Delta P - 11$

SOURCE: Bert C. Henderson, "Predicting Intermodulation Suppression in Double-Balanced Mixers," *Watkins-Johnson Tech-Notes*, vol. 10, no. 4, July/August 1983. (Reprinted by permission.)

tercept of the mixer as being 5 to 10 dB higher than the LO level. For a +13-dBm LO the estimated input third-order intercept is +18 dBm. For an RF input of –10 dBm/tone, the in-band third-order IMD appearing at the IF output will be 56 dB below the desired IF output level.

The above predictions of spurious levels in mixer outputs should be used only to aid in selection of a mixer for a particular application, using actual measurements to make the final choice.

Proper termination of the LO, RF, and IF ports is important when using a diode mixer, the IF and RF ports being most critical. At the IF output, a very broadband termination is particularly important in high-performance mixers in order to prevent the unwanted image frequency and higher-order products from reflecting back into the mixer. These reflected products remix with the mixer inputs, increasing IMD output levels.

Amplifiers

The designer has a wide choice of options in the type of device to use as IF and RF amplifiers. In addition to discrete transistors, packaged IC and hybrid amplifiers are available from manufacturers with a wide range of characteristics. Their advantages include reduced size/board space, controlled characteristics, and reduced circuit design time. They usually are designed for use in 50-ohm circuits, which is compatible with mixer and IF filter termination requirements. Intermodulation distortion, gain, and noise characteristics are controlled parameters, so the designer has a gain block of known characteristics to plug in to his circuit design.

In addition to fixed-gain amplifiers, devices are available with gain controlled by a control voltage. These are candidates for TGC and ALC applications in exciters (and AGC in receivers) provided their IMD performance is adequate over the gain control range. Higher-power amplifiers, commonly used in the cable TV industry, are also suitable for RF output stages at levels to +20 dBm PEP or more.

5.4 Introduction to Transceivers

Exciter and receiver functions are combined into one unit, a transceiver, when size or cost is of great importance. Airborne radios, manpacks, and vehicular radios are just a few applications where the transceiver finds wide application. Radio operation in these cases is typically simplex, that is, transmitting and receiving alternately on a single frequency (occasionally, dual frequencies are used). Some flexibility may be sacrificed in selecting operating modes or conditions, such as independent sideband, extra receiver bandwidths, etc. Performance may also be sacrificed to save power or space.

The receiver and exciter have several functional elements in common such as power supply, frequency synthesizer, frequency standard, control circuits, and front-panel controls. The use of circuit components, such as IF filters, mixers, and amplifiers, in common between the transmit and receive modes is maximized in a transceiver. These and other design considerations unique to transceivers will be discussed.

5.5 Transceiver Architecture

The unique features of transceiver architecture are the functions shared between the transmit and receive operations. The key to a simple, compact transceiver design is to make use of common elements to the maximum extent possible, consistent with performance requirements.

The power supply, synthesizer, and control functions are shared between transmit and receiver operations by using common supply voltages, one frequency scheme for both operations, and similar control methods. Circuit elements can also be shared, such as amplifiers, mixers, and filters. Components that are capable of bidirectional signal flow, such as filters and diode mixers, are particularly useful.

Figure 5.27 shows a block diagram of a typical transceiver which helps to illustrate these points. In the exciter path, the baseband signal enters the left-hand side through the baseband amplifier and processing circuits into the balanced modulator. It then progresses through the rest of the path to the output. Along the way it passes through amplifier stages that are switched into the signal path.

For operation as a receiver, the received signal enters on the right, passing through a broadband "roofing" filter into the mixer M3. This mixer is used as a bidirectional circuit element, with the receiver signal passing through it in the opposite direction of the transmit signal. The signal passes through another bilateral circuit device, the second-IF filter, then into a re-

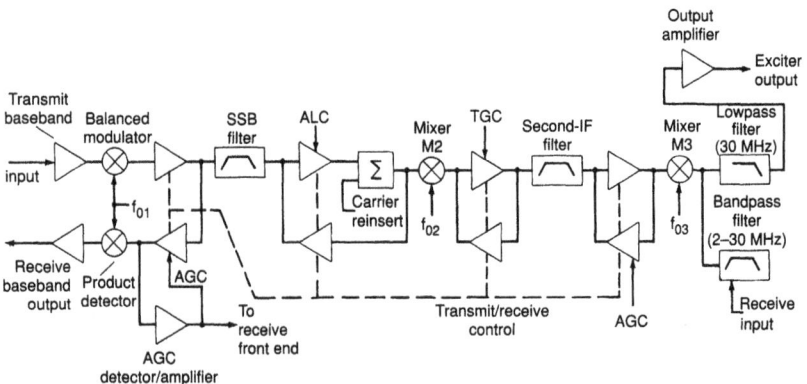

Figure 5.27   Transceiver block diagram.

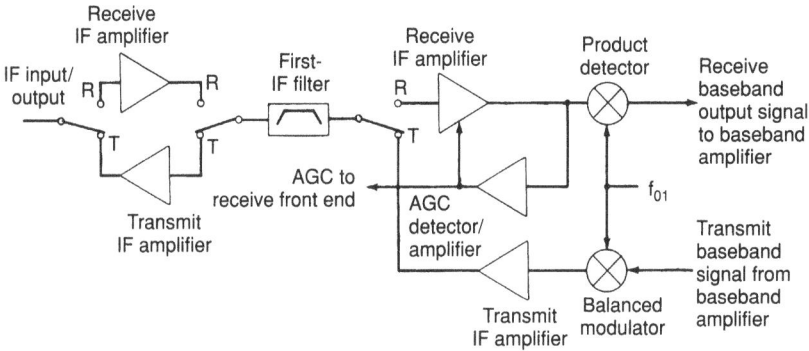

Figure 5.28   Transceiver IF amplifier and filter.

ceive IF amplifier. The mixer M2 is also used as a bilateral device, mixing the receive signal down to the first or lowest IF. Signal flow progresses through the first-IF filter to the receive IF amplifier and product detector M1. The resulting audio signal is amplified and applied to the output terminals.

For detailed design considerations of individual sections of the transceiver, refer to Chapter 4 and previous sections of this chapter. Some considerations unique to transceiver design will be discussed here.

IF filter and amplifier

The IF amplifier and filter section contains the transmit balanced modulator and IF amplifier, receive detectors and IF amplifiers, and the first-IF filter. A block diagram is shown in Figure 5.28.

In the receive signal path, in addition to filtering to select the proper sideband (or other passband), the IF amplifiers must provide enough gain to drive the detectors at the proper level. Gain on the order of 60 dB could be required in the IF amplifier between the IF filter and detectors. Automatic gain control is used in this amplifier to maintain constant levels into the detectors. The transmit IF amplifier, on the other hand, may have little gain, and no AGC requirement. The functions of the two amplifiers are quite different, not lending themselves to sharing functions. The IF buffer amplifiers (to the second mixer) in the transmit and receive paths are both relatively low gain, and are switched on and off to change signal direction.

There are many ways to switch the IF filter into the proper signal path. One simple method is shown in Figure 5.29, where FETs are biased on and off in response to control voltages applied to their gates. Bias voltages V_T and V_R turn on the appropriate FETs in the transmit and receive signal paths. Resistors R_T provide proper terminations for the filter. If FETs are not appropriate to the IF amplifiers being used, diode switches or some other form may be used.

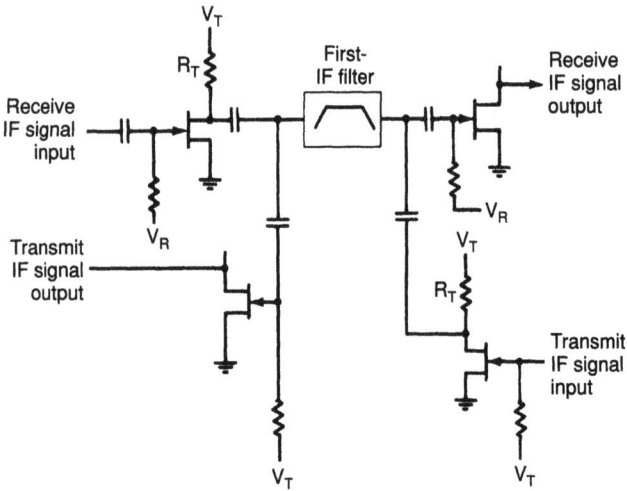

Figure 5.29 A switching method for the first-IF filter.

Although not indicated on the block diagram of Figure 5.28, it may be desirable to use a single circuit element for the transmit modulator and receive product detector. The designer may want to trade off the complexity of switching the circuit against providing two separate circuits. A technique for using the switching-type IC modulator (type 1496/1596) may be considered where the balanced inputs and outputs are separated into baseband and IF inputs and outputs. This is shown in Figure 5.30, where one of the inputs receives transmit baseband audio, and the appropriate output is the

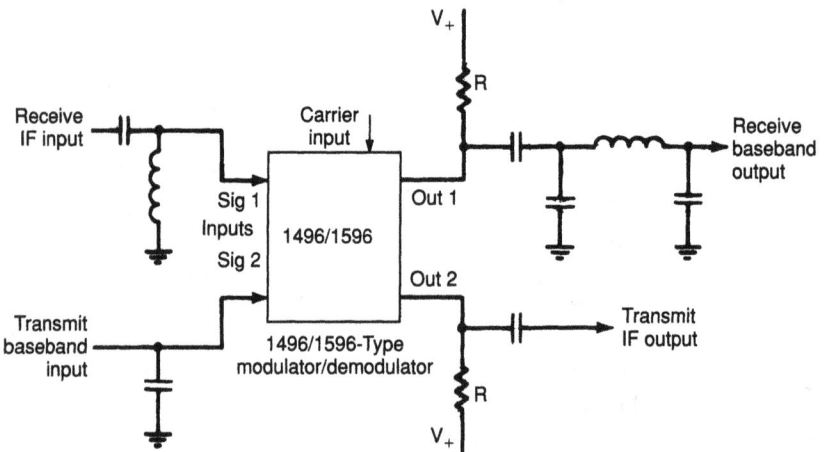

Figure 5.30 Transmit/receive application of 1496 and 1596 modulator/demodulator.

modulator IF output with suppressed carrier. The input has capacitor by-passing, which is effective only at IF, not degrading the baseband signal. The other input to the IC is the receive IF signal. This input has a highpass filter. The appropriate output contains the recovered audio signal and has a lowpass filter to remove the injection f_{01}. Careful layout of circuit components is required, especially in terms of transmit carrier suppression.

It should also be noted that, because of the crowded component layouts common to transceiver design, care should be taken that the receive IF output is well isolated from the IF filter's input (receive-side) circuitry. Because of the high gain usually needed in the receive IF amplifier, it is easy for passband amplitude variations to be increased over that of the filter alone, by output IF signal leaking back and combining with the filter input IF signal, causing partial cancellation or addition. The output IF signal, having passed through the IF filter, undergoes rapid phase shifts with respect to the filter IF input signal across the passband. For example, if the IF output signal leaking back to the filter input is 20 dB lower than the filter input signal, it will introduce an additional 1.7 dB of amplitude variation across the filter passband as it becomes in phase then out of phase with the input signal.

IF-to-RF translator

The IF-to-RF translator has some characteristics that differ from the IF amplifier. Figure 5.31 shows how the translator might be configured to accommodate transmit and receive functions. The mixers and second-IF filter are bilateral elements, with the amplifiers switched to reverse signal flow. The IF amplifiers used here are operating at VHF (around 100 MHz), and so will not be amenable to simply biasing on and off as for the first-IF amplifiers. The inputs and outputs of the amplifiers can be switched in and out of the signal path, perhaps with PIN diodes. One possible configuration is shown in Figure 5.32. The isolation across the switches must be greater than the sum of the amplifier gains at all frequencies within the bandwidth of the amplifiers or in-

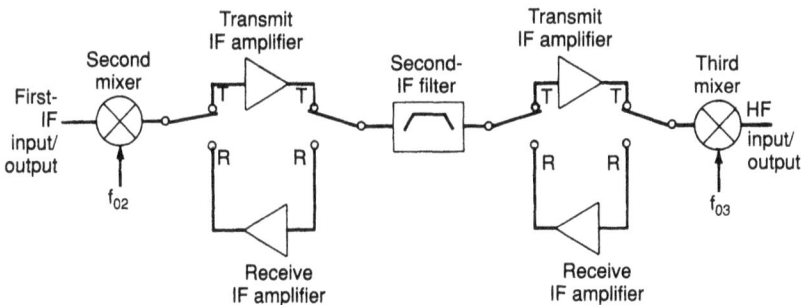

Figure 5.31 Switching IF amplifiers in frequency translator.

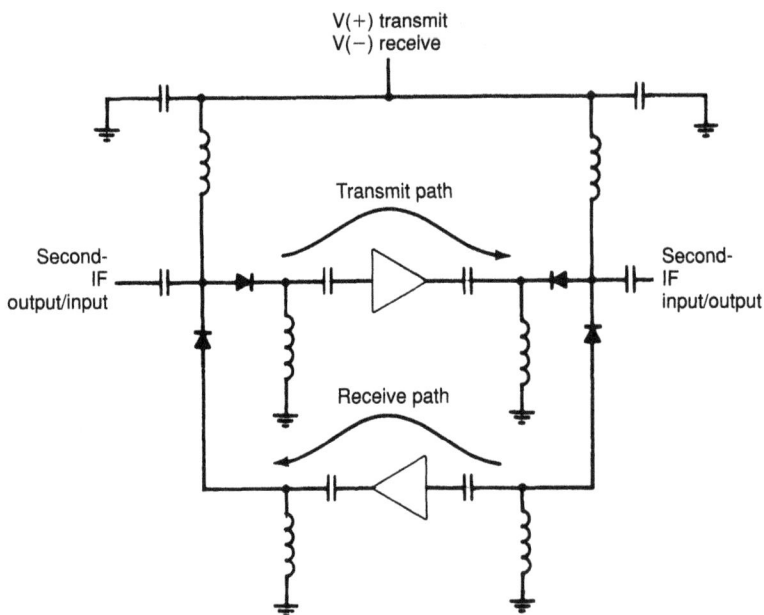

Figure 5.32 Amplifier switching with PIN diodes.

stability will occur. It may be possible to use only one amplifier if the gain each way is nearly the same. A small amount of gain adjustment, in the form of an attenuator, could be combined into one of the diode switches. It should be determined that the amplifiers used are stable under the switched-out condition. If not, the amplifier may need to have power removed when not in use.

The switches used on the HF side of the third mixer M3 will handle higher signal levels, requiring that they have sufficient on current and voltage back bias for the expected signal levels. Since they are switching signal frequencies down to 2 MHz, they must also have sufficiently long carrier lifetimes.

There are alternatives to the switching methods discussed here, of course. One that has been used is shown in Figure 5.33, where the mixer M3 is an active mixer and not bidirectional. The signal flow through the mixers M2 and M3 and the second-IF filter is not reversed. Instead, the inputs/outputs to the translator are switched along with the mixer injections. This method requires that the isolation between the two injection signals across the switches be very high to eliminate generation of spurious signals, as discussed in section 5.2 and Chapter 4.

Filter-saving frequency scheme

If the transceiver has a requirement to operate in USB or LSB, but not ISB, then a frequency scheme using only one first-IF filter may be considered. The scheme uses one SSB filter in the first IF. The desired sideband is selected in the second IF by proper choice of the second injection frequency f_{02}.

Figure 5.34 illustrates the frequency scheme. A USB filter is used in the first IF. In the second mixer, the USB is translated by f_{02} to a USB that falls in the passband of the second-IF filter, and is transmitted as a USB. If the operating mode is LSB, the injection frequency of the second mixer is changed to f'_{02}, which is above the second-IF filter passband. The difference product now falls in the second-IF filter, which is inverted from the original USB, becoming an LSB. It is then transmitted as an LSB. In receive, the same process operates in reverse.

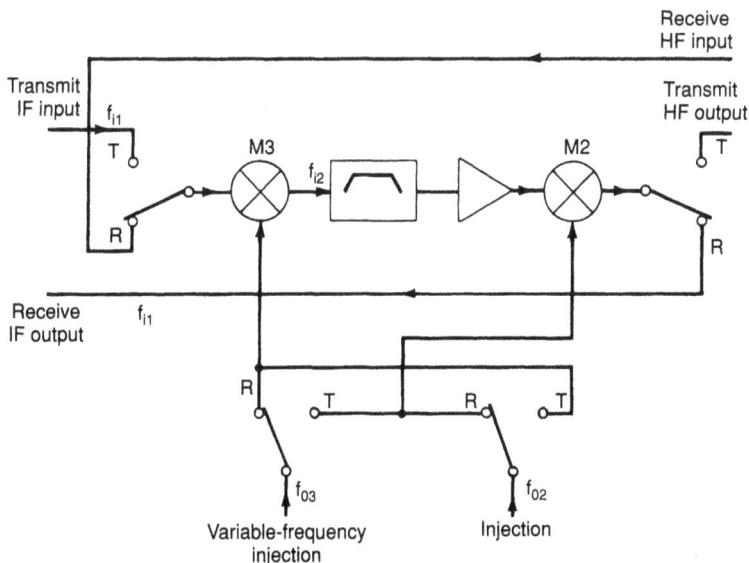

Figure 5.33 Alternative switching scheme for frequency translator.

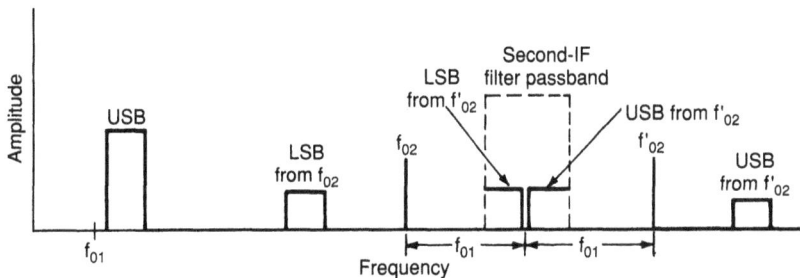

Figure 5.34 Filter-saving frequency scheme.

The advantage of the above scheme is the saving of an IF filter and the associated switching circuits, but at the expense of devising the means to change the injection frequency f_{02}. Note also that different receive and transmit spurious signals will be generated for f_{02} and f'_{02} and should be taken into account.

CW and RTTY modes

Sometimes transmission and reception of CW or RTTY, or both, are required in an SSB transceiver. There are alternative ways to generate the transmit signals and to receive them, and deciding which to use depends on the complexity of the approach.

A CW signal can be generated either by reinserting a keyed carrier at f_{01} or by keying an audio signal. If the emitted signal has to be at an exact assigned carrier frequency, using a reinserted carrier is the preferred approach. If the emission mode allows a keyed audio signal, a keyed audio oscillator can be used. The latter simplifies the receiver operation because it is tuned to the assigned carrier frequency and no BFO is required to receive the keyed audio-modulated signal.

Controlling the CW signal emission bandwidth requires either passing the signal through a filter or shaping the keyed envelope. Envelope rise and fall times of 5 to 10 ms may be sufficient to meet emission limits. An IF filter of approximately 300-Hz bandwidth inserted into the keyed carrier signal path will provide positive bandwidth control.

In the receive mode, if the CW signal is at the assigned carrier frequency, one of the mixer injections in the receiver must be offset to produce an audio output. Shifting either f_{03} or f_{02} will place the received signal in the USB (or LSB) filter, producing audio output. Or, a narrow CW filter could be used in both the receive and transmit modes, centered at f_{01}. In this case, f_{01} would be shifted in the receive CW mode to produce an audio output.

The transceiver is sometimes required to incorporate a single-channel RTTY mode, using an FSK signal. If operation of the RTTY channel in SSB mode is allowed, a two-tone audio generator keyed by the teletype (TTY) mark/space input will suffice. An oscillator that switches from a mark to a space frequency in response to the TTY input will produce the necessary FSK signal in the SSB passband.

If the emission mode requires the shifted frequencies to be on each side of the assigned carrier frequency, the above approach may not be acceptable. A more straightforward approach is to insert the FSK signal in the first IF, shifting each side of f_{01}. A crystal oscillator that is "pulled" between the mark and space frequencies is a method of generating the FSK signals. Figure 5.35 shows a crystal oscillator operating in a frequency range that is more conducive to being pulled, then divided by a fixed ratio to the first-IF frequency. A mark/space input signal is shown that changes the capacitance of a varactor, changing the oscillator frequency. For example, if the mark frequency is 500.085 kHz and the divide ratio is 20, the oscillator frequency is 10.0017 MHz. The required pull range for 170-Hz shift is 340 ppm, well within the capability of such an oscillator. Minor temperature compensation may be required to bring the frequency stability within requirements. If compensated to 5×10^{-6} the frequency error would be 2.5 Hz. Also shown in Figure 5.35 is a filter in the signal path to the IF, which may be required to control spurious emissions.

Another approach to generating the FSK signal, if the frequency synthesizer is capable of it, is to shift the injection signal f_{01}. This has the advantage of not introducing additional frequency error because it is locked to the transceiver frequency standard. The transition time between mark and space frequencies must be fast, however, on the order of 1 ms for data rates of 75 baud.

To receive the FSK signal, it is converted to an audio FSK signal, then detected in a phase-lock loop or filter-type discriminator. If the FSK signal is received in the SSB mode the receiver output is audio and is fed to the FSK discriminator and detected. The appropriate mark/space is then sent to the TTY printer.

If the received signal is a shifted carrier, the required audio is generated by shifting one of the LO injections in the receive path. For example, the variable-frequency injection f_{03} could be shifted by 2 kHz, placing the signal in the USB passband and producing the required audio output. A separate IF filter could be used instead of the normal USB filter, if better selectivity is needed. Or the injection to the product detector f_{01} could be shifted to produce the audio output. In this case, the signal would require filtering by a separate IF filter because it will not pass through the USB (or LSB) filter. Either approach will produce the same result.

Transmit/receive switching

In switching from the transmit to the receive state, the receive circuits must be ready for normal operation quickly so as not to disrupt a received signal.

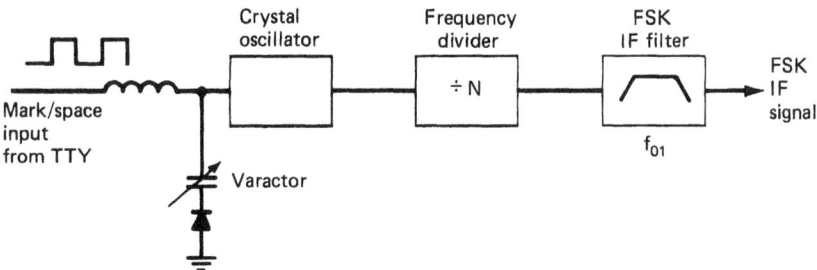

Figure 5.35 Method of generating IF FSK centered at f_{01}.

This means that power must be applied to receive circuits, frequency changes required in any of the injection signals made, and gain of the receiver initialized before a usable baseband output is produced. If the signal to be received is voice, then the time allowed may be 0.1 sec or longer. If FSK or other data is to be received, the allowable time may be on the order of 10 to 20 ms.

A frequent problem concerns the receive AGC circuit. When in the transmit mode, transmit signals are present in the IF amplifier section that will generate high AGC levels. This reduces receive gain when switched back to the receive mode. Disabling the receive IF amplifier when in the transmit mode is usually a necessary step, but other precautions are also necessary. At the top of Figure 5.36 it is shown how the transmit signal is expected to decay in the output of the IF, particularly the IF filter. The filter "rings out" for a period of time t_1, and produces a short signal in the receive IF when the receive circuits are reactivated. The result is that the receive AGC responds to the signal, decreasing the receive gain and desensitizing the receiver for a period of time determined by the AGC decay time. A method of defeating this effect is to "crowbar" the AGC control voltage at the AGC holding capacitor for a period of time t_2, which assures that the filter has completely dissipated its stored energy. Mechanical SSB filters (seven-pole design) with group delay of 900 µs may require t_2 to be on the order of 10 ms, depending on the transmit signal level in the filter.

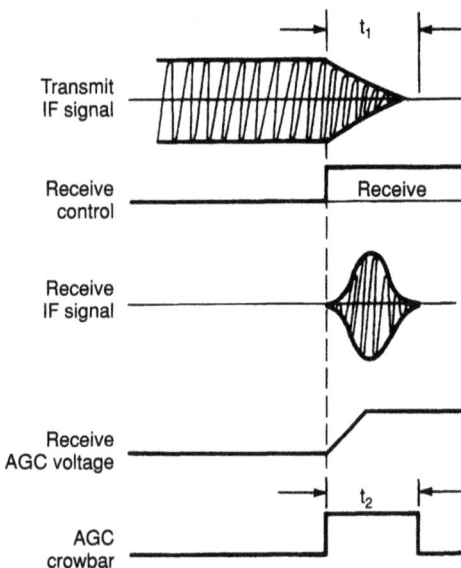

Figure 5.36 Transmit-to-receive timing and AGC crowbar operation.

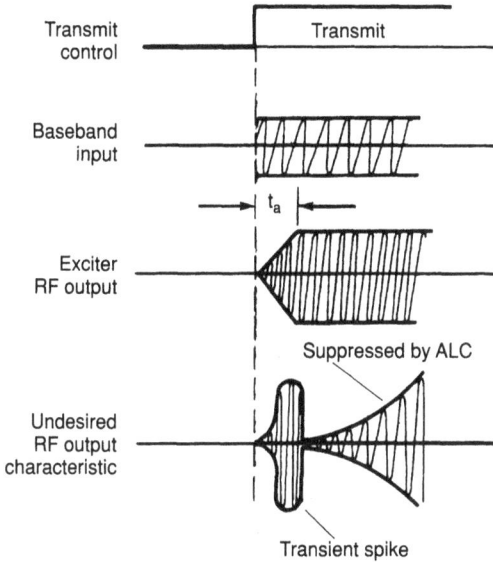

Figure 5.37 Receive-to-transmit switching effect on transmit signals.

Transients caused by changing bias conditions on amplifiers as they are turned on can also generate a receive IF signal that generates AGC voltage. This can be minimized by making coupling capacitors only as large as needed for the IF signals. The AGC crowbar will also nullify the effects of these transients.

In switching from receive to transmit, the transmit circuits must be ready to transmit quickly, because a key signal may be applied simultaneously with the baseband input. If the input is data, keying to the transmit mode may be required in less than 10 ms.

The top of Figure 5.37 shows the application of the key signal, putting the transceiver in the transmit mode. The exciter output is building up during time t_a. It is during this time that circuit gains must stabilize and injection signals change, if necessary, to the correct frequencies. As mentioned previously, the time allowed for t_a may be less than 10 ms. At the bottom of Figure 5.37 is shown an undesirable transient spike in the exciter output caused by the rapid application of bias voltages. Such a transient could cause the transmitter ALC circuit to reduce transmitter gain in the same way that AGC acted to reduce receive gain. These transients can be eliminated by shaping the turn-on voltages and limiting the size of coupling capacitors.

5.6 Modular Radio Communication Systems

What is modular radio?

Modular radio communication systems have come into the marketplace within the past 5 years, and offer the system designer a very flexible alternative for the hardware and software implementation of a complete radio communications system. A modular radio system consists of a family of functional hardware and software modules that share common resources, such as a chassis, power supply, control subsystem, etc. The emergence of modular equipment families has been driven by several key factors:

- Subminiaturization of electronic assemblies
- Evolution of the PC/workstation as a generic mainframe
- Maximization of hardware reuse in the face of rising design costs

A modular system is best described by comparing it with the preceding generation of communication systems.

A traditional communications system (typically) consists of an equipment rack (often a 6-ft high, 19-inch wide EIA standard rack) filled with various radio elements (receiver, exciter, power amplifier, modems, patch panels, etc.) all interconnected by a wiring harness within the rack. The various elements are interconnected to perform different functions and may be reconfigured by remote switches and/or patch panels.

A modular radio communications system typically consists of a chassis (often a 19-inch wide rack chassis) filled with various functional radio modules (receiver, exciter, power amplifier, modem, etc.) all interconnected by a motherboard that is internal to the chassis and augmented by external intermodule wiring as necessary (typically coaxial cables). The various elements are interconnected to perform different functions and may be reconfigured by software, remote switches, and/or patch panels.

Characteristics of a modular radio system

A modular radio system has key characteristics that combine to define the flexibility and functionality:

- All modular elements comply with a selected mechanical, electrical, and software interface standards.
- The family of modular elements includes all functional modules necessary to construct the desired high-level system, often from different vendors and designed to a common standard.
- It has a common control software interface that operates all elements.

A personal computer is an excellent example of a modular computer system, with its accessory card standards and functionality.

When is a modular radio communications system appropriate?

A modular radio system is *not* always an optimum solution. The system designer should carefully weigh the following considerations:

- Overhead (chassis, power supply, CPU, etc.) should be spread over a sufficient number of functions to be economical.

- Multiple functions typically are required to justify a modular system; dedicated single-function systems are rarely economical versus a dedicated single-function radio. These functions include SSB, CW, FSK, AM, ALE, and frequency scanning surveillance.

- For a high-reliability system, redundant, reallocatable, functional modules are integrated into a high-level system.

- Typically a modular system will include at least one and possibly additional radio receivers and transmitters, preselector filters, modems, high-level controllers, CPU, and I/O card(s).

- A complex modular system will generally require an internal "housekeeping" processor and an external "control" processor for the operator interface.

Module examples

Figures 5.38, 5.39, and 5.40 show examples of representative modular radio elements. Figure 5.38 shows a complete four-channel ISB receiver module for an IBM PC/AT (ISA) bus system. This member of the Rockwell International Spectrum 2000 modular radio family features a machined aluminum "clamshell" case, an ASIC for ISA bus interface, MS-DOS memory mapped control and monitor functions, PC/AT module standards, and external connectors in the PC/AT connector zone.

Figure 5.39 shows a 100-W HF/VHF power amplifier module, also from the Rockwell Spectrum 2000 family. This module is of particular note in that it is the highest-power module found in any modular family and demonstrates the limits to which a modular bus system can be pushed. Key features include linear amplification, software-controlled gain and lowpass filters, ASIC for ISA bus interface, remote metering of analog functions, and an unusual heat-sink design tailored to the unique airflow of an ISA motherboard chassis.

Figure 5.38 A four-channel ISB receiver module for an IBM PC/AT (ISA) bus.

Figure 5.39 A 100-W HF/VHF power amplifier module from the Spectrum 2000 family.

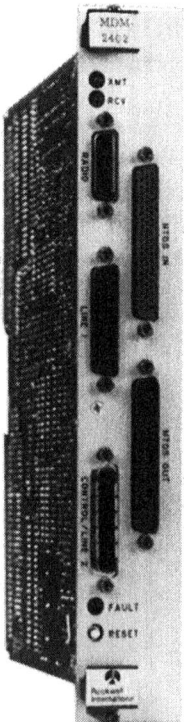

Figure 5.40 VME standard modem module.

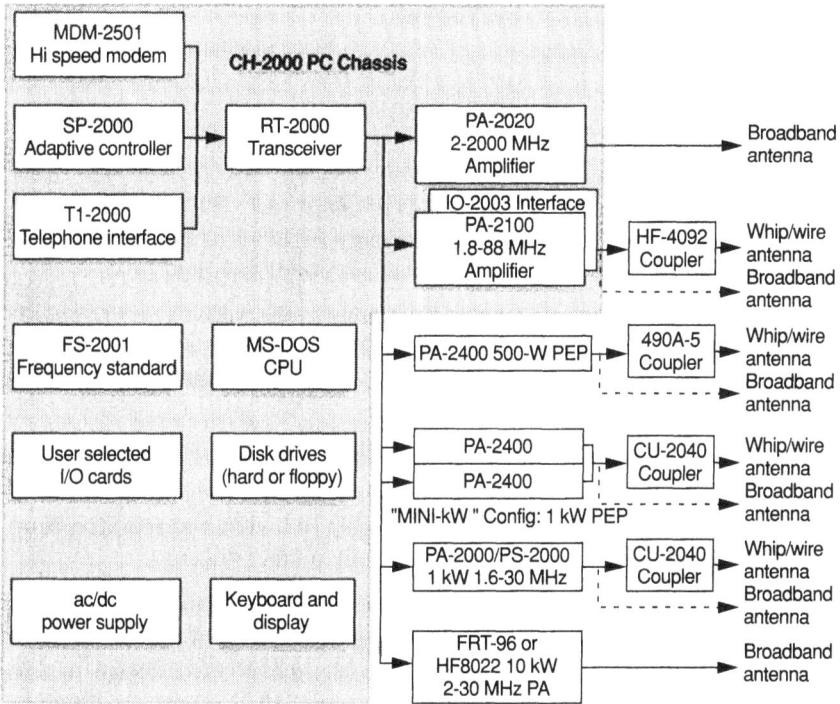

Figure 5.41 Multiradio, multiband Spectrum 2000 system.

Figure 5.40 shows a different variation: a 6U, double-pitch VME standard modem module. This Rockwell modem utilizes digital signal processing (DSP) to provide audio baseband modulation and demodulation of a variety of software-based waveforms.

System examples

Figure 5.41 shows a sophisticated multiradio, multiband system with six radio strings, two modems, two adaptive controllers, and two redundant cross-linked operator control computers. The objective of this compact mobile system is to provide simultaneous communications across a wide spectrum with fail-safe redundancy. A sophisticated, menu-driven graphic operator interface MS-DOS software package gives each operator redundant, easy-to-use access to all communications functions. The instruction manual is a built-in set of comprehensive help screens.

Critical design aspects

The system designer should be aware of several critical aspects of modular radio system design. Among them are:

- Bus EMI from the motherboard or digital functional modules can easily be conducted and/or radiated into sensitive receiver modules! A 5-V logic signal is approximately 130 dB *greater* than a 1-μV signal typically coming into an antenna port! Careful shielding and grounding is required to eliminate such interference, which generally manifests itself as broadband noise. Optical isolation, shutting down CPU buses when commands are not being sent, and using the slowest feasible bus frequency are typical EMI reduction approaches. Newer types of CPUs have clock frequencies well into the VHF range, where receiver sensitivities are considerably less than 1 μV.

- If one of the digital-module clock frequencies falls within a receiver's frequency range, it generally will be heard in the receiver output unless extreme measures are taken to isolate it. Transmitter noise output is also vulnerable.

- Total heat loading within a densely packed chassis must be met with a carefully balanced air flow scheme.

- Sufficient CPU capacity should be available to handle the required command flow necessary. This is often severely underestimated.

In the Spectrum 2000 equipment all of the above problems have been solved through careful engineering and an understanding of EMI techniques. Chapter 16 gives further insight into EMI problems in receiver design. In conclusion, we note that modern designs tend toward dual-conversion receiver-exciter radios rather than the triple-conversion schemes used in earlier years. This is a result of the increasing availability of high-quality, high-frequency crystal filters, at reasonable cost.

References

1. "Part 81—Stations on Land in the Maritime Service and Alaska Pacific Fired Stations," in *Rules and Regulations* (Washington, DC: Federal Communications Commission, U.S. Government Printing Office, 1985).

2. "Appendix 8—Table of Maximum Permitted Spurious Emission Power Levels," in *Radio Regulations* (Geneva: International Telecommunications Union, 1982).

3. Raymond E. Cook, "Cascaded Active Circuits Yield 90-degree Phase Difference Networks," *Electron. Des. News* 18 (April 5, 1973): 52–56.

4. Alan G. Loyd, "90 Degree Phase Difference Networks," *Electron. Des.* 24 (September 13, 1976): 90–94.

5. Bert C. Henderson, "Predicting Intermodulation Suppression in Double-Balanced Mixers," *Watkins Johnson Tech-Notes* 10, no. 4 (July/August 1983).

6. Bert C. Henderson, "Mixers: Part 1," *Watkins Johnson Tech-Notes* 8 (March/April 1981).

6

IF Analog Filters

Joseph A. Vanous

Intermediate frequency filters are used in SSB receivers and exciters to shape the desired signal spectrum and to reduce unwanted signals. Unwanted signals are created by the mixing process when a signal frequency is mixed, or translated, to a different frequency. The desired signal may be shaped by filter design to conserve bandwidth and to prevent adjacent channel interference. Filters also protect receiver stages from overload by strong signals on nearby frequencies. In addition, the filtering must not unduly alter the characteristics of the modulated signal. These alterations show up as distortion in the frequency domain (intermodulation and harmonic distortion) or in the time domain (signal delay differences). These effects must be minimized.

The SSB IF filters differ from other bandpass filters because they are offset from the carrier frequency and have a nonsymmetrical attenuation response. These filters are designed to pass one sideband such as USB and reject LSB. Other IF filters have the signal frequency centered in the passband and are used to reduce spurious signals generated by the mixers. Such filters have a symmetrical response.

6.1 Passband and Stopband Characteristics

Since bandpass filters share certain common characteristics, it is desirable to define the various terms associated with them. A diagram of these is shown in Figure 6.1.

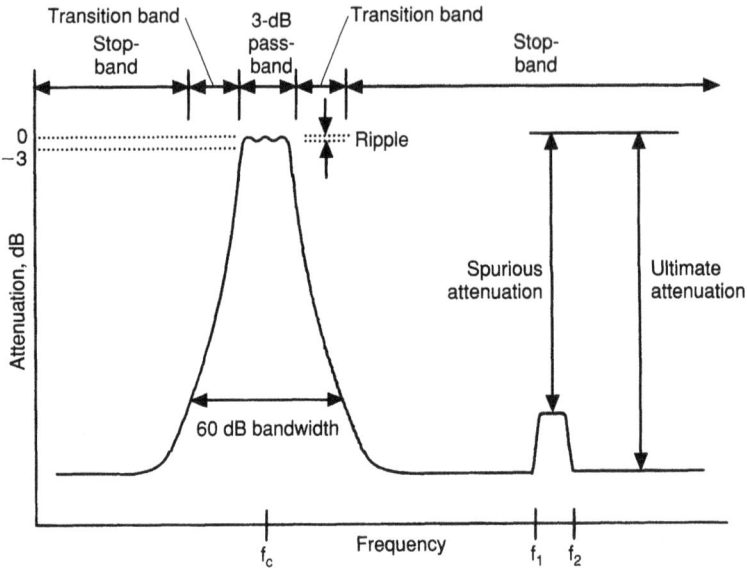

Figure 6.1 Common amplitude characteristics of a bandpass filter.

Passband. The passband defines the limits of the desired signal modulation and is usually defined between the 3-dB points (or half-power points) referenced to the highest peak between these limits. Often, other attenuation points are given, such as 1 or 0.5 dB. This occurs when the overall exciter or receiver 3-dB passband is specified, and the attenuation at 3 dB is then partitioned among all the filters in the signal path.

Passband ripple. This is the variation in signal level or response across the passband. It is defined as the difference in amplitude between the minimum response and the maximum response values. The difference is stated in decibels and can range from a few tenths of a decibel to 3 dB or more.

Transition band. This term is loosely defined but generally includes the region between the 3- and 60-dB points. It is often defined as being monotonic, that is, a smooth transition. It can also be specified as a region where the attenuation is some lesser amount than that existing in the stopband.

Stopband. The stopband is the region beyond the transition band. A specified amount of attenuation is often stated for specific frequencies in the stopband. These usually correspond to images and injection signals. See Chapter 4 for a discussion of image frequencies.

Spurious attenuation. Many filters have spurious responses in the stop-band. The locations of these are sometimes defined by a frequency band and the expected minimum attenuation levels. They are usually above the passband in crystal filters.

Ultimate attenuation. This is specified as an attenuation in the stopband region where the attenuation can be expected to remain greater than this level.

Intermodulation distortion. The in-band IMD is specified for exciters and receivers, and the out-of-band IMD primarily for receivers. This is described in detail in section 6.4. The level of the two tones is specified for a given third-order distortion product level. For example, for a two-tone level of –25 dBm/tone, the third-order distortion product could be specified as no greater than –60 dB down (–85 dBm).

Insertion loss. The insertion loss of a filter is the loss in available power (in decibels) that occurs when the filter is inserted between the source and load. It is defined as $10 \log (P_{available}/P_{load})$ and can be measured as $20 \log (E_1/E_2)$, as shown in Figure 6.2. The maximum amount of power delivered to a load occurs when the source and load are matched. This level can be considered the reference of 0 dB. With the filter inserted, the loss occurring

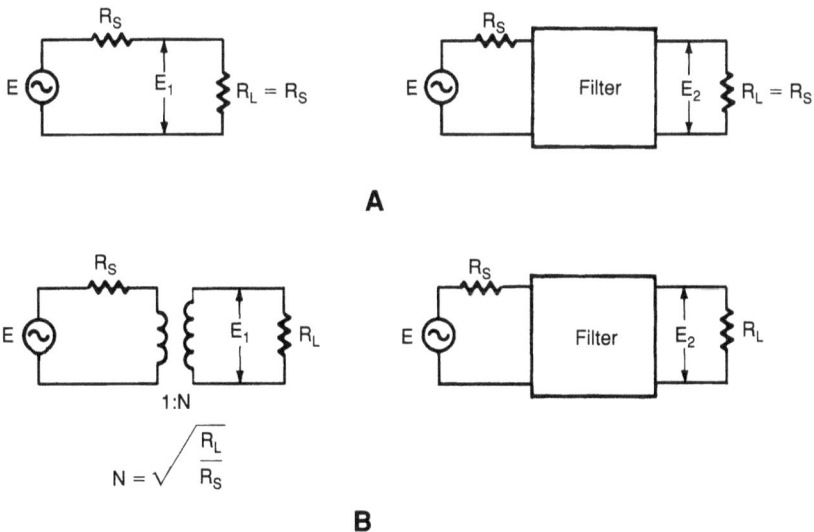

Figure 6.2 Method of determining insertion loss of filters (A) when load and source impedances are matched ($R_L = R_S$); (B) when impedances are different ($R_L \neq R_S$). Insertion loss (in decibels) = $20 \log (E_1/E_2)$.

is the insertion loss (in decibels) and is measured over the specified filter passband. When the source impedance differs from the load, a transformer can be used to obtain the reference voltage E_1.

Shape factor. The shape factor of a filter is a figure of merit that is usually specified as the ratio of the 60-dB bandwidth to the 3-dB bandwidth. At times other numbers are used. For example, a 6-dB bandwidth may be used to provide a 60-to-6-dB ratio.

Delay distortion. Phase linearity is specified as a time delay across the filter passband. Ideally, a constant delay across the passband would provide linear phase shift with signal frequency change. The variation in the time delay is called *delay distortion* and is described further in section 6.5.

Most LC tuned circuits and crystal and mechanical filters have amplitude characteristics that fall within the following major classifications:

1. Butterworth maximally flat passband response

2. Chebyshev (sometimes spelled Tchebycheff) equal ripple response

3. Gaussian response

4. Bessel response

Many other types of responses exist. Another variation of the Chebyshev has equal ripple in the passband and equal attenuation minima in the stopband. These are also called *elliptic-function filters*. This filter has the steepest response in the transition band but at the expense of rising attenuation lobes in the stopband. Other responses include equal-ripple delay, transitional, Legendre, and minimum insertion-loss filters. Each of these has its special characteristics. The majority of filters for general application, however, use responses 1 and 2.

The Butterworth filter has a maximally flat amplitude response at the center of the passband and exhibits a gradual increase in attenuation near the 3-dB point, as shown in Figure 6.3. It provides reasonable selectivity and is monotonic in both the passband and stopband.

The Chebyshev filter has equal ripple in the passband. The response extends closer to the 3-dB point and does not have the gradual attenuation of the Butterworth filter in this region, especially for ripples of 0.1 dB or greater. In the transition region, Figure 6.4A, a sharper monotonic attenuation provides more selectivity for the same number of poles. In the limiting case when the passband ripple is zero, the response becomes a Butterworth filter. It becomes apparent that the allowance of a little ripple in the passband causes a great difference in the selectivity of a filter.

In general, for the Butterworth and Chebyshev filters, the more poles used, the sharper the attenuation transition at the 3-dB point. This type of characteristic results in a higher time-delay variation at that point. Figure

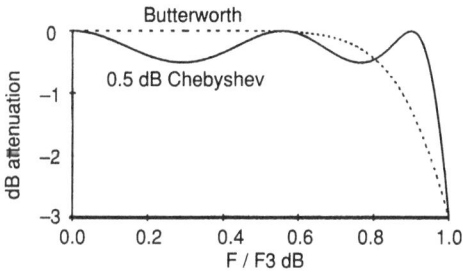

Figure 6.3 Passband comparison between Butterworth and Chebyshev filters, $n = 3$. In-band response of the Chebyshev is wider.

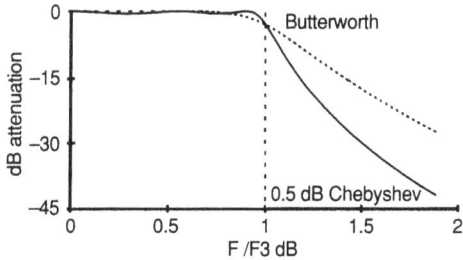

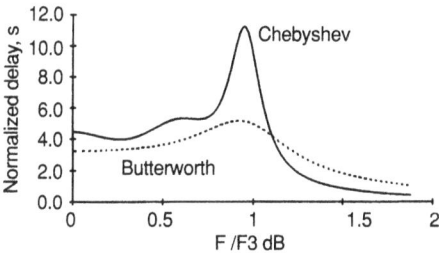

Figure 6.4 Comparison between Butterworth and Chebyshev filters, $n = 5$. (A) Selectivity plot showing steeper Chebyshev response; (B) delay characteristics showing greater delay variation for the Chebyshev. Note: Both (A) and (B) represent half the passband.

6.4B illustrates the difference between delays of a five-pole Butterworth and a five-pole 0.1-dB ripple Chebyshev filter. Because the Chebyshev filter has this sharper attenuation at 3 dB (Figure 6.4A), the delay variation is greater. Both exhibit a saucer-shaped delay across the passband.

The Gaussian filter has a relatively flat delay out to and beyond the 3-dB points, but has a slowly increasing attenuation characteristic in the stopband. The Bessel filter has the flattest delay and has similar attenuation characteristics. These two-filter responses are not widely used because of their poor selectivity. However, in applications where an impulse or step function signal must be preserved with low levels of ringing and overshoot, the constant time delay of these filters may be necessary. See the discussion on noise blankers in Chapter 7.

6.2 Filter Arrangements

In SSB systems, information transmission may occur in the USB, LSB, ISB, or four-channel ISB. In addition, AM may be transmitted as USB with carrier or AME. Since this mode customarily uses the USB filter, it will not be considered separately. More information on this mode is contained in Chapters 2 and 5.

In USB operation, the most widely used form of SSB in the HF band, the USB filter in the exciter or receiver allows the transmission and reception of signals above the carrier frequency, as shown in Figure 6.5. The audio is mixed with the carrier injection in a balanced modulator (for an exciter) producing a double-sideband output that is then filtered to pass only the USB signals. Then this IF is translated once or twice to the desired RF output frequency. In this case, it is presumed that the translation preserves the carrier-to-signal relationship, that is, an increase in AF causes an increase

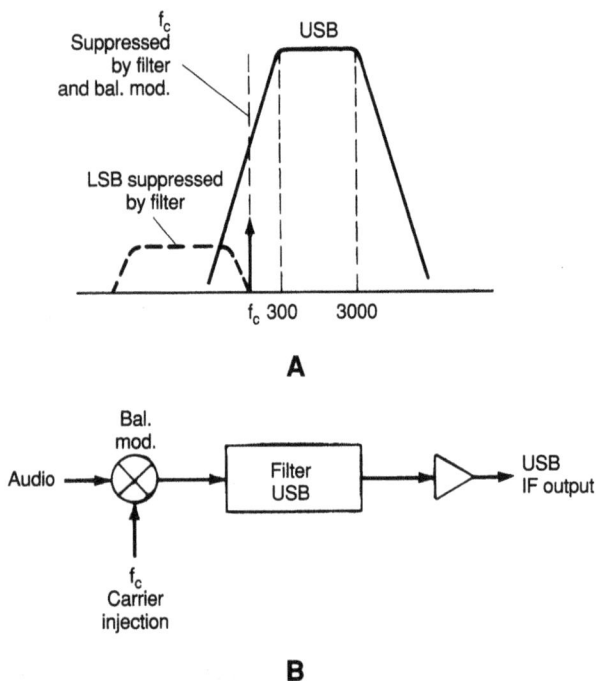

Figure 6.5 Relationship between USB, suppressed carrier, and opposite sideband. (A) Selectivity response; (B) USB exciter circuit block diagram.

in RF. For example, assume that the carrier frequency, although suppressed, is 2.0 MHz. The actual signals transmitted would be 2.000300 through 2.003000 MHz. In some equipment designs, a sideband reversal occurs so that an LSB filter is needed for USB transmission.

An example of bandwidth requirements for equipment can be obtained from a military standard specification (see reference 1). This specification concerns the requirement for an HF tactical digital information link. The overall receiver or exciter passband amplitude response is stated as not more than –2 dB between 450 and 3050 Hz, and not more than –3 dB at 300 Hz. The –60-dB attenuation frequencies are +4400 and –400 Hz. The negative sign indicates that the frequency is on the low-frequency side of the carrier. Since these are overall requirements, the 2- to 3-dB passband response has to be partitioned between all filters in the signal path. The 60-dB selectivity will have to be obtained primarily from the USB filter. The LSB is similarly specified except that the numbers will be negative because they are referenced to the carrier frequency. Typically, 10 to 30 dB of carrier attenuation can be expected of the SSB filter. Since at least –50 dB is usually required, the difference has to come from the carrier null in the balanced modulators.

In the LSB mode, only the frequencies below the carrier are transmitted. It should be noted that LSB is not the predominant mode of transmission in the HF range. When it is used, it is usually in addition to USB. Consequently, when equipment has USB and LSB (without ISB), the same audio and balanced modulator is used for both, with the filter switched to provide one or the other sideband, as shown in Figure 6.6.

Independent sideband uses both the USB and LSB modes simultaneously to transmit and receive two types of signals that can be completely independent of each other. Or the same information can be transmitted in both sidebands simultaneously to enhance the transmission link reliability. The use of ISB requires two separate audio channels, modulators, and filters in the transmitter, and two filters, IF amplifiers, product detectors, and audio amplifiers in the receiver (refer to Figure 6.7).

In the four-channel ISB mode of operation, filters are arranged as shown in Figure 6.8. The USB and LSB modes in the four-channel configuration are the same as described for ISB and use a common carrier frequency injection. The outer channels designated lower LSB (LLSB) and upper USB (UUSB) have additional subcarrier injections that are specified a fixed difference from the carrier injection frequency. The military standard frequency separation is $f_c \pm 6.290$ kHz.

Note that the outer sideband filters have the passband frequencies inverted with respect to the inner sidebands. This method provides the least amount of interference between the four modes of operation when all are in use simultaneously. Four separate audio amplifiers, modulators, and filters are required in the transmitter, and four separate channels are also required in the receiver.

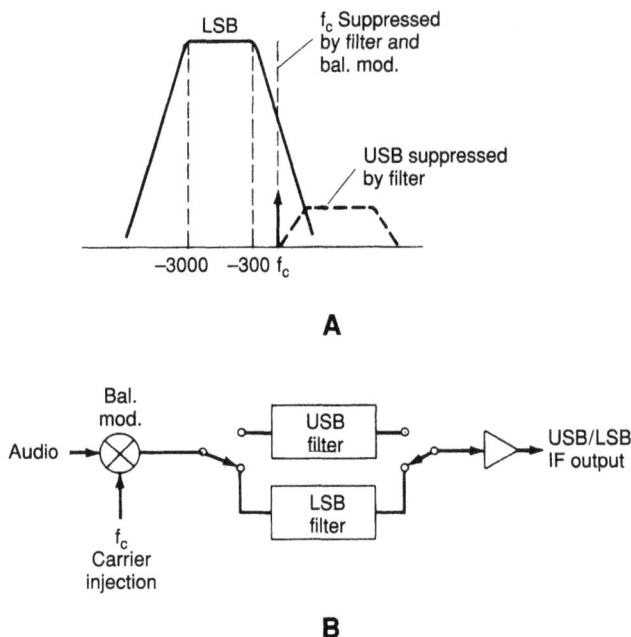

Figure 6.6 Relationship between LSB, suppressed carrier, and opposite sideband. (A) Selectivity response of LSB; (B) block diagram shows filter switching allowing either USB or LSB operation in exciter. Only one is in use at a time.

Another military standard (see reference 2) defines the selectivity characteristics for a four-channel radio as not more than –2 dB between 250 and 3100 Hz and not less than –40 dB at 50 and 3250 Hz. In addition, the attenuation shall not be less than 60 dB at 3550 Hz and higher frequencies and at $f_c - 250$ Hz and lower frequencies. This is stated for USB but applies to the other filters also.

This selectivity is extremely difficult to meet because the transition band is defined at the 40-dB as well as the 60-dB attenuation points.

6.3 Filter Types

Selectivity is a basic requirement for SSB receivers and exciters. Many options appear to be available for the equipment designer when specifying selectivity at the various IF frequencies. For high-performance radios, the range quickly narrows for SSB filtering. The use of mechanical and crystal filters predominates for low IF frequencies up to 500 kHz. Digital signal processing (Chapter 8) will also produce selectivity in this range. However, this method involves complex digital techniques, which is a totally new approach to IF filtering. At the other higher IF frequencies, the crystal filter possesses a clear advantage because of its low insertion loss and high degree of selectivity. Bandpass LC filters have their use also, but are often restricted to lowpass and highpass configurations. Ceramic filters have a low cost advantage in some applications. Surface acoustic wave filter applications generally lie in the VHF/UHF range.

LC filters

The LC bandpass filter is seldom used as an IF filter in HF receivers and exciters. In order to achieve the narrow passband and sharp skirt selectivity required for SSB filtering, extremely high inductor Q's are necessary. Mechanical and crystal IF filters utilize Q's of 10,000 or more, which are not attainable with coils.

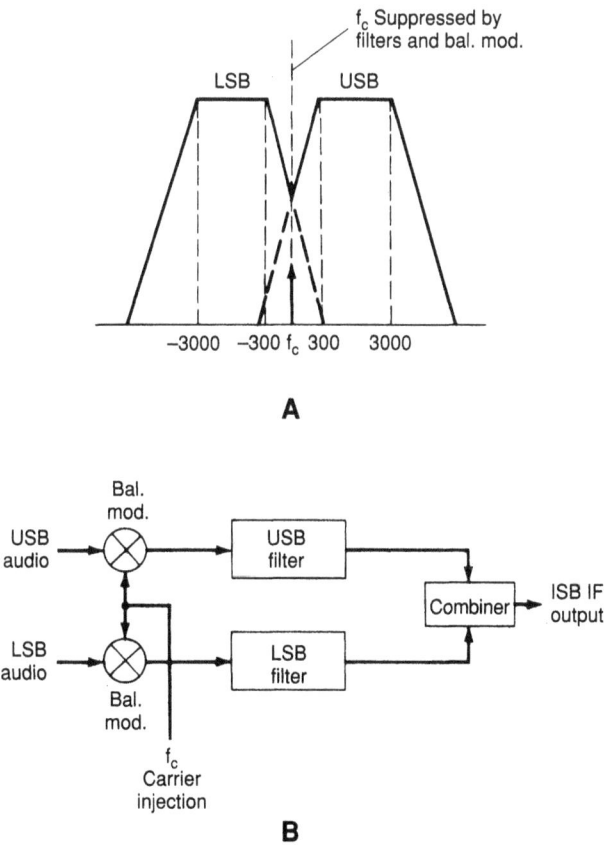

Figure 6.7 Independent sideband operation. (A) Selectivity response of both USB and LSB; (B) ISB exciter block diagram using separate channels allowing use of both simultaneously or each separately.

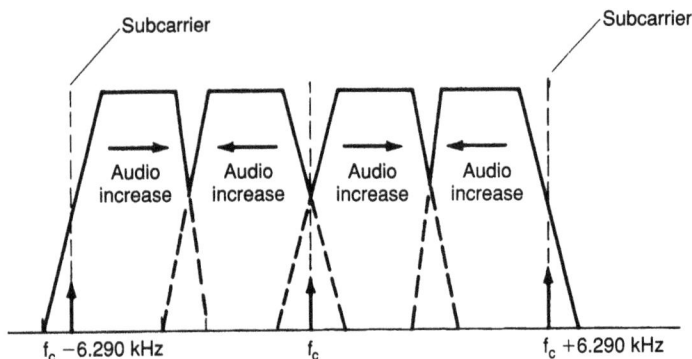

A

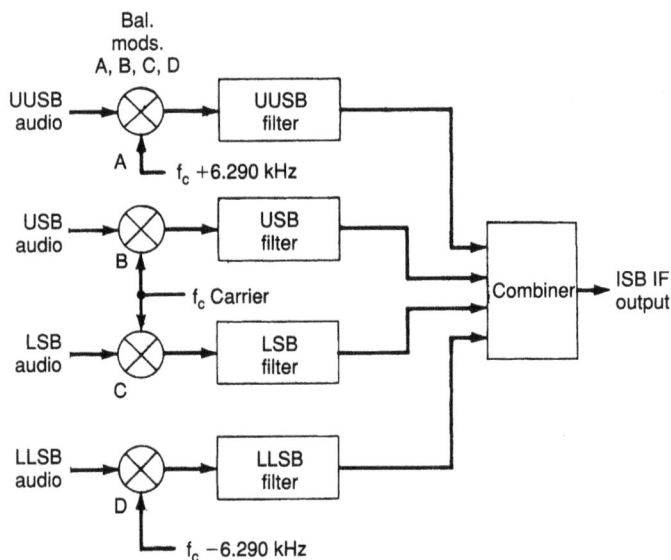

B

Figure 6.8 Four-channel ISB operation. (A) Selectivity response of four filters; (B) block diagram showing four independent audio and balanced modulator circuits.

Nevertheless, single- or double-tuned circuits are often used to provide a high impedance for amplifier stages or to further attenuate some spurious signals. Small single-layer solenoids can provide coil Q's in the 50 to 100 range. The use of iron or ferrite cores can decrease the coil size and increase the Q's into the 100 to 300 range. Toroidal inductors will also exhibit such low losses.

For higher Q's (up to 1000), helical resonators can be used (see reference 3). Although optimal use for these resonators lies in the VHF and UHF ranges, practical use extends down into the HF region. The helical resonator is a single-layer solenoid or helix enclosed in a circular or square highly conductive shield.

The achievable Q for a given size using copper for the coil and shield is determined from the following equation:

$$Q = 60S \, (f_0)^{1/2} \qquad (6.1)$$

where S = length of the square side, in
f_0 = resonant frequency, MHz

For example, a helical coil inside a 2-in. copper shield at 25 MHz can be expected to have a Q of $60 \times 2 \times 25^{1/2}$, or 600. The length of the enclosure would be 2×1.6, or 3.2 in. The large-volume, $2 \times 2 \times 3.2$, or 12.8 in.[3], for a single inductor precludes its use inside a receiver or exciter.

A comprehensive description of LC filters is presented in Chapter 9 which describes their use in preselectors and postselectors.

Mechanical filters

The mechanical filter for radio equipment has been in use since the mid to late 1940s, although the concept was developed earlier. The mechanical filter achieves selectivity by converting the electrical signal to mechanical vibrations that are passed through a network of high-Q resonators and coupling elements which are then converted back to electrical signals. In both the disc-wire type and the newer torsional type of mechanical filter, the disc resonators are coupled together with wires. The electrical analogy of this filter is composed of inductively coupled parallel-tuned circuits in a ladder configuration. Transducers used for the mechanical-electrical conversion are either the magnetostrictive or the piezoelectric type. The result is a highly selective bandpass filter that is stable over a wide range of environmental conditions.

In 1985, torsional resonator mechanical filters for 455-kHz IFs were developed (see references 4 and 5). These filters have resonators in the form of cylindrical rods vibrating in a torsional or twisting mode. Development of the torsional resonator filters resulted in a major step in miniaturization without sacrificing precision or stability. A size of 0.30 in. by 0.50 in. by 1.25 in. makes its size similar to a dual in-line IC. In addition, this approach has allowed a low-cost filter to be produced. Chapter 17 discusses a SPICE subcircuit file for low-cost torsional-mode CW, SSB, and AM mechanical filters.

Mechanical filters provide excellent selectivity characteristics. Their high-Q resonators have values that are typically about 20,000. Figure 6.9A

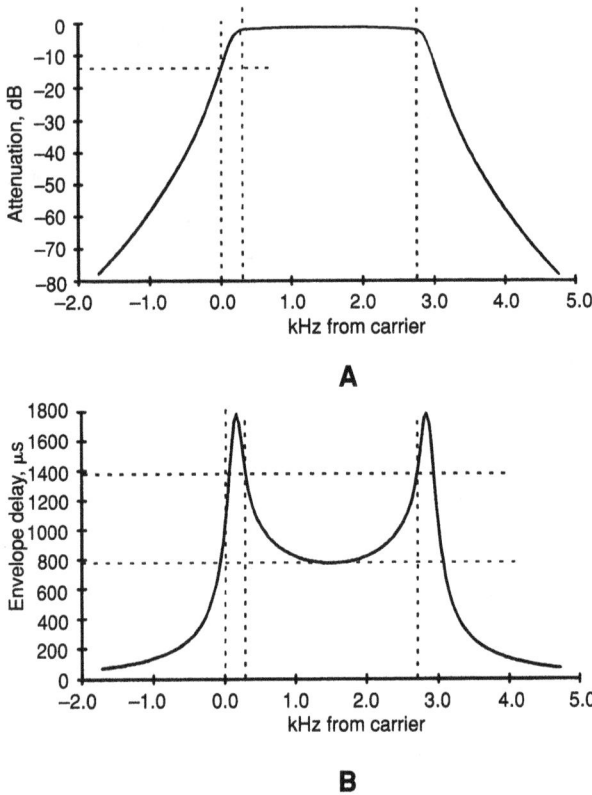

Figure 6.9 Mechanical filter characteristics. (A) Amplitude response; (B) time-delay response.

illustrates the selectivity characteristics of a representative eight-resonator 455-kHz mechanical filter. Note that the carrier frequency is attenuated by 14 dB. Typically, 10 to 30 dB of carrier attenuation can be provided.

The amount of carrier rejection is a function of the 3-dB attenuation frequency on the carrier side and the number of resonators. The stopband selectivity is dependent on the 3-dB bandwidth and, again, the number of resonators in the filter. A measure of selectivity is the filter's shape factor, which is the ratio of the 60-dB bandwidth to the 3-dB bandwidth. A typical 8-resonator filter has a shape factor slightly less than 2:1, whereas a 12-resonator mechanical filter with attenuation poles (transmission zeroes) both above and below the passband has a shape factor of 1.4:1. The attenuation poles are the result of bridging across two resonators. Bridging across one resonator can be used to produce a single attenuation pole to increase the steepness of the response on one side, for instance, the carrier frequency side of the filter. Besides realizing a desired amplitude response, the filter differential envelope delay must often be considered. The resultant filter response may have to be a compromise between the desired amplitude and delay responses.

The envelope delay characteristic is specified as a differential delay between passband frequencies (see also section 6.5). Figure 6.9 shows the amplitude and delay characteristics of a 2.7-kHz bandwidth filter that meets customer specifications. When the delay differential is specified out to the amplitude response edge, the manufacturer is forced to increase the passband limits. Note that in Figure 6.9 the delay begins to increase rapidly near the 3-dB amplitude response points. Consequently, in this application, the width of the passband amplitude response was increased to meet the delay requirement. This usually poses no problem because the passband amplitude response limits are stated as a minimum in most applications. However, if the 60-dB attenuation points have been specified, increasing the passband limits forces the shape factor to be smaller. Reducing the shape factor can be accomplished by increasing the number of resonators or adding attenuation poles.

The design range of the disc-wire and torsional mechanical filters is shown in Figure 6.10. The center frequency can vary from 60 kHz to 525 kHz, with percent bandwidths from 0.05 to 9.0%. The figure applies to both symmetrical and SSB filters. Low-frequency narrowband mechanical filters using flexure-mode resonators composed of iron-nickel alloy bars extend the range down to 3.5 kHz with percent bandwidths ranging from 0.2 to 1.5%.

Intermodulation distortion produced by electromechanical filters (mechanical, crystal, and ceramic) does not follow the normal third-order IMD rules (see reference 6). Instead of the third-order IMD slope being 3:1, the

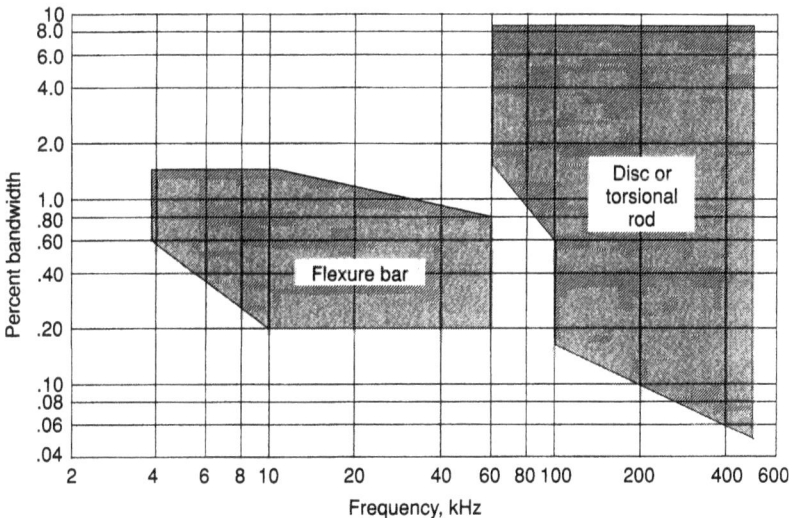

Figure 6.10 Percent bandwidth versus frequency for mechanical filters.

typical slope is 2.5:1. Some filter types have slopes close to 2:1 and others greater than 2.5:1. Filter-to-filter variations are also observed. Sometimes, portions of the third-order IMD curve have zero or even negative slopes. These anomalies are usually due to nonuniform conditions of a resonator surface: a chip, a particle, or bad plating. Although this problem is observed from time to time in all these electromechanical filters, it is most commonly seen in HF (above 40 MHz) crystal filters.

The radio designer has the choice of specifying a minimum attenuation of the third-order IMD over an output voltage range or simply measuring the performance of the filter in the radio. At low frequencies like 455 kHz, it is usually not necessary to require the mechanical filter supplier to measure the IMD on a production basis.

An example of the in-band third-order IMD for a typical 455-kHz SSB of the torsional resonator type is shown in Table 6.1. This is a 2.5-kHz, 3-dB bandwidth SSB filter, as discussed in Chapter 17. The source and load resistances are 2000 ohms shunted with a 30-pF capacitor. The insertion loss of the filter is 1.5 dB.

Disc-wire mechanical filters are remarkably free of spurious responses in the stopband, compared with crystal filters. This is because of the filters' ladder configuration, employing input and output electrical-tuned circuits and transducer resonators that reduce the effects of spurious modes of vibration of the interior resonators. Consequently, the attenuation far-off-frequency can be expected to remain about 70 to 90 dB down and becomes a function of the circuit layout rather than the filter characteristic.

The SSB torsional mechanical filters do not use input or output tuning inductors internally or externally. Nevertheless, the spurious responses 200 kHz on either side of the passband are attenuated below –60 dB, while those farther out are usually of little consequence in receiver design because filters in previous stages effectively attenuate them.

Mechanical filters require source and load terminations consisting of a capacitor in parallel with a resistor. The values of these components depend on the type of transducer used and the bandwidth of the filter. Old filter de-

TABLE 6.1 In-Band Third-Order IM Distortion Measurements on a 455 kHZ Mechanical Filter

Output signal level across 2000 ohms, V/tone	Third-order IMD below each tone, dB	
	Low side	High side
0.500	–50	–50
0.250	–58	–58
0.125	–66	–66
0.0625	–74	–74

signs used magnetostrictive-wire transducers that were self-terminating. These filters still required an external tuning capacitor and a terminating resistor of 100 kohms or greater. They also exhibited high insertion losses of roughly 20 dB. For the disc-wire types, the user must work closely with the manufacturer in specifying the type of termination desired. A wide variety of options are available, including unbalanced and balanced circuits, input and output circuits capable of carrying low-level dc currents (5 mA), circuits with dc blocking, etc. Once the impedance is determined, the user's circuit must duplicate the values to prevent additional insertion loss and, more important, increased passband ripple. The filters are most sensitive to capacitance changes, and tolerances of ±2% are usually specified. The insertion loss of these filters is usually between 3 and 6 dB.

In the case of typical SSB torsional filters, the source and load terminations consist of a 2-kohm resistor shunted with a 30-pF capacitor. The capacitance values are not critical as with the disc-wire type because no internal tuning inductors are used. For example, a 25% change in the capacitance increases the ripple by about 0.10 dB. The torsional filters are more sensitive to terminating resistance than capacitance changes; a 10% change in resistance results in a 0.10-dB increase in ripple.

Drive levels are dependent upon the IM requirements. Although most SSB mechanical filters are capable of being driven up to 10-V rms without damage, the input voltage maximum level for good performance should be less than 1.0-V rms. In low-level applications, the distortion of the amplifier stages may become the deciding factor on the actual signal levels that can be used. In receivers, large amounts of wideband amplification ahead of the filters invite third-order IM problems.

Insertion loss can vary from 1 to 20 dB and is dependent upon the number of resonators, bandwidth, and transducer type. For example, the torsional mechanical filter shown in Figure 6.9 has a 1.5-dB insertion loss.

Mechanical filters possess excellent temperature and aging characteristics. Frequency shifts over temperature are generally parabolic, but some of the newer torsional resonator materials are heat treated to exhibit third-order behavior over a –55°C to +85°C temperature range. Over this temperature range, the frequency variation from the room temperature value is less than ±40 Hz. Over a 20-year period, the filters are stable within 50 ppm. The center frequency shift would be less than 23 Hz for a 455-kHz filter.

Crystal filters

The crystal filter has been used as an SSB filter by the Bell System since the 1930s (see reference 7). That type provided the band selection necessary for the SSB frequency division multiplex voice channels. Their use in radio equipment followed, as SSB equipment came into wide use.

Crystal filters attain selectivity by the use of quartz crystals that act like highly selective tuned circuits. The crystals have extremely high Q's that can range from 10,000 to over 100,000. They utilize the piezoelectric effect of crystalline quartz. Originally, natural or mined quartz was used, but in recent times, cultured quartz has been developed and is used exclusively. The material is cut and precision-ground into blanks. The dimensions and the angle of cut of the material determines the resonant frequency and mode of vibration. This type of filter is characterized by low insertion loss, sharp skirt selectivity, and a high degree of temperature stability. Whereas mechanical filters are limited to approximately 500 kHz as an upper limit, crystal filters range from a few kilohertz to over 250 MHz (refer to Figure 6.11). A new ion-etching technique (see reference 8) enables crystal filters to operate up to 500 MHz. Percent bandwidths can range from 0.01 to 10% of center frequency.

Crystal filters are either the discrete crystal type or the monolithic type. Filters that use discrete crystal blanks require additional inductors, transformers, and capacitors in lattice-type networks to achieve the bandpass characteristic. These are capable of operation from a few kilohertz to 100 MHz. The monolithic filter uses a quartz wafer as a substrate with deposited electrodes to form resonators. This type of construction eliminates the internal components necessary for the discrete type. This unique construction results in a smaller package, lower cost, and higher reliability. Monolithics have been designed to operate from 10 to over 100 MHz.

A single crystal behaves as a tuned circuit, as shown in Figure 6.12. It has a series resonant frequency determined by L and C followed by a parallel resonance caused by the capacitance of the coupling plates or electrodes. Occasionally a single crystal is used as a filter when a fixed frequency, such as a mixer injection, requires a small amount of additional spurious attenuation. Since it is a two-terminal device, it is easily adapted into a circuit.

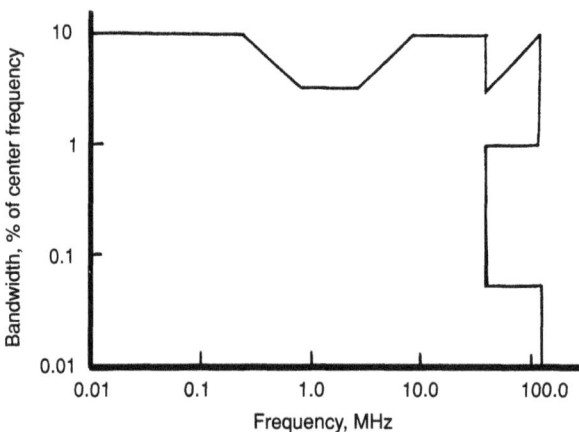

Figure 6.11 Crystal filter percent bandwidth versus frequency.

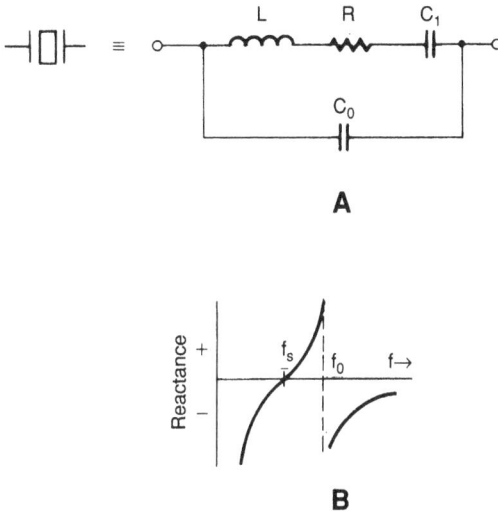

Figure 6.12 Properties of a quartz crystal. (A) Equivalent electrical circuit diagram; (B) frequency response of the crystal circuit. L = motional inductance; C = motional capacitance; R = energy loss resistance; C_0 = capacitance of electrodes and holder.

The source and terminating impedances of crystal filters consist of a resistance and possibly a parallel capacitance. The values of the resistances vary over a wide range, from a few hundred ohms at 50 MHz and above to over 100,000 ohms at 10 kHz. In general, the higher the frequency, the lower the impedance level. The capacitive reactance is greater than the resistance and is specified to accommodate the stray capacitance in the circuit application. The exception is at the 50-ohm impedance level. When filters are designed for use above 30 MHz, it is advantageous to specify a 50-ohm source and load impedance. With such a termination, it becomes a simple task to measure insertion loss and amplitude characteristics in the lab because RF voltmeters and signal generators operate at this impedance level. The filter test fixture requires only input and output RF connectors. More important, circuit design is simplified because diode quad mixers and hybrid amplifiers are usually specified with 50-ohm input and output impedances.

Spurious responses in a crystal filter are usually a function of the number of poles. These occur because of the spurious response modes of the crystals. A two-pole filter will have spurious responses at the –25 to –30-dB level, a few hundred kilohertz above the passband. In an eight-pole configuration the responses may be suppressed below the –70-dB level.

In some circuits, a two-pole filter will provide sufficient selectivity following, for example, a mixer stage. If the injection frequency and image frequency fall on the low side of the filter, the spurious region on the high side can thus be tolerated. If high-side injection is necessary, the injection can be placed outside the spurious area. Most often, frequency schemes are chosen based on other factors, and filters with the desired characteristics are then specified. By noting these filter response characteristics, some advance planning may allow the use of a simpler filter in particular applications.

The third-order IMD in crystal filters does not follow the third-order intercept line. Consequently, the distortion products cannot be predicted at various signal levels based on one distortion measurement. Thus, the intercept equation cannot be used with any validity, and the designer must resort to actual measurements at the signal levels over which the filter is expected to operate. Distortion in crystal filters can occur at low as well as high operating levels. At low drive levels, IM can be due to high starting resistance caused by crystal surface contamination, scratches, plating defects, etc. High drive level causes may include nonlinear resonance and parametric excitation of other modes. Low drive level causes can be controlled in the manufacturing process. However, the high-level problems apparently are not well defined or completely understood at this time.

Nevertheless, crystal filters do provide consistent and predictable results at the design range levels. The examples of signal levels and resulting distortion product measurements on crystal filters given in Tables 6.2 and 6.3 can guide the designer toward reasonable operating levels.

Table 6.2 lists measured values taken on a 24-crystal, 450-kHz, high-performance filter that has a 1-dB passband of 250 to 3100 Hz and a 4-dB insertion loss. The filter source and load impedance is 2000 ohms in parallel with a 33-pF capacitor. The output level is in millivolts per tone across the load impedance. Note that for a 6-dB increase in signal level (30 to 60 mV), the signal-to-distortion ratio remains essentially constant instead of increasing by 12 dB, if the third-order intercept rule were followed. For this filter, the third- and higher-order distortion products are specified not to exceed –60 dB below each tone of a 50-mV/tone or less signal level, measured at the output terminals. The table indicates a distortion margin of approximately 7 dB.

Table 6.3 shows distortion levels measured on a 105-MHz IF crystal filter. This six-pole filter has a 0.5-dB bandwidth of 25 kHz and an insertion loss of 5 dB. The input and output impedances are 50 ohms and the levels are specified in dBm. The tones are 10 kHz apart, centered in the passband. The measurements show evidence of probable phase modulation cancellation as indicated by the level difference between the low- and high-side products.

Out-of-band IMD may be considerably less than the in-band levels. For this type of test, both test frequencies are out-of-band and spaced so that the third order falls inside the passband. The reference signal level is the voltage across a 50-ohm load (maximum available power). Because the test signals are out-of-band, the selectivity of the crystal resonators tends to reduce the levels across them. Therefore, the distortion effects are often decreased. For filters using ferrite impedance transformers, the ferrite material can become a source of distortion, especially at the higher levels.

TABLE 6.2 In-Band Third-Order IMD
Measurements on a 450-kHz Crystal Filter

Signal output level across 2000 ohms, mV/tone	Third-order IMD below each tone, dB	
	Low side	High side
30	–68	–68
40	–68	–69
50	–67	–67
60	–68	–69
79	–68	–68

TABLE 6.3
In-Band Third-Order IMD Measurements
on a Six-Pole 105-MHz Crystal Filter

Signal output level at 50 ohms, dBm/tone	Third-order IMD below each tone, dB	
	Low side	High side
0 (0.22 V/tone)	–50	–55
–5 (0.12 V/tone)	–63	–60
–10 (0.07 V/tone)	–65	–63

For the experimenter or radio amateur who wishes to build his own crystal filters, reference 9 describes a simple design procedure using matched crystals in a shunt-arm ladder configuration. This approach produces an SSB-type filter response with a steep cutoff on one side for carrier rejection and less attenuation on the other side [3].

Another approach [10] uses identical frequency crystals in the series arms of a ladder network. By changing the value of the shunt capacitors, the shape of the filter can be changed from a narrow symmetrical CW response to a wider SSB-type response.

The fabrication of high-performance crystal filters, including prototypes, is normally left to companies that have the resources to design the filter, grind the crystals, and package them to the user's specifications.

SAW filters

The SAW filter is an inherently simple and rugged device that provides a bandpass characteristic from electrodes deposited on a piezoelectric substrate. These electrodes form an interleaved (interdigital) pattern on the substrate and serve as transducers. When an RF signal is applied to one transducer, an electric field is generated, causing acoustic waves at the RF frequency to propagate along the surface to the opposite transducer. The waves generate an electric field that produces an output voltage. The bandpass characteristics are controlled by the choice of substrate material and the electrode interleaved pattern. The fabrication process requires the highly developed manufacturing techniques of the semiconductor industry. Because of the high initial design costs, but low manufacturing costs, the filters are widely used in the high-volume consumer market, such as for TV sets, where the unit cost can be reduced to a few dollars.

Surface acoustic wave filters can be designed to operate from about 10 to over 1000 MHz. Percent bandwidth can vary from a few percent to about 50%. The choice of substrate material affects the temperature coefficient, frequency range, and insertion loss. Widely used materials include quartz and lithium niobate. The insertion loss of SAW filters is high, generally ranging from 10 to 25 dB. However, new techniques have produced filters with losses as low as 2 and 3 dB.

The advantages of the SAW filter include practically constant time-delay characteristics across the passband (linear phase), a flat amplitude response, extremely good shape factor (1.2:1), and small size.

Disadvantages include the presently high insertion loss, difficulty in producing narrow-bandwidth filters (below 1% fractional bandwidth), and design cost. But even though their cost may be high in small volume, the SAW filters' special characteristics find application in radar systems and high-performance communication equipment, primarily in the VHF/UHF range.

6.4 Intermodulation Distortion

Intermodulation distortion is extremely important in radio systems that transmit and receive data. When nonlinearities exist in amplifiers, mixers, and filters, distortion products are generated. If a single tone is transmitted in-band, only harmonics are generated, which in the case of IF frequencies are out-of-band and easily attenuated. If two or more tones are transmitted, in-band IMD products are created around the transmitted tones (in addition to out-of-band products around the harmonics). Since data transmission usually involves closely spaced audio tones, predicting distortion levels becomes a necessity in radio design.

In general, the nonlinearities in a transfer function of an amplifier can be expressed as a power series around a zero-signal operating point:

$$i = K_0 + K_1e + K_2e^2 + K_3e^3 + K_4e^4 + K_5e^5 \qquad (6.2)$$

K_0 and K_1e represent the linear transfer. The terms with powers of 2 to 5 represent the nonlinearities that create the distortion products.

A two-frequency signal is represented by

$$e(t) = A_1 \cos \omega_1 t + A_2 \cos \omega_2 t \tag{6.3}$$

where A_1 and A_2 = amplitudes of the signals
  $\omega_1 = 2\pi f_1$ and $\omega_2 = 2\pi f_2$
  f_1 and f_2 = signal frequencies

Substituting for e in Equation 6.2 produces the following terms which are grouped in terms of distortion product orders.

1. The fundamentals components are

$$\tag{6.4}$$

$(K_1 A_1 + \tfrac{3}{4} K_3 A_1^3 + \tfrac{3}{2} K_3 A_1 A_2^2 + \tfrac{5}{8} K_5 A_1^5 + \tfrac{15}{4} K_5 A_1^3 A_2^2 + \tfrac{15}{8} K_5 A_1 A_2^4) \cos \omega_1 t$

$(K_1 A_2 + \tfrac{3}{4} K_3 A_2^3 + \tfrac{3}{2} K_3 A_1^2 A_2 + \tfrac{5}{8} K_5 A_2^5 + \tfrac{15}{4} K_5 A_1^2 A_2^3 + \tfrac{15}{8} K_5 A_1^4 A_2) \cos \omega_2 t$

2. The second-order components are

$(K_2 A_1 A_2 + \tfrac{3}{2} K_4 A_1^3 A_2 + \tfrac{3}{2} K_4 A_1 A_2^3) \cos(\omega_1 \pm \omega_2) t \tag{6.5}$

$(\tfrac{1}{2} K_2 A_1^2 + \tfrac{1}{2} K_4 A_1^4 + \tfrac{3}{2} K_4 A_1^2 A_2^2) \cos 2\omega_1 t$

$(\tfrac{1}{2} K_2 A_2^2 + \tfrac{1}{2} K_4 A_2^4 + \tfrac{3}{2} K_4 A_1^2 A_2^2) \cos 2\omega_2 t$

3. The third-order components are

$(\tfrac{3}{4} K_3 A_1^2 A_2 + \tfrac{5}{4} K_5 A_1^4 A_2 + \tfrac{15}{8} K_5 A_1^2 A_2^3) \cos(2\omega_1 \pm \omega_2) t \tag{6.6}$

$(\tfrac{3}{4} K_3 A_1 A_2^2 + \tfrac{5}{4} K_5 A_1 A_2^4 + \tfrac{15}{8} K_5 A_1^3 A_2^2) \cos(\omega_1 \pm 2\omega_2) t$

$(\tfrac{1}{4} K_3 A_1^3 + \tfrac{5}{16} K_5 A_1^5 + \tfrac{5}{4} K_5 A_1^3 A_2^2) \cos 3\omega_1 t$

$(\tfrac{1}{4} K_3 A_2^3 + \tfrac{5}{16} K_5 A_2^5 + \tfrac{5}{4} K_5 A_1^2 A_2^3) \cos 3\omega_2 t$

4. The fourth-order components are

$(\tfrac{1}{2} K_4 A_1^3 A_2) \cos(3\omega_1 \pm \omega_2) t \tag{6.7}$

$(\tfrac{3}{4} K_4 A_1^2 A_2^2) \cos(2\omega_1 + 2\omega_2) t$

$(\tfrac{1}{2} K_4 A_1 A_2^3) \cos(\omega_1 \pm 3\omega_2) t$

$(\tfrac{1}{8} K_4 A_1^4) \cos 4\omega_1 t$

$(\tfrac{1}{8} K_4 A_2^4) \cos 4\omega_2 t$

5. The fifth-order components are

$(\tfrac{5}{16} K_5 A_1^4 A_2) \cos(4\omega_1 \pm \omega_2) t \tag{6.8}$

$(\%K_5 A_1^3 A_2^2) \cos (3\omega_1 \pm 2\omega_2)t$

$(\%K_5 A_1^2 A_2^3) \cos (2\omega_1 \pm 3\omega_2)t$

$(\%_{16}K_5 A_1 A_2^4) \cos (\omega_1 \pm 4\omega_2)t$

$(\%_{16}K_5 A_1^5) \cos 5\omega_1 t$

$(\%_{16}K_5 A_2^5) \cos 5\omega_2 t$

The fundamental components can be considered the first order; the second-order components contain the second harmonics $2f_1$ and $2f_2$ (dropping the 2π constant) and the sum and difference of the fundamentals $f_1 + f_2$ and $f_1 - f_2$. The third order contains the third-order harmonics of the signals, $3f_1$, $3f_2$, and the third-order products $2f_1 + f_2$, $2f_1 - f_2$, $f_1 + 2f_2$, and $f_1 - 2f_2$, each of which represents a single frequency. The same applies to the fourth and fifth orders.

A considerable amount of information can be obtained from this expansion. The primary concern, although not the only one, has to do with the in-band components. However, out-of-band components are of interest too, especially in broadband amplifier applications. Grouping the frequencies in increasing frequency order shows which components are in-band and which are out-of-band. To show the actual frequency positions of the various orders, numbers are assigned to the fundamental pair. Let $f_1 = 2.0$ MHz and $f_2 = 2.1$ MHz. Although separated by 100 kHz, let us assume that the frequencies are in-band. Table 6.4 shows the frequency positions.

A number of observations can be made from Table 6.4 for the parameters assumed for this example:

1. All the frequencies shown, except the fundamentals, are caused by the nonlinearity of the device (K_2, K_3, K_4, and K_5 coefficients).

2. Only the odd orders (third, fifth) show up around the fundamental frequencies and only these are in-band. Although in this case the frequency separation was exaggerated to simplify the table, the frequency spacing would normally be from 100 Hz to 3 kHz.

3. The frequency difference Δf between all the distortion components in any harmonic region is the same and is equal to the Δf frequency spacing of the fundamental signals.

4. Note that the second-order product falls between the two second harmonic frequencies. The relative level of the second order and second harmonic can be calculated.

5. Note also that the two third-order products fall between the two third harmonics. The relative levels of the third-order products and the third harmonics can be calculated.

6. The orders fall between the harmonics in the fourth and fifth harmonic regions.

7. Knowing the sequence of the distortion products can be very helpful when attempting to identify the products on a spectrum analyzer.

The preceding analysis is based on a simple analytical model where only AM distortion occurs and only one stage is involved. If the signal undergoes a small phase shift in addition to the amplitude change, then phase modulation distortion will occur also. In AM distortion, all the distortion

**TABLE 6.4 Distortion Products
Illustrating the Sequence of Orders and Harmonics**

Region	MHz	Frequency components	Orders/harmonics
Audio and VLF region	0	dc	
	0.1	$f_2 - f_1$	Second order
	0.2	$2f_2 - 2f_1$	Fourth order
Fundamental region (in-band)	1.8	$3f_1 - 2f_2$	Fifth order
	1.9	$2f_1 - f_2$	Third order
	2.0	f_1	Fundamental
	2.1	f_2	Fundamental
	2.2	$2f_2 - f_1$	Third order
	2.3	$3f_2 - 2f_1$	Fifth order
Second harmonic region	3.9	$3f_1 - f_2$	Fourth order
	4.0	$2f_1$	Second harmonic
	4.1	$f_1 + f_2$	Second order
	4.2	$2f_2$	Second harmonic
	4.3	$3f_2 - f_1$	Fourth order
Third harmonic region	5.9	$4f_1 - f_2$	Fifth order
	6.0	$3f_1$	Third harmonic
	6.1	$2f_1 + f_2$	Third order
	6.2	$2f_2 + f_1$	Third order
	6.3	$3f_2$	Third harmonic
	6.4	$4f_2 - f_1$	Fifth order
Fourth harmonic region	8.0	$4f_1$	Fourth harmonic
	8.1	$3f_1 + f_2$	Fourth order
	8.2	$2f_1 + 2f_2$	Fourth order
	8.3	$f_1 + 3f_2$	Fourth order
	8.4	$4f_2$	Fourth harmonic
Fifth harmonic region	10.0	$5f_1$	Fifth harmonic
	10.1	$4f_1 + f_2$	Fifth order
	10.2	$3f_1 + 2f_2$	Fifth order
	10.3	$2f_1 + 3f_2$	Fifth order
	10.4	$f_1 + 4f_2$	Fifth order
	10.5	$5f_2$	Fifth harmonic

products have a positive sign. In low-level phase modulation, the same odd-order components can be generated, but they have a negative sign (180° out-of-phase) for the lower-sideband products and a positive sign (in-phase) for the upper-sideband products. Consequently, the resulting sidebands will be unequal, with the lower-sideband amplitudes smaller and the upper-sideband amplitudes larger. Generally, in low-level class A stages, this effect is minor.

The same effect can result as the signal passes through tuned circuits, filters, and additional stages. Phase shifting of the distortion products will create unequal product amplitudes.

Product levels

The relative amplitude levels between the second harmonic and the second order can be determined from Equation 6.5. The equation for the second-order frequency is

$$(K_2 A_1 A_2 + \tfrac{3}{2} K_4 A_1^3 A_2 + \tfrac{3}{2} K_4 A_1 A_2^3) \cos (\omega_1 \pm \omega_2) t$$

and for the second harmonic:

$$(\tfrac{1}{2} K_2 A_1^2 + \tfrac{1}{2} K_4 A_1^4 + \tfrac{3}{2} K_4 A_1^2 A_2^2) \cos 2\omega_1 t$$

Assume that the amplitude of the two signals (Equation 6.3) are $A_1 = A_2 = 1$. The ratio of the amplitude coefficients of the second order and the second harmonic, after combining the K_4 terms, becomes

$$\frac{\text{Second-order amplitude}}{\text{Second harmonic amplitude}} = \frac{K_2 + 3K_4}{\tfrac{1}{2} K_2 + 2 K_4} \tag{6.9}$$

If K_2 is much greater than K_4, that is, the second order predominates over the fourth, then the ratio reduces to

$$\frac{\text{Second-order amplitude}}{\text{Second harmonic amplitude}} = \frac{K_2}{\tfrac{1}{2} K_2} = 2 \tag{6.10}$$

Thus it is seen that the second-order amplitude is twice the amplitude of the second harmonic (6 dB greater).

The relative amplitude levels of the third-order and third-harmonic components can also be determined in the same manner using Equation 6.6. As before, let $A_1 = A_2 = 1$, and combining the K_5 terms together, the ratio becomes

$$\frac{\text{Third-order amplitude}}{\text{Third harmonic amplitude}} = \frac{\tfrac{3}{4} K_3 + \tfrac{25}{8} K_5}{\tfrac{1}{4} K_3 + \tfrac{25}{16} K_5} \tag{6.11}$$

If K_3 is much greater than K_5, then the ratio reduces to

$$\frac{\text{Third-order amplitude}}{\text{Third harmonic amplitude}} = \frac{\frac{3}{4}K_3}{\frac{1}{4}K_3} = 3 \tag{6.12}$$

In this case, the third-order component is three times the amplitude of the third harmonic (9.5 dB greater).

These relative levels are shown graphically in Figure 6.13. Note that the in-band third order and out-of-band third order have the same magnitude. This is evident from Equation 6.6, where the third-order frequencies $2\omega_1 \pm \omega_2$ and $\omega_1 \pm 2\omega_2$, which represent the in-band and out-of-band third orders, have the same amplitude coefficients.

The levels of the second- and third-order products with respect to the fundamental frequency levels can be calculated by using the intercept point equation. It should be noted that this analysis represents an ideal condition and does not take into account possible phasing alterations that would change these relationships.

Intercept point

The intercept point method is used for predicting third-order IM product levels for nearly linear circuits, such as class A amplifiers, mixers, etc. The third-order output intercept is the point where the two-tone output signal level crosses over the third-order distortion that is created. This is a theo-

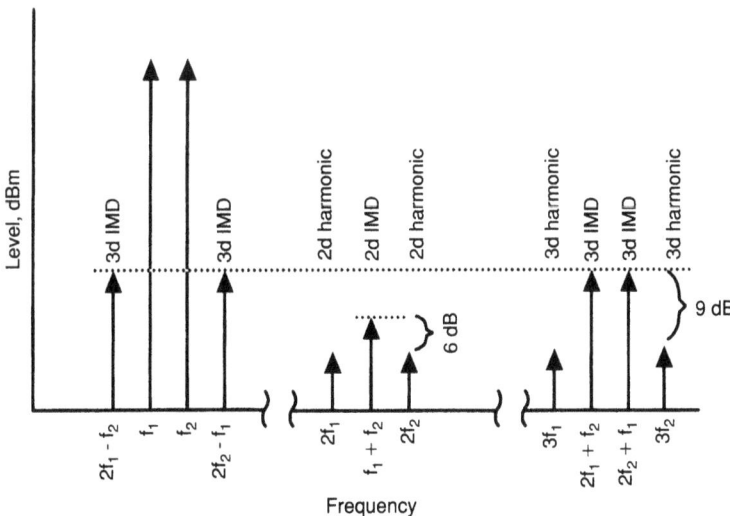

Figure 6.13 Frequency spectrum of a two-tone signal showing the relative levels between second order and second harmonic, and third order and third harmonic.

retical point and lies above the device's compression point (where the level begins to depart from linearity and starts to limit). This is described in Chapter 4. Because the third-order products are created by the power-of-3 coefficients, a 10-dB change in the two-tone signal output will cause a 30-dB change in the third-order distortion. However, referenced to the desired output level, the change in distortion is only 20 dB. When a device has distortion products that follow the intercept line, it is considered "well behaved."

The graph shown in Figure 6.14 illustrates the intercept concept (see reference 11). Another useful tool is the equation for the third-order intercept, which can be derived from Figure 6.14.

$$OPI^3P = P_0 + \frac{IMD}{2} \tag{6.13}$$

where OPI^3P = third-order output intercept, dBm
P_0 = output level per tone, dBm/tone
IMD = third-order IMD ratio below each output tone, dB

Rearranging the equation for IMD, which is the quantity usually desired, produces

$$Third\ IMD = 2\,(OPI^3P - P_0) \tag{6.14}$$

For example, if an amplifier has an OPI^3P of 30 dBm and is driven to produce an output level of –5 dBm/tone, the third-order IMD is $2[30 - (-5)]$ or 70 dB below either –5-dBm tone.

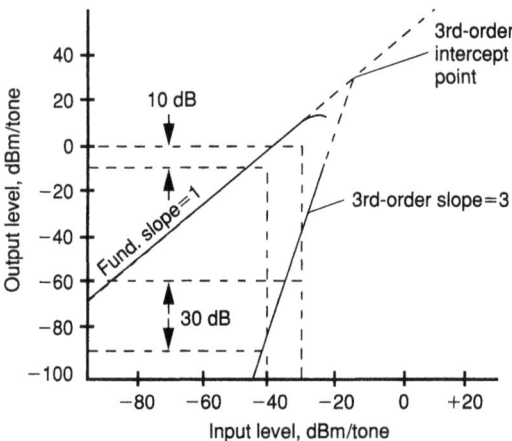

Figure 6.14 Third-order intercept line. A 10-dB change in output (or input) produces a 30-dB change in third-order distortion.

The intercept point can be referenced to the input signal instead of the output. Circuits that have losses, such as mixers, have input intercept points (IPI^{3}P) specified. The IMD calculated from the input signal level will be the IMD at the output. The output level will be the input level reduced by losses of the circuit.

The second-order intercept point is sometimes specified for mixers and amplifiers. It becomes useful in calculating the second-order out-of-band signal for broadband circuits that have no selectivity. The equation for the second-order intercept point is

$$OPI^2P = P_0 + IMD \tag{6.15}$$

or

$$IMD = OPI^2P - P_0 \tag{6.16}$$

where OPI^{2}P = second-order output intercept point, dBm
   P_0 = output level per tone, dBm/tone
   IMD = second-order IMD below each output tone, dB

Since the second order is created by the second-power coefficients, a 10-dB change in signal level will cause a 20-dB change in the second-order distortion. Referenced to the output level, the distortion increase is only 10 dB, as shown by the equation.

6.5 Envelope Delay Distortion

Distortion through time delay

The SSB signal can be distorted by a varying time delay across the passband as well as by a changing amplitude characteristic. When a signal passes through a filter, a time delay occurs because of the phase shift caused by the circuit reactances. If this delay is not constant with frequency, signal waveform distortion of the signal will result. This is made obvious by examining a plot of two frequency components as shown in Figure 6.15. In Figure 6.15A, both the fundamental and the third harmonic start at time zero, resulting in the composite signal shown. If both signals are delayed an equal amount of time, this relationship will not change. If the third harmonic is delayed an amount Δt with respect to the fundamental, as shown in Figure 6.15B, the composite signal is distorted with respect to the original waveform. Consequently, in order to maintain envelope integrity, the delay for all modulated signal frequencies must be the same.

The envelope delay can be expressed as an equation relating phase shift to frequency. Mathematically the delay is equal to $-d\phi/d\omega$ where ϕ is in ra-

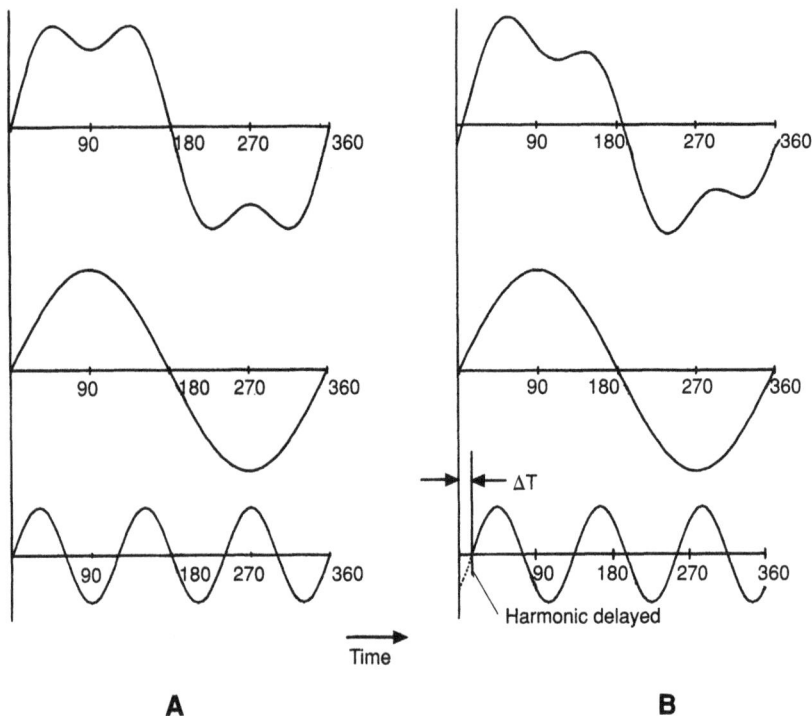

Figure 6.15 Effect of signal delay on composite waveform. (A) Fundamental and third harmonic in phase; (B) harmonic delayed by amount Δt. Top curve: resultant waveform; middle: fundamental; bottom: third harmonic.

dians and ω is in radians per second. For incremental differences, the following equation applies and the minus sign is omitted:

$$T = \frac{\Delta\phi}{360\Delta f} \tag{6.17}$$

where $\Delta\phi$ = change in phase, deg
Δf = change in passband frequency, Hz
T = envelope delay, sec

In order to obtain a constant envelope delay T, the phase shift must be directly proportional to the frequency change. Any deviation from linearity will cause a change in the time delay or delay distortion. Figure 6.16 helps to illustrate the equation. The steeper the phase-shift slope, the greater the total delay.

The term *group delay* is sometimes used in place of envelope delay. Group delay is a telephone circuit term that applies to the delay of groups of audio frequencies. In RF circuit applications the terms envelope delay and group delay are used interchangeably.

Envelope delay distortion is usually specified as a differential time delay between tones in the passband. In multitone data transmissions, a certain amount of delay distortion is tolerated. For example, a 16-tone digital data transmission system in common use allows a 500-µs differential delay between passband tones of 815 and 3050 Hz for the receiver and also for the transmitter.

Figure 6.17 illustrates a typical delay characteristic of a USB filter. The differential delay is the difference between the minimum point of the delay and the delay at some specified passband frequencies. In the graph, this is shown as a differential delay boundary. The actual delay curve is shown to be less than the specified amount. If the delay curve is unsymmetrical, the worst-case frequency is used to determine the Δ.

The total delay for a receiver or exciter is specified as a maximum within the passband frequencies. In Figure 6.17, the total delay can be determined for any frequency because the y ordinate is scaled as total delay. The other filters in the signal path cannot be neglected because they also contribute delay. However, in SSB equipment the greatest delay, both differential and total, occurs in the low-frequency SSB filters. The other IF filters in the signal path can be considered to contribute about 10% of the overall amount. Consequently, the delay of the other filters must be subtracted from the delay characteristics that are specified for the SSB filter. Total delay is ob-

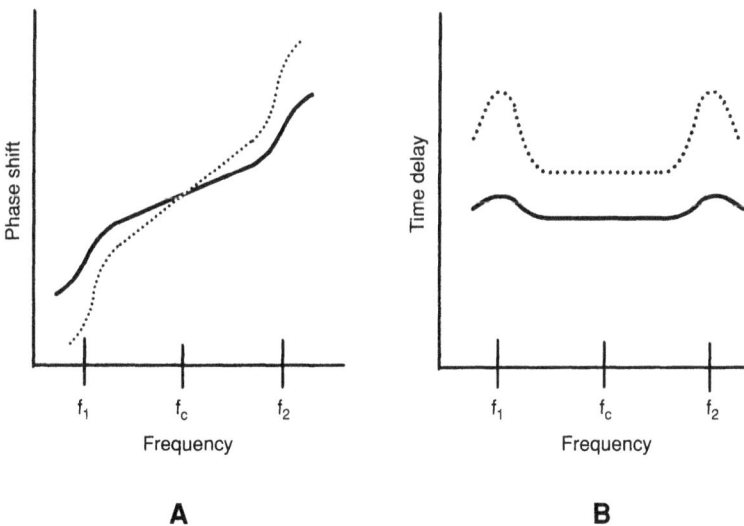

Figure 6.16 Phase-delay characteristics. (A) Phase-shift versus frequency response; (B) steeper phase curve produces a greater delay.

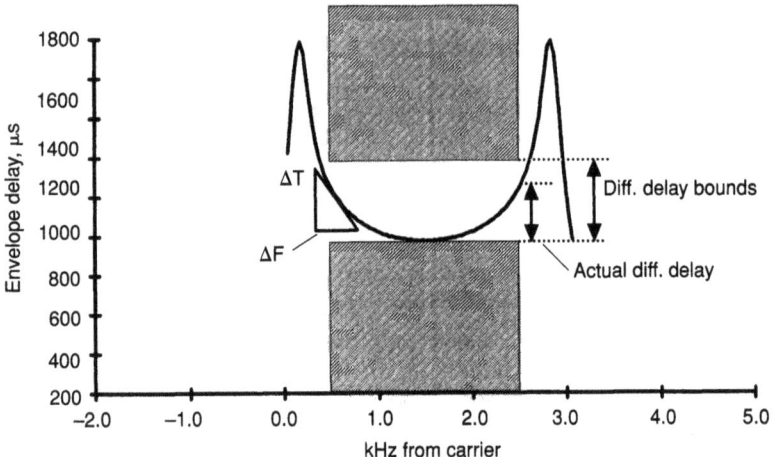

Figure 6.17 Typical delay characteristics of a USB filter with differential delay specified across the passband frequencies.

tained from Equation 6.17, provided the measurement is made between the audio input and the RF of an exciter or the RF input and the audio output of a receiver.

Another specification that occasionally must be taken into consideration is the delay slope distortion. When specified, it is stated as a ratio of Δ delay to Δ frequency that must be met between the differential delay frequencies. In Figure 6.17 the delay distortion could be specified not to exceed 150 µs for any 100-Hz frequency increment between 600 and 3000 Hz. The intent of this requirement, probably, is to control the delay ripple. For most filters the in-band delay ripple is minor, but the problem is created at the passband edges. Here the delay can have a sharp slope that exceeds the slope specification but still meets the differential delay. The problem can be resolved two different ways. One choice for the designer is to widen the passband and thereby lower the slope. This approach forces a decrease in the shape factor ratio because the 60-dB attenuation bandwidth is probably specified. The result is an increase in filter complexity and cost. The alternative is to equalize the delay to produce a relatively flat delay response over the passband frequencies. This approach requires additional circuitry either at the IF or audio frequencies, but is often used.

Delay equalization

The total delay through a receiver or exciter is the sum of the individual delays, most of which occur in the bandpass filters. This additive characteristic of the delay makes it possible to compensate any undesirable variation by adding a delay with an opposite slope characteristic as shown in

Figure 6.18. Equalizing the delay increases the total delay through the equipment. The total delay is not a critical element and in most systems can be tolerated. Some difficulties may be encountered in receiver AGC loops that contain filters with large amounts of time delay. Loop instability can occur and must be corrected.

A very convenient method of equalization utilizes the allpass filter as a delay equalizer. This type of filter provides a flat amplitude frequency response and a prescribed phase-shift frequency or delay characteristic. This is precisely the characteristic that is needed for compensation.

The allpass filter delay equalizer can be a first- or a second-order design. The first-order equalizer has a delay that is maximum at dc and consequently has limited usage. It can be used to compensate a lowpass filter, for example. The second-order equalizer has a delay that can be positioned anywhere in the passband at the IF or audio frequencies. In addition, the amplitude response can be made at the IF or audio frequencies, constant and independent of the delay chosen. The characteristics of a first- and second-order equalizer are shown in Figure 6.19. The width of the delay curve in Figure 6.19B is a function of the peak delay time. The greater the delay, the narrower the delay response becomes. The delay curve can be considered to possess a Q factor similar to tuned circuits.

The second-order allpass transfer function in the complex frequency plane S is given by (see reference 12):

$$T(S) = \frac{S^2 - \omega_r S/Q + \omega_r^2}{S^2 + \omega_r S/Q + \omega_r^2} \tag{6.18}$$

where ω_r = pole resonant frequency, rad/sec
Q = pole Q

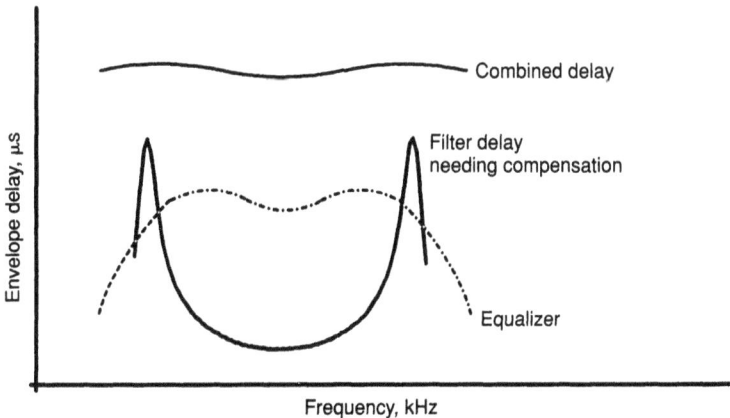

Figure 6.18 Delay equalization of a bandpass filter.

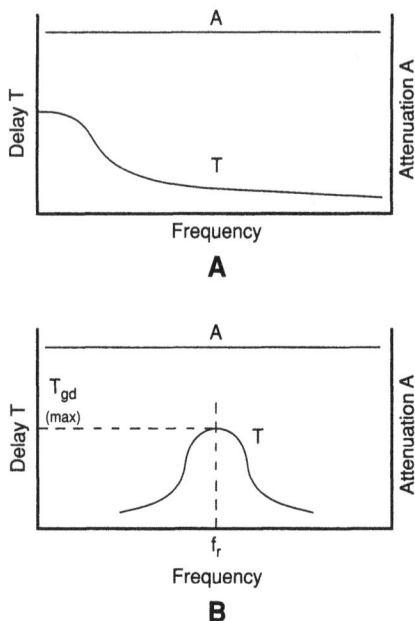

Figure 6.19 Delay and amplitude characteristics. (A) First-order allpass equalizer; (B) second-order allpass equalizer.

From this equation, the group delay is derived as

$$T_{gd} = \frac{2Q\omega_r(\omega^2 + \omega_r^2)}{Q^2(\omega^2 - \omega_r^2) + \omega^2\omega_r^2} \tag{6.19}$$

By letting $\omega = \omega_r$ in Equation 6.19, the delay at ω_r becomes

$$T_{gd, max} = \frac{4Q}{\omega_r} = \frac{2Q}{\pi f_r} \tag{6.20}$$

$T_{gd, max}$ is in seconds, f_r is in hertz, and Q is a dimensionless number. When the Q is greater than 2, the peak delay occurs at ω_r (or f_r) for all practical purposes. For Q's less than 2, the peak value will be slightly lower in frequency. Most of the Q's will range from about 1 to 20.

The two variables, T_{gd} and f_r, are known quantities. For example, a certain amount of delay is described at a specific frequency. Using Equation 6.20, a value for Q is calculated. Then this value of Q is used in Equation 6.19 to determine the delay of other values of f (converted to ω by multiplying the frequency by 2π) in the passband. Thus, one delay response is obtained with its peak at f_r.

More than one delay equalizer section is often required. The number of equalizer sections required to compensate a particular delay curve can be roughly approximated from (see reference 12)

$$N = 2\,(\Delta\text{BW})(\Delta T) + 1 \tag{6.21}$$

where ΔBW = bandwidth of interest, Hz
ΔT = delay distortion over ΔBW, sec

For example, to compensate a delay distortion of 500 µs in a 2700-Hz bandwidth may take $(2)(2.7 \times 10^3)(0.5 \times 10^{-3}) + 1$, or 4 equalizer sections.

Equalizing a delay curvature is a tedious task. The delay to be compensated is first plotted as a function of the passband frequency. Then, starting at the point of minimum delay, the additional Δ delay required to make this point equal to the delay at the delay specification edges is determined from the graph. The frequency at this point is also noted. Then these two values are used in Equation 6.20 to calculate Q. This Q value is used in Equation 6.19 to determine the delay at other frequencies in the passband. These values are then plotted on the graph, and a composite curve is drawn by adding the new delay to the original. At one of the resulting curve dips another estimate is made as to the amount of delay necessary to bring its level up with the delay just calculated. The frequency of this new delay point is noted and the computation process is repeated. This process continues until a satisfactory composite delay curve is obtained. Figure 6.20 illustrates

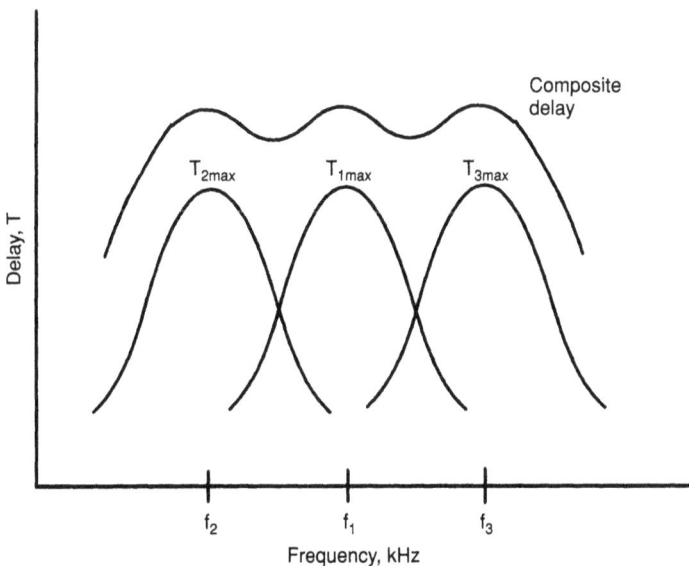

Figure 6.20 Equalizer delay curvature obtained from three equalizer sections.

the equalizer delay curvature based on three equalizer sections. An iterative process will be necessary varying the f_r and Q of each section to create a better fit.

The preceding method can be simplified by normalizing the delay and the frequency. Tables of normalized values for various Q's can then be established to speed the graphics. Programmable calculations can also be utilized to solve Equation 6.19 for the desired frequencies and Q's.

Once the resonant frequencies f_r and the corresponding maximum delays T_{gd} are determined, the equalizer sections must then be designed and added to the radio equipment. The equalization can be accomplished either with passive or active filters. For passive circuits, a bridged T network functions as a delay equalizer. This type of circuit, shown in Figure 6.21, exhibits a second-order delay response. Its allpass characteristic is a function of inductor Q and may require additional amplitude correction. When a number of such sections are necessary, it can be difficult to implement because of interaction between sections and the need for trimming adjustments.

The use of active RC filters at audio makes the design approach much simpler. In this type of circuit the delay is determined by a choice of resistors and capacitors. The flat amplitude response is independent of the delay. One version of the second-order active filter is shown in Figure 6.22 (see reference 12). The values of R and C are arbitrarily chosen, and A corresponds to the gain. The maximum delay ($Q > 2$) is related to the components by (see reference 12)

$$R_2 = \frac{T_{gd,\,max}}{2C} \tag{6.22}$$

and the resonant frequency f_r corresponding to the maximum delay by

$$R_{1b} = \frac{R_2}{(\pi f_r T_{gd,\,max})^2 - 2} \tag{6.23}$$

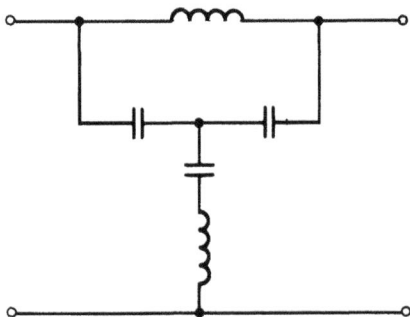

Figure 6.21 Bridged T network that has second-order delay characteristics.

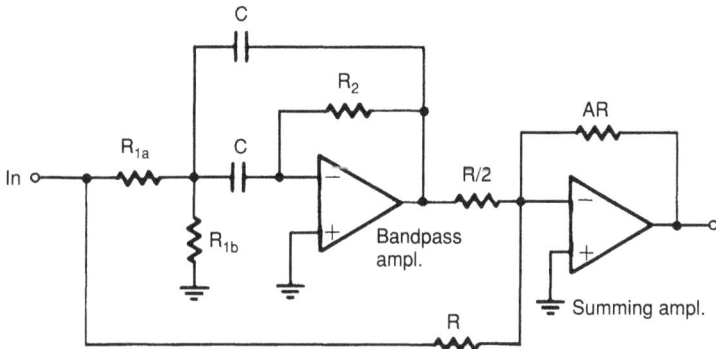

Figure 6.22 Active allpass delay equalizer section $(0.7 > > Q > > 20)$.

and

$$R_{1a} = \frac{R_2}{2}$$

Other active RC allpass circuits can also be used. The circuit shown in Figure 6.22 illustrates the relative simplicity of designing an active equalizer.

In order to compensate for passband delay, a number of such sections must be used in cascade. Each section is designed for one $T_{gd, max}$ and its corresponding f_r. The overall gain can be made unity by adjusting each section for a gain of 1. The equalizing sections can then be added to the receiver or exciter audio amplifier stages at some convenient location.

For the designer who wants to bypass the trial-and-error method of equalizing delay, it should be noted that commercial equalizers are available. Some manufacturers of mechanical and crystal filters also design delay equalizers at audio and IF frequencies to the user's specifications. Computer programs for delay equalization are also available (see reference 13).

Acknowledgment. The author wishes to acknowledge the valuable suggestions of Robert A. Johnson, Rockwell Corporation, who kindly reviewed the sections of the chapter on filter types and delay.

References

1. *Subsystem Design and Engineering Standards for Tactical Digital Information Link (TADIL-A)*, MIL-STD-188-203-1 (Philadelphia: Naval Publications and Forms Center, September 10, 1982).

2. *Standards for Long Haul Communications*, MIL-STD-188-317 (Philadelphia: Naval Publications and Forms Center, March 30, 1972).

3. A. I. Zverev, *Handbook of Filter Synthesis* (New York: John Wiley & Sons, 1967).

4. William J. Domino and Robert A. Johnson, "Miniature Precision Bandpass Filters Solve IF Design Problems," *RF Design* (October 1991).

5. William E. Sabin, "The Mechanical Filter in HF Receiver Design," *RF Design* (March 1993): 43–54.

6. Robert A. Johnson, *Mechanical Filters in Electronics* (New York: John Wiley & Sons, 1983).

7. T. H. Simmons, Jr., "The Evaluation of the Discrete Crystal Single-Sideband Selection Filter in the Bell System," *Proc. of IEEE* 67 (January 1979): 109–115.

8. B. d'Albaret and P. Siffert, "Recent Advances in UHF Crystal Filters," *36th Annual Frequency Control Symposium*, Argenteuil, France, 1982. (Available from Defense Technical Information Center, Alexandria, VA, Doc. AD/A-130.811.)

9. John Pivnichny, "A Different Approach to Ladder Filters," *Communications Quarterly* (Winter 1991).

10. Wes Hayward, "Designing and Building Simple Crystal Filters," *QST* (July 1987).

11. Franz C. McVay, "Don't Guess the Spurious Level," *Electron. Des. 3* 15 (February 1, 1967): 70–73.

12. Arthur B. Williams, *Electronics Filter Design Handbook, Second Edition* (New York: McGraw-Hill, 1988).

13. *Super Filsyn Version 4.5* (Santa Clara, CA: DGS Assoc., 1980). (Available from DGS Associates, 1353 Sartia Way, Santa Clara, CA 95051.)

7

Speech Processing, Squelch, and Noise Blanking

William E. Sabin

7.1 Speech Processing

The human voice does not utilize the SSB power amplifiers very well. The 15-dB peak-to-average ratio, or 24-dB dynamic range, means that a high-power amplifier that handles the peaks well is just "loafing" most of the time. Unless the amplifier and power supply are marginal, much greater power can be transmitted if the deficiencies of speech can be improved. The usual case is that amplifiers that will handle the peaks well are capable of higher average levels than natural speech provides. The central idea is that the weaker components which we will emphasize in this section contribute much to the intelligibility.

Audio processing

The two methods of audio processing are clipping and compression. In the clipper the audio signal, after 10 to 20 dB of additional amplification, is simply sliced off at the required level. It is "memoryless" in that the action at any one time does not affect subsequent actions. The strong syllabic peaks, which occur from 5 to 10 times per second, are thus instantly reduced. The weaker parts which lie between the peaks are clipped much less and are therefore relatively stronger. The resulting truncated wave is then high-/lowpass filtered to remove low-/high-frequency distortion products, and then converted to SSB. The clipper generates significant amounts of audible harmonic and IM distortion, but emphasizes weaker speech elements effectively. Many of these are high-frequency components which are important. The distortion reduces the effectiveness appreciably since it degrades the improvement in intelligibility.

In a compressor the peaks are limited by gain reduction, and the behavior after a peak differs from the clipper. The amplifier gain does not restore immediately but is allowed to recover more slowly in a controlled exponential manner. It has memory. On subsequent peaks the amount of gain reduction required may therefore be much less. The result is that the compressed signal has less distortion but tends to reduce the weaker elements that lie between peaks, especially those that immediately follow the peaks, and therefore tends to be less effective. Some of this reduction in effectiveness is ameliorated by the lower distortion. Compressors are widely used to maintain a constant output level despite variations in voice level. Figure 7.1 illustrates the differences in the two methods.

There are alternative ways of performing audio processing of the memoryless variety that reduce the distortion by eliminating harmonic distortion. This reduction in distortion increases the effectiveness. These methods will be treated in detail after other subjects have been introduced.

Speech clipping

To put the subject of speech clipping into perspective, we discuss the most important basic factors that influence the design and use of such circuitry. The actual circuit designs will be covered in later sections. Of particular interest here is the speech waveform and its relationship to the SSB signal. This relationship changes in some rather peculiar ways as a result of the speech processing operations. It is necessary to understand these changes so that the SSB system can be designed to function properly with the speech processing operational. The central idea is that the transmitter must be kept linear while processing is going on, to prevent splatter, and must be as free as possible of excessive background noise and intelligibility degradation caused by distortion. But we find, in a PEP-limited system, that certain SSB waveform effects can reduce the effectiveness of the speech processing because they introduce envelope peaks that partially restore the original peak-to-average ratio.

The equation of a USB signal is given as

$$f(t) = g(t) \cos \omega_c t - h(t) \sin \omega_c t \qquad (7.1)$$

where $g(t)$ = audio input waveform
$\quad h(t)$ = Hilbert transform of the audio waveform
$\quad \omega_c$ = RF signal frequency, rad/sec

(Chapter 2 discusses this signal in much more detail.)

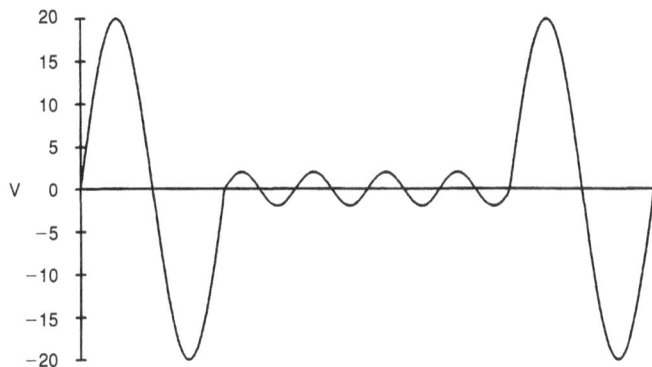

A

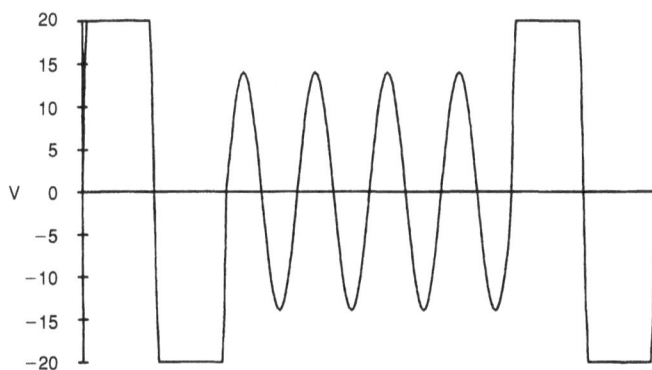

B

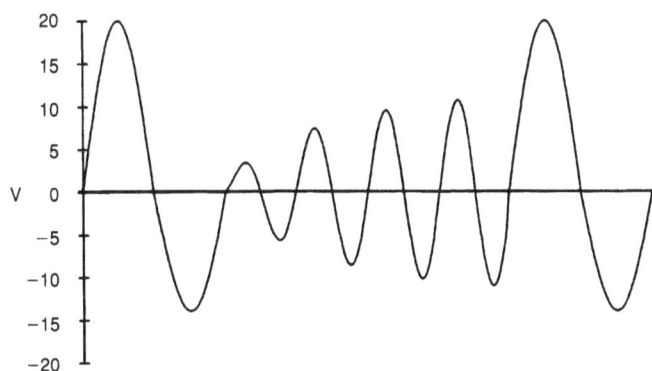

C

Figure 7.1 Computer-generated graphs showing (A) Small high-frequency components between two large low-frequency components, (B) 10 dB of clipping, and (C) 10 dB of compression.

To find the envelope of this SSB wave, we draw a line through the peak values of Equation 7.1 as discussed in Figure 1.2. This envelope is given by

$$f(t)(\text{env}) = \sqrt{\frac{g(t)^2 + h(t)^2}{2}} \qquad \text{RMS} \qquad (7.2)$$

Of particular interest at this point is the shape of the $h(t)$ wave. It is derived from $g(t)$ in the following manner:

1. Find all the spectral components of $g(t)$. In terms of complex exponentials there are elements at positive frequencies and similar elements at negative values of frequency, as shown here:

$$A \cos \omega t = \frac{A}{2} e^{j\omega t} + \frac{A}{2} e^{-j\omega t} \qquad (7.3)$$

2. Change the phase angle of all positive-frequency components by –90°.

3. Change the phase angle of all negative-frequency components by +90°.

4. Convert this modified spectrum back to the time domain to get $h(t)$.

In terms of analog circuit design, we operate on the analog audio signal and apply a 90° phase lag. Circuitry to do this is discussed in Chapter 5. Digital implementations are discussed in Chapter 8.

The operation described above produces waveform distortion that can result in detrimental peaking of the envelope. To see how this happens, consider Figure 7.2, which shows a computer-generated plot of a square-wave input signal that has been subjected to various degrees of lowpass filtering and then subjected to the Hilbert transform operation. The SSB envelope is then generated using Equation 7.2. The bandpass filter in the SSB exciter would eliminate harmonics above 3000 Hz. As the waveform $g(t)$ is made less square, the peakiness of $h(t)$ and therefore of the SSB signal is reduced, as shown in the illustration. As the peaking is reduced, the tendency to overdrive the linear amplifiers is reduced and the average power in a peak-limited system is at the same time increased.

The importance of this discussion, in the present context, is that audio speech clipping can produce the results just described. The speech signal contains strong components at the lower audio frequencies, and clipping of this signal can produce approximately the kind of square-wave signal described above. Speech compression, on the other hand, produces a more "rounded" signal that is less prone to this effect.

The long-term spectrum of speech shows that above 500 Hz there is a rolloff of high-frequency power spectral density of about 6 dB/octave. Preemphasis of these frequencies at 3 dB/octave has been shown to improve intelligibility [1]. Also, as mentioned previously, speech has a peak-to-

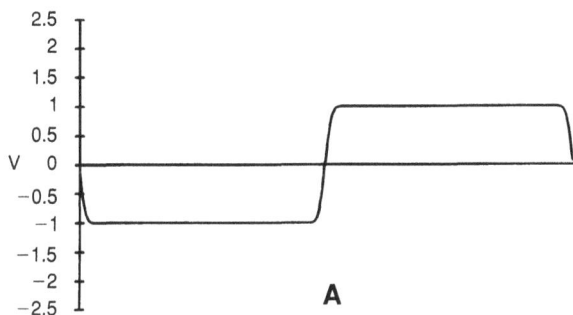

Figure 7.2 Computer-generated graphs showing (A) a square wave that is slightly lowpass filtered.

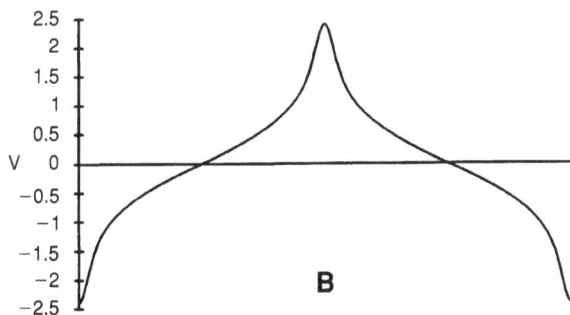

Figure 7.2 (B) its Hilbert transform.

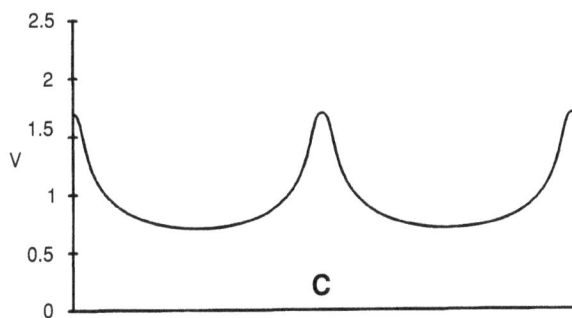

Figure 7.2 (C) the positive side of the SSB envelope, using Equation 7.2.

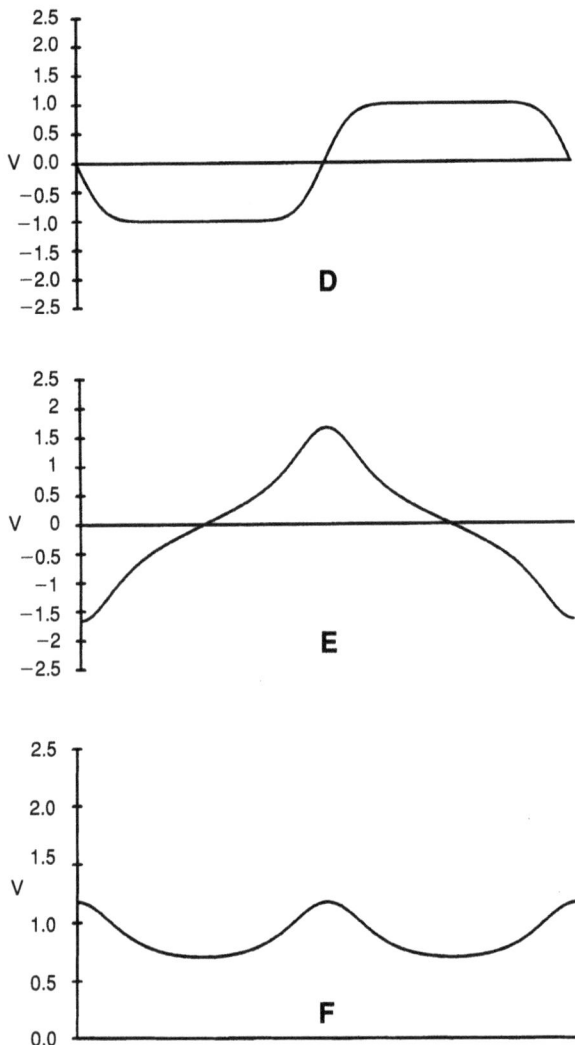

Figure 7.2 (D), (E) and (F) repeat (A), (B), and (C) when the square wave is subjected to additional filtering. The SSB envelope in (F) shows much less peaking effect.

average ratio of about 15 dB. It happens that preemphasis increases the peak-to-average ratio somewhat, partially offsetting the advantage of preemphasis (Figure 7.3A). However, it has been found experimentally [1] that preemphasis, followed by speech clipping, leads to an overall improvement, especially if deemphasis is used at the receiver. Clipping tends to suppress the higher-frequency formants (harmonics of certain speech sounds) that are needed; preemphasis helps to restore them.

After the speech clipper, a lowpass/highpass filter attenuates distortion products at low and high frequencies. This filter may introduce nonlinear phase shift of the speech spectrum and therefore some waveform distortion in the audio signal. But a consideration of the trigonometry involved in Equation 7.1 shows that no additional distortion of the SSB wave is produced by these filter phase shifts. This assumes, of course, that audio stages and modulators are not overdriven by the possibly peaky audio waveform. Also, if the filters exhibit overshoots to a fast-changing signal, that could add additional peaks to the waveform.

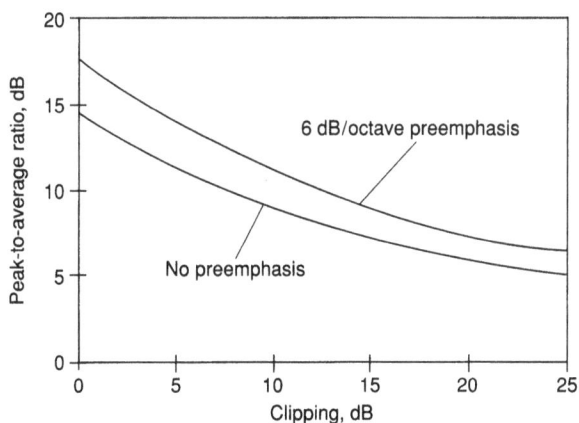

A

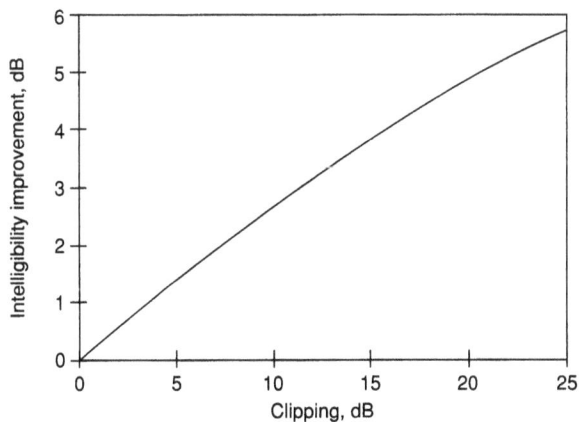

B

Figure 7.3 Computer-generated graphs showing performance of an audio clipper. (A) The peak-to-average ratio of the audio signal; (B) the improvement in intelligibility in white noise.

However, it is possible that the processed audio wave will produce envelope peaking of the SSB wave because of its squareness, as previously described. Therefore we filter the audio signal at low and high frequencies to no more than the necessary bandwidth to minimize this degrading effect.

An additional point is that a squared-off low-frequency audio signal, upon being filtered, is subject to a "repeaking" effect. That is, if a square wave has amplitude equal to 1.00, the fundamental component out of the filter has amplitude 1.27, or $4/\pi$ (2.1 dB). This effect has an important influence on gain settings for the entire transmitter because the voice dynamics are quite different from the steady tone used to make transmitter adjustments. As a result, serious overloading of linear amplifiers can result. Transmitter design can easily guard against this effect, and others previously described, using a concept—automatic level control (ALC)—mentioned in earlier chapters and discussed later. Still, the repeaking effect increases the peak-to-average ratio, and this degrades the speech processing effectiveness.

The curves of Figure 7.3 show the peak-to-average ratio of the audio signal (not the SSB envelope) and the intelligibility improvement of the SSB signal. These results are quite useful and inexpensive to achieve using simple circuitry. The distortion and the SSB envelope peaking are the main drawbacks. These can be greatly reduced at a somewhat higher cost, as will be described later. The block diagram of the processor is shown in Figure 7.4.

Speech compression

The discussion of audio clipping was used to introduce some of the basic considerations of speech processing. This section on audio compression will build on that base and show how the two methods differ. As mentioned before, audio compressors are used mostly to provide an output that is invariant from one talker or microphone distance to another. However, they can also be used to improve somewhat the intelligibility of speech. Figure

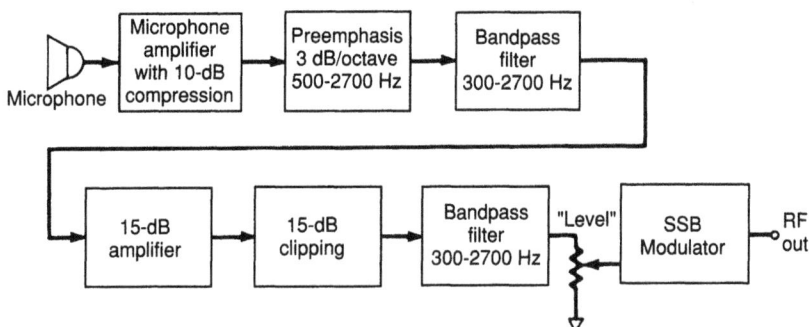

Figure 7.4 Block diagram of an audio speech processing system employing compression and clipping. Careful filter design reduces distortion products and SSB envelope peaking.

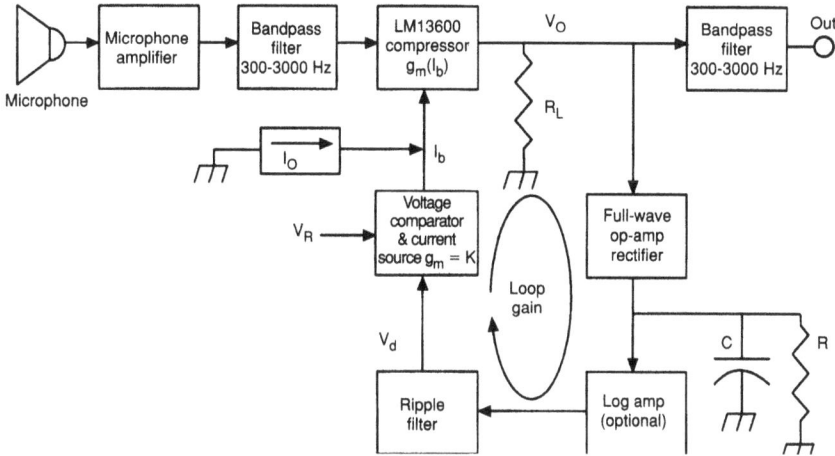

Figure 7.5 Block diagram of a speech compressor. A control current i_b reduces amplifier gain. Full-wave rectification catches both positive and negative excursions. The loop gain path is indicated.

7.1 showed the difference in the way the signal is modified by the two methods. We now consider the operation of the compressor.

Figure 7.5 shows a block diagram of a speech compressor. The compression is usually accomplished by a variable-gain IC amplifier such as the National LM13600, used here as an example. The output voltage is full wave rectified (so that both positive and negative excursions of speech will be detected) using op-amp detector circuits, and if v_d is greater than a reference voltage V_r, then current i_b, which controls transconductance g_m, decreases and the gain is reduced in the manner shown in Figure 7.6A. This action occurs while the magnitude of the input signal continues to increase. But when the input level starts to decrease, the diodes go out of conduction because of the voltage v_d stored on the capacitor C. The gain control loop is then in an open state, and voltage v_d falls exponentially with time constant RC until such time that the diodes start to conduct again. During this fall time the amplifier gain increases, and if v_d becomes less than V_r, the maximum gain is restored. Thus there are two situations to consider, the closed-loop state and the open-loop state.

For the closed-loop case the output signal is found from the following equation:

$$v_o = v_i g_m (i_b) R_L \qquad (7.4)$$

where $g_m (i_b)$ is a function of a bias current i_b as shown in Figure 7.6A. Note the log-log relationship. Using the points marked on the graph, the equation for the straight line is given by

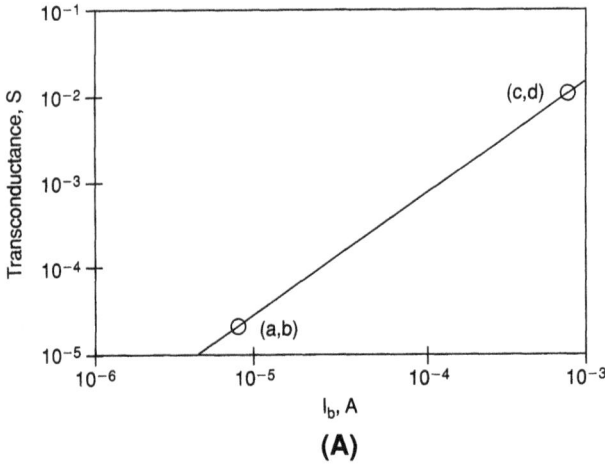

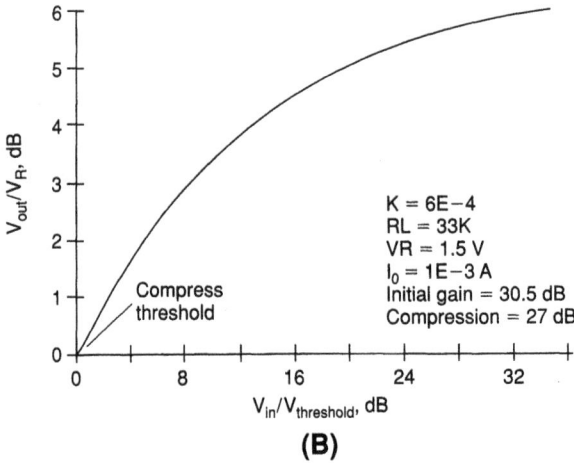

Figure 7.6 Computer-generated graphs showing gain control characteristics. (A) The almost logarithmic effect of i_b on the transconductance of a National LM16300 amplifier. (B) The audio output versus input of the compressor circuit, showing the output flattening at large input values, indicating high loop gain.

$$\log g_m = \left[\frac{d-b}{c-a}\right] \log i_b + \left[\frac{bc-ad}{c-a}\right] = M \log i_b + B \tag{7.5}$$

Using Equations 7.4 and 7.5 and Figures 7.5 and 7.6, we find the output in logarithmic form as follows:

$$\log v_o - M \log [I_0 - K(v_o - V_R)] = \log v_i + B + \log R_L \tag{7.6}$$

where $v_o > V_r$ (the closed-loop mode). Figure 7.6B shows the response, using realistic circuit values as shown. The response slope is not constant but flattens out, indicating an increase in the loop gain path indicated in Figure 7.5. This effect is shown also in Equation (7.7), which shows the loop gain increasing as v_o increases (large input signal).

$$\frac{d(\log v_o)}{d(\log v_i)} = \frac{1}{1 + \dfrac{MKv_o}{I_o - K(v_o - V_R)}}$$

$$= \frac{1}{1 + (\text{loop gain})}; \quad v_o > V_R$$

(7.7)

The output changes 6 dB for a 33-dB input change (the compression is 27 dB). To achieve a very small output change, a very large value of loop gain (high K) would be needed. This can create difficult problems with loop stability, especially at large values of input signal. If there is some time delay (phase shift) in this loop due to the filtering of the control signal, overshoots and instability could result. *Gulping* is a term used to describe a situation where an overshoot in i_b shuts off the output signal until v_d can decay (via RC) sufficiently to recover. Also, any slight audio signal on i_b will modulate the output signal and produce some distortion; therefore the v_d line must be adequately filtered. So the design goals are as follows:

1. Restrict, by filtering, the low- and high-frequency ends of the input signal. Restricting low frequencies makes the v_d line easier to filter.

2. Make the loop response fast enough to follow the syllabic rises in the voice signal (as filtered in step 1) so that saturation of the amplifier does not occur. To aid in this, design the amplifier with adequate "headroom," or large-signal handling ability.

3. Lowpass filter the v_d line to reduce ripple, but not to such an extent that step 2 is violated or to such an extent that gulping occurs.

4. Postfilter the audio output signal to remove as many low- and high-frequency distortion products as possible.

These four requirements obviously interact and possibly conflict, and a combination of experimental, analytical, and simulation analyses must be performed to achieve a working design.

A better design can be obtained, at greater cost, if a logarithmic amplifier is inserted as suggested in Figure 7.5. This reduces loop-gain variations and improves dynamic response. Op-amp design literature for these circuits is ubiquitous (see reference 2, for example).

After a peak has occurred, v_o falls off because of the discharge of RC. As Figure 7.1 shows, the speech that immediately follows the peak is compressed, but the compression decreases with time. This effect becomes more noticeable as RC becomes small. As a result, the weaker speech elements are enhanced. At the same time, the signal becomes more distorted because of the modulation of signal amplitude by v_d. As RC becomes small, the compressor acts more like a clipper.

It has been found [1] that with a 0.5-sec recovery time and 10 dB of compression, a 1- or 2-dB improvement in intelligibility is obtained. Compare this with Figure 7.3B. The combination of compression (10 dB) and clipping (20 dB) has been found to be useful (the compressor helps to maintain a more constant average level into the clipper). Finally, reference 3 notes that if the gain control is applied in a balanced manner to a push-pull gain control amplifier, the effect of ripple voltage on v_d is reduced (but some IM effects can still be created).

RF clipping

We have seen that there are two problems with audio processing. One is the amount of distortion that accompanies audio clipping (a more effective method than compression for intelligibility improvement), and the other is the peakiness of the SSB envelope due to the Hilbert transform phenomenon. Radio frequency clipping, and a form of audio processing that is mathematically equivalent to RF clipping, help to resolve these problems.

We first discuss true RF clipping as shown in the block diagram of Figure 7.7. A high-speed diode circuit slices the RF cycles in a memoryless manner. The diodes are in shunt with a high-impedance tuned circuit and the amplifier is essentially a current source. The peak signal currents are diverted to the low-impedance diodes. High-conductance RF diodes are needed to get a flat envelope when clipping (see reference 4). Any Hilbert peaks present are also clipped. There are some initial conditions at the beginning and end of diode conduction that involve the charge and current in the tuned circuit, but at RF frequencies these are quickly resolved and cause no problems in the circuit performance. The degree of clipping (in decibels) is the decrease in output of a single-tone signal when the diodes are introduced. An exception to this for a speech or multitone signal will be noted shortly. The Hilbert peaks are not re-created after the clipping process if the envelope delay after clipping is fairly constant.

Because the nonlinear clipping occurs at an RF frequency, any harmonic distortion products that are generated are at multiples of the RF frequency and can easily be filtered out. Thus, only IM products of those voice frequencies that are simultaneously present and which are within the output filter passband are present. This filter must be capable of rejecting IM products (mostly of higher order) that are close to the desired passband and which

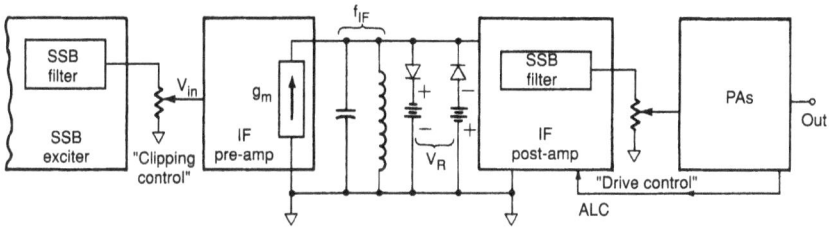

Figure 7.7 Block diagram of an RF clipper circuit.

would constitute adjacent channel splattering. For heavy clipping (20 dB or more) the third-order IM products are about 9 dB below each tone of a two-equal-tone signal. Speech signals can often approximate such a two-tone signal for short periods of time. The absence of harmonic distortion greatly improves the intelligibility of the processed signal. Figure 7.8 shows this improvement, which should be compared with the effects of audio clipping.

Figure 7.7 also reveals some of the main problems with RF clipping. One is that the circuit must be added to the RF signal path. Another is that an additional high-quality bandpass filter (crystal or mechanical) must be added. These make this approach expensive and also somewhat difficult to retrofit to existing equipment. The additional RF gain added to the signal path can create vulnerability to various kinds of spurious signals and instability unless special care is taken to shield and filter the RF clipper assembly.

Another problem is that when the clipped envelope is bandpass filtered, a repeaking effect just like the one described in audio clipping occurs. If a two-equal-tone signal is infinitely clipped, the peak envelope output level after postfiltering is 2.1 dB greater than the clipped envelope. Of course, the

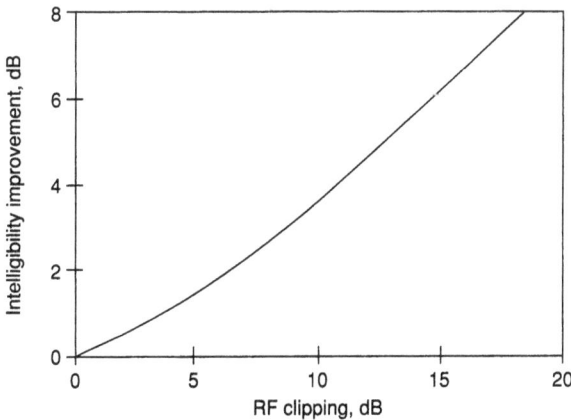

Figure 7.8 Computer-generated graph showing improvement in speech intelligibility for RF clipping.

amount of repeaking depends on the voice constituents present at any one time and is therefore a variable quantity. This means that the amplifiers after the clipper can flattop unless they are protected by ALC. For this reason, ALC should be considered an important ingredient in any speech processor approach. Of course, the ALC should be applied to a stage that follows the clipper, rather than precedes it, in order to be effective. Also, if a single-tone signal is used to set up the RF amplifier signal levels, this adjustment is not correct for the speech signal dynamics encountered because of repeaking. See reference 4 for further discussion. The results shown in the plot of Figure 7.8 are corrected for this effect. Measurements verify that with 20 dB of clipping, the peak-to-average ratio of speech, including the repeaking effect, is about 9 dB.

Experience with RF clipper units has shown that if a high-quality second filter is used, and if the amplifier stages after the clipper are carefully designed, using ALC, to prevent even very slight flattopping, a very high-quality narrowband signal can be transmitted with 20 dB of clipping and very high "talk power" and excellent intelligibility. The second-IF filter contributes to this cleanliness. The two filters must have well-matched passbands for best quality of speech sound. The overall response should be sufficiently flat to minimize response peaks. The IMD does not create a "mushy" sound, but rather a crisp and clean effect that is pleasant sounding, even when the received signal is strong. Preemphasis of the audio signal at 3 dB/octave is sometimes suggested to improve the result. It is desirable, though, to use a noise-canceling microphone.

Two adjustments are required: a gain control ahead of the clipper to set the degree of clipping and one after the clipper to set the output level. The correct output level is that which produces the required ALC action on voice peaks. The ALC recovery time constant should be short (for example, 100 ms), so that the average power will be reduced as little as possible. The ALC reduces the effectiveness of RF clipping slightly, but this cannot be helped.

RF compression

The most common form of RF compression is ALC. It is widely used to control the peak levels in power amplifiers and is discussed thoroughly in the power amplifier chapters. We will mention briefly one way that ALC is used to provide speech processing. The discussion will perhaps suggest similar methods to the reader.

As indicated in Figure 7.9, a control signal, proportional to the SSB envelope, is developed in a power amplifier, which is applied to an early stage in such a way that the peak envelope level is regulated. Think of it as envelope feedback. This control is usually developed on voice peaks. If the control has a short recovery time constant, some enhancement of the speech signal occurs between the syllabic peaks. Figure 7.9 shows a diode/RC circuit that has been used to maintain a slowly varying average level with a 3.0-sec time

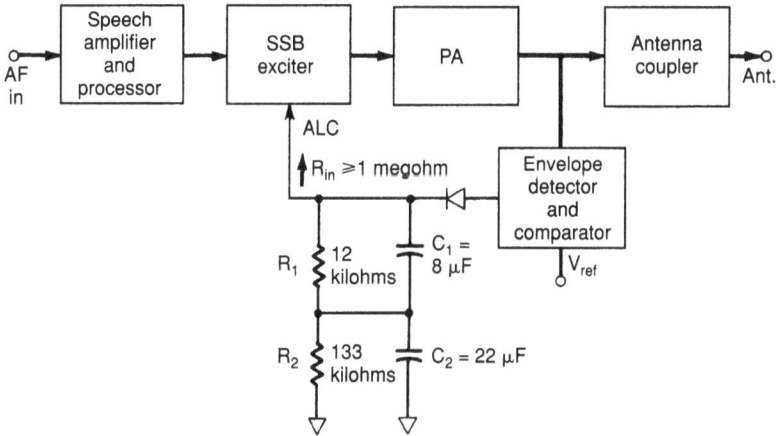

Figure 7.9 Block diagram and simplified schematic diagram of an ALC circuit that provides a small measure of speech intelligibility enhancement.

constant (R_2, C_2) and also a faster varying level (R_1, C_1) with a 0.1-sec time constant. The voltage across the short time constant decays between syllabic peaks, allowing the SSB envelope to increase, and provides a small amount of speech processing.

The proportion of control voltage that is fast-varying is determined by the ratio $R_1/(R_1 + R_2)$. Choose C_1 and C_2 to satisfy the time-constant requirements. This ratio should be such that about 3 dB of gain fluctuation occurs between syllabic peaks. The slowly varying voltage across C_2 constitutes a simplified transmit gain control (TGC) voltage (see Chapter 5 for a discussion of TGC). In order to determine the quantities involved, it is necessary to make a plot of RF output versus control voltage for the intended system. The driver circuit for the RC network should have a very low output impedance so that the capacitors can be charged quickly and the control voltage can therefore follow the fast-rising syllabic peaks. Of course, there should be very little time delay within the complete control loop so that overshoots (gulping) do not occur. Also, if the control loop is not designed properly, transient flattopping, with splatter, can occur on voice peaks because the envelope peaks are not properly restrained. Just as we found with the audio compressor, wide variations in ALC loop gain can be difficult to deal with. The gain control function should be free of abrupt changes and should be as logarithmic as possible.

Baseband envelope clipping

We have seen that RF clipping is superior to audio clipping for two reasons: there is less harmonic distortion (therefore greater intelligibility), and the envelope peaks created by the Hilbert transform are also clipped, thereby increasing the average power. With certain precautions and at a slightly greater cost and complexity than audio clipping, the same results can be obtained in an audio processor.

If an SSB envelope that has been clipped and filtered is frequency translated to baseband, the resulting audio signal may then be fed into an SSB modulator to create an SSB signal that has the same envelope as the first SSB signal. That is, the SSB envelope is invariant with respect to frequency translation, even if the translation is down to baseband and back up to some different RF frequency. However, the baseband signal does not have an "envelope," strictly speaking, because the RF wave associated with an envelope is not present. But recall that in Equation 7.2 the audio signal and its Hilbert transform accomplish this quite well. From previous discussion regarding the Hilbert transform it can be seen that if a signal is accurately Hilbert transformed repeatedly, the signal changes from $g(t)$ to $h(t)$ to $-g(t)$ to $-h(t)$ and so forth, with the respective wave shapes preserved. So if an SSB envelope is demodulated and remodulated, the envelope shape is reproduced. The requirement is that the Hilbert transform be done accurately. This unit can be external to the transmitter and plug into the microphone jack.

Figure 7.10 shows a block diagram of an RF clipper that clips at some low RF frequency. A single local oscillator performs the upconversion and downconversion. The relative phase of the two injections is not important for correct operation. However, the audio output must not overload the amplifiers and SSB modulators in the transmitter. Because large peaks have been removed by the clipper, this problem is well solved.

Figure 7.11 shows another, cheaper approach. An SSB envelope is generated at baseband by creating the square root of the sum of the squares of the audio signal and its Hilbert transform (see Equation 7.2). The resulting envelope voltage is used to limit the desired audio $g(t)$, or I (in-phase) component, in a gain control amplifier (LM13600). The resulting audio is mathematically equivalent to that produced by the circuit shown in Figure 7.10. This circuit has been patented by R. L. Craiglow and F. W. Werth for the Rockwell-Collins Co. [5] and is used in production line equipment.

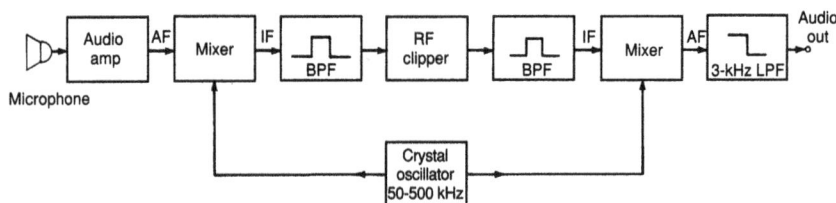

Figure 7.10 Block diagram of a baseband processor that uses RF clipping. The audio signal is converted to SSB, clipped, filtered, and converted back to baseband. Upon subsequent conversion to SSB, the envelope characteristics of the RF clipper are retained.

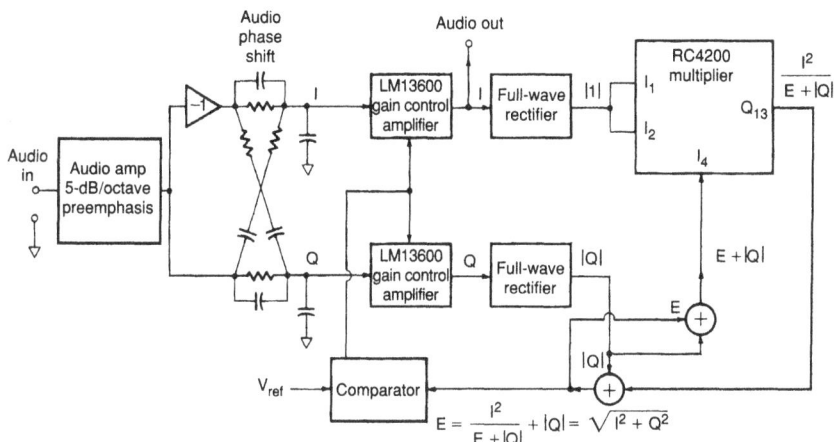

Figure 7.11 Baseband processor that generates the envelope shape of an SSB wave by taking the square root of the sum of squares of the audio signal and its Hilbert transform. The envelope waveform then controls the gain of an amplifier that provides the audio output signal.

LINCOMPEX processing

A major complaint regarding speech processing is that the natural amplitude variations that one expects to hear are "washed out," creating an unpleasant effect for nontechnical users. The LINCOMPEX system (for Linked Compression and Expansion, developed by the British Post Office in the 1960s; see reference 6) encodes the audio amplitude levels at the transmitter and sends them on an auxiliary channel. The receiver decodes this information and reconstructs the original signal. However, the improvement in intelligibility is preserved during the process and the received signal is "telephone quality." Since the auxiliary channel contains information regarding the absolute short-term amplitude of the speaker's voice, the receiver output also has the same amplitude if the equipment is properly adjusted. Therefore, the received signal is transparent to fading and path loss. Receiver AGC helps this process and prevents receiver overload. The control information is usually sent at reduced amplitude so that more speech power will be available in a peak power-limited system, but of course some "talk power" is sacrificed. Because of the companding, the background microphone noise usually associated with speech processors is reduced. This permits larger amount of compression (30 to 40 dB) to be used.

Figure 7.12 shows a block diagram of a LINCOMPEX transmitting unit that uses FM at about 2900 Hz for the amplitude information. The VCO frequency is proportional to the amplitude (in decibels) and has a sensitivity of about 0.5 dB/Hz. Therefore the transmitter and receiver must both have very excellent frequency stability to prevent excessive amplitude drop-off at the receiver. Doppler shift due to relative motion of the transmitter and receiver would also

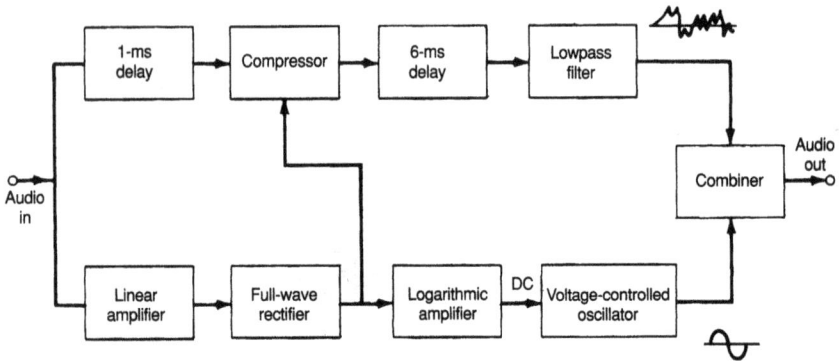

Figure 7.12 Block diagram of a LINCOMPEX modulator. The VCO provides the FM auxiliary signal that contains the original amplitude variations.

create problems. The time delays assure that the envelope information is well coordinated with the compressed speech. Digital implementations of LIN-COMPEX have been suggested [7] that reduce the frequency stability requirements and provide some automatic calibration at the receiver.

Figure 7.13 is the receiver demodulator. The constant-volume amplifier helps (along with AGC) to eliminate amplitude variations prior to volume expansion. The antilog amplifier complements the log amplifier at the transmitter. Under weak-signal conditions the FM channel can be contaminated with noise and some degradation can occur. But since the channel is very narrowband, this effect is reduced.

Amplitude compandoring for SSB

Another method of speech processing is called amplitude compandored single-sideband (ACSSB) [8]. Compandor ICs are available from several vendors and are used in numerous audio signal processing applications, including telephone and stereo systems. In SSB applications a reduced amplitude (10 dB below PEP) pilot carrier is often added at 3.1 kHz.

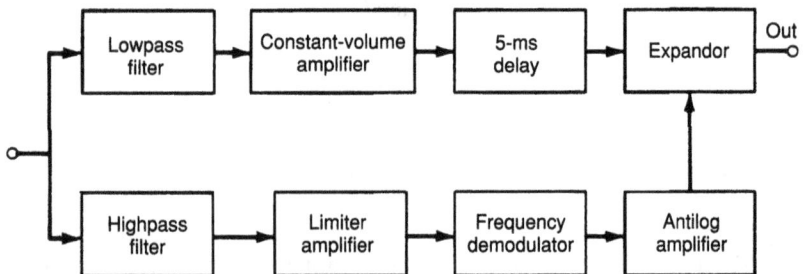

Figure 7.13 Block diagram of a LINCOMPEX demodulator. The expander restores the original amplitude variations, as conveyed by the subband FM signal.

The pilot carrier is used in the receiver for frequency locking by a phase-locked loop and for AGC and squelch. By using the pilot carrier for AGC a main objection of SSB is solved. In conventional SSB the fluctuating speech signal creates AGC. In speech pause intervals the background noise level greatly increases because AGC falls to zero. The pilot carrier remains much more constant.

A compandor normally operates in a 2-to-1 mode. The compressor at the transmitter compresses each 2-dB change in the input level to a 1-dB change in the output level. The expandor at the receiver does the opposite. If the same type of IC is used at both locations the receiver output matches the original voice amplitude, but in SSB a 10- to 15-dB improvement in signal-to-noise ratio is achieved. Figure 7.14A shows how the compressor increases weak speech components so that they rise above the noise level. Figure 7.14B shows how the expandor suppresses the noise level and reconstructs the original speech dynamic range.

If the average received signal approaches the noise level the effectiveness of the system is greatly reduced. But if the average is several decibels above the noise level the resulting quality of the receiver output is significantly improved.

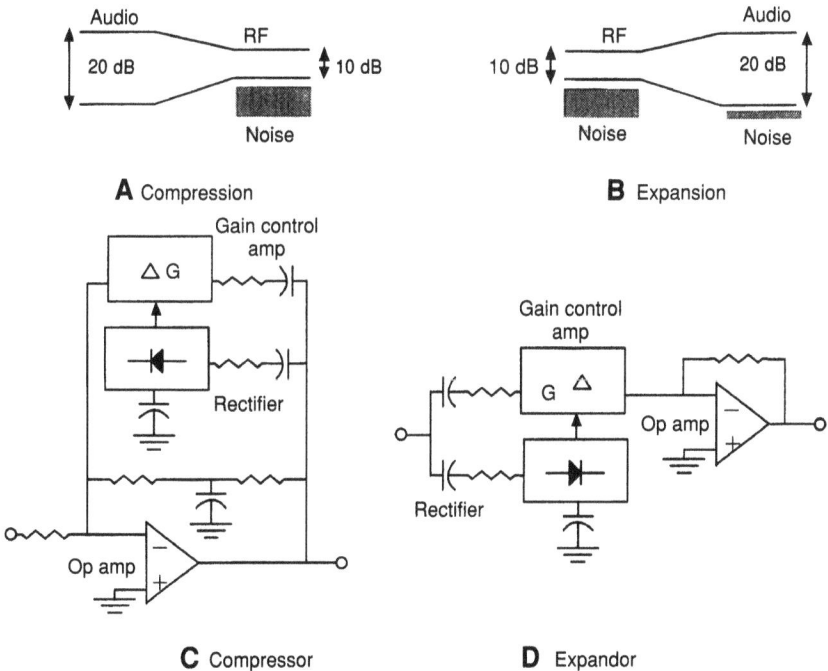

Figure 7.14 Implementation of amplitude-compandored SSB. (A) Compression at the transmitter. (B) Expansion at the receiver. (C) Block diagram of the compressor. (D) Block diagram of the expandor at the receiver.

The SSB receiver IF AGC loop must maintain a constant *average* audio level into the expandor so that it is getting the intended level of input. The receiver itself must not compress or otherwise "color" the received voice signal. The pilot carrier is useful in assuring this and is recommended for ACSSB. The AGC loop should be capable of responding to slow fades in order to maintain the constant output level, but fast fades or "flutters" may be too rapid to accomplish this. In this situation the quality of the speech output can suffer.

Figure 7.14C shows an NE571 IC in a compressor. The NE571 consists of a rectifier unit and a variable-gain transconductance multiplier. The NE571 is used as a feedback device for an op-amp. Figure 7.14D shows the same device in an expandor circuit.

Correct operation requires that the two devices track each other in amplitude and frequency response. This is done by carefully selecting the supporting component values, as described in the vendor's application notes and data sheets.

The NE571 can also be used as an ALC in the receiver output to assist the AGC in its task of maintaining the constant average level that the expandor requires. A more elaborate set of functions is available in the NE5750 and NE5751 ICs. The NE5750 contains a microphone amplifier, a VOX circuit, and a speaker driver amplifier. The NE5751 contains a 300- to 3000-Hz switched capacitor BPF, mute switching, and other functions for FM cellular telephone operations. These can easily be adapted to SSB applications. The main attraction for compandoring is its simplicity. The ICs are widely used and are low cost.

The design of the phase-locked loop (PLL) for the pilot carrier must accommodate some initial frequency offset in the received signal. It must also reject the SSB speech modulation, which constitutes a strong "noise" level that competes with the pilot carrier during the phase-locking process. This means that the lock range of the PLL is quite narrow. Auxiliary frequency sweep methods, at a fairly slow sweep rate, may be needed to assure reliable lockup. The frequency resolutions and stabilities of the transmitter and receiver, and any possible Doppler shifts, must be coordinated with the PLL design. However, some relaxations with respect to reference frequency standard specifications at both the transmitter and receiver (and therefore their cost) are feasible. One approach that has been used to expedite the PLL lockup is to transmit the pilot carrier at full PEP for a brief time, such as 1/4 sec, before applying speech modulation. Chapter 2 contains a more detailed discussion of the system design considerations for pilot-carrier SSB systems and also further notes on compandoring.

The receiver IF filter bandwidth must be wide enough to contain the pilot-carrier offset frequency, even when it is more than the nominal 3.1 kHz. The high values of differential group delay at the edge of the IF filter can present PLL problems. The transmitter output must have low splatter and noise at the pilot-carrier frequency. Out-of-band transmitter interference can be caused by IM between the pilot carrier and the speech modulation. The need for a clean, high-quality design at both ends of the link is clearly indicated.

An intriguing approach to the receiver design would be to use digital signal processing (DSP) to perform all of the necessary compandoring, output leveling, AGC, and PLL functions. A study of Chapter 8 should suggest some ideas along those lines.

7.2 Squelch

Many users of SSB equipment want the receiver to be activated only when the desired signal is present, even though it may be noisy and contaminated with interference. In conventional AM and FM the presence of a carrier signal can be determined, and this can be used to operate a squelch gate. The carrier conveys no speech information, but it is a 1-bit message that says "signal present" or "signal not present." In SSB, where there is usually no carrier, the voice signal itself must be recognized unambiguously in the presence of noise and interference. The time needed to recognize a voice signal must be short enough that very little speech information is lost but long enough that false triggering on transient noise pulses is minimized. These conflicting requirements have led to a great deal of effort to discern the unique characteristics of SSB speech that a squelch can recognize and which will optimize its performance in the presence of noise, fading, and interference. Also, quite often the squelch must be tolerant of frequency mistuning of, say, 100 Hz or so.

One useful property of speech is its long-term spectral distribution, which shows a strong peak in the area of 500 to 600 Hz and a 6-dB/octave rolloff. However, the short-term spectrum over a period of 100 ms or so will at different times show a somewhat different distribution.

A comparator is adjusted for balance with "white," or uniform, noise. The voice signal then unbalances the comparator. Once triggered, the audio remains gated on for about 2 sec so that annoying rapid on/off toggling will not occur. This is improved by providing hysteresis so that the pull-in level is 6 to 10 dB higher than the dropout level. A problem with this approach is that a beat note at low frequency causes false trigger. A high-pitch note will prevent, or lock out, desired response to a speech signal. Certain kinds of unvoiced, or fricative, speech sounds have mostly high-frequency components that do not properly actuate the squelch. If these are the opening sounds, the squelch will tend to miss them. Also, certain types of pulses may not have a flat spectrum, causing false trigger. In addition, it is necessary to be able to preset both transmitter and receiver frequencies to within 100 Hz or so to be sure the squelch will work properly. With synthesizer control, this is no problem. Despite its faults, this approach has been used very successfully in production equipment.

A second property of speech is the low-frequency "syllabic" variations. These are in the range of 1 to 10 Hz or so. Figure 7.15 shows a schematic diagram that utilizes these variations. The signal is amplitude limited with diodes to reduce sensitivity to signal level, bandpass filtered at 600 Hz, envelope detected, and then bandpass filtered to recover the low-frequency syllabic variations that amplitude modulate the 600-Hz speech band. That is, the bursts of speech spectral energy within the 600-Hz bandpass filter vary at the syllabic rate. Noise and CW carriers are relatively weak in these syllabic frequency components. Also, slight mistuning of receiver frequency is less detrimental than some other approaches. This simple circuit is used in a production receiver. Because of the limiting action, all of the desired information is contained in the time fluctuations in the zero crossing—in other words, in the phase variations. A more complex circuit also detects the syllabic variations in a passband centered at 2200 Hz and compares the results of the two bands. The two sets of variations tend to be either in phase or 180° out of phase most of the time, that is, either correlated or uncorrelated. The full-wave rectification corrects for this likelihood. The use of correlation reduces the sensitivity to signals other than desired speech signals because noise and many types of interference rarely have this kind of coincidence.

In mobile SSB systems a pilot carrier has been found to be a reliable way to detect the signal [9]. A special filter passes only the carrier frequency. Since the bandwidth of this filter is very narrow, the pilot carrier can be 10 to 15 dB below PEP.

7.3 Noise Blanking

The elimination of impulsive interference in an SSB receiver contributes substantially to the quality of communication in situations where the source of interference cannot be controlled, such as in a mobile environment and near power lines and machinery. Certain pulse-generating transmitters, such as HF over-the-horizon radars, send complex pulse packets that present a somewhat special problem [10] and have become an international nuisance. There has been a great deal of effort to perfect the various means of making a receiver transparent to these kinds of interference, which not only degrade speech intelligibility but affect the AGC and thereby desensitize (paralyze) the receiver. In this section the discussion is directed mostly at noise blanking, in which the receiver signal path is turned off for the duration of the pulse. The absence of a carrier in SSB systems makes this method especially effective.

Impulse noise differs from ordinary random noise in two respects. Its short-term spectrum is phase coherent and has an amplitude variation and autocorrelation, as shown in Figure 7.16, that are a function of its width and its edge slopes. As a result, the received voltage is nearly pro-

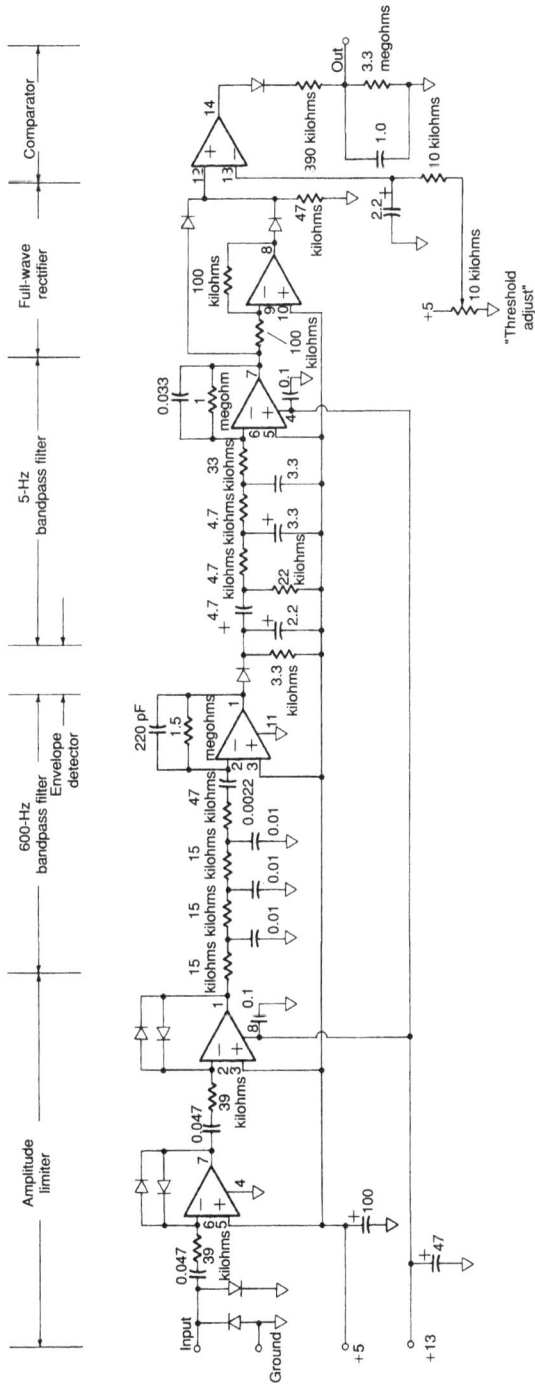

Figure 7.15 Schematic diagram of a squelch that detects syllabic modulation of a speech signal.

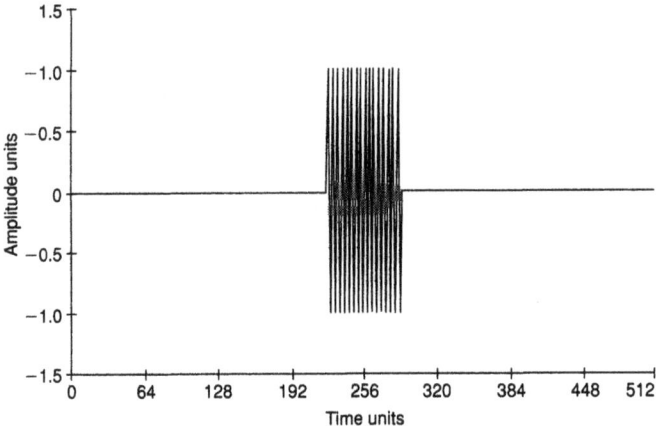

A

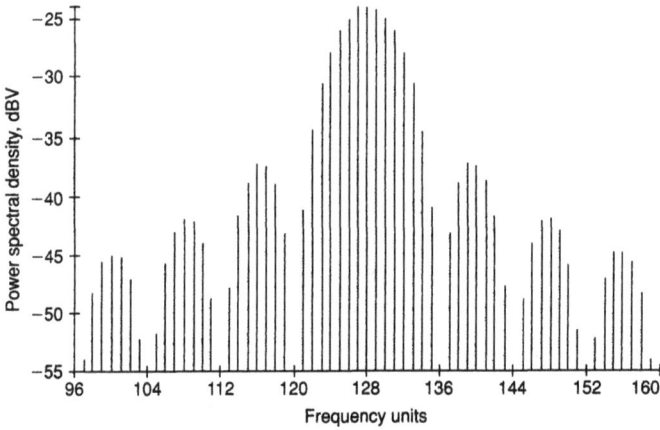

B

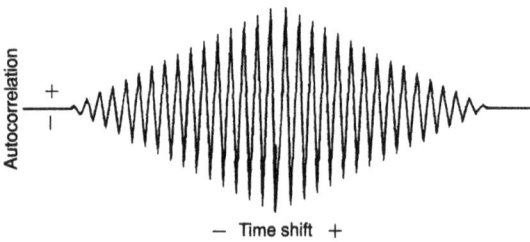

C

Figure 7.16 Computer-generated graphs showing (A) a narrow repetitive RF noise pulse, (B) its one-sided spectrum, and (C) its autocorrelation function.

portional to bandwidth (within the limits indicated in the figure) rather than to the square root of bandwidth. Also, the pulse power is very high for a short time and then falls quickly to zero. The duty cycle of these pulses is usually less than, say, a few percent or so. For example, automobile ignition pulses are from 3 to 10 ns in duration (at the source; they are stretched in the process of being radiated by the vehicle), and for an eight-cylinder four-stroke engine the pulse rate is roughly rpm/15 pulses per second. Other artificially produced sources emit pulses in the 0.1- to 10-μs range. These characteristics are sufficiently different from the speech signal and noise that they may be exploited by the noise blanker in ways that do not seriously degrade the desired signal. Speech can be interrupted for about 10% of the time with an acceptable (usually) loss of articulation, as discussed in Chapter 2. Other types of interference, such as lightning and pulse packets, are sufficiently similar to speech both in time and spectrum that they are more difficult to blank. Amplitude limiting or clipping can help somewhat in this case.

A signal or pulse entering a receiver encounters at some point a bandpass filter that rejects undesired signals and interference. If an RF pulse of narrow but finite width is passed through this filter, the time output is the convolution of the pulse shape and the impulse response of the filter, or equivalently, the output spectrum is the product of the pulse spectrum and the frequency response of the filter. If the pulse spectrum is essentially flat across the filter passband, the time output is approximately the impulse response of the filter. This impulse response causes the output pulse energy to be delayed by an amount equal to the slope of the filter phase shift, to be elongated in time to about twice the reciprocal of the bandwidth, and to be reduced in amplitude (the time-power product is fixed by the energy of the pulse). Also, multiple-response echoes, called *ringing*, occur which are separated in time by the reciprocal of the filter bandwidth and which must often also be blanked. The relative amplitudes of these responses are determined by the shape of the filter transition band and its differential phase delay (the two are related). Figure 7.17 illustrates these points and also shows that a rounded-response, linear-phase filter is a better choice for the bandpass filtering of RF impulses. This filter may not be optimal in terms of eliminating undesired signals, though, so a conflict appears that must be resolved in the receiver design employing a noise blanker.

At this point the discussion centers on a particular technique for noise blanking, called the in-band blanker, because the noise pulse energy is in the frequency band closely adjacent (10 to 50 kHz) to the desired signal. The emphasis is not on circuit details but on general principles that can be applied in many different ways by the designer. The discussion is for the most part qualitative, because a detailed and very involved analysis requires more space (and rigor) than is feasible here. Figure 7.18 shows the generic block diagram to be discussed.

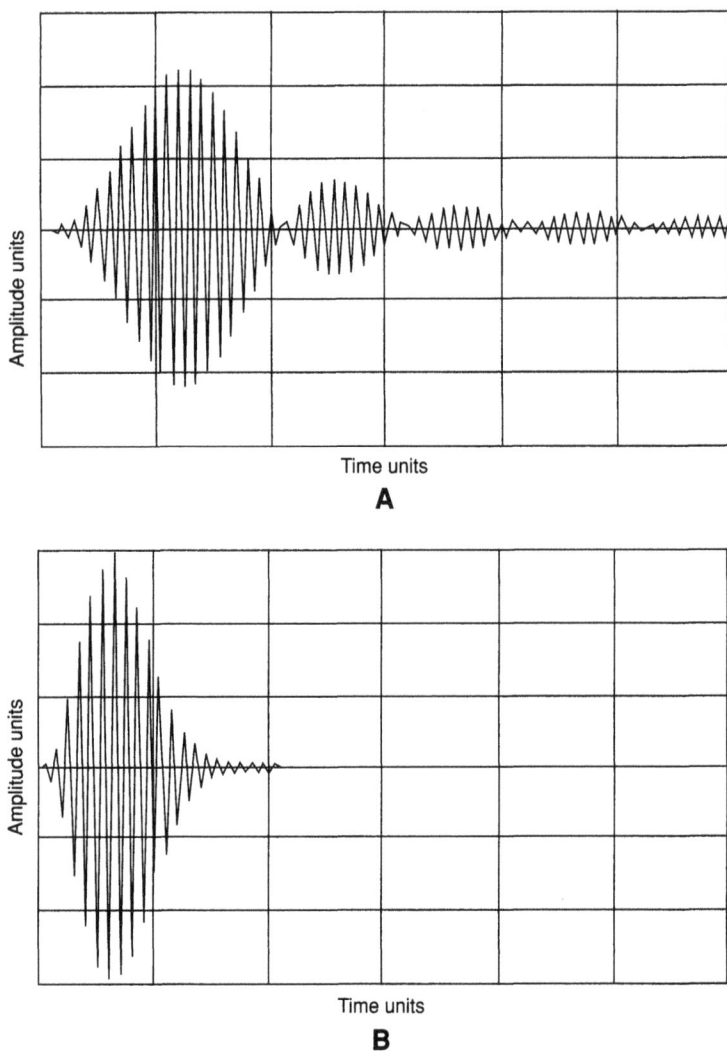

Figure 7.17 Computer-generated graphs showing response of a bandpass filter to a narrow pulse of RF energy. A Chebyshev and a Bessel filter are compared: (A) four-pole, 0.5-dB Chebyshev; (B) four-pole Bessel.

The noise amplifier follows the RF translator and is preceded by a filter that is much wider than that required for the desired speech or CW signal. This filter should have as much bandwidth as possible and minimal ringout responses. Because of the wide bandwidth and poor shape factor of this filter, the signal path IF stages preceding the narrow filter must have as little gain and as much signal-handling ability as possible in order to minimize the chances of spurious signal generation in these stages. Refer to Chapter 4 for a detailed discussion of this subject. A further important point is that the input impedance of the noise amplifier must be constant with respect to signal and noise spike levels so that IM will not be generated by a nonlinear impedance. There are two ways to help this: use a MOSFET input stage and operate its gate terminal at a very low impedance level (this reduces signal levels at the input gate). The noise figure for this stage is relatively less important, so considerable noise mismatch is tolerable. The wide bandwidth of the filter magnifies the amplitude of the noise pulses and makes them much narrower than would be the case for a narrow filter, as previously discussed, and this helps to get more noise spike into the noise amplifier. The signal IF amplifiers that precede the gate switch, however, must handle the large noise spikes and must have a low noise figure. This conflict creates a difficult design problem that can be solved only by minimizing the gain and by using a power amplifier with a low noise figure. Amplifiers after the gate switch do not have the noise spike, and this eases the design problems of these stages. The use of the wide filter does make the signal path more difficult, and a degradation of receiver noise figure may have to be accepted if strong signal performance is a major specification. Very often this is a mild penalty. From this discussion, the trade-offs are readily apparent. For example, a narrower filter will widen the noise pulses (this may be quite acceptable in a speech SSB application) and make the circuit design easier.

Following the noise amplifier, an envelope detector provides a pulse that actuates a retriggerable monostable flip-flop which provides an output as long as the noise pulse lasts and can be quickly ready to provide another pulse if it is needed. This method has been found to be very helpful in production equipment in reducing interference from HF radar pulse packets. We can now estimate the required gain in the noise channel. Suppose we

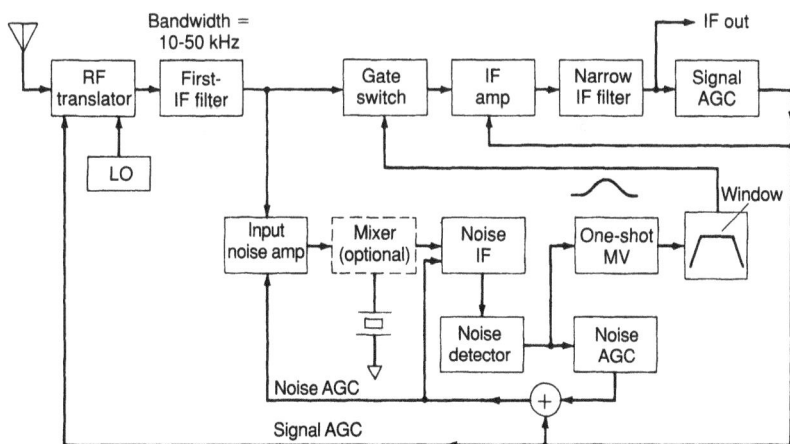

Figure 7.18 Block diagram of an in-band noise blanker.

want to blank a spike whose power in a 3-kHz bandwidth at the antenna is equal to a desired –115-dBm signal. If the RF translator has 15-dB gain and if the first-IF filter has a 50-kHz bandwidth, the spike level at the noise amplifier input is $-115 + 15 + 20 \log (50/3) = -76$ dBm, or 112 μV at 500 ohms. The flip-flop requires about 3 V to trigger, so the voltage gain needed, including rectification, is 27,000. Three stages at 30 per stage will suffice. Some of the gain can be at dc after rectification. Often, especially in upconversion receivers (see Chapter 3), a frequency conversion to a much lower IF frequency is performed, as indicated in the block diagram.

The gate switch attenuates the signal path by at least 50 dB. Figure 7.19 shows a typical example. It is singly balanced with respect to the blanking pulse so that the spectrum of the blanking pulse itself will introduce less noise into the signal path. In addition, the blanking pulse should be *windowed*, or shaped as indicated in the figure, to attenuate IF frequency components. Despite these precautions the blanking pulse can intermodulate with undesired signals to produce interference in the desired passband. Figure 7.20 shows how the product of a blanking pulse and an off-frequency carrier creates the problem. The windowing of the blanking pulse reduces this effect; for an excellent discussion of windowing see reference 11.

A serious problem with the in-band noise blanker is that undesired signals within the passband of the wide filter can trigger the gate circuit, cutting off the signal path. To minimize this, AGC is derived from the impulse detector and applied to the noise amplifier as shown in the Figure 7.18 block diagram. Normal signals will then be restricted to levels that do not affect the gate switch. This AGC must not come on too quickly, though, because noise spikes will make AGC and will not attain enough amplitude to operate the gate. For this reason, the AGC attack time is made very slow,

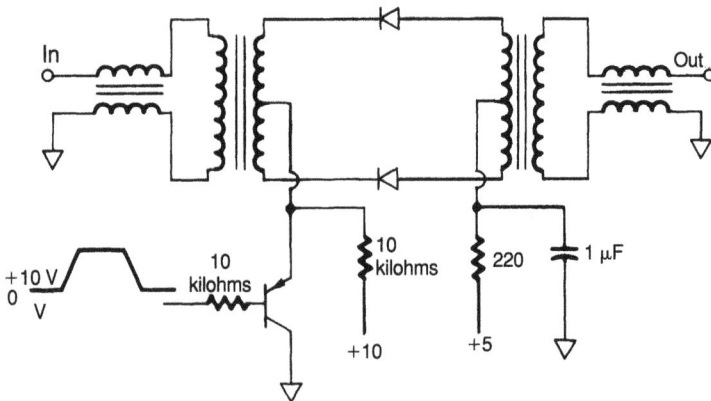

Figure 7.19 Schematic diagram of a noise blanker gate switch. The switch is singly balanced to reduce the leakage of gate pulse energy into the signal path.

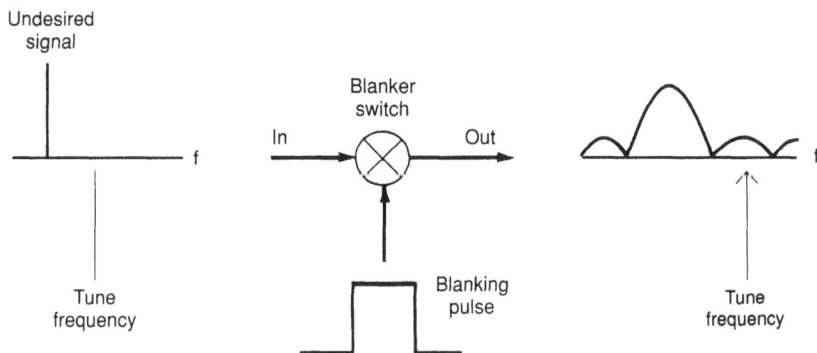

Figure 7.20 The IM of a blanking pulse and an out-of-band undesired signal produces an in-band interference.

perhaps 10 to 20 ms, so that the noise impulses will produce the desired output. The leading edges of normal speech signals, both desired and undesired, will also often trigger the switch momentarily, but a considerable amount of this can be tolerated, if it is less than about 5%. The rejection of speech signals can be much improved by postdetection filtering; a filter that rolls off below 3 kHz will be less affected by the greater low-frequency speech energy. Also, the AGC recovery must be fairly fast, say 50 ms or so, so that rapid repetitions of pulses can be blanked. The dual time-constant AGC idea described in the section on ALC would work well in a blanker.

A further problem is that strong undesired signals make noise channel AGC and therefore reduce the effectiveness of the blanker for weaker spikes. This can be improved by deriving at least part of the noise channel AGC from the normal AGC that is generated after the narrow filter. Since this filter greatly attenuates, stretches, and delays impulses, the impulse is blanked before the normal AGC can react significantly. This approach was patented by S. L. Dawson for Rockwell Corp. [12]. To prevent excessive blanking by out-of-band signals, as described in the previous paragraph, some of the noise AGC should be derived in the manner previously discussed. Also, as the desired signal becomes large, the need to blank is reduced; this method aids in this if it is implemented properly.

In order for the pulse to be properly blanked, the gate switch must be fully opened at the time that the pulse arrives at the switch in the signal path. This timing becomes more critical as the first-IF bandwidth becomes wider, and also as the blanking pulse is windowed, as we have already discussed. The windowing creates a time delay in the blanking pulse that must be matched in the signal path. One author [13] uses a glass acoustical delay line. Another approach for short delays is to use a cascaded multiple-resonator filter. The value of the delay can be determined from the lowpass prototype from which the tuned filter is derived [14].

The methods discussed so far are adequate for SSB speech systems. For high-speed data systems the blanking width must be much less to prevent data loss. To accomplish this, a separate wideband noise-sensing receiver is often used as the noise channel. This receiver is tuned to a relatively unused frequency. Its wide bandwidth assures a very large noise spike, relative to the incidental signals, which may also be present in the noise-sensing receiver. The wide bandwidth greatly reduces the amount of noise amplification needed [15]. The noise and signal path time delays are carefully coordinated.

References

1. R. L. Craiglow et al., *A Study of the Effects of Elementary Processing Techniques on the Intelligibility of Speech in Noise*, Collins Radio Res. Rep. CCR 201, 1963.

2. D. F. Stout and M. Kaufman, *Handbook of Operational Amplifier Circuit Design*, Chapter 17 (New York: McGraw-Hill, 1976).

3. E. W. Pappenfus, W. B. Bruene, and E. O. Schoenike, *Single Sideband Principles and Circuits*, p. 330 (New York: McGraw-Hill, 1964).

4. W. E. Sabin, "RF Clippers for SSB," *QST* 51, no. 6 (July 1967): 13–18.

5. R. L. Craiglow and W. F. Werth, "Speech Processor for Processing Analog Signals," U.S. Patent 4,410,764 (October 18, 1983).

6. *Proceedings* III, International Radio Consultative Committee (CCIR) XIII Plenary Assembly, Geneva, 1974.

7. S. M. Chow et al., *Syncompex, a Voice Processing System for Low Cost HF Radio Telephony* (Ottawa, Canada: Communications Research Centre, Canadian Department of Communications, 1980).

8. Lusignan, B. B. "Use of Amplitude Compandored SSB in the Mobile Radio Bands," Stanford Electronic Laboratory, Stanford University, July 1980.

9. Barry Manz, "SSB Technology Fights Its Way into Land-Mobile Market," *Microwave & RF* 22 (August 1983): 72–80.

10. Bradley Wells, "The Russian Woodpecker: A Continuing Nuisance," *Ham Radio* 17 (November 1984) 37–45.

11. F. J. Harris, "On the Use of Windows for Harmonic Analysis with the Discrete Fourier Transform," *Proc. of IEEE* 66 (January 1978): 51–83.

12. S. L. Dawson, Noise Blanker, U.S. Patent 4,479,251 (1984) (assigned to Rockwell International).

13. W. Gosling, "Impulsive Noise Reduction in Radio Receivers," *Radio & Electron. Eng.* 43 (May 1973): 341–347.

14. H. J. Blinchikoff and A. I. Zverev, *Filtering in the Time and Frequency Domain* (New York: John Wiley & Sons, 1978).

15. R. H. Sternowski, "Noise Blanking at the Antenna Input of a Communication Receiver," Master's thesis, Iowa State University, Ames, IA, 1977.

8

Digital Signal Processing

Richard A. Groshong

8.1 Introduction

Single-sideband radio communication equipment has traditionally been designed using analog signal processing techniques and circuitry. In the past,
the inherent advantages offered by a digital implementation were offset by
the higher cost and greater power dissipation of the digital hardware. The
limited dynamic range of available analog-to-digital (A/D) converters and
sample-and-hold (S/H) devices also prevented the digital implementation
from realizing its full potential. The advent of high-speed, low-cost digital
signal processing (DSP) devices has changed all that.

Over the past 10 years, microprocessor technology has produced several
generations of highly integrated DSP chips primarily targeted at the telecommunications industry, but aptly suited for SSB radio communication applications. During this same period, A/D converter manufacturers have made the
transition from technology that was directed mostly at the data acquisition
market to the DSP market. The current generation of highly linear, sampling
A/D converters, combined with techniques such as bandpass sampling, are
now making digital radio equipment design the architecture of choice.

This chapter will discuss SSB digital receiver and exciter concepts, algorithms, and architectures that primarily utilize IF bandpass sampling techniques. Additional discussions will center on critical digital receiver/exciter
design considerations and on DSP device characteristics that set or limit
the performance of a particular design.

There are several key concepts that are used in digital SSB radio design.
Since they are essential to understanding the radio implementations to fol-

low, we shall briefly review them. It is not the purpose here to provide a general treatment of DSP techniques, and it is assumed that the reader is familiar with sampling techniques, Z transforms, and digital filtering theory (see references 1 through 5). Only techniques that are directly applicable to SSB are reviewed here.

One of the key concepts in DSP as applied to radio communications is that of analytic or complex signals. An ordinary or real signal, if represented in the frequency domain, has symmetrical positive and negative frequencies. For example, a signal $e_1 = A \cos 2\pi f_0 t$ has a spectrum point at both f_0 and at $-f_0$. If one displays the signal on a spectrum analyzer, of course only the positive portion is displayed. The negative component is nevertheless present, and if we mix the signal with another signal $e_2 = B \sin 2\pi f_1 t$, both sum and difference frequencies will be generated. In complex signal representation it is possible to obtain signals with one-sided or nonsymmetrical positive and negative sideband structures. It is also possible to keep track of which components are positive and which components are negative. The complex signals that exist in the computer, while represented in complex mathematical form, behave precisely as predicted by the mathematical theory and can be processed, translated, phase-shifted, etc., without the attendant restrictions of real analog signals. Once the signals have been processed or manipulated as desired, they can be converted into real signals by D/A converters or used directly by other digital processors.

Mathematics of complex signals

The area of DSP that relates to complex signals and the concept of positive and negative frequencies is often not well understood and leads to much confusion in signal processing. An attempt will, therefore, be made here to convey a sufficient understanding to the reader so that the techniques described subsequently can be understood. A real signal, say, $f(t) = \cos \omega_0 t$, is normally considered to be a positive frequency and can be displayed on a frequency spectrum analyzer. If it is in the RF region it can be transmitted and occupies a definite position in the electromagnetic spectrum. A real signal of this type can be represented by two rotating vectors in the complex plane, as shown in Figure 8.1.

Since the vectors rotate in opposite directions, the imaginary component resulting from the sum of the vectors is always zero. Therefore, the signal can be transmitted on a single circuit. The Fourier transform of this signal is, of course, composed of identical positive and negative components of the form

$$F[\cos \omega_0 t] = \pi \delta(\omega - \omega_0) + \pi \delta(\omega + \omega_0) \tag{8.1}$$

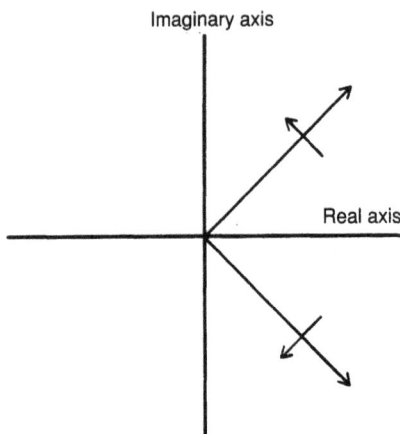

Figure 8.1 Phasor representation of a real signal.

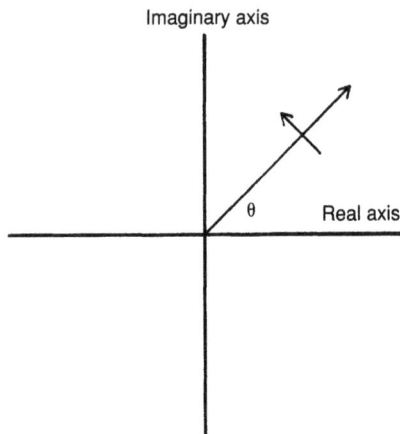

Figure 8.2 Phasor representation of a positive frequency.

A real signal is always composed of symmetrical positive and negative frequencies, although in some cases they may be of opposite sign; for example,

$$F[\sin \omega_0 t] = j\pi\delta(\omega + \omega_0) - j\pi\delta(\omega - \omega_0) \tag{8.2}$$

It is possible to represent a signal that has only positive or only negative frequencies if we allow two circuits and represent it in the form $f(t) = I(t) + jQ(t)$. This is shown in Figure 8.2. In this case $I(t) = \cos \omega_0 t$ and $Q(t) = \sin \omega_0 t$. We note here that the $I(t)$ component taken by itself still has a positive and a negative component as before, as does the $Q(t)$ component. The explanation is that when taken together in the form $I(t) + jQ(t)$, the negative spectral components cancel each other while the positive frequency components reinforce each other. Complex signals of this type are referred to as *analytic signals*.

It is often desirable in SSB equipment to begin with a real two-sided signal, such as the audio from a microphone, and convert it to a one-sided signal by eliminating either the positive or the negative components. The signal can then be frequency converted without producing unwanted sidebands.

A real two-sided signal, say, $f(t)$, can be converted into a one-sided signal. First, all the components of $f(t)$ are shifted in phase by 90°. Let this signal be represented by $\hat{f}(t)$. If the shifted component is multiplied by j (where $j = \sqrt{-1}$) and added to the original signal, an analytic signal is created that has only the positive frequencies of the original function. Thus if $f(t)$ has a frequency spectrum $F(\omega)$ as shown in Figure 8.3A, the complex signal $f(t) + j\hat{f}(t)$ has a spectrum as shown in Figure 8.3B. Interestingly, the signal $f(t) - j\hat{f}(t)$ has a spectrum consisting of the negative frequencies of $f(t)$. We shall see more of positive and negative frequencies later. In general, we represent a complex signal in the form $I + jQ$, where I can be thought of as the in-phase component and Q the quadrature component.

Another very useful concept in DSP is that of frequency translation. This is based on the Fourier frequency translation theorem which states that if a signal $f(t)$ has a frequency spectrum $F(\omega)$, the signal $e^{j\omega_0 t}f(t)$ has a frequency spectrum $F(\omega - \omega_0)$. Thus all components in the original signal are translated up in frequency by a value f_0 (where $\omega_0 = 2\pi f_0$). The theorem holds whether the signal is real or complex. If it is real, both the positive and negative sidebands are increased by a value f_0 and multiplication by $e^{j\omega_0 t}$ results in a complex signal.

From the Euler identity

$$e^{j\omega_0 t} = \cos \omega_0 t + j \sin \omega_0 t \qquad (8.3)$$

the output of a frequency translator has the form

$$e^{j\omega_0 t}(I + jQ) = (\cos \omega_0 t + j \sin \omega_0 t)(I + jQ) \qquad (8.4)$$

$$= (I \cos \omega_0 t - Q \sin \omega_0 t) + j(I \sin \omega_0 t + Q \cos \omega_0 t) \qquad (8.5)$$

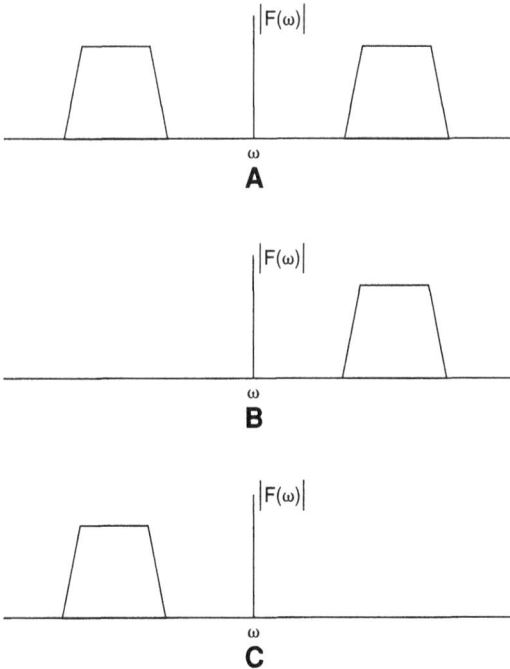

Figure 8.3 (A) Frequency spectrum of a real audio signal. (B) Frequency spectrum of signal $f(t) + j \hat{f}(t)$. (C) Frequency spectrum of signal $f(t) - j \hat{f}(t)$.

Thus a complex mixer can be implemented as shown in Figure 8.4.

A third technique is that of shifting the phase of a signal by 90°. This is normally done on a real signal. The phase shifter delays all components by 90°; consequently all positive frequencies are shifted by –90° and all negative components are shifted by +90°. Such a phase shifter is also called a *Hilbert transformer*. For a signal with only one frequency, for example, $e = A \cos \omega_0 t$, the phase shift can be realized by providing a delay of one-quarter cycle, as shown in Figure 8.5A. If the signal consists of a band of frequencies, however, the phase shift is a perfect 90° only at the center frequency. A better phase shifter can be obtained using the implementation shown in Figure 8.5B. Here each block T represents one sample time delay in a digitized signal. The signal multipliers are h_0, h_1, . . . , etc. The phase shifter can be made as broadband as necessary by using additional delays. The values for the h's are optimized for the bandwidth and sample rate used. Section 8.4 discusses this form of finite impulse response (FIR) filter in more detail; however, it is interesting to note that the coefficients have odd symmetry about the center. If the total number of taps is odd, the even tap weights are normally zero.

Complex signals are sometimes translated in frequency so that the center of the frequency spectrum is at zero. This allows the lowest possible sample rate to be used since the analytic signal is then composed of both positive and negative frequencies. After filtering, the signal can again be translated to the normal position in the audio spectrum.

Digital receiver/exciter concepts

The concepts discussed in the previous paragraphs will now be used to illustrate the design of both digital SSB exciters and receivers. We shall re-

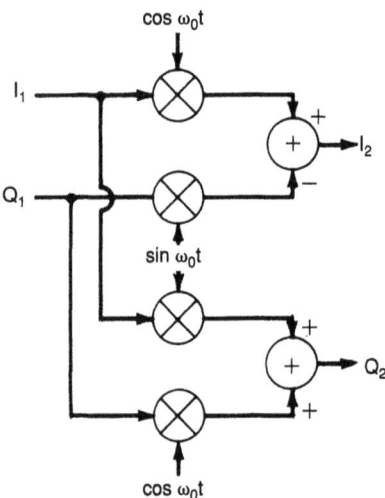

Figure 8.4 Block diagram of a complex frequency translator.

A

B

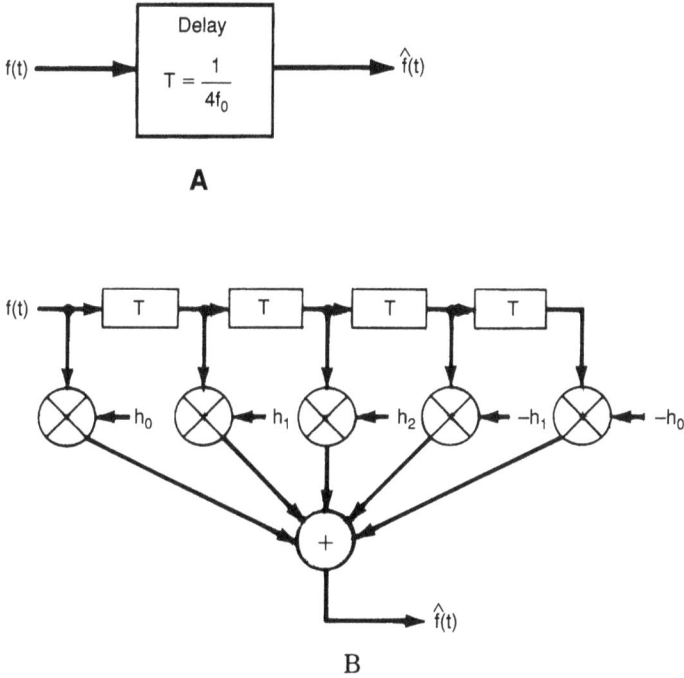

Figure 8.5   (A) Narrowband phase shifter; (B) 90° phase shifter.

turn later to those ideas and develop them in more mathematical detail, particularly as related to limitations imposed on digital radio performance. There are, of course, many different ways in which an exciter or receiver can be configured, and only a few are illustrated here to show the concepts.

The block diagram of a digital SSB exciter is shown in Figure 8.6. The audio signal must be lowpass filtered to remove frequencies greater than half the sample frequency of the A/D converter. A sample rate in the range of 16 to 32 kHz may be found to be appropriate. The digitized audio signal can then be compressed or limited using a fast envelope detector in a digital speech processor, or analog processing can be used prior to the A/D converter. The signal is then passed through a bandpass filter to form the I component of a complex signal and through a similar bandpass filter having a built-in 90° phase shifter to form the Q component. The filters are identical except for the values of the multipliers. The resulting complex signal $I + jQ$ has a one-sided frequency spectrum. This signal is translated to an IF frequency by multiplying by

$$e^{j\omega_0 t} = \cos \omega_0 t + j \sin \omega_0 t \tag{8.6}$$

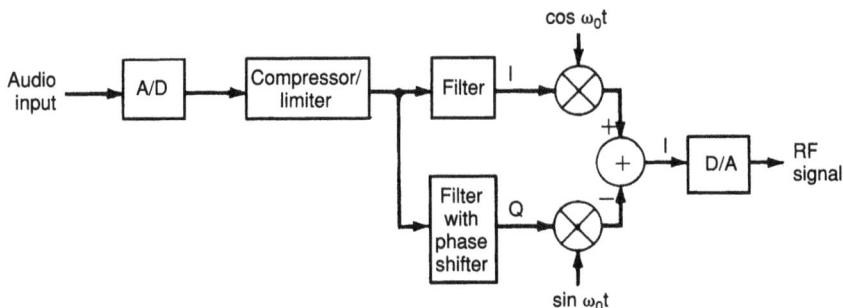

Figure 8.6 Block diagram of digital exciter.

The resulting signal is shown in Figure 8.7. This signal is then converted to an analog signal by the D/A converter. In this case only half the complex mixer is used because it is not necessary to compute the Q component. The case shown results in USB transmission. If the sum of the two components from the sine and cosine multipliers had been used, the lower sideband would have resulted.

An independent sideband exciter can be constructed as shown in Figure 8.8. The LSB and USB circuits are shown independently for clarity. These functions can be combined by moving the adder and subtracter to the outputs of the compressor/limiters. The resulting configuration is shown in Figure 8.9.

The sampling process results in the spectrum repeating at harmonics of the sampling frequency. The harmonic spectra must be removed by an analog filter following the D/A converter. In reality the D/A converter does not

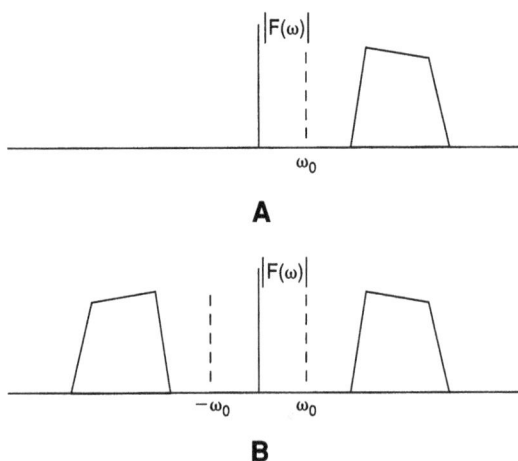

Figure 8.7 (A) Frequency spectrum of complex translated signal $I + jQ$. (B) Frequency spectrum of I component only.

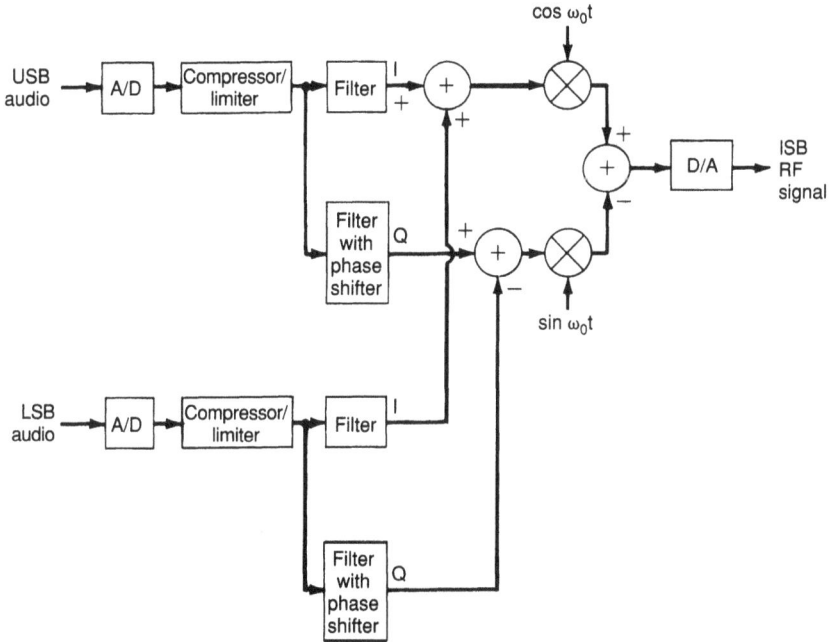

Figure 8.8 Block diagram of an ISB exciter.

provide a series of impulses equal to the sample values, but rather holds the sample value until the next sample time. This results in the sample and hold function which acts as a filter.

If the sample rate is too low, this filtering action may result in attenuation of the upper portion of the audio band. For this reason as well as to reduce the requirements of the analog filter, which eliminates the harmonic spectra, it may be desirable to add an interpolation filter ahead of the D/A converter to increase the sample rate.

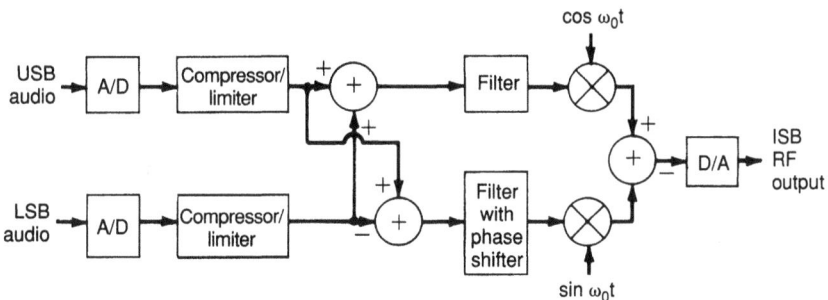

Figure 8.9 Block diagram of an ISB exciter with common filters.

An interpolation filter is a filter operating at an output (higher) sample rate that should be integrally related to the input sample rate. The additional required input samples are taken to be zero. The bandwidth of the interpolation filter is chosen to provide sufficient attenuation of the harmonic spectra in the original signal.

The design of digital SSB exciters requires a considerably greater depth of knowledge than presented here. However, it is intended that this introduction will give the reader a basic idea of how DSP techniques can be used to perform the function of a digital exciter. There are, of course, many different ways of developing a digital exciter. For example, the filtering rather than phasing technique can be used to generate the SSB signal. The signals can then be filtered at an intermediate digital frequency rather than at baseband. The latter implementation does not require the use of 90° phase shifters, so the use of infinite impulse response (IIR) filters is more convenient. Several techniques that can be used to develop signals for a four-channel SSB exciter are discussed in section 8.9.

A digital SSB receiver can be designed in many different ways, and the reader is cautioned not to assume that the methods described here are optimum for all applications. Again, the approaches can be divided into phasing methods and filtering methods to separate the sidebands. As with the digital exciters the phasing methods become more practical using digital techniques, because good broadband phase shifters can be realized and amplitude balance is not upset by the digital system. Systems using FIR or IIR filters can be used; however, FIR filters are easier to design if a 90° phase shifter is to be included. Finite impulse response filters can also be designed without differential time delay, which may be an advantage for data transmission.

A block diagram of one implementation of a digital SSB receiver is shown in Figure 8.10. The function of A/D conversion using IF sampling is discussed in section 8.2, and will not be described here other than to indicate that IF sampling allows the sampling rate to be determined by the bandwidth of the signal rather than requiring it to be twice the actual IF frequency. The digitized signal is then translated to baseband using the translation theorem discussed previously in this section. Since the sampled signal is real, translation requires only two multipliers:

$$x(n)e^{j\omega_0 t} = x(n)\cos\omega_0 t + jx(n)\sin\omega_0 t \tag{8.7}$$

Thus $I = x(n)\cos\omega_0 t$ and $Q = x(n)\sin\omega_0 t$, where $t = nT$. The value ω_0 is chosen to beat the center of the desired sideband to zero (not the frequency of the carrier). It is convenient to choose the IF frequency to be one-fourth the sample rate. Then $\cos\omega_0 nT$ takes on only the values 1, 0, –1, 0, etc., while $\sin\omega_0 nt$ takes on the values 0, 1, 0, –1, etc. The multipliers can then be replaced by inverters and multiplexers.

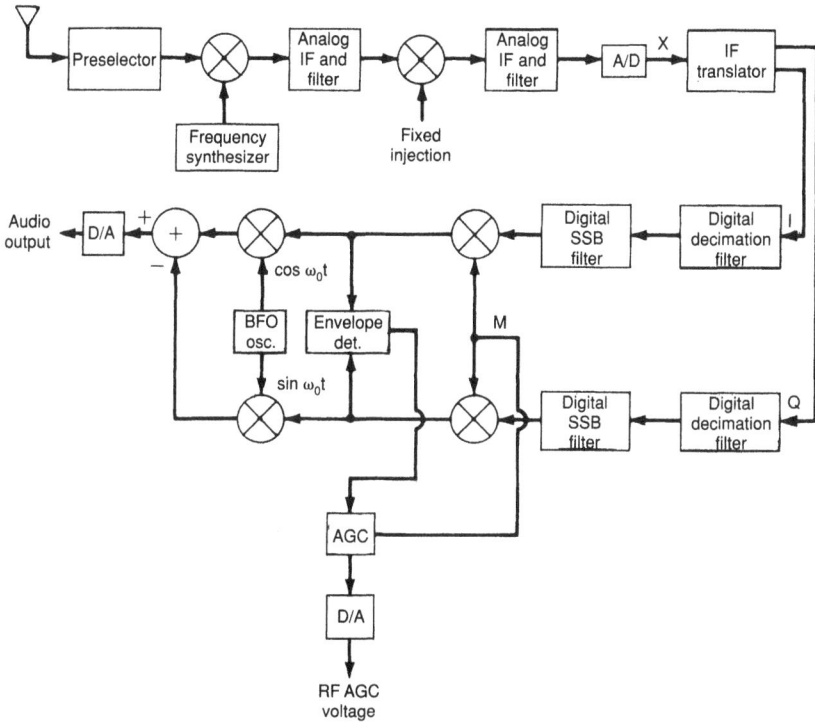

Figure 8.10 Block diagram of a digital SSB receiver.

The sampling rate should now be reduced to the lowest possible value prior to narrowband filtering. This minimizes the number of taps required in the SSB filters. This is accomplished by the decimation filter, which reduces the bandwidth to perhaps 8 kHz. A 16-kHz sample rate can then be used in the narrow SSB filters. If FIR filters are used, approximately 64 taps are required to obtain a good passband response. In this case the I and Q filters are identical because no 90° phase shifter is required. The level of the signal is next adjusted by the AGC multiplier M, and the BFO beats the signal up in frequency by the amount that it was beat below zero in the IF translator. A value on the order of 1700 Hz is appropriate for voice bandwidth signals. Because a real output signal is required, again, only two multipliers are required for the translator. The digital BFO oscillator must generate both cos ωnT and sin ωnT. An accumulator keeps track of the accumulated phase value $\theta(n + 1) = \theta(n) + \Delta$, where $\theta = \omega nT$. Δ determines the BFO frequency. A lookup table for the sine and cosine along with an interpolation routine can be used to find the actual values. The AGC may be quite complex because of the limited dynamic range of the A/D converter, as discussed previously.

The block diagram in Figure 8.10 shows a single-channel SSB receiver. Recall that in this case the frequency synthesizer is programmed to beat the desired sideband below baseband so that the signal occupies a frequency range of about ±1.4 kHz, and the I and Q filters are identical. Another implementation is also possible in which the carrier frequency is at zero and the USB corresponds to positive frequencies while the LSB corresponds to negative frequencies. This implementation is more conducive to building an ISB receiver.

A block diagram of an ISB SSB receiver is shown in Figure 8.11. Here the SSB filter in the Q channel includes a 90° phase shifter. The outputs of the filters can then be subtracted to extract the USB and added to extract the LSB. The AGC is not shown here and is considerably more complex than the signal-channel AGC. This results because two AGC detectors are desirable with separate digital gain multipliers. A single RF gain control voltage must be used, which is determined by the larger of the ISB signals.

It should be pointed out that the information presented here is tutorial in nature and that considerably more in-depth knowledge is required to design digital SSB equipment. Indeed, unexpected results often occur as a result of aliasing, quantization effects, nonlinearities, etc., and a thorough analysis and simulation is recommended prior to undertaking the design of digital SSB equipment.

The sections following in this chapter will explain more thoroughly some of the concepts, limitations, and advantages of DSP.

8.2 Sampling and Data Conversion

Since the input and output signals of most radio communications systems are analog, the first and last steps in a digital radio must be converting between

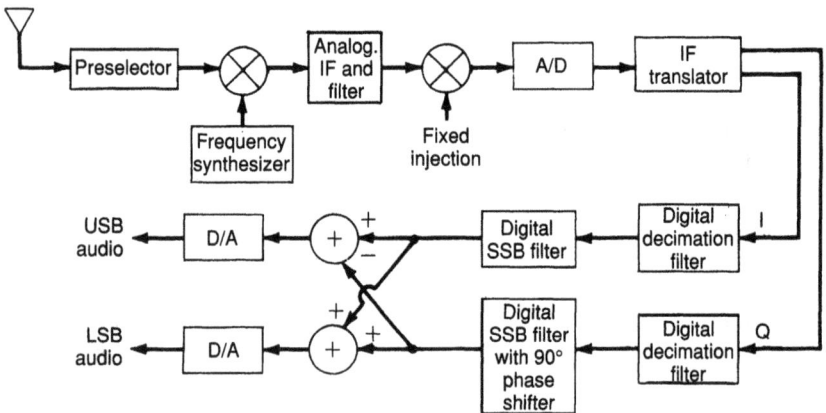

Figure 8.11 Block diagram of an ISB receiver.

analog and digital forms. At the RF side of this transformation, there are stringent demands for linearity and dynamic range. In the transmitter, the modulated digital RF or IF signal must be converted to analog form with very low distortion and noise so as not to contaminate the RF spectrum with unwanted emissions that destroy the usefulness of other channels. In the receiver, the wideband analog RF or IF signal must be converted to digital form without introducing noise and distortion which will fall back on the desired signal channel and make it unusable. To maintain good orthogonality between assigned frequency channels, both in the transmitter and the receiver, requires extremely good linearity and resolution in the RF or IF A/D and D/A converters. Indeed, the linearity and resolution of high-speed A/D and D/A converters presently limit the achievable performance of digital radios. They are by far the most critical subsystems in digital radio implementations, and it is therefore often necessary to reduce the bandwidth and range of signal amplitudes at the RF/IF A/D interface by analog RF/IF signal conditioning.

The sampling theorem

Sampling refers to the process of periodically extracting the value of the analog waveform. Once the analog value is extracted, it must be converted to a digital value or number (A/D conversion) that can be processed mathematically.

In order to understand the sampling theorem, we consider a special type of sampling, namely, impulse sampling. Impulse sampling replaces the continuous analog signal by a train of impulses occurring at the sample times, with the area or *weight* of each impulse equal to the original signal amplitude at those sample times. This process can be described by the following equation:

$$s(t) = x(t)i(t) \tag{8.8}$$

where $x(t)$ is the original continuous signal, and $i(t)$ is the impulse train given by

$$i(t) = \sum_{k=-\infty}^{\infty} \delta(t - kT) \tag{8.9}$$

The sampled signal is simply the original signal multiplied by a train of unit impulses, and results in the convolution of the two corresponding frequency spectra. Figure 8.12 shows the frequency spectrum of a unit impulse train consisting of equispaced frequency δ functions, an arbitrary band-limited signal spectrum, and the resulting spectrum from the convolution of the two. We can see that the band-limited signal spectrum repeats at every multiple of the sample frequency which is the reciprocal of impulse train period. The information contained in any one of these frequency bands is sufficient to describe the input signal.

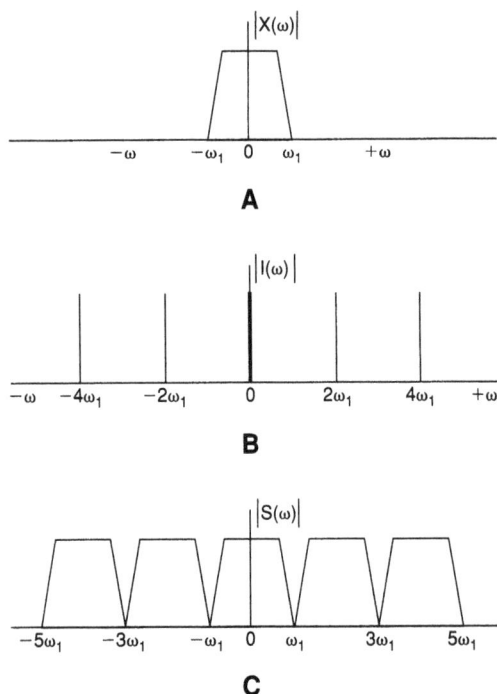

Figure 8.12 Frequency domain illustration of the impulse sampling process: (A) band-limited signal spectrum; (B) unit impulse train spectrum; (C) spectra resulting from the convolution of (A) and (B).

If the sampling frequency in the example shown in Figure 8.12 is reduced, a distortion due to aliasing results. Figure 8.13 illustrates this effect. Aliasing distortion occurs when the corresponding frequency spectra overlap. This distortion cannot be removed once it has occurred, and must be avoided or reduced to an acceptable level in practice.

This brings us to the result of the sampling theorem. That is, the sampling frequency must be at least twice the bandwidth of the desired signal to avoid aliasing distortion and loss of information due to the sampling process. Sampling at twice the bandwidth of the desired signal is said to be *sampling at the Nyquist rate*. In practice, it is usually necessary to sample at a rate somewhat higher than the Nyquist rate to allow for finite transition bandwidths in the antialias filters.

Bandpass and harmonic sampling

Digital radio implementations typically require the sampling of a receiver IF signal. Assuming the IF has been band limited by a filter of some type, as indicated in Figure 8.14A, it is not necessary to sample this IF at twice the highest frequency component of the signal, although there may be advantages in doing so. If the IF band is located between adjacent integer multiples of the total IF bandwidth, the sample frequency is only required to be twice the bandwidth of this signal (which can be considerably less than the highest-frequency component). Under these conditions bandpass or harmonic sampling can be used to efficiently digitize the IF signal. If the IF band does not fall within the above constraint, bandpass sampling can still be used, but the sample frequency will necessarily be greater than twice the IF bandwidth to prevent aliasing.

Bandpass or harmonic sampling is simply the result of the convolution of the IF frequency spectrum with the spectrum associated with the repetitive sampling impulse, as shown in Figure 8.14B. The sampled baseband output spectrum will be an undistorted representation of the IF signal provided there is no aliasing. Aliasing is avoided as long as the IF spectra of Figure 8.14C do not overlap.

Sampling clock noise

It is important to remember that the sampling impulse spectrum is determined by the A/D and D/A converter sample clock. Therefore, the sample

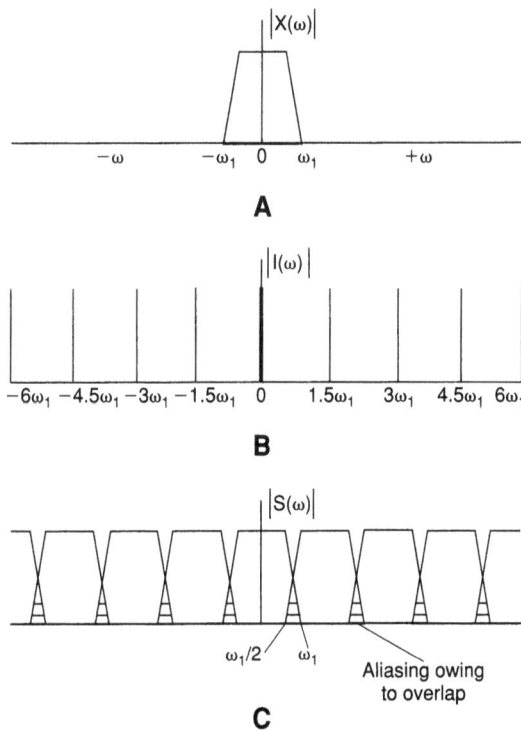

Figure 8.13 Frequency domain illustration of aliasing: (A) band-limited signal spectrum; (B) unit impulse train spectrum; (C) spectra resulting from the convolution of (A) and (B).

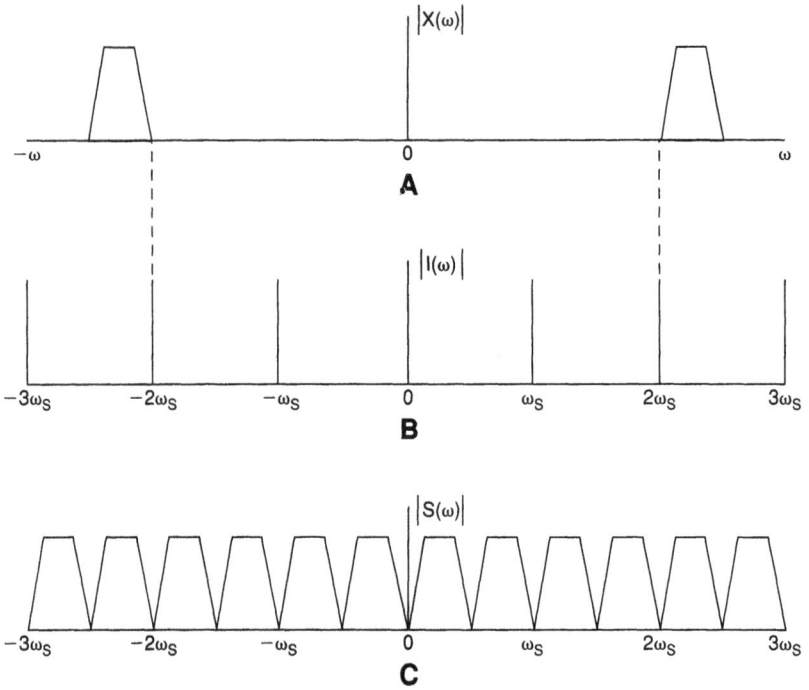

Figure 8.14 Frequency domain illustration of the harmonic sampling process: (A) band-limited IF signal spectrum; (B) sampling sequence spectrum; (C) spectra resulting from the convolution of (A) and (B).

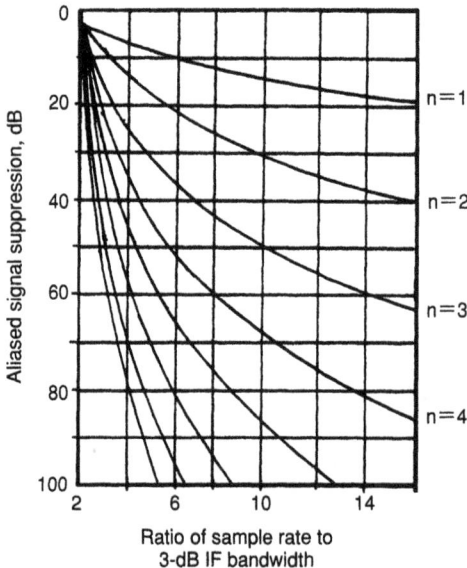

Figure 8.15 Minimum aliased signal suppression versus the ratio of sampling frequency to 3-dB lowpass filter bandwidth. The parameter n is the number of poles required for a 1-dB ripple Chebyshev filter.

clock must be as "clean" and jitter-free as any local oscillator or mixer injection signal in an SSB receiver or exciter. If isolation or buffering is inadequate, the spurious noise picked up on the sample clock can prove disastrous.

Antialias filter requirements

An antialias filter is required to band limit the input signal prior to sampling. A reconstruction filter is also required following D/A conversion. In a digital transceiver, these filters may be a bidirectional IF filter and an active audio output or microphone input filter. In order to avoid aliasing completely, a filter with infinite out-of-band rejection would be required. With all practical filters, only finite out-of-band rejection can be obtained. A curve of minimum suppression of an out-of-band aliased signal as a function of the sample frequency and the number of poles in the antialias filter is shown in Figure 8.15. Note that even with an eight-pole, 1-dB ripple, Chebyshev filter, the sample frequency must be four times the 3-dB bandwidth or twice the Nyquist rate for 80-dB suppression. Thus, regardless of how good the digital filtering is that follows, the filter must be very good if the sample frequency is to be kept within a factor of two of the Nyquist rate.

Data converters

The A/D converter is a device or functional block that accepts an analog value and produces its digital equivalent value. The D/A converter, on the other hand, performs the reverse operation. Many types of data converters are available to perform these operations, each with varying degrees of resolution, speed, and distortion.

As discussed in this chapter, the A/D converter is assumed to be of the sampling-type which includes a sample-and-hold (S/H) amplifier or the equivalent function. Normally, for signal processing, an S/H amplifier is required in one form or another to "acquire" and "hold" the analog input signal voltage until it can be completely converted to a digital value by the A/D converter.

Noise and distortion sources in A/D and D/A converters

Analog-to-digital and D/A converters introduce the following types of noise and distortion:

- Quantizing noise due to ideal quantization
- Phase noise due to sampler aperture jitter
- Harmonic and IM distortion due to nonuniform quantizer step sizes
- Harmonic and IM distortion due to slew rate limiting or nonlinear transfer curve in amplifiers (frequency dependent)

- Harmonic and IM distortion due to nonlinear impedance (flash converters)
- Harmonic and IM distortion due to amplitude-dependent aperture jitter (diode bridge samplers)
- Spurious noise and distortion due to system noise pickup (already discussed)
- Spurious noise and distortion due to aliasing (already discussed)

Quantizing noise. Quantizing noise is the most fundamental type of distortion in a data converter. The total quantizing noise Nq for an ideal converter is

$$N_q = \frac{\Delta^2}{12R} = \frac{(V_{pp}/2^b)^2}{12R} \text{ W} \tag{8.10}$$

where Δ = voltage step size
R = input resistance
V_{pp} = peak-to-peak voltage range
b = number of bits of resolution

Ideally, this noise is spread uniformly over the Nyquist bandwidth of $f_s/2$ where f_s is the sample rate and the quantizing noise density N_{0q} is therefore

$$N_{0q} = \frac{N_q}{f_s/2} = \frac{(V_{pp}/2^b)^2}{6f_sR} \text{ W/Hz} \tag{8.11}$$

The decrease in noise density with increasing sample rate is shown in Figure 8.16. Thus, the quantizing noise density decreases by 6 dB for every bit of resolution added to the converters and by 3 dB for every doubling of the sample rate. The signal-to-noise ratio can be increased by increasing the sample rate and then digitally filtering the signal back to the bandwidth corresponding to a lower sampling rate. This is called *oversampling*.

The maximum sine wave signal power S_{max} that the converter can handle without overload is

$$S_{max} = \frac{\frac{1}{2}(V_{pp}/2)^2}{R} = \frac{\frac{1}{8}V_{pp}^2}{R} \text{ W} \tag{8.12}$$

The maximum signal-to-noise ratio is therefore

$$\frac{S_{max}}{N_q} = \frac{3}{2}\, 2^{2b} \tag{8.13}$$

$$= 6.02b + 1.75 \quad \text{dB}$$

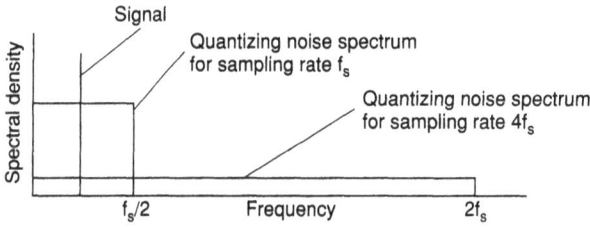

Figure 8.16 Reduction in quantizing noise density by oversampling.

A useful measure of normalized dynamic range is the maximum sine wave signal-to-quantizing noise density ratio given by

$$\frac{S_{max}}{N_{0q}} = \tfrac{3}{4}\, 2^{2b} f_s \tag{8.14}$$

$$= 6.02b + 10 \log_{10} (f_s) - 1.25 \qquad dB$$

The conventional definition of dynamic range is the normalized dynamic range (S_{max}/N_{0q}) divided by the IF passband bandwidth. Figures 8.17 and 8.18 show the sine wave signal-to-quantizing noise ratio and the normalized dynamic range for ideal converters.

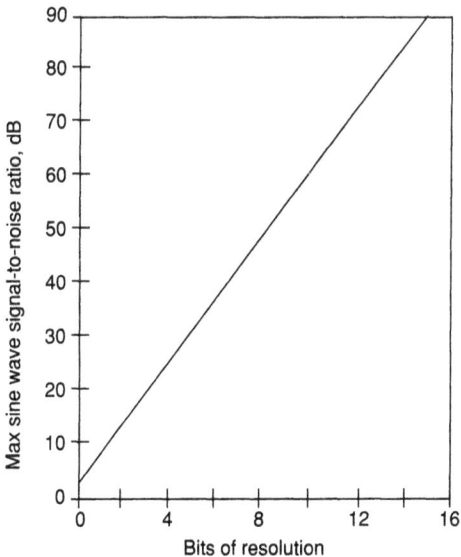

Figure 8.17 Maximum sine wave signal-to-quantizing ratios for ideal data converters.

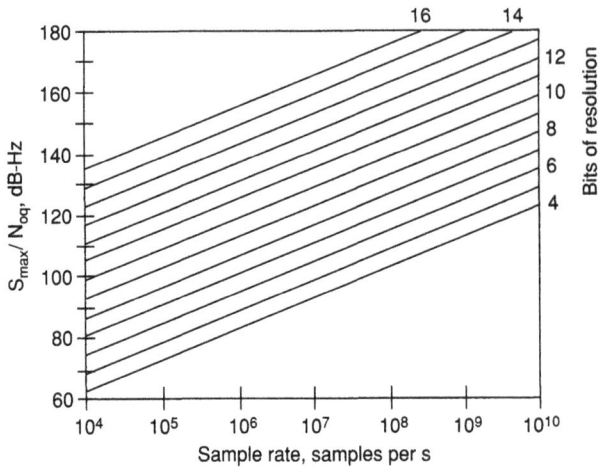

Figure 8.18 Normalized dynamic range versus sampling rate for various sizes of data converters.

This discussion assumed that the quantizing noise was spread uniformly over the Nyquist bandwidth. In practice, this is only the case if the signal being converted is large and traverses many states or quantizing levels of the converter, or if the signal at the converter is itself sufficiently random. In the case of the SSB receiver application, the designer may have to ensure that sufficient front-end or out-of-band dither noise is present at the input of the IF sampling A/D converter when sensitivity level receive signals are being sampled. This consideration is presented in more detail in section 8.10.

Nonlinearities. In practice, data converter quantizing steps are not of equal size, and this nonlinearity increases the quantizing noise and introduces discrete distortion products. Not only does this increase the amount of noise, but much of the distortion-producing noise will be concentrated in discrete low-order distortion products. The type and amount of distortion depend on the type of converter and on the tolerances of the various converter components.

At low frequencies, the data converter transfer curve can be considered linear with small random quantizing errors distributed throughout the converter's full scale range. Therefore, increasing the signal's amplitude doesn't necessarily increase the level of the distortion products. Instead, the distortion due to nonuniform quantization is normally fixed relative to the full scale range of the converter.

At high frequencies, various parasitic effects such as slew rate limiting, signal feedthrough, nonlinear behavior of resistors and capacitors, and decreased open-loop gain in the data converter's amplifier result in harmonic and IM distortion that increases with the signal level. This frequency-dependent nonlinearity is extremely difficult, if not impossible, to linearize. Therefore it is very important to ensure that the full power bandwidth of the S/H amplifier is adequate to support the signal frequency being sampled, especially in IF bandpass sampling applications.

Nonlinearity correction. The nonlinear A/D converter distortion can be reduced by using lookup table correction at the A/D converter output so that the true mean or center value of the analog quantizing step is outputted when the input voltage falls in this range. This arrangement is shown in Figure 8.19. Note that the lookup table must be driven by all the bits out of the A/D converter, but that the table output words need to have greater resolution than the A/D converter.

The least significant bit must have less weight than the least significant bit of the A/D output, while the most significant bit must have enough weight to correct the largest error. In general, the lookup table must be continuously updated to correct for temperature-dependent errors and can be implemented by random access memory (RAM).

Aperture lifter. In addition to the noise and distortion caused by the quantizing amplitude errors, there is noise induced by jitter at the exact instant of sampling. This source of distortion is called *aperture jitter*, and causes, in effect, phase modulation of the received waveform. If the jitter or error in the sampling instant is independent from sample to sample, a sine wave component in the signal generates a white noise component in the output. The ratio of the sine wave signal power C to the aperture noise density that it generates N_{0a} is

$$\frac{C}{N_{0a}} = \frac{f_s}{8\pi^2 f_0^2 \sigma_a^2} \tag{8.15}$$

where f_s is the sampling frequency, f_0 is the signal frequency, and σ_a is the rms time jitter in the sampling (see reference 6). This is also the form of the normalized dynamic range (usually expressed in dB-Hz). Note that this nor-

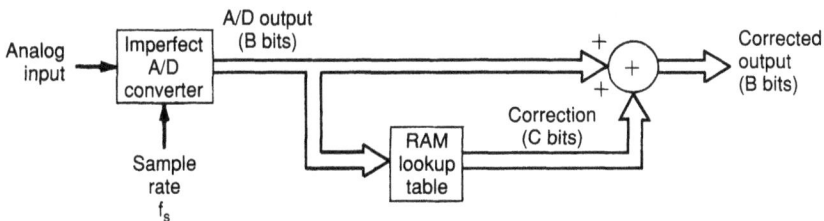

Figure 8.19 Block diagram of an A/D converter with linearity correction.

malized dynamic range is inversely proportional to the square of both the signal frequency f_0 and the rms aperture jitter σ_a. That is, the aperture jitter noise density decreases by 6 dB for every halving of the signal frequency, signal amplitude, or aperture jitter.

Parametric distortion. In flash A/D converters, there are $2^b - 1$ voltage comparators driven in parallel by the input signal, where b is the number of bits of resolution. These comparators have an input capacitance that varies with the applied signal voltage. Because there are so many comparators in parallel, the input capacitance is quite large and can distort the input voltage waveform significantly unless the driving impedance of the flash A/D converter-driving amplifier is quite low. Therefore, low-impedance A/D converter-driving amplifiers are generally required.

Many high-performance S/H amplifiers use a high-speed analog switch, constructed using fast quad diodes in a ring or bridge configuration, much like today's diode mixers. In a track mode, diodes are forward biased to connect the switch input to the output. In the hold mode, the control waveform changes state, reverse biasing the diodes and disconnecting the input from the output. Because the diodes switching times are not instantaneous and the diode bias point is affected by the input signal voltage, the actual switching time can vary depending upon the input signal voltage at the time of the switch. As a result, this amplitude-dependent jitter results in harmonic and IM distortion that is very difficult to compensate.

Converter noise. For high-resolution A/D converters, the smallest quantizing step size is frequently so small that the noise figure of the voltage comparators becomes an issue. For example, consider an A/D converter with 16 bits of resolution, a peak-to-peak range of 2 V, and an input impedance of 50 ohms. The mean square noise contributed by the smallest quantizing step is –88 dBm. If the comparators have a bandwidth of 50 MHz and a noise figure of 10 dB, the thermal noise at the comparator input is –83 dBm or just equal to the quantizing noise. Thus, it is very difficult to design high-speed, single-stage A/D converters with high resolution, even considering only the noise figure problems in the voltage comparators. This problem is partially solved by the use of multistage A/D converters.

D/A converter zero-order hold distortion. Most D/A converters are zero-order hold devices. That is, the analog output of the converter is held at a level corresponding to a digital value applied to the converter for an entire sample period. A zero-order hold D/A converter has a frequency response of

$$|H(\omega)| = \left| \frac{\sin(\omega T_s/2)}{\omega T_s/2} \right| \tag{8.16}$$

where T_s is the hold period or the reciprocal of the sample frequency. This results in a gradual output frequency rolloff that may not be desirable in certain instances. For example, if the output passband width is one-fourth the sample frequency, a passband tilt of approximately 1 dB must be tolerated. This rolloff can be eliminated by adding a digital compensation filter prior to D/A conversion. The compensation filter frequency response must be

$$\frac{1}{|H(\omega)|} = \left| \frac{\omega T_s/2}{\sin(\omega T_s/2)} \right| \qquad (8.17)$$

over the desired frequency range.

Impulse sampling of the D/A converter output can also be used to reduce the hold period and consequently the rolloff. Although the rolloff in the desired passband region is reduced, this method increases the high-frequency content of the D/A converter output and places more stringent requirements on the output filter.

Converter saturation. Every A/D and D/A converter application must prevent the converter from saturating on the peak values of those signals being converted. If saturation occurs, large harmonic and IM products will be generated that can alias many times and fall back in-band, effectively corrupting the signal being converted. In order to prevent saturation, adequate *headroom* must be allocated at the A/D and D/A converter to support the desired signal peaks as well as any undesired signal peaks.

For a typical speech signal, the peak-to-rms voltage ratio may reach as high as 15 to 20 dB, requiring the designer to allocate that much headroom. If the signal to be converted has the Gaussian distribution, allocating 10 dB of headroom would result in saturation occurring 0.1% of the time, while 15 dB of headroom would result in saturation only 0.0000002% (2×10^{-9}) of the time (refer to the Gaussian or normal probability distribution curve).

Sigma-delta A/D converters

Analog-to-digital and D/A converters utilizing "sigma-delta (Σ, Δ)" processing techniques possess many characteristics which are attractive in digital radio design. Sigma-delta converters use oversampling, noise shaping, digital filtering, and sample-rate decimation and interpolation to obtain high dynamic range and high resolution. Also, because they use oversampling, antialias and reconstruction filtering are reduced to simple low-order designs.

A sigma-delta A/D converter consists of two basic processing blocks: an analog noise-shaping modulator and a digital decimation filter, as shown in Figure 8.20.

An analog input signal enters the modulator where it is sampled at a rate much greater than the Nyquist rate. It is called a modulator because the sig-

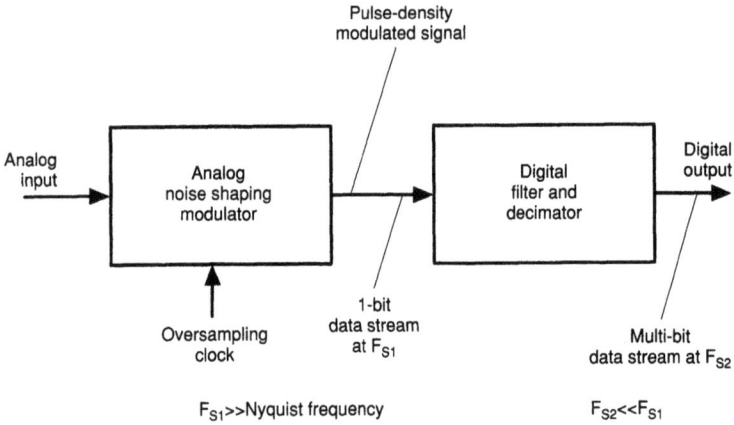

Figure 8.20 Sigma-delta A/D converter.

nal is pulse density modulated. In other words, the density of the output pulses over a given time period is approximately equal to the mean value of the input signal over the same time period. The modulator generates a 1-bit output data stream, which is spectrally shaped by the modulator so that most of the quantizing noise power is at high out-of-band frequencies. The digital filter then removes the out-of-band quantization noise prior to decimation so that it does not alias back in-band.

One of the advantages of the sigma-delta converter over a multibit linear converter is that the 1-bit quantizer can theoretically be perfectly linear, eliminating one of the nonlinearities that limit dynamic range. In addition, the sigma-delta converter is more amenable to VLSI because only a small amount of analog circuitry is required.

A disadvantage is that the sigma-delta modulator must generally be a rather high order to keep the oversampling rate practical. High-order, stable modulators are difficult to design, especially if the converter must work with high signal frequencies, at which device parasitic effects become significant.

8.3 Sampling Rate Reduction

Sampling at a high rate is often advantageous because it reduces the requirements of the antialiasing filter preceding the A/D converter. It also reduces the quantization noise density from the A/D converter. In cases of this type it is advantageous to reduce the sample rate prior to narrowband filtering. It is also often advantageous to perform filtering in several stages, reducing the sample rate between stages as the bandwidth is reduced.

Reducing the sampling rate is often referred to as *decimation*. The rate of reduction is normally done in integer values and is equivalent to resampling the signal at a lower rate. This results in the spectrum being repeated at harmonics of the lower sampling frequency, and consequently the bandwidth must be reduced to half the lower sampling frequency prior to decimation. This is illustrated in Figure 8.21. Figure 8.21B shows the input spectrum, which repeats at intervals of $3f_0$. The decimation filter, which operates at the higher sample frequency, eliminates the frequency components above $f_0/2$ so that aliasing will not occur after the sample rate is reduced. The output signal can now be further filtered with sample rate f_0, which requires less computation. Decimation is particularly effective for FIR filters since only the output samples that are actually used need to be computed.

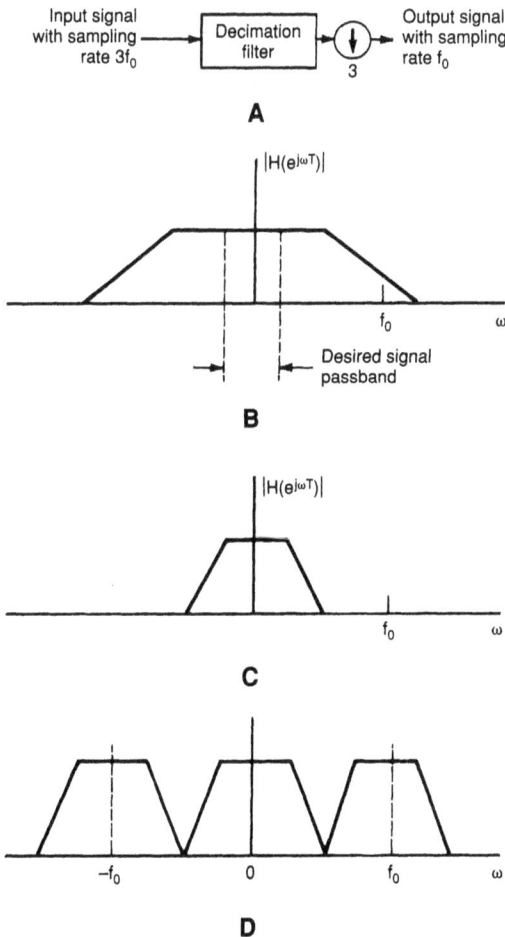

Figure 8.21 (A) Block diagram of a decimation filter; (B) frequency spectrum of input signal; (C) frequency response of decimation filter; (D) frequency response of output signal.

Occasionally it occurs that a signal is grossly oversampled to reduce the quantization noise associated with an A/D converter. A very simple FIR decimation filter can then sometimes be used in which all the multiplier coefficients are unity. Only an adder, accumulator, and counter are required to implement a filter of this type, called a *boxcar filter*, as shown in Figure 8.22.

The Z transform of a boxcar filter is given by

$$H(Z) = \sum_{k=0}^{N} h(k)Z^{-1} = \frac{Z^{-N} - 1}{Z^{-1} - 1} \tag{8.18}$$

and the frequency response is

$$|H(e^{j\omega T})| = \left| \frac{\sin (\omega NT/2)}{\sin (\omega T/2)} \right| \tag{8.19}$$

where T is the input sample period.

8.4 Digital Filtering

The use of digital filtering is one of the major reasons for using DSP techniques in SSB equipment, and digital filters offer several significant advantages over analog filters. One of the most obvious is that once the filter is designed each unit performs identically, which eliminates the tuning, tweaking, and testing often required for analog filters. In addition, because the characteristics of the digital filters are determined by the values of the multipliers in the program, it is possible to store the values for many filter bandwidths in a given processor or even to download the coefficients when it is desired to change the filter bandwidth.

Another advantage, in some cases, is if FIR filters are used the filters can be designed with no differential time delay. This is discussed further in the following section. Finally, because there is no production variation, it is possible to consider filter types that are difficult to handle with analog techniques because of component tolerances.

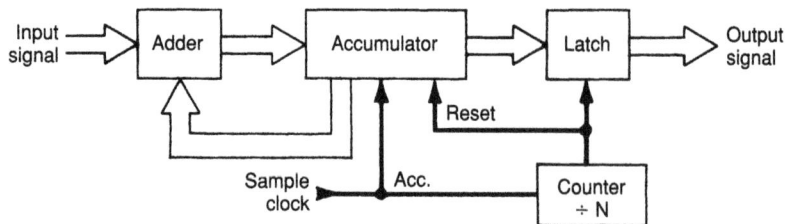

Figure 8.22 Block diagram of a boxcar filter.

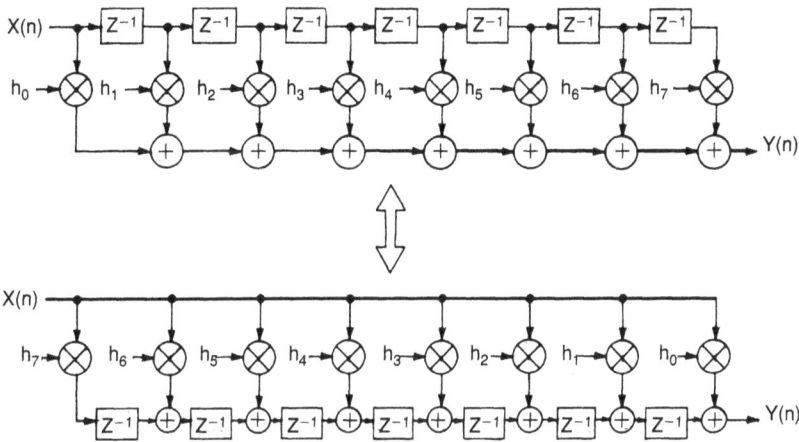

Figure 8.23 Equivalent FIR filter structures.

FIR filtering

A finite impulse response filter transfer function consists of only zeros, compared with the pole-zero structure of infinite impulse response filters. The transfer function of the FIR filter has the form

$$H(Z) = \sum_{n=0}^{N-1} h(n)Z^{-n} \tag{8.20}$$

where $h(n)$ is the impulse response represented by N coefficients of the filter. A block diagram of two equivalent forms of FIR filters is shown in Figure 8.23. Since the current filter output depends only on a finite number of past and present input values, it is sometimes referred to as a *tapped delay line filter* or a *transversal filter*.

Finite impulse response filters have several advantages that make them a good choice in digital transceivers. Computer tools are available to design FIR filters with arbitrary frequency or time response characteristics (see reference 7). The filters are normally designed to have symmetrical or anti-symmetrical coefficients about the center of the filter. This is a sufficient condition to produce a filter with no delay distortion. Since there is no feedback, the filters are absolutely stable and there are no limit cycles. Furthermore, the absence of feedback results in a considerable computational savings if the sample frequency is to be decimated following the filter, because only the output samples required at the decimated rate need be computed.

It is relatively easy to incorporate a Hilbert transformer in an FIR filter so that the phase of the output is shifted by 90° from a similar filter without a Hilbert transformer. This allows phasing of the signals to generate or demodulate SSB signals.

It can be shown that if the impulse response of a lowpass filter is $h_{LP}(t)$, with a frequency response $H_{LP}(f)$, the frequency response of a filter with an impulse response $h_{LP}(t) \cos \omega_0 t$ or $h_{LP}(t) \sin \omega_0 t$ is given by

$$\left| H_{BP}(f) \right| = \tfrac{1}{2} \left| H_{LP}(f - f_0) \right| + \tfrac{1}{2} \left| H_{LP}(f + f_0) \right| \qquad (8.21)$$

which is a bandpass filter centered at f_0. The phase response of the filter obtained by multiplying the impulse response by $\sin \omega_0 t$ is 90° out-of-phase from the filter obtained by multiplying by $\cos \omega_0 t$. Since the impulse response of an FIR filter consists of N coefficients, $h(n)$, these values can be multiplied by

$$\sin \omega_0 \left(n - \frac{N}{2} + \frac{1}{2} \right) T_s \quad \text{or} \quad \cos \omega_0 \left(n - \frac{N}{2} + \frac{1}{2} \right) T_s$$

for $n = 0, 1, \ldots, N - 1$, to perform the transformation where T_s is the sample time.

Thus, a pair of filters can be designed for the I and Q channels in a digital transceiver that includes a Hilbert transformer. Using a design program, such as that described in reference 7, it is also possible to design an FIR filter simply to implement a Hilbert transformer.

The primary disadvantage of FIR filters is that a large number of taps can be required to achieve a sharp cutoff. The approximate number of taps required by a particular design can be estimated (see references 5 and 8) by the equation

$$N = \frac{-10 \log_{10}(\delta_1 \delta_2) - 15}{14 F_T / F_s} + 1 \qquad (8.22)$$

where N = number of taps
   F_T = transition bandwidth
   F_s = sample rate
   δ_1 = passband peak per unit ripple voltage
   δ_2 = stopband voltage attenuation

After the number of taps is estimated and the full precision filter coefficients are computed by the design program, overflow scaling and quantization effects must be evaluated. Overflow scaling is required in most filter designs implemented in fixed-point signal processors. For example, a typical processor might be capable of representing numbers from +1 to −1 in 16-bit 2's complement fractional form. In these implementations, care must be taken to avoid intermediate or final values in the overall computation that exceed ±1. Even though a particular filter design produces a maximum sine wave gain of 1 throughout the passband, the filter output can often exceed a value of ±1 for arbitrary input waveforms. This is especially true in very high-performance sharp-transition bandwidth filters, such as SSB filters. In general, the worst-case gain is computed by summing the absolute value of all the filter coefficients:

$$\text{Worst-case gain} = \sum_{n=0}^{N-1} \left| h(n) \right| \tag{8.23}$$

To prevent overflow, all the coefficients can be scaled so that the worst-case gain is slightly less than unity, or the filter input value can be scaled by the reciprocal of the worst-case gain. Both of these methods result in some degradation of the overall signal-to-noise ratio because of quantization effects, since the maximum sine wave signal power through the filter has been reduced.

There are three primary quantization effects that can affect the performance of the FIR filter. Input quantization noise is determined by the input data word length and the signal level through the FIR filter delay registers. This noise is white and uniformly distributed over the Nyquist bandwidth. It is computed in the same way that the A/D converter noise is computed. That is, the total input quantizing noise σ_i^2 is

$$\sigma_i^2 = \frac{\Delta_i^2}{12R} \, W \tag{8.24}$$

where Δ is the resolution determined by the value of the least significant bit of the fixed-point word and R is the assumed resistance. This noise is directly affected by the overflow scaling method.

Roundoff noise is another quantization effect and is a result of errors in the filter computation due to truncations following the coefficient multiplications and the final filter output accumulation. It is uniformly distributed white noise and is, therefore, computed as above. For example, if each multiplication is truncated, in an N-tap filter, the total roundoff noise is

$$\sigma_r^2 = \frac{N\Delta_r^2}{12R} \, W \tag{8.25}$$

where Δ_r is the minimum step size following the truncation.

Many digital multiplier/accumulators today provide a very large accumulator word length so that full precision products can be accumulated. In this case, σ_r^2 is zero and the roundoff noise corresponds to a single filter output truncation such that

$$\sigma_a^2 = \frac{\Delta_a^2}{12R} \, W \tag{8.26}$$

The maximum signal-to-quantization noise ratio of the FIR filter is then S_{max}/N_q, where S_{max} is the maximum sine wave signal power determined by the input level and the gain of the overflow scaled filters, and N_q is the total noise power consisting of σ_a^2 or σ_i^2 and that portion of σ_i^2 which is not attenuated by the filter.

The final quantization effect, coefficient quantization, is a function of many filter design variables and is difficult to predict. Coefficient quantization does not add to the overall noise power of the filter, but does affect the overall response of the filter.

Since the filter coefficients are represented by a finite number of bits in a particular machine implementation, the filter's impulse response and consequently the frequency response will be affected. In some cases, a high-performance SSB filter's stopband requirements may not be met in a fixed-point implementation without factoring the filter into smaller cascade filters requiring less stopband attenuation in each of the cascade sections.

The frequency response of a 64-tap SSB filter, operating at a 16-kilosample/sec rate and using 16-bit coefficient values, is shown in Figure 8.24. Figure 8.25 shows the same filter response when the filter coefficients are reduced to 13-bit precision. Note that the original 60-dB stopband requirements cannot be met by the reduced coefficient filter.

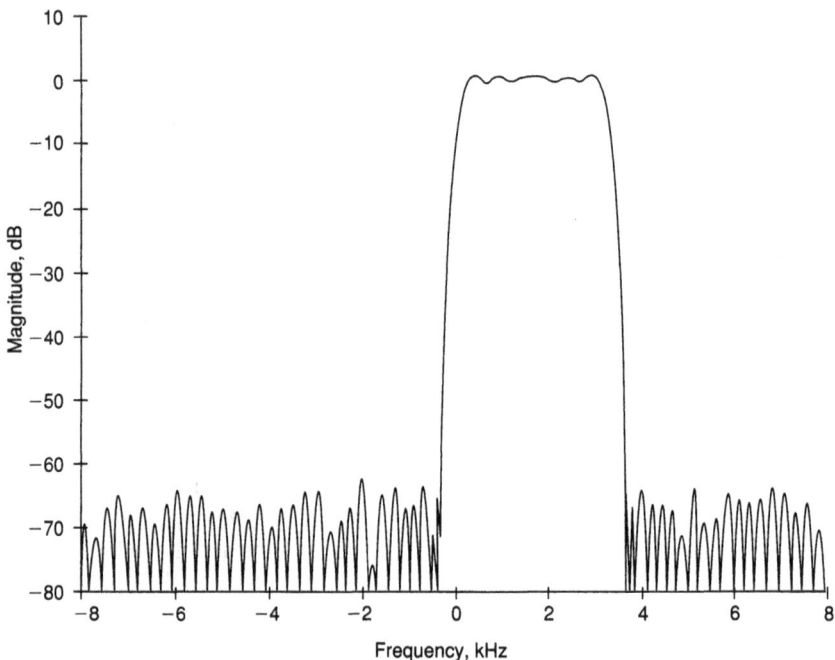

Figure 8.24 Computer-generated graph showing 64-tap sideband FIR filter response using 16-bit coefficients.

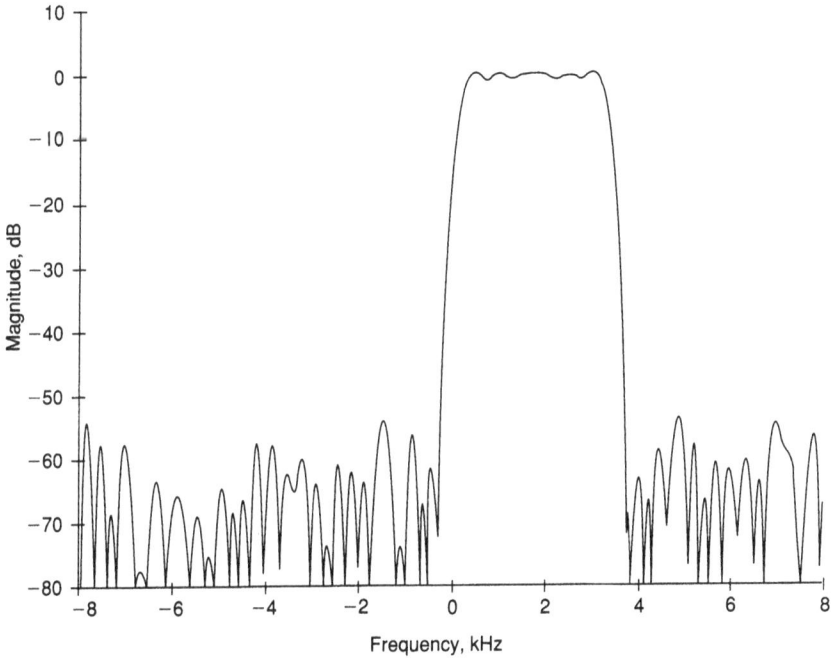

Figure 8.25 Computer-generated graph showing 64-tap sideband FIR filter response using 13-bit coefficients.

If the original N-tap filter transfer function is factored so that zeros can be optimally paired to break the original filter into two or more cascade sections, the coefficient quantization effect can be reduced with a slight increase in roundoff noise resulting from an additional accumulator truncation for each section. In general, the polynomial of order $N - 1$ is factored using a good root-finder program. The roots are then grouped to evenly distribute each filter section's gain and response requirements and the new coefficients are calculated. The factoring procedure results in an overall increase of one tap computation for a total of $N + 1$ taps compared to the original filter's N taps.

IIR filters

Infinite impulse response filters are distinguished from FIR filters in that feedback is present. This results in the presence of poles in the transfer function that are not present in FIR filters. Normally IIR transfer functions are derived by converting the transfer functions of analog filter types to the digital domain. Thus the design of an IIR filter often approximates the frequency response of a Butterworth, Chebyshev, or elliptic filter. An IIR filter can be designed to have a sharper cutoff frequency for a given order than an FIR filter; however, it has generally the same phase-delay characteristics as the analog filter from which it was derived. Infinite impulse response filters tend to require more careful scaling to prevent overflow, and they are also subject to oscillation if any pole is outside the unit circle in the Z plane. The presence of limit cycles must also be carefully studied. Limit cycles are low-level oscillation usually in the order of the least significant bits of the digital word. Infinite impulse response filters can be equalized if necessary to minimize the delay distortion. However, if this is required, the complexity may well exceed that of an FIR filter designed to perform the same function.

A number of excellent IIR design programs are available in the marketplace. These programs generally begin by inputting the desired filter characteristics and determine the filter type and order by interacting with the operator. Some of the programs allow the operator to determine the effects of truncated coefficients and arithmetic as well as to scale the filter to prevent overflow. Several programs also allow the operator to designate the order of the filter sections to minimize the scaling required to prevent overflow. Limit cycles are also identified in some cases.

It is not the purpose of this discussion to describe specific filter design programs or to redevelop the theory of IIR digital filter design, which is treated in a number of excellent texts (see, for example, references 1 through 3) on the subject. A brief description of a common implementation of digital IIR filters is presented here, along with an overview of the bilinear transform method of converting an analog filter transfer function to a digital transfer function.

An analog lowpass filter transfer function of the Butterworth or Chebyshev type can generally be written in the form

$$H(S) = \frac{K}{S_p^n + a_1 S_p^{n-1} + a_2 S_p^{n-2} + \ldots + a_n} \tag{8.27}$$

Tables are available in the literature listing the coefficient values [1]. Standard methods are also given to convert the transfer functions to bandpass or bandstop filters.

Several methods can be used to approximate the digital transfer function $H(Z)$ from the analog transfer function $H(S)$. The most common of these are the impulse invariant method and the bilinear transform. The impulse invariant method guarantees that the digital filter will have the same impulse response as the analog filter at the sample times. Thus the digital filter also has the same phase response as the analog filter. This method is quite good so long as the frequencies of interest are well below one-half the sampling frequency. As this condition is approached, the frequency response of the digital filter deviates considerably from that of the analog filter. For this reason, the bilinear transform method is more widely used.

The bilinear transform produces considerably better results and is implemented by making the following substitution for S in the analog transfer function:

$$S = \frac{2}{T} \frac{Z-1}{Z+1} \qquad (8.28)$$

It can also be shown that the digital frequency corresponding to a given analog frequency value is given by

$$\omega_d = \frac{2}{T} \tan^{-1} \frac{\omega_a T}{2} \qquad (8.29)$$

where T = sampling time
  ω_d = digital frequency
  ω_a = analog frequency

Thus, if a specific cutoff frequency is desired in the final digital filter, the analog filter from which the digital filter is realized must have a cutoff frequency given by

$$\omega_a = \frac{2}{T} \tan \frac{\omega_d T}{2} \qquad (8.30)$$

A digital filter can be realized in many different configurations. In general the overflow characteristics are different for each configuration and it may be worthwhile to investigate the scaling requirements of several configurations. Among the configurations used are the direct form, the canonic form, the cascade form, and the parallel form. Block diagrams for these forms are given in the references cited. The transfer functions are of the form

$$H(Z) = \frac{a_0 + a_1 Z^{-1} + a_2 Z^{-2} + a_3 Z^{-3} + a_4 Z^{-4}}{1 + b_1 Z^{-1} + b_2 Z^{-2} + b_3 Z^{-3} + b_4 Z^{-4}} \qquad (8.31)$$

Often the transfer function is factored into two-pole sections, as given below, and implemented as cascaded bi-quad sections such as shown in Figure 8.26 using the canonic form (this configuration tends to be more tolerant of truncated arithmetic than a single realization of the entire filter):

$$H(Z) = \left[\frac{a_{10} + a_{11} Z^{-1} + a_{12} Z^{-2}}{1 + b_{11} Z^{-1} + b_{12} Z^{-2}} \right] \left[\frac{a_{20} + a_{21} Z^{-1} + a_{22} Z^{-2}}{1 + b_{21} Z^{-1} + b_{22} Z^{-2}} \right] \qquad (8.32)$$

A bi-quad section requires five multiplications and four additions that must be performed during each sample time. It should be noted that arithmetic overflow is a serious problem because, with 2's complement arithmetic,

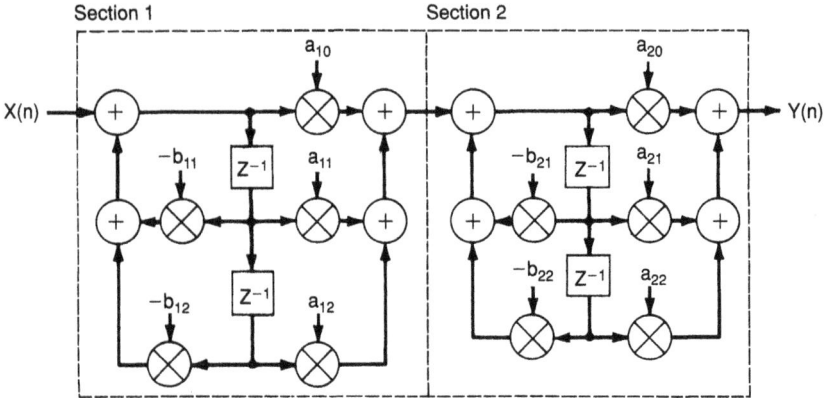

Figure 8.26 Block diagram of cascaded bi-quad filter sections.

overflow from the largest positive number results in the largest negative number.

Optimum pairing of the poles and zeros of the transfer function can reduce the scaling required between sections to prevent overflow.

It should also be noted that since IIR filters utilize feedback, oscillation is possible if the poles are outside of the unit circle on the Z plane. Even for a stable filter, however, limit cycle oscillations can occur with truncated arithmetic. These low-level oscillations, which may be in the order of the least significant bits, result because for specific low-level signals the truncation results in a pole apparently moving to the unit circle, causing oscillation. Information concerning limit cycles is given by several of the IIR filter design programs.

Adaptive digital filtering

An adaptive digital filter is an FIR or IIR digital filter with time varying coefficients. Adaptive filters have been used to implement adaptive notch filters (see reference 9), adaptive noise cancellers (see reference 10), and adaptive channel equalizers (see reference 11). All of these functions can be applied to SSB radio communications equipment.

A block diagram of a generalized adaptive filter is shown in Figure 8.27. The input signal $x(n)$ is filtered to produce an output $y(n)$. This output is compared to a reference signal $r(n)$, and the resultant error signal $e(n)$ is used by the coefficient update algorithm to adapt the coefficients in such a manner as to reduce the error.

In order to determine the updates required for the N coefficients, it is necessary to generate a sensitivity vector that consists of the partial derivatives of the error signal power with respect to each coefficient.

Sensitivity filters are required to generate each of the partial derivatives. These sensitivity filters can be determined using Tellegen's theorem (see reference 12). The transfer function of the sensitivity filter, for the ith coefficient, is the transfer function from the adaptive filter's input to the input of the ith coefficient multiplier, multiplied by the transfer function from the output of the ith coefficient multiplier to the adaptive filter error signal output.

Once the sensitivity vector is computed, a number of different algorithms can be used to update the coefficients. Some of the more popular ones are the Gauss-Newton, Newton-Raphson, or steepest descent methods (see references 13–15). Regardless of the algorithm used, the coefficients are adapted to minimize the error signal and force all sensitivity filter outputs to zero.

In real-time applications the "gradient" or steepest descent algorithm is often used in order to determine the coefficient updates. In this case the update formula is

$$h_i(n) = h_i(n-1) - Ke(n)s_i(n), \ i = 1, 2, \ldots, N \qquad (8.33)$$

where K is an adaptation rate parameter that is chosen to guarantee stability and also control the rate of adaptation. The quantity $e(n)$ is the error output of the adaptive filter as described above, and $s_i(n)$ is the output of the sensitivity filter for the ith coefficient.

To better illustrate the DSP involved, we will examine the adaptive FIR filter. If Tellegen's theorem is used as described above, we find that the sensitivity filters degenerate to a simple multiplication of the delayed input sig-

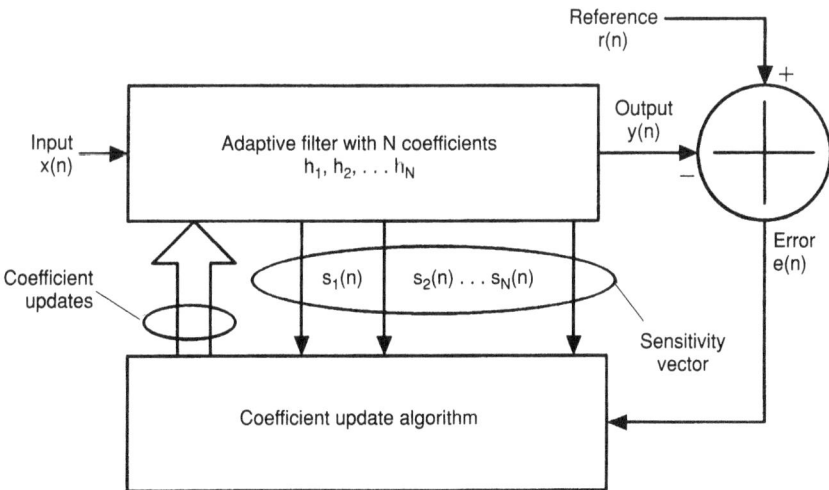

Figure 8.27 Block diagram of a generalized adaptive filter.

nal values by the error signal output, as shown in Figure 8.28. For an adaptive IIR filter, the sensitivity filters will necessarily be more complex because of the feedback.

The gradient coefficient update processing is also illustrated in Figure 8.28. Note that for purposes of more clearly illustrating the gradient algorithm, each sensitivity filter output is being multiplied by the adaptation rate parameter K. Since the processing is linear there is no reason why the multiplication could not have been done only once on the error signal, before multiplication by the delayed input signal values. In most implementations, the multiplication is moved to the error signal output.

Adaptive notch filter

Typically, an adaptive notch filter uses the input signal $x(n)$ as the reference signal $r(n)$, as in Figure 8.28. Its coefficients are then adapted to minimize the difference (or error) between the filter's input signal and the filter's output signal. Also, the error signal $e(n)$ is taken as the notch filter output rather than the adaptive filter output $y(n)$. Consequently, if the input signal is correlated, the adaptive filter will be successful in forcing the error signal, or notch filter output, to zero. On the other hand, if the input signal is not correlated, the filter will not be able to adapt and the input signal will be passed through to the notch filter output. The notch filter can therefore be viewed as a decorrelating filter or *whitening* filter.

This form of adaptive notch filtering works very well for removing highly correlated signals, such as tone interferers, from uncorrelated signals. If the

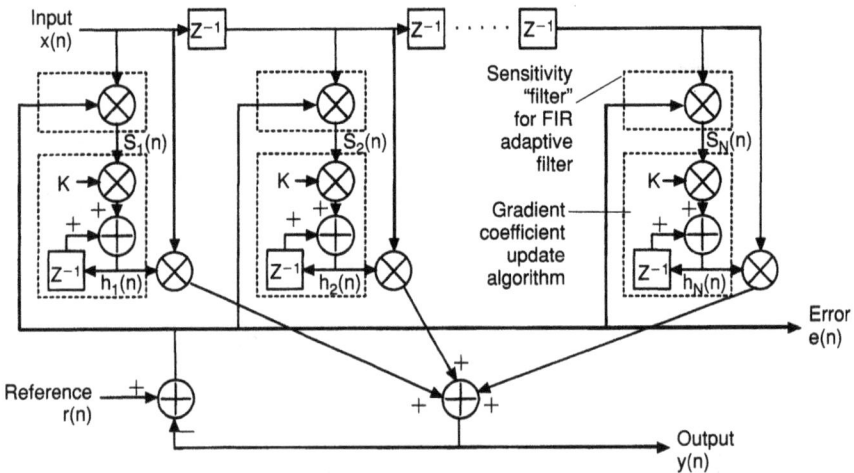

Figure 8.28 Adaptive FIR filter using the gradient algorithm.

desired signal is somewhat correlated, as speech is for example, the adaptive notch filter may have to be modified or "tweaked" to prevent it from being too successful at canceling the desired signal as well as the interferer. The adaptation rate parameter K can be reduced to prevent the filter from adapting to the desired signal while still allowing it to adapt to the interferer. Another technique is to insert a delay between the input signal and the adaptive filter input, but not delay the reference signal input [9]. This delay will then serve to decorrelate signals such as speech, which are correlated only over a short time interval.

An estimate of the required number of taps in an adaptive FIR notch filter can be obtained using Equation 8.22.

8.5 SSB Detection

Product detectors

Conventional analog product detectors are heterodyne detectors (mixers) that combine two input signals (IF and BFO) to produce a difference frequency at audio or baseband. The sum frequency, IF, and BFO frequency are removed from the mixer output by relatively simple lowpass filters.

A single frequency component of an SSB signal, at complex baseband, can be represented as

$$Y(t) = A e^{j\omega_0 t} = A(\cos \omega_0 t + j \sin \omega_0 t) = I(t) + jQ(t) \qquad (8.34)$$

and as quadrature components

$$I(t) = \mathrm{Re}\{Y(t)\} \qquad (8.35)$$

$$Q(t) = \mathrm{Im}\{Y(t)\} \qquad (8.36)$$

In one implementation of a digital receiver, the desired sideband is translated to baseband and selected by filtering. The digital product detector must take the filtered complex baseband input, perform the BFO translation, and convert the complex result to real audio.

Complex signal translation can be accomplished by multiplying the input SSB signal $Y(t)$ by a complex BFO signal $e^{j\omega b t}$ for USB or $e^{-j\omega b t}$ for LSB and outputting only the real or imaginary part, as shown in Figure 8.29. If $Y(t)$ is sampled at $t = nT_s$ at a rate $f_s = 1/T_s$:

$$f(nT) = [A e^{j\omega_0 n T_s}][e^{-j\omega b n T_s}] \qquad (8.37)$$

$$= A e^{j(\omega_0 \pm \omega_b)n T_s} = A[\cos(\omega_0 \pm \omega_b)nT_s + j\sin(\omega_0 \pm \omega_b)nT_s] \qquad (8.38)$$

where

$$\text{Re}\{f(nT)\} = A \cos (\omega_0 \pm \omega_b) nT_s \qquad (8.39)$$

The analysis shown above illustrates the cancellation of the undesired mixer product (either the sum or difference) and assumes perfect amplitude and phase balances of the input and the BFO. However, there are always distortions that produce some errors in the amplitude or phase balance of both. This error results in imperfect cancellation of the undesired mixer product which shows up as a spurious distortion product. This distortion (see reference 16) can be expressed as the following ratio:

$$\frac{\text{Undesired product}}{\text{Desired product}} = \frac{a^2 + b^2 - 2ab \cos (\Delta + \delta)}{a^2 + b^2 + 2ab \cos (\Delta - \delta)} \qquad (8.40)$$

where a and b are proportionality constants such that the in-phase input I can be modeled as aI and the quadrature component Q can be modeled as bQ (these constants take into account amplitude imbalance in both the input signal and the digital BFO signal), Δ is the BFO phase error (relative to 90°), and δ is the input phase error between I and Q components (relative to 90°).

In general, the amplitude and phase balances of the complex input can be made nearly perfect by proper digital design. However, any error in the BFO algorithms will contribute to amplitude and phase imbalances, and the algorithm complexity will increase as the required distortion levels decrease. Two BFO algorithms are presented below. Each has advantages and disadvantages in terms of speed, memory requirements, and accuracy that must be weighed in each machine implementation.

BFO generation

One method of implementing a complex oscillator for digital BFO generation uses a phase accumulator driving a sine and cosine algorithm or generator. The output of the sine and cosine algorithm is the sin $\omega_b nT$ and cos $\omega_b nT$ values which are used in the product detector, as shown in Figure 8.29.

The phase accumulator represents the $\omega_b nT$ modulo 2π quantity used to compute sin $\omega_b nT$ and cos $\omega_b nT$. The value is obtained by simply incrementing a phase angle θ by an amount Δ at each sample time. If the sample frequency is f_s,

$$\Delta = \frac{2\pi f_b}{f_s} \text{ radians} \qquad (8.41)$$

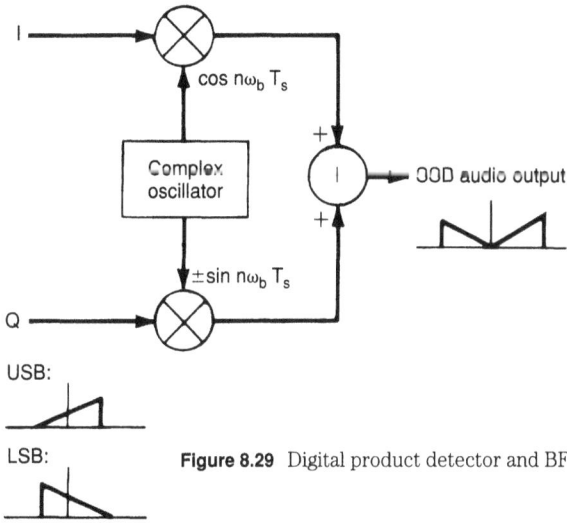

Figure 8.29 Digital product detector and BFO.

A sine and cosine generator can be implemented using lookup tables and interpolation. The following algorithm utilizes the trigonometric identities below to provide the required complex oscillator outputs:

$$\sin (A + \delta) = \sin A \cos \delta + \cos A \sin \delta \tag{8.42}$$

and

$$\cos (A + \delta) = \cos A \cos \delta - \sin A \sin \delta \tag{8.43}$$

If δ is made small, the following approximations can be made:

$$\sin (A + \delta) \approx \sin A + \delta \cos A \tag{8.44}$$

and

$$\cos (A + \delta) \approx \cos A - \delta \sin A \tag{8.45}$$

The sine and cosine generators use the above approximation and identity to calculate

$$\sin (X) = \sin (Y) + \delta \cos (Y) \tag{8.46}$$

and

$$\cos (X) = \cos (Y) - \delta \sin (Y) \tag{8.47}$$

where

$$X = \omega_b n T \tag{8.48}$$

$$Y = \text{nearest quantized lookup table value less than } \omega_b n T \qquad (8.49)$$

and

$$\delta = X - Y \qquad (8.50)$$

In general, it is desirable to limit the size of the lookup table. The algorithm presented normalizes the input angle ($0 \le \theta \le \pi/2$) and corrects the sign of the sine or cosine after computation by using the following rule:

$$\text{Sign of } \sin(\omega_b n T) = \text{sign of } \omega_b n T \qquad (8.51)$$

and

Sign of $\cos(\omega_b n T)$

$$= \text{sign of } \left\{ \frac{\pi}{2} - \text{ABS}(\omega_b n T) \right\}; \; -\pi < \omega_b n T \le \pi \qquad (8.52)$$

where ABS is the absolute value. A single table can be used for both sine and cosine functions since $\cos \theta$ is equal to $\sin(\pi/2 - \theta)$ over the range $0 \le \theta \le \pi/2$.

The distortion performance for several table sizes are shown in Table 8.1. The distortion values listed were computed assuming a complex translator implementation where both sine and cosine envelope and phase error values are taken into account.

The sine and cosine generator can also be implemented without large lookup tables by using an algorithm based on a series expansion. The method is characterized by an increase in accuracy as the number of terms in the series increases, and requires a normalization of the input angle and a sign correction after the sine and cosine are computed, as described previously.

TABLE 8.1 Distortion Performance of Translator Using Trigonometric Approximated Sine and Cosine Generators ($\pi/2$ Normalization)

sin/cos Table length N	Total translator distortion	
	dB	%
4	−28	4
8	−40	1
16	−52	0.25
32	−64	0.06
64	−76	0.016
128	−88	0.004

TABLE 8.2 Distortion Performance of Translator Using MacLaurin Series Sine and Cosine Generators ($\pi/2$ Normalization)

Order of polynomial represented by series		Total translator distortion	
Sine	Cosine	dB	%
3	4	–28	4
5	6	–52	0.25
7	8	–82	0.008

The most common series expansion is the Maclaurin series, which is just a special case (expansion about zero) of the familiar Taylor's power series as illustrated:

$$\sin (x) = x - \frac{1}{3!}x^3 + \frac{1}{5!}x^5 - \frac{1}{7!}x^7 + \ldots \tag{8.53}$$

$$\cos (x) = 1 - \frac{1}{2!}x^2 + \frac{1}{4!}x^4 - \frac{1}{6!}x^6 + \ldots \tag{8.54}$$

As the angle increases, the error increases rapidly. The distortion performance for several expansion lengths was calculated for a complex translator implementation and is summarized in Table 8.2.

8.6 Digital AGC Methods

Automatic gain control provides the same overall function in a digital receiver as in a purely analog receiver. However, there can be significant differences in the design approach used, because of receiver hardware differences. For instance, more gain is generally required before the digital IF filter in the digital receiver in order to drive the A/D converter at a satisfactory level. This means that appreciable AGC must be applied before the digital IF filter. In addition, because of the envelope delay associated with a high-performance digital filter, special techniques must be used to provide an AGC with adequate speed, while still controlling overshoot. A digital receiver block diagram illustrating a typical AGC processor interconnection was shown in Figure 8.10.

The AGC processor operates on the digital IF signal components to provide a desired gain control value to a D/A converter, which in turn controls the attenuators in the IF translator, and consequently the signal level at the A/D converter. It also provides a digital gain multiplier to control the signal level at the audio output D/A converter.

The desired receiver gain distribution, or analog versus digital receiver gain, is selected on the basis of a system level analysis. The maximum analog translator gain must be great enough so that the quantizing noise of the A/D converter does not degrade the receiver noise figure or sensitivity below the desired limit. As the signal level increases above the sensitivity level, the analog gain should not be decreased by AGC action until an adequate signal-to-noise ratio is obtained. As the signal level increases still further, the analog AGC should hold the signal level at the A/D converter essentially constant. In addition, adequate headroom at the A/D converter and D/A converter must be provided to avoid saturation during normally high peak-to-average voltage ratio periods of the desired signal.

Strong signals that fall outside the narrowband digital filter bandwidth, but inside the analog IF translator bandwidths, can overload or saturate the A/D converter. This results in the generation of in-band IMD products and can result in significant degradation of the desired signal. If large signal levels are detected at the A/D converter, the receiver gain may have to be redistributed by reducing the preconversion analog gain and increasing the digital gain to maintain the desired signal output level. This will, however, reduce the desired signal-to-quantization noise ratio.

8.7 Digital Squelch

Chapter 7 presented various methods of implementing the SSB squelch function. Nearly any squelch that can be implemented with analog circuitry can also be implemented digitally because the bandwidths are relatively low and the filter requirements are generally modest. One of the advantages of DSP may well prove to be the ability to realize a signal delay while the squelch decision is being made. This should allow a longer averaging time in the squelch detector without losing the first syllable of the message while the squelch decision is being made.

One of the more widely used squelch circuits for SSB compares the energy in the lower and upper portions of the audio band, as indicated previously. A block diagram of this circuit, implemented digitally, is shown in Figure 8.30. A two-pole lowpass filter with a 1.2-kHz cutoff may be appropriate, while the highpass filter has a cutoff in the order of 1.8 kHz. Two FIR filters may be used. Finite impulse response filters have the advantage that the delayed signal is also available in the filter, eliminating the need for a separate delay. The Hilbert transformer produces a 90° phase shift for the detectors. A transformer with 8 to 12 taps may suffice. The detectors perform an approximation of the function $\sqrt{I^2 + Q^2}$ by taking the absolute value of the larger of I or Q and adding 0.4 times the absolute value of the smaller. A smooth window function such as a Hamming window can be used as the squelch gate, or a simple on/off threshold can be used.

8.8 Speech Compression

Over the years there have been many approaches used to increase the ef-

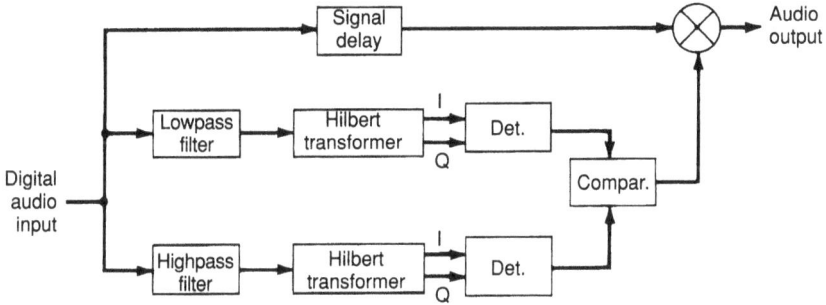

Figure 8.30 Block diagram of digital squelch.

fective "talk power" of SSB transmissions. Chapter 7 presented various methods of speech processing that can be used to accomplish this. Each approach has distinct advantages and disadvantages. From a performance viewpoint, RF speech processing has often provided the best results. A digital compressor can be implemented to operate on the instantaneous speech modulation envelope and provide the same effect as RF processing. The compressor time constants can be varied to provide speech compression or clipping. A block diagram of the compressor is shown in Figure 8.31. The real audio input signal is phase shifted 90° in the Hilbert transformer so that the instantaneous envelope of the signal can be computed from the in-phase I and quadrature Q components. The envelope $E(n)$ is then used to generate a multiplier $M(n)$ which scales the delayed signal on a sample-by-sample basis.

A signal path delay is provided to prevent hard limiting during transient responses and thereby eliminate "clicks" normally encountered in closed-loop or analog systems.

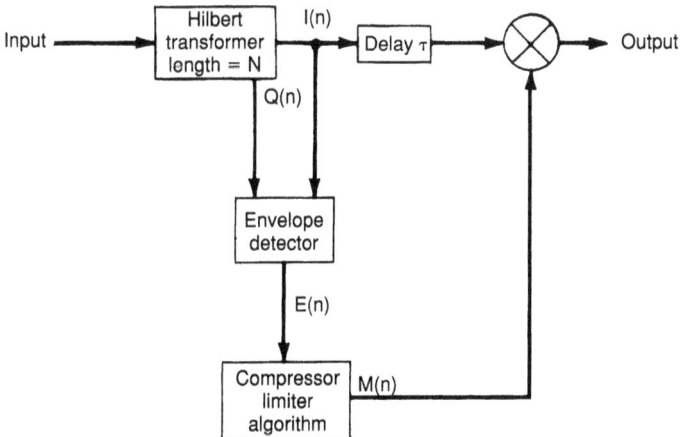

Figure 8.31 Block diagram of speech compressor/limiter processing.

The envelope detector can approximate the $\sqrt{I^2 + Q^2}$ by taking the absolute value of the larger of I or Q and adding 0.4 times the absolute value of the smaller. The compressor algorithm can adjust the multiplier value using the following recursive equation:

$$M(n) = [1 - K]M(n - 1) + \frac{KE_{\text{Desired}}}{E(n)} \tag{8.55}$$

where K = positive constant less than 1
   $M(n - 1)$ = previous multiplier value
   E_{Desired} = desired output envelope level

The time constant of the compressor is then

$$\text{TC} = \frac{T_s}{\ln(1 - K)} \tag{8.56}$$

where T_s is the sample period or $1/f_s$.

Thus the time constant can be varied by changing the factor K. As K tends toward 1, the time constant is reduced and the compressor will operate as a clipper. In general, the time constant must be somewhat larger than the Hilbert transformer envelope delay of $N/2$ sample periods. A speech compressor requires a fast-attack, slow-release characteristic, and, thus, two values of K are used depending on whether the compressor multiplier is increasing or decreasing.

This example simply illustrates the advantage of DSP in that flexibility is obtained by using sophisticated mathematical algorithms that are difficult to implement with analog circuitry.

8.9 SSB Modulation

Single-sideband modulation techniques are discussed in detail in Chapter 5 and briefly in section 8.1. These techniques as well as several others used to produce four-channel SSB are examined in more detail here.

Single-sideband signals are often generated by filtering techniques in the analog domain because of the difficulty in obtaining wideband phase shifters and maintaining exact amplitude balance as required by phasing techniques. These limitations are largely overcome with digital systems, and consequently, phasing methods for SSB generation tend to dominate with DSP techniques.

The filtering techniques can, of course, also be implemented with digital filters; however, the processing load may be higher than for the phasing techniques. One of the basic methods of SSB generation is shown in Figure 8.6. Here the signal is passed through two bandpass filters, one of which contains a 90° phase shift. Typical bandwidth characteristics are shown in Figure 8.32A. The amplitude response of a real filter is the same for either positive or negative frequencies, and both halves are shown in the drawing. If FIR filters are used, it is convenient to begin with a lowpass filter as shown in Figure 8.32B and transform it to a bandpass filter as described in section 8.4. This can be accomplished by multiplying each of the lowpass coefficients $h(n)$ by a cosine function such that

$$h_c(n) = h(n) \cos \omega_0 \left(n - \frac{N}{2} + \frac{1}{2} \right) T_s$$

$$n = 0, 1, \ldots, N-1$$

(8.57)

where ω_0 is the desired angular frequency shift. In this case $\omega_0 = 2\pi(1500)$. A companion filter that has a 90° phase offset relative to the first can be obtained by using a sine function multiplication, such that

$$h_s(n) = h(n) \sin \omega_0 \left(n - \frac{N}{2} + \frac{1}{2} \right) T_s$$

$$n = 0, 1, \ldots, N-1$$

(8.58)

Care must be exercised in the low-frequency region if the filters are not used as a complex pair, because the overall characteristic of each filter is produced by the overlapping positive and negative responses and the response at −300 Hz should be in the order of 60 dB down. The re-

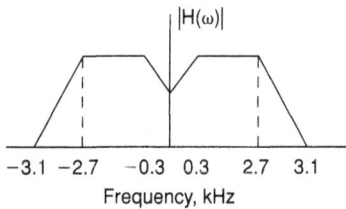

A

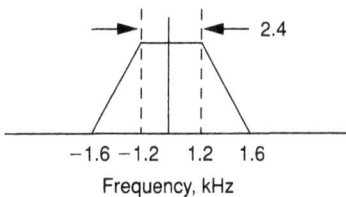

B

Figure 8.32 (A) Typical SSB filter characteristics; (B) lowpass source filter frequency response.

sulting I and Q outputs from the filters represent a positive-frequency spectrum and, as shown in Figure 8.6, may be translated to the digital IF frequency.

Another technique, known as the Weaver method, translates the audio signal down by about 1600 Hz, forming a complex signal with unsymmetrical positive and negative parts (Figure 8.33). The resulting signal can then be lowpass filtered and no 90° phase shifter is required in the filters. This technique results in a low computation rate and is also more amenable to the use of IIR filters. The second full-complex mixer translates the SSB signal to the digital IF. The circuit shown in Figure 8.33 can also be used to generate the upper USB or the lower LSB for a four-channel exciter by using an ω_1 that is $2\pi(6100 \text{ Hz})$ greater or lower than that used for the LSB and USB translators, and all four signals are summed prior to applying to the D/A converter. The filter characteristics will, in general, be more severe for a four-channel system, however. The spectrum generation using the Weaver method is shown in Figure 8.34.

As shown in section 8.1 (see Figure 8.9), a considerable savings can be realized in a two-channel ISB exciter using the phasing method. This results because for a 90° phase-shifted pair derived from a real signal the representation $I + jQ$ gives the positive frequencies and $I - jQ$ gives the negative frequencies. An examination of the block diagram shown in Figure 8.8 reveals that by interchanging the order of filtering and summation, one pair of filters can be eliminated. We first consider the component from each of the audio inputs. In each case, the audio signal is

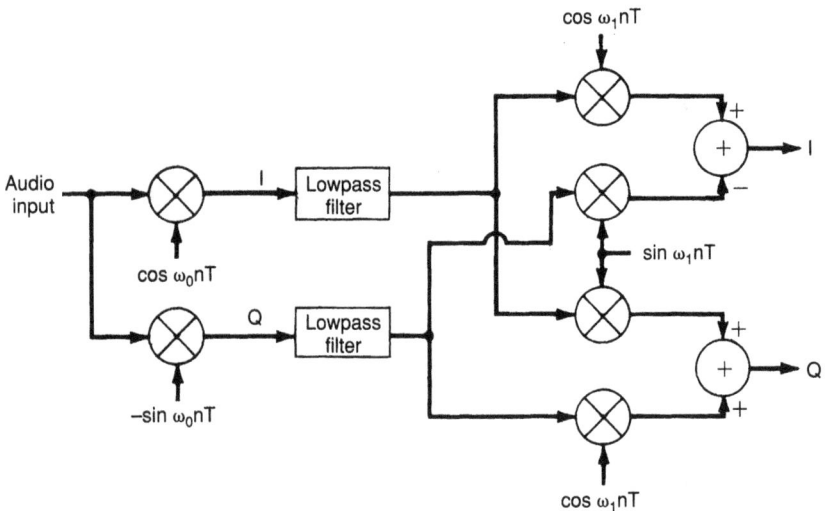

Figure 8.33 Weaver method for SSB generation.

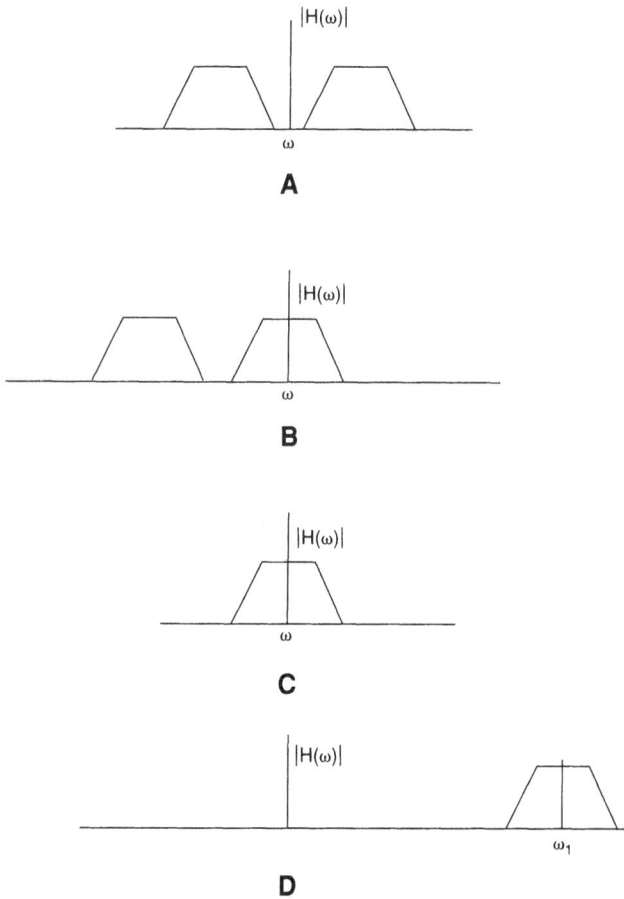

Figure 8.34 (A) Input audio spectrum; (B) audio spectrum of complex signal after first mixer; (C) audio spectrum of complex signal after filters; (D) frequency spectrum of signal after final mixer.

passed through a filter and the outputs of the two filters are added together. Obviously the signals could have been added first and then passed through a single filter.

In the case of the Q components each audio input is passed through a filter incorporating a 90° phase shifter. The outputs of the filters are then subtracted. An equivalent result could have been obtained by first subtracting the signals and then passing them through a single filter that incorporates a 90° phase shifter. The results of these modifications are shown in Figure 8.9.

A brief discussion of the conversion of complex signals into real signals at the D/A converter will now be given. In general, if one component (either I

or Q) of a complex signal is applied to a D/A converter, all the positive-frequency components of the complex signal are also duplicated on the negative side and all the negative-frequency components of the complex signal are duplicated on the positive side. This can occasionally lead to interesting results if the complex signal has low-level residual spurious signals on the undesired side. These effects can be avoided by using a circuit, such as shown in Figure 8.35, that includes a full-complex mixer, as compared with the circuit shown in Figure 8.6 which uses only half the mixer.

The circuit shown in Figure 8.35 selects only the positive-frequency components of the complex spectrum and outputs a real signal with the positive components duplicated on the negative side. In this case, any residual negative-frequency components in the complex spectrum are not outputted.

The D/A converter normally operates as an S/H circuit, holding one value until the next sample is available. This can cause a significant drop in the upper frequency response of the audio spectrum. A low sample rate also

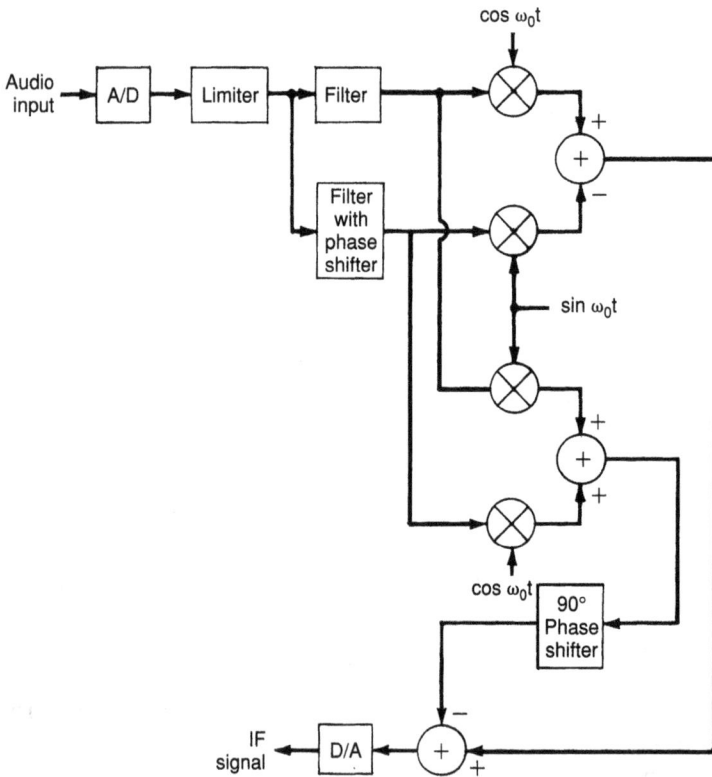

Figure 8.35 Block diagram of digital exciter with complex frequency translator.

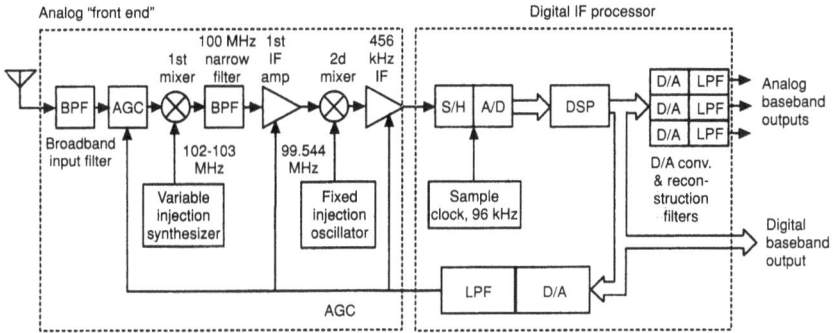

Figure 8.36 Example HF digital SSB communication receiver block diagram.

limits the maximum frequency to which the signal can be translated by the digital mixer. Consequently it is often necessary to increase the sample rate by interpolation prior to translation or application to the D/A converter. This can be accomplished by supplying an integral number of zero samples between the actual signal values and following the process with a lowpass filter operating at the higher sample frequency. The filter removes the components of the signal caused by harmonics of the lower sample rate and provides a clear spectrum for application to the D/A converter.

8.10 Digital SSB Receiver Design Example

Previous sections of this chapter have concentrated mostly on the basic DSP issues and concepts. In this section, a digital SSB receiver design example is presented to pull some of the concepts together in a system-level design.

In this example, IF bandpass sampling will be used to implement a 2- to 30-MHz HF digital SSB receiver, as illustrated in Figure 8.36. A maximum front-end bandwidth of 16 kHz has been selected, to allow up to four 3-kHz -wide ISB channels to be processed by the DSP. This bandwidth also allows the DSP to perform receiver fine frequency tuning over at least a 1-kHz range, simplifying the variable injection synthesizer.

The first IF will be 100 MHz, which is high enough to allow the broadband input filter to suppress the IF and image response to 80 dB, but low enough to allow a low-cost crystal filter to provide the 16-kHz-wide front-end filtering.

The variable injection synthesizer will tune the range of 102 to 130 MHz in 1-kHz tuning steps. The DSP can provide the 1-Hz fine-tuning steps required. This "high-side" first-IF mixer injection results in a passband reversal or "flip" in the mixer, which will be corrected in the A/D converter sampling process.

Following the first-IF crystal filter, the signal is amplified to make up for losses in the first-IF mixer and filter, and to maintain a front-end noise figure of 15 dB. The signal is then mixed with a fixed oscillator signal of 99.544 MHz. This mix produces a second IF of 456 kHz, which is high enough to allow the second-mixer image response at 99.088 MHz to be attenuated by 80 dB in the first-IF filter, but low enough to be sampled by the A/D converter.

Providing the S/H bandwidth of the A/D converter is greater than 456 kHz, we can harmonically sample the second-IF bandpass signal efficiently using a sample rate of 96 kHz. This sampling process is illustrated in Figure 8.37. Note that when the second-IF spectrum is convolved with the sampler impulse spectrum, the IF signal is effectively frequency translated, with a passband reversal or "flip" to a lower IF of one-fourth the sample frequency, or 24 kHz. This passband reversal will be used to counteract the passband reversal in the first-IF mixer. Of course, the resultant sampled IF spectrum repeats at every multiple of the sample frequency as expected.

From Figure 8.37 it can be seen that if the second-IF bandwidth increases, aliasing can occur due to an overlap in the sampled IF spectrum. Although aliasing cannot be totally eliminated using a real-world IF filter, due to the fact that only finite stopband attenuation is possible, it can be reduced to an acceptable level. In this design, a passband width of 16 kHz must be alias protected to 60 dB.

If we examine the IF sampling process in more detail, we can now specify the bandpass requirements of the first-IF filter (refer to Figure 8.38).

Using the selected 96-kHz sampling frequency and 456-kHz second IF, the first-IF filter must provide a maximum 80-kHz wide-stopband and a minimum 16-kHz-wide passband, corresponding to a 5:1 shape factor. The filter must have at least 60 dB of attenuation for signals 40 kHz away from the passband center of 100 MHz. An additional 20 dB of attenuation is also required at the image response frequency of 99.088 MHz, and a typical passband ripple specification might be 1 dB. These requirements could possibly be satisfied using a four- or five-pole crystal filter design. The differential group delay in the passband should be small enough that it does not degrade the excellent phase response within the passbands of the digital ISB IF filters. The first-IF passband should be wide enough that the group delay variations at the edges are avoided.

As a general rule, the sum of the filter passband width and the stopband width must be less than or equal to the sample frequency for IF bandpass sampling designs. This will be true if the IF is an integer multiple of the sample frequency plus or minus one-quarter of the sample frequency, as in this example. If this is not the case, the stopband width will have to be reduced, resulting in a more stringent IF filter design.

Following the second-mixer, the second-IF bandpass signal must then be amplified to a sufficient level to drive the A/D converter. The gain required is, of course, a function of the selected A/D converter and also various re-

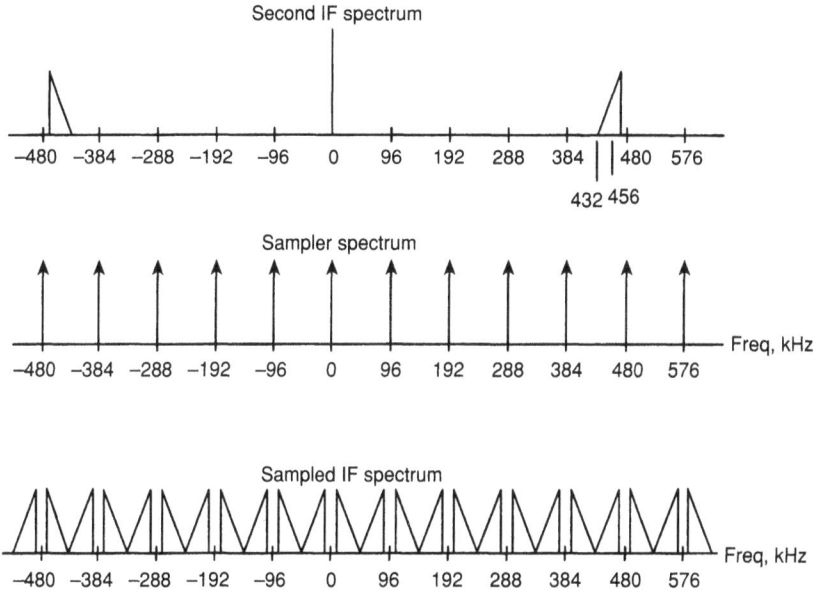

Figure 8.37   Intermediate frequency sampling process illustration.

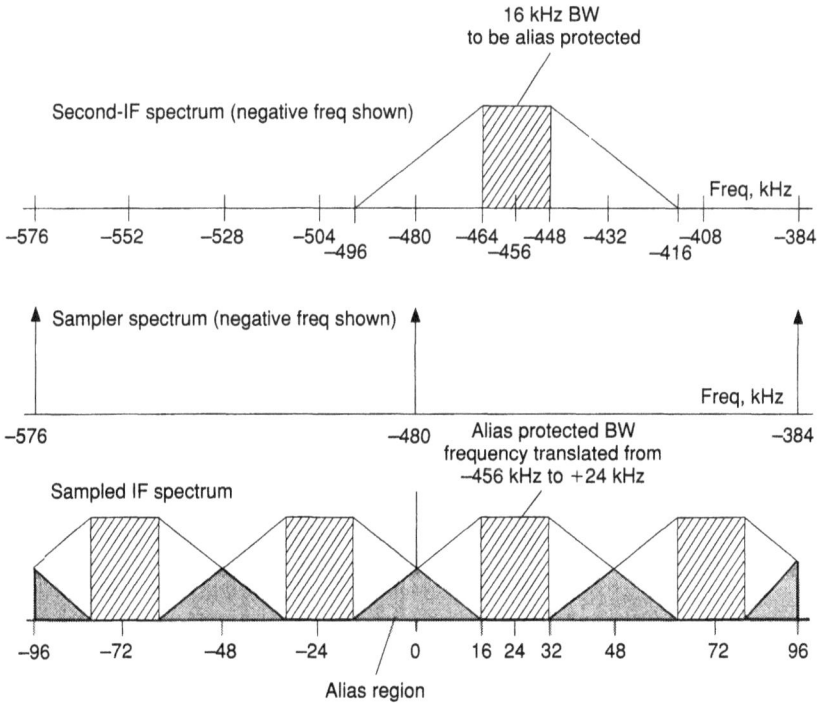

Figure 8.38   Detailed IF sampling process illustrating required bandpass response.

ceiver performance trade-offs. It must be evaluated on the basis of a system-level analysis.

Using a 14-bit A/D converter with a 10-V (peak-to-peak) range and a sampling frequency of 96 kHz, the theoretical quantizing noise density, from Equation 8.11, is 1.2935×10^{-14} W/Hz or –108.9 dBm/Hz, assuming 50-ohm impedance normalization. The thermally generated receiver noise for a 15-dB noise figure receiver front-end is –174 + 15 = –159 dBm/Hz.

In the receiver front-end, the minimum weak signal gain must be large enough so that the weakest usable signal plus receiver noise is greater than at least one A/D converter quantizing level in order to be recovered at all. And, as discussed earlier, several quantizing levels must be traversed if the A/D converter quantizing noise is to be uniformly distributed. The best way to ensure that the quantizing noise will be uniformly distributed is to provide enough noise to dither across several quantizing levels. This can be achieved by amplifying the in-band receiver thermal noise or by adding out-of-band dither noise that will be filtered out later in the DSP.

Although it may not be the best method, this example will discuss the simple approach of amplifying the receiver thermal noise until it bridges an adequate number of quantizing levels. Normally, if the noise is Gaussian distributed, and the rms level of the noise at the A/D converter is greater than or equal to the level of a sine wave which just bridges a single quantizing level, an adequate number of quantizing levels will be bridged to guarantee uniformly distributed quantizing noise.

For the 14-bit A/D converter in this example, one quantizing level is 610 mV peak-to-peak. The equivalent sine wave level is then 216 mV rms, or –60.3 dBm into 50 ohms. Assuming a receiver front-end or pre-A/D noise bandwidth of 16 kHz, the gain required to amplify the receiver thermal noise is then

$$\text{Gain} = \text{quantizing level} - \text{thermal noise density} - 10\log_{10}\text{BW}$$
$$= -60.3 \text{ dBm} + 159 \text{ dBm/Hz} - 10\log_{10} 16{,}000 \text{ Hz}$$
$$= 56.7 \text{ dBm}$$

A spreadsheet program can be used to perform these calculations and others in order to examine interactively the receiver performance using various A/D converters, sample rates, bandwidths, and gain distributions. One example of a typical spreadsheet program output, for the case discussed above, is shown in Table 8.3.

Note that in this example the analog gain is large enough to allow thermal noise to adequately bridge several quantization levels. Because this approach is used, the receiver noise performance is dominated by thermal noise and not A/D quantization noise. At sensitivity level, the thermal noise is 7 dB greater than the quantization noise as measured in the 3-kHz information bandwidth. Out-of-band dither noise could have been added to the A/D converter input to reduce the front-end gain required to meet the sen-

TABLE 8.3 Digital Receiver Performance Spreadsheet Output for Thermal Noise-Limited Example

DSP Receiver Gain Distribution and Noise Performance

Analog RF/IF noise figure	15.0 dB
A/D converter resolution (bits)	14.0 bits
A/D converter full-scale level (V-peak)	5.0 V-peak or 23.98 dBm
A/D converter quantization level	-60.3 dBm
A/D converter sample frequency (Hz)	96.0 kHz
A/D converter input bandwidth (BW1)	16.0 kHz
Information bandwidth (BW2)	3.0 kHz
Analog AGC threshold	-73.0 dBm
Analog AGC range	90.0 dB

Output S/N, dB vs antenna level, dBm

Antenna signal level (dBm)	Analog RF/IF gain (dB)	A/D signal level (dBm)	Noise at A/D input in BW1 (dBm)	Noise at A/D input in BW2 (dBm)	Quantizing noise of A/D in BW2 (dBm)	Total noise in BW2 (dBm)	Output S/N ratio in BW2 (dB)	Digital proc. gain (dB)	Output signal level (dBm)	Antenna overload level (dBm)
-113	57.0	-56.0	-60.0	-67.2	-74.1	-66.4	10.4	60.0	4.0	-33
-103	57.0	-46.0	-60.0	-67.2	-74.1	-66.4	20.4	50.0	4.0	-33
-93	57.0	-36.0	-60.0	-67.2	-74.1	-66.4	30.4	40.0	4.0	-33
-83	57.0	-26.0	-60.0	-67.2	-74.1	-66.4	40.4	30.0	4.0	-33
-73	57.0	-16.0	-60.0	-67.2	-74.1	-66.4	50.4	20.0	4.0	-33
-63	47.0	-16.0	-70.0	-77.2	-74.1	-72.4	56.4	20.0	4.0	-23
-53	37.0	-16.0	-80.0	-87.2	-74.1	-73.9	57.9	20.0	4.0	-13
-43	27.0	-16.0	-90.0	-97.2	-74.1	-74.1	58.1	20.0	4.0	-3
-33	17.0	-16.0	-100.0	-107.2	-74.1	-74.1	58.1	20.0	4.0	7
-23	7.0	-16.0	-110.0	-117.2	-74.1	-74.1	58.1	20.0	4.0	17
-13	-3.0	-16.0	-120.0	-127.2	-74.1	-74.1	58.1	20.0	4.0	27
-3	-13.0	-16.0	-130.0	-137.2	-74.1	-74.1	58.1	20.0	4.0	37
7	-23.0	-16.0	-140.0	-147.2	-74.1	-74.1	58.1	20.0	4.0	47
17	-33.0	-16.0	-150.0	-157.2	-74.1	-74.1	58.1	20.0	4.0	57
										See text

sitivity specification. On the other hand, if the front-end gain is not reduced, the dither noise could be used to guarantee that an adequate number of quantizing levels are traversed with smaller input signals. This improves the sensitivity of the receiver.

Note from the spreadsheet in Table 8.3 that the output SNR is better than 10 dB at an antenna signal level of –113 dBm. The digital AGC algorithm holds the D/A output signal level at 20 dB below full scale (+4 dBm), while the antenna signal level increases, until the analog RF AGC threshold is reached. The output SNR continues to increase with the antenna signal level until the level at the A/D reaches 40 dB below full scale (–16 dBm). At this point the analog RF AGC holds the A/D input level and D/A output level constant. The output SNR continues to rise to 58 dB until the A/D quantization noise dominates the receiver noise; it then rises no further.

The antenna overload level (column 11) listed in Table 8.3 is the level of an undesired signal at the antenna input that can pass through the 16-kHz bandwidth front-end to the A/D converter, therefore saturating it. This undesired signal would be outside the 3-kHz information bandwidth and would not affect the normal AGC setting. As the desired signal level increases, AGC reduces the front-end gain. This allows larger undesired signals to be tolerated without overloading the A/D converter. As the level of the undesired signals increases, at some point it causes the front-end gain to go into compression or saturation. This results in unacceptable nonlinearity for the desired signal. For very large undesired signals, some form of damage protection may be needed.

The reason for maintaining nearly 40 dB of headroom on the A/D converter input and only 20 dB on the DSP output was mentioned earlier. Only in-band desired signals should be present after the final narrowband filtering in the DSP. Consequently only 20 dB of headroom is required to support the normal speech signal level peaks at the D/A output. But at the A/D converter, large out-of-band undesired signals may be present. Therefore it is necessary to have as much headroom as possible to support the undesired signal peaks.

In our case the maximum SNR for large desired signals, without interference, is 58 dB. Suppose an out-of-band undesired signal is present at the A/D converter at a level of +4 dBm, or –20 dB from the saturation level. The A/D converter will generate harmonic and IM distortion products that will alias and potentially fall within the desired signal bandwidth. If these products are 70 dB below the level of the undesired signal, or at a level of –66 dBm (+4 – 70), and the in-band desired signal level at the A/D converter is –16 dBm, the ultimate SNR or signal-to-noise and distortion (SINAD) ratio will be limited to 50 dB (–16 + 66). So in order to maintain a good output SNR or SINAD under real-world conditions, it is very important to have an extremely linear A/D converter in a digital receiver design.

Suppose the A/D converter's S/H has an rms aperture jitter of 200 ps. Using our 96-kHz sampling frequency and 456-kHz IF, the carrier-to-aperture noise power density ratio, from Equation 8.15, is 1.46×10^{11}, or 111 dB-Hz. Assuming the jitter noise is white, the ratio will be 35 dB less in the 3-kHz information bandwidth or 76 dB. With the same +4-dBm undesired signal present at the A/D converter as in the example above, the noise power due to aperture jitter will be at a level of –72 dBm (+4 – 76). Again, if the in-band desired signal level at the A/D converter is –16 dBm, the ultimate SNR will be limited to 56 dB (–16 + 72). Therefore, it is also very important to have an extremely low S/H aperture jitter in digital receiver design.

References

1. William D. Stanley, Gary R. Dougherty, and Ray Dougherty, *Digital Signal Processing* (Reston, VA: Reston Publishing, 1984).

2. Andreas Antoniou, *Digital Filters Analysis and Design* (New York: McGraw-Hill, 1979).

3. Alan V. Oppenheim and Ronald W. Schafer, *Digital Signal Processing* (Englewood Cliffs, NJ: Prentice-Hall, 1975).

4. Ronald E. Crochiere and Lawrence R. Rabiner, *Multi-Rate Digital Signal Processing* (Englewood Cliffs, NJ: Prentice-Hall, 1983).

5. Lawrence R. Rabiner and Bernard Gold, *Theory and Application of Digital Signal Processing* (Englewood Cliffs, NJ: Prentice-Hall, 1975).

6. Walter Kester "Test Video A/D Converters under Dynamic Conditions," *EDN* 27 (August 18, 1982): 103–112.

7. Digital Signal Processing Committee of the IEEE Acoustics, Speech, and Signal Processing Society (eds.), *Programs for Digital Signal Processing* (Piscataway, NJ: IEEE Press, 1979).

8. L. R. Rabiner, J. F. Kaiser, O. Herrmann, and M. T. Dolan, "Some Comparisons between FIR and IIR Digital Filters," *Bell Syst. Tech. J.* 53 (February 1974): 308.

9. S. Reyer and D. Hershberger, "Using the LMS Algorithm for QRM and QRN Reduction," *QEX* (September 1992): 3–8.

10. B. Friedlander, "System Identification Techniques for Adaptive Noise Canceling," *IEEE Trans. Acoustics, Speech, & Signal Processing* ASSP-30, no. 5 (1982): 699–709.

11. M. Frerking, *Digital Signal Processing in Communication Systems*, pp. 464–485 (New York, Van Nostrand Reinhold, 1994).

12. R. Seviora and M. Sablatash, "A Tellegen's Theorem for Digital Filters," *IEEE Trans. Circuit Theory* CT-18 (1971): 201–203.

13. B. Widrow, "Adaptive Filters," in *Aspects of Network and System Theory*, R. Kalman and N. Declaris (eds.), pp. 563–587 (New York: Holt, Rhinehart, and Winston, 1971).

14. W. Press, B. Flannery, and S. Vetterling, *Numerical Recipes in Pascal: The Art of Scientific Computing*, pp. 309–374 (Cambridge, UK: Cambridge University Press, 1989).

15. T. Kwan and K. Martin, "Adaptive Detection and Enhancement of Multiple Sinusoids Using a Cascade IIR Filter," *Proc. 1988 IEEE Int. Symp. Circuits & Systems* (June 1988): 2647–2650.

16. E. W. Pappenfus, W. B. Bruene, and E. O. Schoenike, *Single Sideband Principles and Circuits* (New York: McGraw-Hill, 1964).

9

Preselectors and Postselectors

Bill D. Hart

Although wideband receiver front ends are in increasing use, they cannot handle situations with very strong off-channel signals. Passive tuned circuit filters are still needed to protect against cross-modulation, IM, damage, and other interference effects. The trend is toward separately packaged preselector filter units using sophisticated tuning and control methods. Selectivity, noise figure degradation, and maximum interference voltage are important constraints. It is possible to design filters with an optimal trade-off of these parameters for given component limitations. With some forethought in the filter design, several preselectors can be driven by the same antenna in a multicoupling configuration.

The same filters or similar circuits can be employed as postselectors to filter the transmitter signal before final power amplification, thus reducing its out-of-band noise and spurious emissions that could interfere with nearby receivers.

9.1 Purpose and General Description

Receiver preselector filters

A receiver preselector is a passive, tuned filter used between the antenna and the receiver input. It provides additional selective filtering that prevents or reduces the numerous interference and damage problems resulting from collocation of transmitters and receivers, as discussed in detail in Chapter 2. The interference problems that preselectors reduce include IM, cross-modulation, reciprocal mixing, desensitization, spurious and image responses, and circuit overload damage. These problems occur when adequate antenna separation is not available, such as aboard aircraft, ships, vehicles, and transportable communication shelter units, or in crowded fixed-station sites. It is common to find isolation between two HF antennas on an aircraft of less than 6 dB at many frequencies. Figure 9.1 shows that coupling between two whip antennas is significant at appreciable distances, perhaps as great as 1000 ft (300 m). When these users need multiple radio installations, or when several users are in close proximity, simultaneous transmitter and receiver operation (SIMOP) nearly always requires the selectivity of tuned circuits.

Receivers have historically included tuned circuit stages at the input to attenuate out-of-band undesired signals. Research has improved mixer dynamic range (Chapter 4) and synthesizer noise levels (Chapter 10) to the point where receivers are now produced with wideband input stages. They cover their entire frequency ranges with no retuning of filter elements and provide adequate performance for many users. Omitting tuned circuit filtering is economical and permits the operating frequency to be changed rapidly. The dynamic range of the best wideband receivers is still not adequate, however, for sites where there are strong interfering signals.

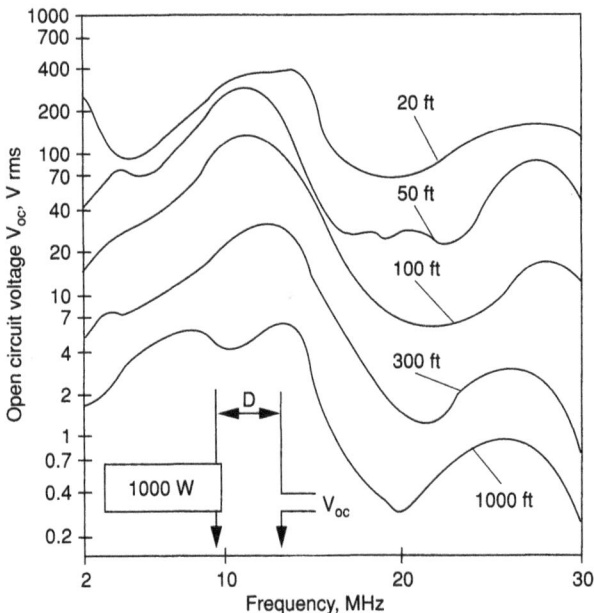

Figure 9.1 Voltage on receiving antenna at varying distances from transmitter. Antennas are 32-ft vertical whips over "good" ground; $s = 10^{-2}$ S/m.

Exciter postselector filters

Exciter postselectors are very similar to receiver preselectors. They filter the transmitter signal between the low-level exciter stages and the power amplifier to reduce emissions that could interfere with nearby receivers. They reduce transmitted broadband noise, spurious signals, and in some cases harmonics.

Postselector design is in many respects similar to that for preselectors. A filter can be configured for use as both a preselector and a postselector in a transceiver installation. Postselectors differ from preselectors in that the available desired signal level is much higher, so the emphasis is on in-band IM instead of noise figure. There are no large out-of-band signal voltages to stress components.

Adding a postselector has a small impact on transmitter design. The postselector may include an amplifier so no net loss is incurred and the power amplifier drive level is maintained. An allowance for postselector gain variation versus tuned frequency must be made in the design of ALC or TGC loops, either by having sufficient margin for variation or by correcting the gain from a lookup table. A postselector's tuning time may delay the tuning of the power amplifier or antenna coupler while they wait for a stable RF signal.

Research continues into improvements in noise and linearity in transmitters. Such improvements could eventually reduce the need for postselectors, but at the current level of technology a filter still improves collocation performance.

Typical preselectors

Modern preselectors typically have one to four bandpass resonators automatically tuned to the radio operating frequency. They are usually packaged as separate units because it is not economical to provide the basic transmitting and receiving equipment with the additional selectivity that, although sometimes vital, is required only in a minority of installations.

The principal performance specifications for two tunable preselection filters are given in Table 9.1. The values include an allowance for variation over the operating frequency range, temperature, and other environmental conditions.

A 3-dB bandwidth of 1 or 2% of the tuned frequency will provide considerable attenuation to undesired signals 5 or 10% away from the tune frequency. A bandwidth less than 1% may cause problems with tuning accuracy and stability, resulting in an unacceptable insertion loss.

Suboctave bandpass filters

A bank of fixed-tuned suboctave-width bandpass filters is a limited form of preselector and postselector filtering. The appropriate filter is typically

TABLE 9.1 Example Preselector Specifications

Characteristic	Unit A	Unit B
Frequency range, MHz	2–32	2–30
Number of bands	4	2
Number of resonators	3	2
Tuning method	Capacitor servo	Diode switch
Tuning time (typical), sec	1	0.01
Impedance (nominal), ohms	50	50
Impedance out-of-band	50–150 pF	2 or 0.5 µH
Operating RF input @ 10% frequency separation, V_{rms}	200	140
Protection limit, V_{peak}	500	500
Selectivity @ 10% frequency sepration, dB	47	40
Minimum passband, kHz	12	6
@dB points, dB	2	1
Filter loss, dB	6	10
Loss with amplifier, dB	0	
Total noise figure, dB	12	12
Transmit IM below each tone, dB	46	40
IM test tone level, V_{rms}	1.58	0.56

diode switched or relay switched into the signal path by the receiver tuning control circuits.

Half-octave filters (high/low passband frequency ratio = 1.4) can usually provide adequate suppression of second-order IM interference because, as shown in Figure 9.2, one undesired signal producing the IM must be at least a full octave away (no closer than half or twice the desired frequency). A second-order IM product will be reduced 1 dB for each decibel of attenuation the filter has at f_2.

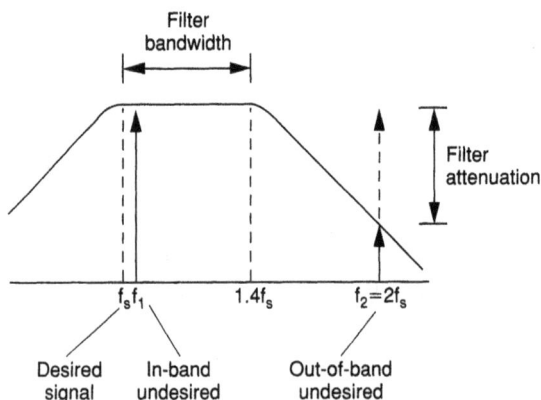

Figure 9.2 Worst-case frequencies for second-order IM in half-octave filter. Second-order product is $f_2 - f_1 = f_s$.

The 2- to 32-MHz band is covered by eight half-octave filters, which are typically smaller than a tunable filter covering the band. Narrower filters may be desired, but the number of filters required goes up rapidly and a tunable design soon becomes advantageous. The number of fixed filters needed, in terms of the band limits to be covered and filter corner frequencies, is

$$N = \frac{\log\,(\text{band high/band low})}{\log\,(\text{filter high/filter low})}$$

For example, with 15% bandwidth filters, 20 filters would be required to cover the 2- to 30-MHz band.

The design of half-octave filters is covered in references 1 and 2. Chapter 17 of this book describes a personal computer program that accompanies this book. It designs a fifth-order half-octave bandpass filter. The filter is simple and very effective in eliminating second-order IM interference.

Relationship between second- and third-order IM

Broadband receiver front ends, such as the upconversion design with 2- to 30-MHz coverage, are confronted with the possibility of a second-order IM problem that is potentially more serious than the third-order problem. Figure 9.3 shows that as the strength of each of the two equal interfering tones increases, the second-order product due to $I_2{}'$ may rise above the noise level prior to the third-order product due to I_3. The question then occurs: What must the second-order intercept I_2 be, such that its second-order IM is not worse than third-order IM due to I_3? For a given quality of mixer and amplifier I_2 performance, a preselector filter can then be specified that assures this objective.

The third-order spurious-free dynamic range (SFDR) is given by

$$\text{SFDR}_3\,(\text{dBm}) = 0.67(I_3\,(\text{dBm}) - kTBF\,(\text{dBm})) \qquad (9.1)$$

The second-order SFDR is given by

$$\text{SFDR}_2\,(\text{dBm}) = 0.5(I_2\,(\text{dBm}) - kTBF\,(\text{dBm})) \qquad (9.2)$$

If we equate SFDR_2 and SFDR_3, in other words make the level of second- and third-order IM products all equal to the noise level for the same pair of equal input tones, then the following relationship derives:

$$I_2(\text{dBm}) = \frac{4I_3\,(\text{dBm}) - kTBF\,(\text{dBm})}{3} \qquad (9.3)$$

From the discussion in connection with Figure 9.2 we can now find the minimum attenuation of the $2f_s$ tone in that figure, using a half-octave filter, that achieves equality of the second- and third-order dynamic ranges. In practice we would like to do much better than this, if it is convenient.

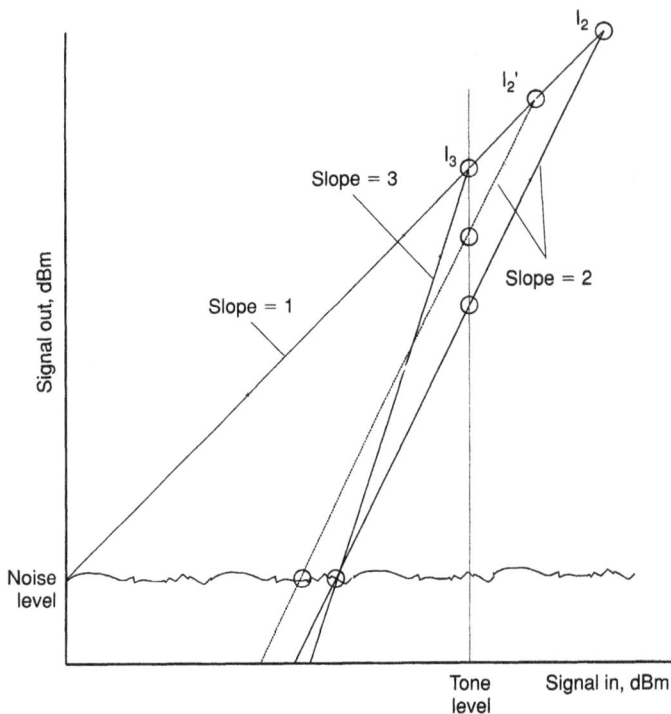

Figure 9.3 Relationship of second-order and third-order IM interference.

A similar situation occurs if one strong tone in Figure **9.2** is near the upper edge of the passband and the other tone is f_s less than this. This case and the previous case are usually the worst-case situations. In all other combinations of frequencies, both tones are outside the passband, in which case the second-order intercept point is considerably greater. However, in some cases I_2, when the two tones "straddle" the passband $(f_{hi} - f_{lo} = f_s)$, may also have to be evaluated if the filter is marginal.

9.2 Design Considerations for Preselectors

A preselector contains many circuits besides the basic RF filter. This section describes some of the supporting circuitry associated with overload protection, signal amplification, and frequency tuning, and some practical limitations on the components and circuits that the filter employs.

Overload protection

The preselector may be exposed to several sources of electrical energy on its input terminals, which it should be able to withstand even if operation is temporarily interrupted. These include large RF signals inside or outside the filter passband, static charges, and transients induced by nearby lightning strikes. Several techniques should be considered to protect the filter from these overloading voltages. Some of these are illustrated in the design examples (see Figures 9.12 and 9.13).

A static discharge resistor of 100 kohms or more should be placed across the input terminals if there is no other dc path. Spark-gap or ionization-tube surge protectors may be effective in shunting the transients induced by nearby lightning strikes. Semiconductor surge protectors, if used, should be very carefully checked for IMD at the largest specified level of undesired signals.

Protecting the filter from overload by large RF signals usually requires a detector and monitor circuit. If the monitor senses an overloading signal, it can disconnect the filter from the input or alter the filter response to prevent absorption of excess power. Shunting a filter resonator with a resistance will lower the resonant Q factor. With most filter designs the resulting mismatch will lower the power absorbed from the antenna and the shunting resistor does not have to dissipate very much undesired power. A discharge device such as a neon lamp can perform this same function if the maximum filter voltage levels happen to coincide with its breakdown voltage and if the resulting output level is acceptable.

Both rapid and long-term protection is required. An overload monitor sensor will have some actuating time constant and the disconnect device will take time to operate. A fast relay, for example, will require at least 1 ms to operate. Surge protection from zeners, voltage clamp circuits, or discharge devices is necessary to protect amplifiers and the receiver during this interval. On the other hand, disconnecting or detuning is essential to prevent overheating of surge protection devices.

Both in-band and out-of-band RF signals can overload the filter. Out-of-band signals develop a large voltage across the input without actually dissipating much power because the filter input impedance is mostly reactive at out-of-band frequencies. In-band signals cause overload at much lower voltages because the power is transferred with low loss to all filter sections and to the output.

It is desirable for the protection circuit to restore normal operation when the overloading signal goes away. This is a great convenience for the operator, since occasional momentary overload conditions do not disable the equipment or require manual reset.

The filter should be automatically disconnected from the antenna when power is turned off and the automatic overload circuits can no longer protect it.

Amplifiers

A passive preselection filter's loss adds directly (in decibels) to the receiver noise figure (neglecting output mismatch). A better system noise figure can be achieved if the filter is followed by a low-noise amplifier, which restores the received signals to near their original level or higher. If the overall pre-selector unit has unity gain, then the receiver AGC threshold and S-meter calibration are undisturbed. The noise figure of the amplifier must be much lower than that of the receiver to minimize the system degradation.

The amplifier must also have good dynamic range in order to handle un-desired signals within the preselector bandwidth. These could cause odd-order IM products to fall within the receiver passband. The filter attenuation curve reduces the desired signal slightly and out-of-band undesired signals much more. The net effect is to improve the third-order intercept point, rel-ative to the amplifier alone. The gain of the amplifier, of course, causes larger signals to be applied to the receiver.

In very lossy filters, it may be advantageous to place an amplifier between sections of the filter to compromise between noise and distortion. It must be located where considerable selectivity has been achieved, to protect the amplifier, but before too much loss has occurred.

Proper design of the filter and amplifier require a systematic analysis of the system noise figure and IM intercept, considering the cascaded system of filter, amplifier, and receiver. These topics were treated in more detail in Chapter 4 and are illustrated in Example 9.2 of section 9.4.

Tuning methods

Passive filters may be tuned by varying an inductance or capacitance, or both. The most common tuning elements are switched capacitors, servo-driven variable capacitors, and varactor (variable capacitance) diodes. Tunable cavities are often used at VHF and UHF for high performance, particularly for very low-loss designs, but become huge at lower frequencies and are al-most never used in the HF band.

Because of the large ratio of high- to low-frequency limits often encoun-tered, it is usually not practical to tune the entire range in one band. For ex-ample, the 2- to 30-MHz HF band would require variable elements with more than a 225:1 ratio and have a 15:1 variation in impedance. Band switching of inductors is used with variable tuning capacitors, or vice versa, to obtain a realistic tuning ratio while reusing the tuning elements. One-octave or two-octave bands are commonly used, so that the 2- to 30-MHz unit uses two to four bands.

In servo-tuned units, wafer switches or relays are suitable for band switch-ing. High-reliability vacuum relays are available that operate in milliseconds and can switch kilovolts of RF. Fast-tuning units used in ECCM applications will need fast relays or diode switches. These units will accumulate a huge number of retuning cycles and may need special relay designs to obtain ad-equate life. Diode switches can provide faster band switching, but stray ef-fects may be hard to control when switching inductors.

Variable inductors are sometimes used for tuning but are generally limited to slow manually or electromechanically tuned designs. Two common forms of mechanically variable inductors are those with movable tuning slugs and roller coils. Rapid-tuning, electronically variable inductors based on gradual magnetic saturation have problems with IM and temperature drift that prevent their widespread use.

The filter shown in Figure 9.4 uses switched capacitors for tuning. If each capacitor is half as large as the previous one (binary weighting), equal steps of capacitance are available up to the total. A larger number of tuning elements will provide finer tuning steps. The tuning control must convert operating frequencies to switch combinations that add up to the right capacitance.

Although the scheme is conceptually simple, the control algorithm or the actual component values or both will have to be adjusted for stray effects. The inductors will be found to have a significant amount of capacitance effectively in parallel with them, determined by their self-resonant frequency. Great care must be taken to keep the lead length of the capacitors small or the lead inductance will cause a frequency-dependent error in the effective capacitance values. Lead inductance is a greater problem in low-reactance resonators. The switch circuit can also cause an alteration of the effective capacitance. All these effects must be minimized or the tuning algorithm becomes quite involved with corrections to the capacitance values at each new tuned frequency.

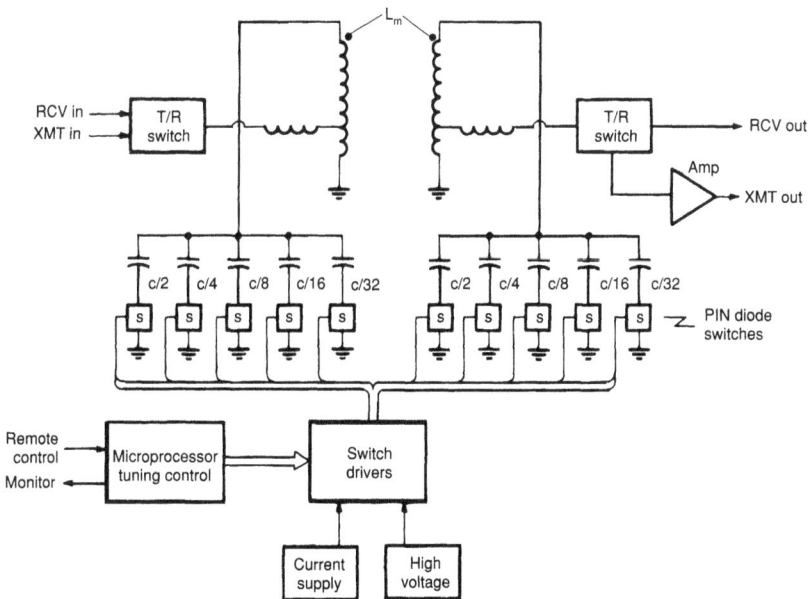

Figure 9.4 Block diagram of PIN switch-tuned preselector.

Filters could also be tuned with switched inductors, but this is less commonly done. Switched inductors are harder to design because they add in series, requiring switches to operate at floating RF potentials. The open-state capacitance of the switches and their capacitance to ground can resonate with unused inductor sections and affect the filter tuning. Inductors are usually large and expensive compared to capacitors, and must be optimized for low dissipation rather than for ease of switching. Inductor switching is usually restricted to band changes.

PIN diode switching

Positive intrinsic negative layered diodes are commonly used to switch RF circuits. The relevant diode specs are on-state resistance, off-state (reverse-bias) capacitance, reverse-bias resistance, noise, intrinsic carrier lifetime, and breakdown voltage.

A possible PIN diode switch for tuning capacitors is illustrated in Figure 9.5. The driver circuit switches between a low-voltage, high-current supply that turns the diodes on, and a high-voltage, low-current supply that back-biases the diodes to turn them off. Two diodes are usually required to provide the switching current path and to isolate the bias circuits, which could

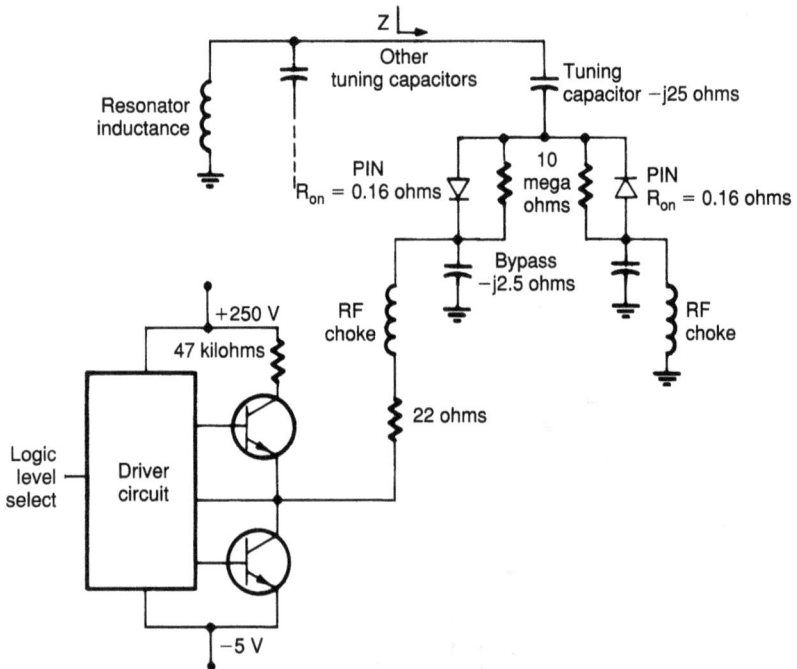

Figure 9.5 PIN diode switch with the driver and bypassing.

create undesired resonances. A switched inductor may not always need the second diode.

A diode will not be properly switched off if it has a significantly lower back-bias resistance than the other one, and therefore much less of the bias voltage. This can cause unacceptable mistuning and loss in the resonator. Large resistances across the diodes or a single resistor connected to a supply of half the total voltage is usually needed to assure equal voltage division. These resistors increase the resonator dissipation and must be made as large as practical. It should be noted that high-value carbon resistors have a lower effective resistance to RF (the value in parallel with the equivalent capacitance) than to dc. This is due to the distribution of capacitance between carbon granules which effectively finds a shorter path through the granular array and partially shunts the resistive contacts.

The second diode is effectively in parallel with the first for the RF signal current and can be utilized to lower the overall switch resistance, with proper precautions as discussed below. The bypass capacitors must be large with respect to the tuning capacitor, but do not need stable RF characteristics. It is sometimes possible to series resonate the bypass capacitor in the lowest band and thus achieve a low impedance where it is needed most. The effect of this frequency-dependent bypass on the total tuning reactance should be checked.

Because of the large switching voltages, it is often impractical to make the bypass capacitor reactance smaller than the diode resistance, which leads to a somewhat surprising requirement. Both ends of the diode switch must have the same bypassing impedance if we are to benefit from the conduction of both diodes. For example, in the circuit shown in Figure 9.5 the total impedance presented to the tuned circuit is $0.08 - j26.25$, representing a capacitor with $Q = 328$. If we replace one bypass circuit with a direct ground connection, the impedance becomes $0.16 - j25.01$ and the Q drops to 158. This is because essentially all the RF current is flowing in the grounded diode and we no longer have the second diode resistance in parallel.

The switch diodes can generate noise that is significant in receiver preselectors because it degrades the noise figure. This noise level is usually related to the back-bias leakage current of the diodes. The current and the noise increase very rapidly because of avalanche multiplication as the breakdown voltage is approached, and it is usually necessary to use a large voltage derating factor.

Varactor tuning

Varactors are semiconductor diodes that have been optimized for a variation in the effective capacitance of the junction as the dc reverse-bias voltage is changed. A typical range of tuning capacitance per diode is 5 to 25 pF for small varactors which are commonly used at VHF and UHF frequencies.

Some very large varactors are being made for the low-HF frequency range, although their loss goes up at higher frequencies. The frequency tuning range is set by the ratio of maximum to minimum capacitance (including strays).

Varactor-tuned filters are used for their small size and low control (tuning) power in circuits that have moderate signal levels and a moderate tuning range. Very high-frequency and UHF receivers have used them extensively, but they are less common in the HF band.

A fundamental limitation on varactor-tuned filters is the IM caused by the variation in junction voltage when a large signal is applied. The varactor is inherently and necessarily a nonlinear device. Even-order distortion is greatly reduced by the back-to-back circuit shown in Figure 9.6, where the signal increases the voltage across one diode while decreasing that on the other. In general, all distortions are reduced by using more varactors in series or parallel combinations so that the signal swing on each varactor is reduced.

The impedance feeding the tuning voltage to the varactors affects the performance. A resistor will lower the resonator Q. A high resistance has been found to cause more IMD than a high reactance feed (such as a resonant choke). Large out-of-band signals tend to detune the resonator because of their effect on the varactor voltages. This can lead to oscillation at a low frequency, determined by the time constant of the tuning voltage feed circuit.

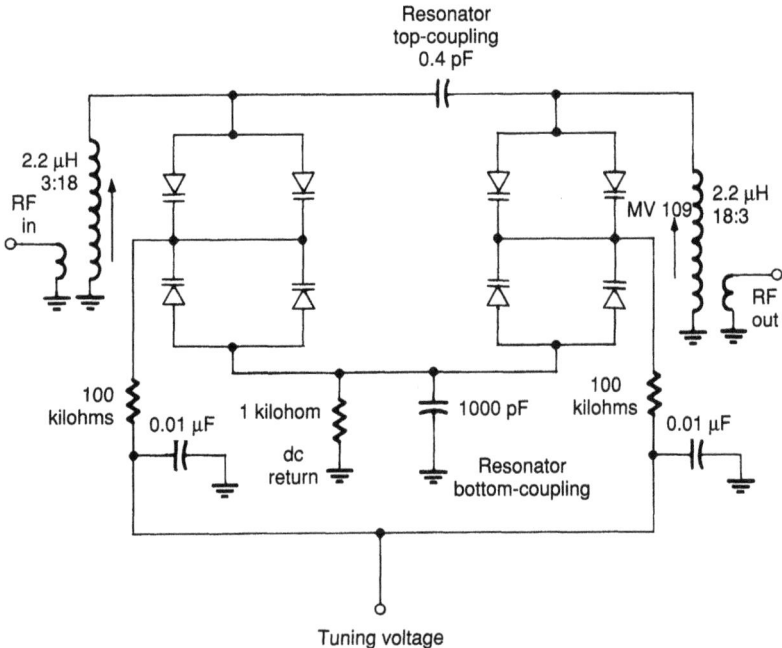

Figure 9.6 Varactor-tuned filter. Frequency range: 20 to 30 MHz; loss: 2 to 3 dB; tuning voltage: 5 to 27 V; 3-dB bandwidth$/f_0$: 6.5 to 7.5%.

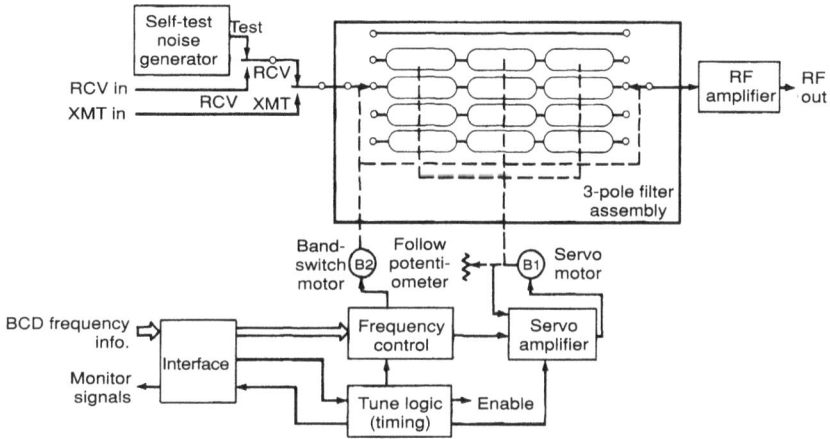

Figure 9.7 Block diagram of servo-tuned preselector.

As a rule of thumb, the circuit should be expected to operate with a peak-to-peak resonator RF voltage no more than half the minimum dc tuning voltage. If all the tuning range is not needed, the minimum tuning voltage can be increased. The out-of-band available power should be less than 25 mW/varactor when the small 25-pF varactors are used.

It is usually necessary to buy varactors in sets matched at three or more points on their capacitance-versus-voltage curve, or at least to use parts that are from the same lot. This is to ensure that the tuning of all sections of the filter track together as the tuning voltage varies.

Control

A tunable preselector filter must be kept tuned to the same frequency as its associated receiver. Synthesized receivers will have logic level binary or binary coded decimal (BCD) frequency information. The preselector will take this information and convert it to the necessary switching or servo positioning signals its tuning elements need. Conversions and calculations can be done by a microprocessor.

In electromechanical servo-tuned filters it is necessary to obtain a feedback of the tuning element position as shown in Figure 9.7. Potentiometers or shaft encoders may be used. The digital operating frequency information must be converted into a code or voltage specifying where the elements are to be positioned. The servo system drives the tuning elements to the position where the feedback matches the desired value.

Dither sensing tunes the filter to a sample of the desired RF frequency. It senses the direction to tune by rapidly varying the filter's tuned frequency back and forth (dithering) a small amount and determining which direction increases the level of the RF signal. When properly tuned, no increase is found in either direction. Tuning dither can be provided by a varactor or small diode-switched capacitor. Dither tuning is not a widely applicable method because it cannot be used with receive-only equipment and is relatively slow.

Switched-element tuning also requires conversion of the frequency information (Figure 9.4) to select the proper tuning element switches for each frequency. This conversion is sufficiently complex to require a microprocessor or a very large lookup table. In the simplest case, with binary-weighted tuning elements, the square of the frequency must be calculated. In practice, there may be several stray effects to compensate for in the selection of elements. Often it is simpler to perform interpolation in the microprocessor between entries of a lookup table that has the frequency-squared curve and the major stray effects precomputed.

9.3 Filter Design

Resonator loss

The most important limitation on the performance of a filter design is the resonator dissipation. A low dissipation (high Q) allows the designer to improve other design trade-offs. It is generally constrained by the volume available for the resonator and the materials suitable for the operating frequency range. Reference 3 is a good introduction to RF component limitations.

At frequencies up to a few megahertz, low-permeability ferrites are attractive for physically small designs. High-permeability ferrites are not sufficiently temperature stable and cause unacceptable tuning drift. At the top of the HF band, powdered iron and similar materials are suitable. Toroidal and cup-core forms are popular. Any magnetic material can cause distortion due to nonlinearities in its magnetization curve, particularly if operated at high flux densities, as would be done to obtain physically small inductors. A test should be made for IMD at the largest operating levels of received interfering signals or in-band transmit signals. Refer to Chapter 16 for a discussion of IM tests.

Helical resonator cavities [4, 5] are carefully proportioned coil and cavity assemblies that provide the highest unloaded Q for an air-core inductor in a given volume. They can be operated as high-Q coils below their self-resonant frequency. Their equivalent parallel capacitance must be included in the filter design. The Q, when optimum proportions are used, is estimated from

$$Q_u = 60Sf_0^{1/2}$$

where S = side of the square cavity, in
f_0 = frequency, MHz

Dielectric materials in a high-Q resonator, including structural pieces, should be chosen with caution because they can increase the dissipation. The choice of materials generally becomes more critical with increasing frequency. The loss depends on the product of the dissipation factor (loss tangent) of the material and the fraction of the resonator capacitance that is effectively within that material. Coil supports should avoid, when possible, high-field-strength regions, including the inside of the coil. For a given mechanical configuration and operating frequency, the product reduces to the dielectric constant times the dissipation factor, both of which can be found in handbooks [6]. For example, glass-epoxy board, nylon, phenolics (Bakelite), and any kind of hygroscopic materials should be avoided. Teflon, polyethylene, and polystyrene are usually good materials, if suited to the environment. Ceramics, any organic material, and other circuit board materials should be chosen with caution and carefully specified, because good materials are available but apparently similar materials may have much higher dissipation.

Capacitor losses, while lower than those of most inductors, can still significantly increase resonator dissipation. Switching of tuning elements introduces some resistance, particularly if PIN diode switching is used.

Topology

The choice of a filter topology is strongly influenced by the requirements for switching and controlling the tuning elements. Most variable elements and switching circuits should have one side referred to chassis ground; for example, diode switch biases, varactor control voltages, and mechanically tuned capacitors. If both ends of a tuning element are RF hot, additional isolating RF chokes are required. The stray capacitance to ground of these isolation circuits and of the component mountings must be absorbable into the filter topology. If the strays are not in parallel with capacitors of the filter model, they can form parasitic resonances and undesired lowpass/highpass sections, and the expected response may not be obtained.

Grounded elements rule out lattice and series-arm, shunt-arm ladder filters but fit well in the coupled-resonator model. It places one end of each resonator inductor and capacitor near RF ground, and all nodes can have a capacitor to ground, which means that stray capacitances to chassis and component mountings are easily absorbed into the design. The coupled-resonator form is suitable for filters with a fractional bandwidth less than about 10%, which covers most preselector and postselector applications.

Mathematically, coupled-resonator filters are usually derived from a lowpass prototype ladder filter, although it is also possible to derive exact equations for the response of low-order coupled-resonator filters. Converting a lowpass prototype to a "classical" bandpass gives a series-arm, shunt-arm ladder filter with a response having geometric symmetry (not arithmetic symmetry). The total bandwidth at any attenuation is exactly proportional to that of the prototype, but the high and low frequencies at a given attenuation will not have the same separation from f_0.

Conversion from the classical ladder form to coupled-resonator implementation uses impedance transformation networks to change alternate series and parallel resonators into all-series or all-parallel resonators with coupling elements. Several forms of coupling can be used, as discussed in the next section. These transformations are exact only at one frequency but are usable over a frequency range that is suitable for most preselectors.

Resonator coupling methods

The design procedures for coupled-resonator filters give the desired resonator coupling coefficients rather than element values. The coupling can be implemented in many ways, such as those shown in Figure 9.8 or reference 7. The most common are mutual inductance, top C, and bottom L. Mutual inductance may be provided by a link winding or an aperture between resonator compartments. End resonator loading to the source and load terminations may be provided in a similar manner, as shown in Figure 9.9. The form with the termination coupled by a large reactance to the top of the resonator tank is preferred over the tapped divider form when large off-resonance input or output impedance is needed for multicoupling applications (see section 9.7).

The equations in Figure 9.9 are exact for parallel resonance. Fractional errors less than $1/Q_t^2$ occur if the equations are used for a resonator with a series load on the other side. Variations will occur if the resonator loss is better modeled as a series (wire resistance) than as a shunt (dielectric loss) resistance. The variations are usually negligible for simulations because the detuning is a small part of the passband width.

Inductive coupling models may be converted using the equations given in Figure 9.10. Input and output couplings to a given resonator should not be made through the same element (for example, both being taps on the coil), as this couples the signal past the resonator and its selectivity is reduced.

Each resonator in Figure 9.8 must have its component values altered from the nominal used in the resonant frequency equation in order to account for the detuning effects of the coupling or end-loading reactances on both sides. The corrected value for one coupling is shown in each figure. For example, the center resonator in a three-resonator top-C coupled filter must be corrected for each coupling capacitor and uses a main capacitor of value $C - 2C_t$.

There is a frequency variation inherent in each of these methods that shows up in response plots as a passband tilt and a greater stopband attenuation on one side of resonance than the other (after accounting for geomet-

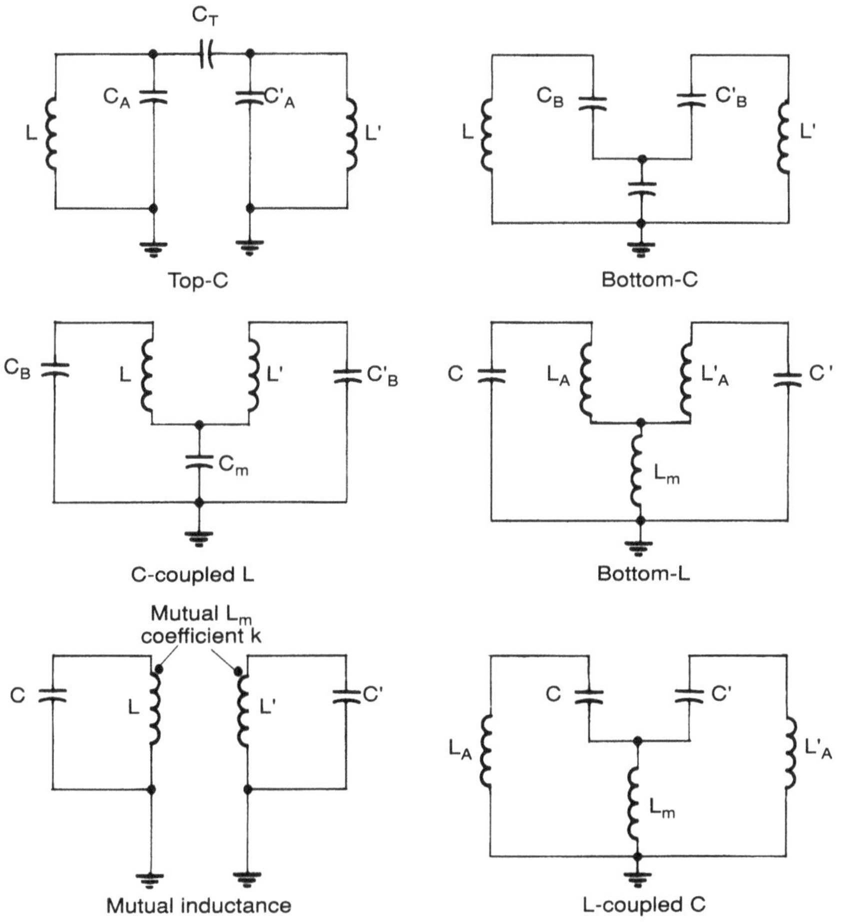

C_T
C_A
C'_A
L
L'
Top-C
C_B
C'_B
L
L'
Bottom-C
C_B
L
L'
C'_B
C_m
C-coupled L
C
L_A
L'_A
C'
L_m
Bottom-L
Mutual L_m
coefficient k
C
L
L'
C'
Mutual inductance
C
C'
L_A
L'_A
L_m
L-coupled C

$$f_0 = \frac{1}{2\pi\sqrt{LC}} = \frac{1}{2\pi\sqrt{L'C'}}$$

$$k = \frac{L_m}{\sqrt{LL'}} = \frac{C_T}{\sqrt{CC'}} = \frac{\sqrt{CC'}}{C_m}$$

$$C_A = C - C_T \qquad C'_A = C' - C_T$$

$$L_A = L - L_m \qquad L'_A = L' - L_m$$

$$C_B = \frac{C_m C}{C_m - C} \qquad C'_B = \frac{C_m C'}{C_m - C'}$$

Figure 9.8 Resonator coupling methods.

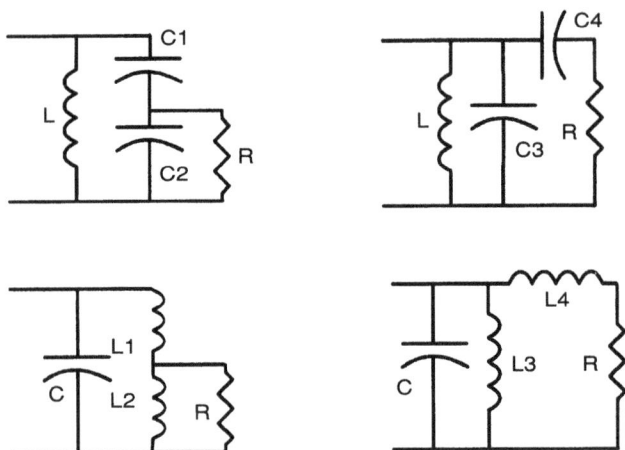

$$X = 2\pi f_0 L = \frac{1}{2\pi f_0 C}$$

$$\frac{X}{Q_t} < R < X Q_t$$

$$\alpha = \frac{R}{X}\left(Q_t + \frac{1}{Q_t}\right) - 1 > 0$$

$$\beta = \frac{X Q_t}{R} - 1 > 0$$

$$X_2 = \frac{R}{\sqrt{\alpha}}$$

$$X_4 = R\sqrt{\beta}$$

$$X_1 = X\left(\frac{Q_t^2}{Q_t^2 + 1}\right) - X_2\left(\frac{\alpha}{\alpha + 1}\right) > 0$$

$$\frac{1}{X_3} = \frac{1}{X} - \frac{1}{X_4}\left(\frac{\beta}{\beta + 1}\right) > 0$$

Analysis (if X_1 resonates circuit):

Analysis (if X_3 resonates circuit):

$$Q_t + \frac{1}{Q_t} = q = \frac{X}{X_2}\left(\frac{X_2}{R} + \frac{R}{X_2}\right)$$

$$Q_t = \frac{X_4}{X}\left(\frac{X_4}{R} = \frac{R}{X_4}\right)$$

$$Q_t \approx q - \frac{1}{q}; Q_t \geq 5$$

Figure 9.9 Resonator input and output loading methods. Note: (1) Matching range $X/Q_t >> R >> X Q_t$; (2) X is the nominal total resonator reactance, either inductive or capacitive; (3) all subscripted X_i must have signs opposite X; (4) equations are approximate, with fractional errors on the order of $1/Q_t^2$; (5) error partially depends on whether resonator loss is actually where shown.

ric symmetry). Each capacitive coupling adds 6 dB/octave tilt and each inductive coupling subtracts 6 dB/octave tilt. The tilt may not be significant within a narrow passband. The asymptotic behavior at very high and low frequencies is determined entirely by the coupling methods between resonators and at the input and output [8]. It is usually necessary to choose a combination of coupling methods throughout the filter to obtain the best compromise stopband tilt for the application. For some ranges of values the end-loading circuits and couplings may give a significant "shelf" effect where the response flattens out at only 15- to 20-dB attenuation for a decade of frequency before beginning the 6 dB/octave decrease.

A performance loss occurs in tunable filters if the optimum coupling and end-loading values cannot be maintained as the resonator is tuned across the band. Couplings of the same type as the tuning element (for example, top-C between variable capacitors) will cause this problem. It also arises when the resonator Q_u varies appreciably over the band, as with helical resonator inductors. The ratio of load resistance to resonator reactance will determine which of the end-loading methods shown in Figure 9.9 will have the most acceptable variation in Q_t. Overcoming these variations can tax the ingenuity of the designer. The effect can sometimes be alleviated with more complex matching networks at the ends of the filter, as shown in Figure 9.4, with tailored-curve variable capacitors such as shown in Example 9.1 of section 9.4, or by combinations of resonator coupling methods, as in Figure 9.7.

Filter types

The best prototype for most preselector applications is the Cohn minimum-loss (min-loss) filter. Other prototypes are useful when they provide some characteristic that is not controlled by the min-loss design. The min-loss, with its simple design procedure, should be calculated first in any case to provide a basis for comparing with other designs.

The min-loss prototype is appropriate when designing for the lowest center frequency loss consistent with a specified stopband width. The theory

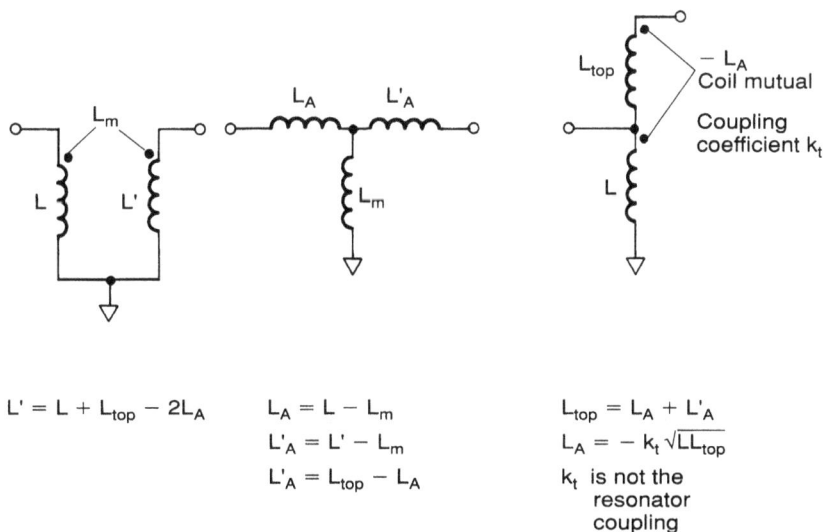

$$L' = L + L_{top} - 2L_A \qquad L_A = L - L_m \qquad L_{top} = L_A + L'_A$$
$$L'_A = L' - L_m \qquad L_A = -k_t \sqrt{LL_{top}}$$
$$L'_A = L_{top} - L_A \qquad k_t \text{ is not the}$$
$$\text{resonator}$$
$$\text{coupling}$$

Figure 9.10 Equivalent forms of coupled coils.

developed by Cohn and others [9–13] explicitly trades off resonator dissipation factor (or Q), selectivity at some point on the skirt (for example, 40 dB), and midband filter loss to give a design that is optimum for these three parameters. It does not try to control passband shape, but generally gives good results in preselector applications. The min-loss has other good properties, as discussed below, which may make it a good choice even when the lowest loss is not required.

Lossless filter prototypes such as Chebyshev, Butterworth, and Bessel give their theoretical response shapes when the elements have infinite Q. Filters implemented with the lossless prototype values but lossy elements are called *pseudoexact* designs, and their response shapes vary with the element loss. With plots of these responses [14] or a good computer program for network analysis, a designer can iteratively find a trade-off between selectivity, ripple, passband width, and midband loss. Equation 6.3-15 in reference 15 provides a useful way to estimate the trade-off between loss and normalized q (element Q_u divided by bandwidth) for any pseudoexact prototype. Selectivity and loss are closely tied to passband width and ripple in this design method. If the passband is not a critical aspect of the design, its central place will hinder the design trade-off process.

Predistorted Butterworth and Chebyshev designs are available to maintain the theoretical shape in the presence of element loss. This is done with impedance mismatch (high-voltage standing wave ratio [VSWR]) and increased passband loss, and is rarely applicable to preselector design.

A min-loss design may not be chosen in an application requiring a relatively wide and flat passband. The passband width is not specified in the min-loss design process. It has a ripple that is greatest at the edge of the passband and becomes quite large for filters with several resonators and low loss. If these characteristics are not satisfactory, then a low-ripple (for example, 0.1-dB) Chebyshev or Butterworth pseudoexact design may be the best choice. The loss obtained will not be the lowest achievable, but the penalty is often small. A side benefit is sometimes a better VSWR.

Each resonator of the min-loss filter has equal loss (in decibels). For an equally terminated filter, this results in an equal-element lowpass prototype. In the case of a one- or two-pole doubly terminated filter, the Cohn filter is also a Butterworth pseudoexact filter. The Cohn has some advantages over many others. All resonators have nearly equal voltages for a midband signal, and it has a large in-band power-handling capability. No resonator is more sensitive to tuning adjustment than the others, which helps with temperature stability, ease of alignment, and tracking in a tunable filter.

When optimizing a receiving system for a very low noise figure, it is sometimes found that improvement is obtained by allowing some impedance mismatch between the filter and its load (an RF amplifier, for instance) and reoptimizing the selectivity/loss trade-off. This optimum noise matching has been taken into account in an extension of the Cohn theory [12, 13].

The improvement usually turns out to be a fraction of a decibel and is unimportant for typical HF preselector designs, but can be more significant at higher frequencies.

In many cases the coupling values for any of the filter types discussed will differ by 10% or less. Then the unavoidable variations inherent in a tunable filter, as discussed earlier, make the choice of prototype a moot question. In practice, the calculated design is a simplified model that omits many real component effects. The designer must use it as a starting point, then measure and adjust the experimental model until its response is acceptable according to the performance specifications.

Min-loss design procedure

The design of equally terminated min-loss filters is covered in reference 15. A design procedure for doubly terminated coupled-resonator filters where all resonators have the same unloaded Q is summarized here. It is not required that the resonators be implemented with the same element values.

The following symbols are used:

- f_0: resonant frequency

- f_H: a frequency above resonance with specified attenuation

- f_L: frequency below resonance with same attenuation as f_H

- f_0^2: $f_L f_H$

- B: $f_H - f_L$ bandwidth at the specified attenuation

- N: number of resonators

- L_0: midband insertion loss (in decibels)

- Q_u: unloaded resonator Q

- Q_t: "terminal Q," the Q of a lossless resonator loaded only by the terminating resistance on one side

- Q_L: $Q_u Q_t / (Q_u + Q_t)$, resonator Q with its loss and termination

- r: Q_t / Q_u, dissipation to loading ratio; also the prototype filter's series element normalized resistance

- k: resonator-to-resonator coupling

The filter will have geometric symmetry, except for the effects of the coupling as discussed earlier, and thus only two of the four frequencies (f_0, B, f_H, and f_L) at a given attenuation level may be chosen arbitrarily. The fractional bandwidth at the selectivity specification point is needed:

$$\frac{B}{f_0} = \frac{f_H}{f_0} - \frac{f_0}{f_H} = \frac{f_0}{f_L} - \frac{f_L}{f_0}$$

The design procedure begins with two of these three specifications: selectivity requirement, insertion loss, and unloaded resonator Q_u. The designer selects a filter with an acceptable number of resonators, midband loss, and off-frequency attenuation. The curves shown in Figures 9.11A–C represent the insertion loss response of doubly terminated Cohn filter prototypes (equal-element pseudoexact) on a normalized frequency scale and provide a quick means of determining the design trade-offs. The prototype curves can be extended at $6N$ dB/octave, but a coupled-resonator implementation eventually departs from the prototype shape.

Then the following approximate equations are used to design the couplings (they are accurate for filters with low midband loss):

$$r = \frac{L_0}{4.343N} \quad \text{(accurate if } L_0 < 2N, \text{ in decibels)}$$

$$Q_t = rQ_u$$

$$k = \frac{1}{Q_t}$$

Implementation consists of setting the component values to obtain these coupling and loading values, as shown in Figures 9.8 through 9.10. All res-

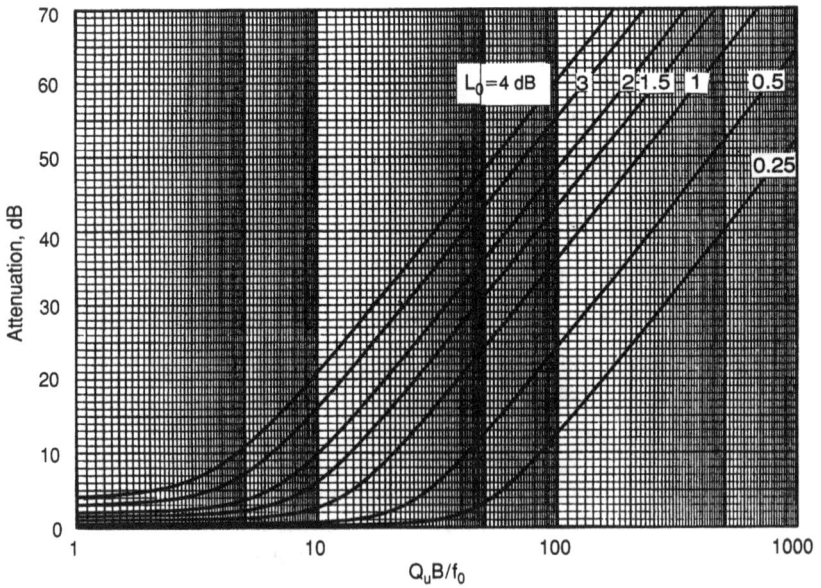

Figure 9.11 (A) Attenuation of two-resonator Cohn filter.

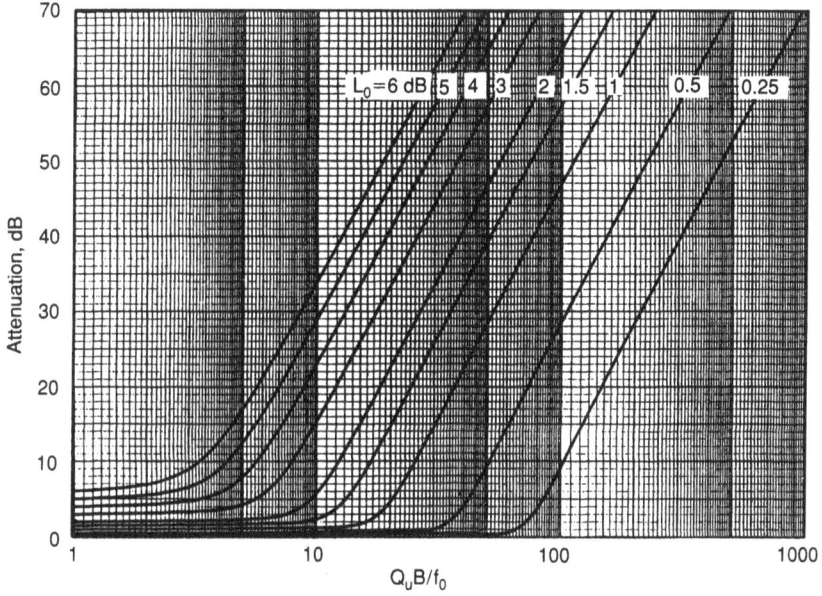

B

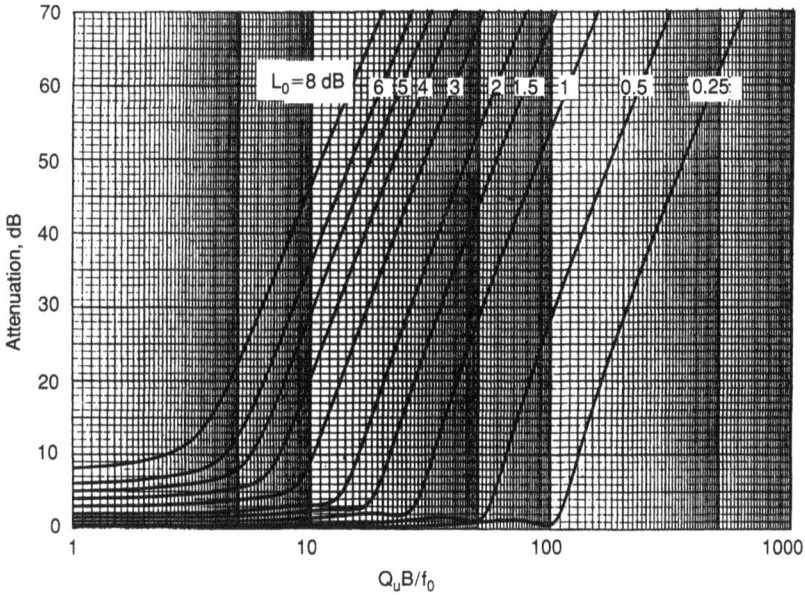

C

Figure 9.11 (B) Attenuation of three-resonator Cohn filter. (C) Attenuation of four-resonator Cohn filter.

onator-to-resonator coupling coefficients are k and each end termination is chosen to obtain Q_t.

After k and Q_t are fixed the skirt bandwidth on the straight-line portion of the response is independent of Q_u. The actual Q_u determines the midband loss and passband shape.

9.4 Design Examples

The following design examples illustrate the use of the filter design formulas and curves and the resonator coupling techniques. The second example is also used to illustrate the design trade-offs between noise figure and distortion that the preselector design engineer must make. These examples are intended as first-pass studies which the designer would use for initial feasibility and trade-off studies, but which do not include margins for variations in temperature and production part tolerances.

Example 9.1: Tunable preselector

We are designing a receiver and need a tunable preselector to offer as an accessory. It should provide 20 dB of absolute attenuation for signals 5% away from the tuned frequency. The space available requires us to use toroid inductors, and their unloaded Q_u will be 200.

Solution. We start with the following equations:

$$B = \left(1.05 - \frac{1}{1.05}\right)f_0 = 0.098f_0$$

$$Q_u\frac{B}{f_0} = 200 \times 0.098 = 19.5$$

Referring to Figure 9.11A we find that a two-resonator filter will give us the required 20-dB attenuation with a midband loss of about 2.0 dB. Then we calculate

$$r = \frac{2.0}{4.343 \times 2} = 0.230$$

$$Q_t = 0.230 \times 200 = 46.0$$

$$k = \frac{1}{46.0} = 0.022$$

The filter is illustrated in Figure 9.12. The input coupling will be provided by a capacitance to the top of the resonator in order to obtain a suitable out-of-band input impedance for multicoupling. In order to maintain reasonably constant input loading and therefore the required Q_t, we will make this a

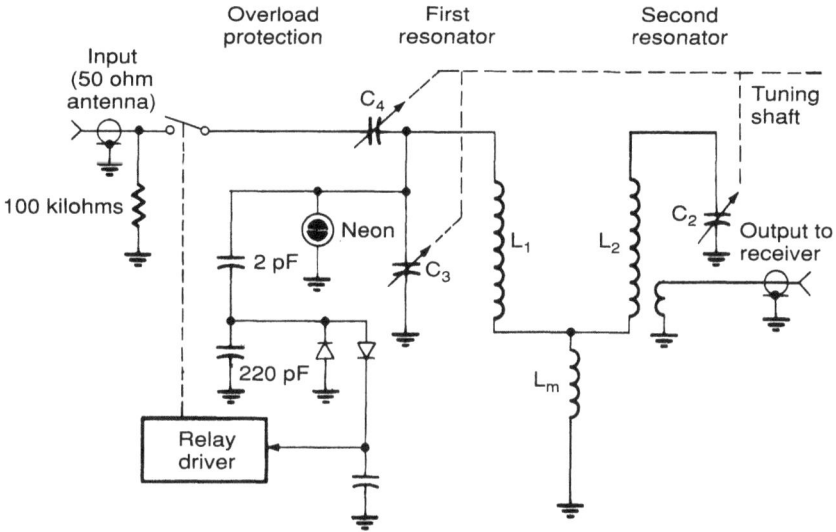

Figure 9.12 Two-resonator tunable filter of Example 9.1.

$$L = \frac{1}{(2\pi f_0)^2 C} = \frac{1}{(2\pi\, 4 \times 10^6)^2\, 50 \times 10^{-12}} = 31.7\ \mu H$$

$$L_m = kL = (0.022)\,(31.7) = 0.7\ \mu H$$

$$L_m = L_2 = L - L_m = 31.0\ \mu H$$

	2 MHz	3 MHz	4 MHz
C, pF	200	88.9	50
X, ohms	398	597	796
X_4, ohms	955	1171	1352
C_4, pF	83.3	45.3	29.4
X_3, ohms	681	1215	1931
C3 + stray, pF	117	43.6	20.6

variable capacitor section tied to the tuning shaft. The capacitance versus rotation of the sections of this capacitor can be tailored to obtain the desired loading. The resonator coupling will be "bottom L." The output loading will be provided by a link winding on the second inductor, with a turns ratio as needed to obtain Q_t, in conjunction with the receiver input impedance. We will choose a maximum total resonator capacitance of four times the minimum resonator capacitance in order to tune octave frequency bands. Pick a minimum capacitance, including strays, of 50 pF at 4 MHz. The inductors can be calculated as shown in Figure 9.12. The capacitors are calculated from the equations for X_3 and X_4 in Figure 9.9. The capacitors must be calculated at several frequency points within the band to establish their tracking curves. The inductors and input capacitor may be switched to cover additional frequency bands, while the main tuning capacitors are reused.

Example 9.2: Fixed preselector

An 18-MHz receiver needs a filter to protect it from a nearby foreign broadcast transmitter only 2% from the center of the band. The receiving antenna signal is 50 V and the receiver can operate with 1.0-V undesired signals. Resonators with $Q_u = 525$ can be obtained by using cavities containing helical resonator coils operated below their self-resonance. The receiver has a 14-dB noise figure, and a low-noise pre-amp (noise figure 5 dB) is being considered because the antenna is not efficient and there is loss incurred in distributing the signal to several receivers. The preselector should be designed for multicoupling.

Solution. The 2% separation gives a normalized frequency variable of

$$\frac{Q_u B}{f_0} = 525 \left(1.02 - \frac{1}{1.02} \right) = 20.8$$

We need 20 log (50/1) = 34-dB overall reduction in the undesired signal, which is the total of filter attenuation and amplifier gain. This gain will depend on the noise figure goal. We will guess the filter in-band loss to be 4.5 dB and try for a 12-dB system noise figure. The Friis noise figure equation [16] is used, as discussed in Chapter 4 and reference 17 of this chapter, to find the noise figure of the cascaded stages. See also the computer program NFIIP in Chapter 17. All values must be converted from decibels to power. Remember that the filter noise figure is equal to its loss (neglecting mismatch effects) because the filter contributes thermal noise, as well as attenuating the signal power. We start with

$$F_R = 25 \quad \text{(14-dB receiver noise)}$$

$$F_A = 3.2 \quad \text{(5-dB amplifier noise)}$$

$$F_F = \frac{1}{G_F} = 2.8 \quad \text{(4.5-dB filter noise)}$$

$$F_{\text{sys}} = 15.8 \quad \text{(12 dB)}$$

$$= F_F + \frac{F_A - 1}{G_F} + \frac{F_R - 1}{G_F G_A} = 2.8 + \frac{3.2 - 1}{0.35} + \frac{25 - 1}{0.35 G_A}$$

Solving for the gain gives $G_A = 9.8$ or about 10 dB. Then we need a filter attenuation of 34 + 10 = 44 dB.

Filters with various numbers of resonators and 44-dB attenuation at this frequency are checked on the plots of Figure 9.11. It is decided that three resonators give the best hardware/loss trade-off:

n	L_0 (dB)
2	8 (out of valid range)
3	4.3
4	3.7

We note that the passband is fairly flat out to a normalized bandwidth of 3 or so. This indicates that we may be able to use a fixed-tuned filter with a tunable receiver if our band of interest is no wider than

$$B = \frac{3f_0}{Q_u} = \frac{3 \times 18}{525} = 0.103 \text{ MHz}$$

This possibility should be checked by a circuit analysis program before committing to the design. Passband tilt could be a problem, depending on the coupling methods.

We next calculate the filter coupling and loading:

$$r = \frac{4.3}{4.343 \times 3} = 0.330$$

$$Q_t = 525 \times 0.330 = 173$$

$$k = \frac{1}{173} = 0.0058$$

The filter is illustrated in Figure 9.13. Figure 9.14 shows the insertion loss response of the filter. Coupling is obtained by an aperture in the wall between each resonator and the next. The aperture dimensions can be determined experimentally using the coupling measurement technique discussed in section 9.5. The input and output end-loading is obtained with a small capacitor or a probe inserted into the top of the resonator cavity. This provides capacitive coupling to balance the inductive coupling between resonators and improves the stopband symmetry. A slight tilt remains, but the result is very close to the original design values. This form of coupling also presents a high input impedance for good multicoupling, as shown in Figure 9.15.

The Friis equation is now used to check the system noise figure.
Filter:

$$F_F = 2.7 \quad (4.3 \text{ dB})$$

$$G_F = 0.37 \quad (-4.3 \text{ dB})$$

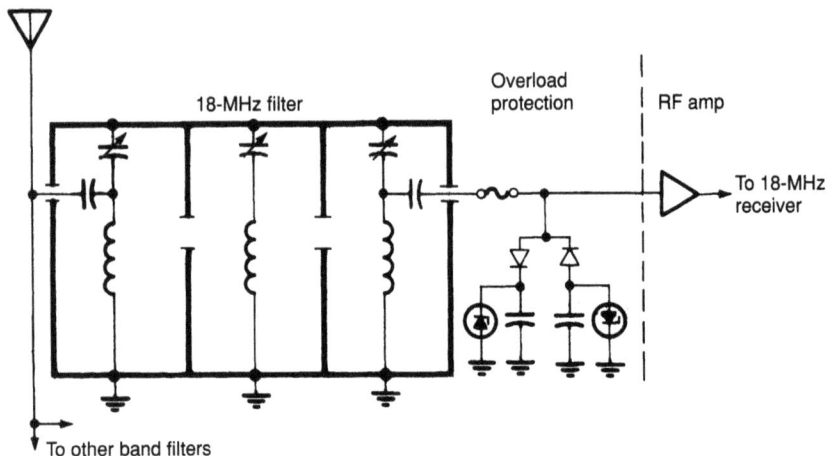

Figure 9.13 Three-resonator aperture-coupled fixed filter of Example 9.2.

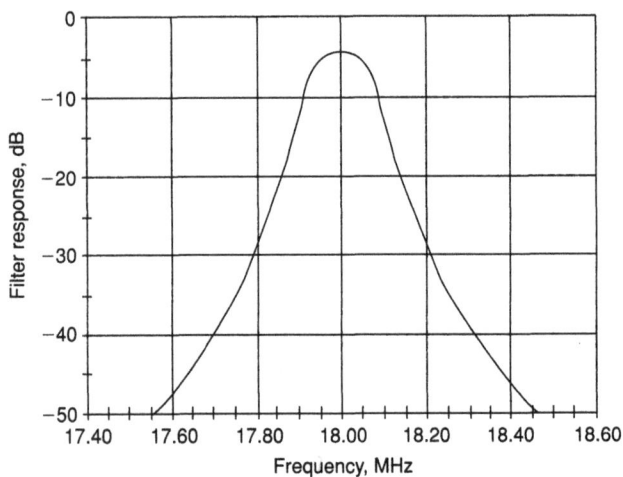

Figure 9.14 Frequency response of fixed filter of Example 9.2.

Amplifier:

$$F_A = 3.2 \quad (5 \text{ dB})$$

$$G_A = 10 \quad (10 \text{ dB})$$

Receiver:

$$F_R = 25.1 \quad (14 \text{ dB})$$

System noise figure:

$$F_{sys} = F_F + \frac{F_A - 1}{G_F} + \frac{F_R - 1}{G_F G_A}$$

$$= 2.7 + \frac{3.2 - 1}{0.37} + \frac{25.1 - 1}{(0.37)(10)} = 15.0 \quad (11.8 \text{ dB})$$

The new system noise figure is 11.8 dB, which is a significant improvement over the original receiver noise figure if the system is used with an inefficient antenna system or in a low-noise location.

Next we should consider what we have done to the IM intercept point. Assume the receiver's third-order intercept point is +15 dBm and the amplifier input intercept is +13 dBm. As discussed in Chapter 4 and references 17 and 18, the intercept points should be adjusted one to one by the gains and losses in order to refer them to a new point in the system, and when expressed in milliwatts, they will, in the worst case, combine like resistors in parallel. In this example, we will find the combined intercept point of the receiver and amplifier at the amplifier input:

$$I_{sys} = \left[\frac{1}{I_A} + \frac{G_A}{I_H} \right]^{-1} = \left[\frac{1}{20} + \frac{10}{32} \right]^{-1} = 2.7 \text{ mW or } 4.4 \text{ dBm}$$

Referring this to the filter input gives an input intercept of 4.4 + 4.3 = 8.7 dBm for signals near the peak of the preselector passband. It is assumed that the filter has no distortion because it uses no ferrous materials. The amplifier is not contributing much distortion, but its gain significantly in-

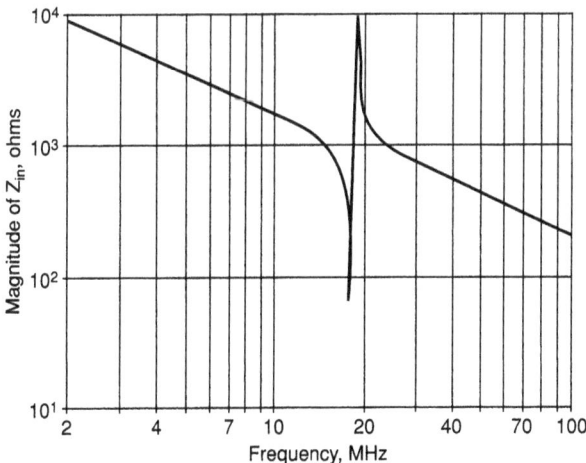

Figure 9.15 Filter input impedance in Example 9.2.

creases the distortion in the receiver for signals within the preselector pass-band. This situation would improve if we let the system noise figure equal the original receiver noise figure and used less amplifier gain.

The out-of-band distortion performance is considerably improved. If the interfering signals are at least 2% off-frequency, and the third-order distortion of the receiver follows the expected 3:1 rate of change, the filter will increase the intercept point at a rate of 3/2 dB for each decibel of relative selectivity, obtaining

$$8.7 + \tfrac{3}{2}(44 - 4.3) - 68 \text{ dBm}$$

which is a very respectable value. The value will increase for greater frequency separations. Actually, one interfering tone will have to be separated twice as far as the other from the operating frequency in order to produce an in-band third-order product (for example, $f_0 = 2f_2 - f_1$). It will be attenuated much more, improving the IM performance much more than that calculated above.

The numbers in Example 9.2 illustrate the trade-offs the receiving system designer must go through to arrive at an acceptable system design, considering the sources and types of interference, noise, and distortion which are most critical in each application.

9.5 Implementation Considerations

Operating voltages

The midband rms voltages in the min-loss filter can be estimated from the loaded Q_L values, reactance at resonance X_0, and power P flowing through the filter:

$$Q_L = \left[\frac{1}{Q_t} + \frac{1}{Q_u} \right]^{-1} = Q_u \frac{r}{1 + r}$$

$$V^2 = Q_L X_0 P$$

This estimate is approximate and is valid only at resonance. It will usually be found that some resonator has more voltage at frequencies near the edge of the passband. For farther out-of-band signals the selectivity improves through each resonator. The equal-loading property of the min-loss filter assures that the in-band voltages will be approximately equal. Other filter designs such as Butterworth may have some resonators with extremely high in-band voltages. It is advisable to use a circuit analysis program to determine the voltage-versus-frequency curve for each resonator before setting component ratings. Such a plot for Example 9.2 is shown in Figure 9.16, for an antenna signal of 1 V into a matched load.

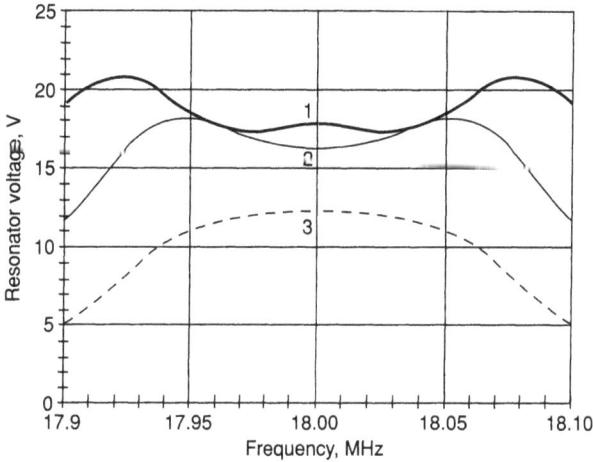

Figure 9.16 Resonator voltages in Example 9.2.

Measuring Q and coupling

It is desirable to measure the assembled unloaded resonator's Q_u to verify that all the losses have been accounted for, even if the components were measured. For comparison and combining components, it is best to work with measurements as dissipations ($d = 1/Q$) because the dissipations add numerically. This leads more quickly to an intuitive grasp of where the losses are than does working with Q values.

The in-place measurement is based on Dishal's methods [19, 20]. The simplest equipment setup uses a stable RF generator having a fine-tune adjustment, a frequency counter, and a sensitive RF voltmeter. A network analyzer with a sweep display is more convenient, if available. The generator and voltmeter are each very lightly coupled to the resonator with a "sniffer" probe, loop of wire, or one of the filter-terminating methods at an extremely high Q_t. The insertion loss when the resonator is coupled between generator and meter should be no less than 30 dB to avoid lowering the Q being measured, or else their contribution to dissipation calculated according to the coupling used should be subtracted out. Adjacent resonators and terminations are disconnected (opened or shorted as appropriate) when measuring a single resonator.

The signal generator is tuned for a peak voltmeter reading, and then tuned above and below to find the frequency separation between the 3-dB (71% voltage) frequencies. We use the equation

$$d = \frac{1}{Q} = \frac{F_{3H} - F_{3L}}{F_0}$$

End-resonator loaded Q_L can be measured by the same method, with the termination reconnected and other resonators still disconnected. Then Q_t can be calculated from the measurements to check the accuracy of link-coupled or tapped-coil terminations.

A vector impedance meter provides another means for determining Q by reading the 45° phase points, which are nearly coincident with the 3-dB points. The high impedance of this instrument makes it easy to use with less instrument loading on the resonator.

The coupling between two resonators can be measured in a similar manner by lightly coupling the generator to one and the meter to the other. All other resonators or terminations are removed (open or short circuit as appropriate for the topology). The two resonators are carefully tuned to the same frequency. For useful values of coupling between the resonators, a double-resonance peak will be obtained. The separation between the peaks is measured and k is computed:

$$k = \frac{F_H - F_L}{F_0}$$

If the coupling is low, relative to the resonator dissipations, the dip between peaks will be shallow and the measurement will yield too small a value for k, which can be approximately corrected according to Table 9.2.

9.6 Computer Aids

Several general-purpose circuit analysis programs, such as SPICE, Touchstone, and Harmonica help to verify the design by making frequency response plots. Monte Carlo simulations of component variations will also give an indication of the effects caused by component tolerances.

Spreadsheets (EXCEL, QUATTRO, Lotus 1-2-3) and mathematical evaluation tools such as Mathcad and MATLAB can also be used to evaluate the design equations and solve for elements where algebraic solution is difficult.

TABLE 9.2
Corrections for k Measurements

dip (dB)	$k\,Q_u$	k/k (measured)
2	2.0	1.15
3	2.4	1.10
4	2.8	1.07
6	3.7	1.04

Programs on disk

These programs are discussed in Chapter 17. The ones most relevant to this chapter are

- Half-octave filter design program. This program aids in designing suboctave filters and generates a SPICE text file that can then be edited to evaluate the effects of stray capacitance and component tolerances. Optimization of inductor values when the nearest standard capacitor values are used is also a valuable option.

- SPICE file for Example 9.2 (see section 9.4). This is set up to plot the frequency response of Example 9.2. Lines are included as comments which can be used to plot input impedance and resonator voltages.

- Spreadsheet for min-loss design equations. This file is set up with values for Example 9.2, but may have the entries changed to repeat the calculations for any design.

Considerations for simulation

Simulating a narrow filter such as a preselector requires that the resonances be accurate within a small fraction of the passband width. Otherwise a de-tuned filter is simulated and the response does not represent the design.

Such accurate tuning requires good precision in the element values, just as trimmer adjustments are used in the lab to obtain precise resonances. For a filter with a small percentage bandwidth the elements should be correct to at least three significant figures. The principal elements must be corrected for the coupling and end-loading elements, as noted in previous sections. The stray capacitance values at each node and for each inductor should be estimated or measured and included in the simulation.

The resonant frequency of each resonator can be verified by a preliminary simulation, with the other resonators shunted by resistances to lower their Q and allow the resonance of interest to be seen. Circuit losses should be added only after the lossless circuit is properly tuned.

Much additional information can be obtained in addition to the basic frequency response. The voltages and currents at all points in the filter are readily accessible and they can help to set component ratings. These values can be examined over frequency to determine how well the filter components will withstand very strong signals at nearby frequencies.

9.7 Multicoupling

The need frequently arises to operate several receivers from one antenna. Signal preamplifiers and power splitters allow this to be done with minimal degradation of noise figure, but such devices may not be able to handle large interfering signals. If a preselector is needed on a single receiver, it is also needed on a multiple-receiver connection. A single preselector cannot be used unless the receivers always operate on very closely spaced frequencies. Therefore we must connect several preselector inputs to the antenna and are faced with the problem of assuring satisfactory performance.

This situation is very difficult to analyze because the input impedance of each preselector filter varies as it is measured across the frequency band and is also different for each frequency it may be tuned to. The two methods of attack are worst-case analysis and statistical simulation. Worst-case analysis can often suggest a configuration that looks promising, although the worst-case degradation is large. Then statistical analysis of many randomly chosen combinations of tuned frequencies can tell whether 90 or 99% of the cases are acceptable.

As an example, consider a preselector tuned to some frequency in the middle of the HF band and which has an input impedance above its operating frequency modeled by a 150-pF capacitor (Figure 9.17). Let us analyze the degradation it causes a second unit tuned to 30 MHz with an input impedance on frequency of 50 ohms. At 30 MHz the capacitance has a reactance of 35 ohms. If there is a 1.5-ft (0.46-m), 50-ohm, 66% velocity coax between the capacitance and the common connection point, the transmission line effect transforms the 35 ohms to 9 ohms of reactance. It is shunt-

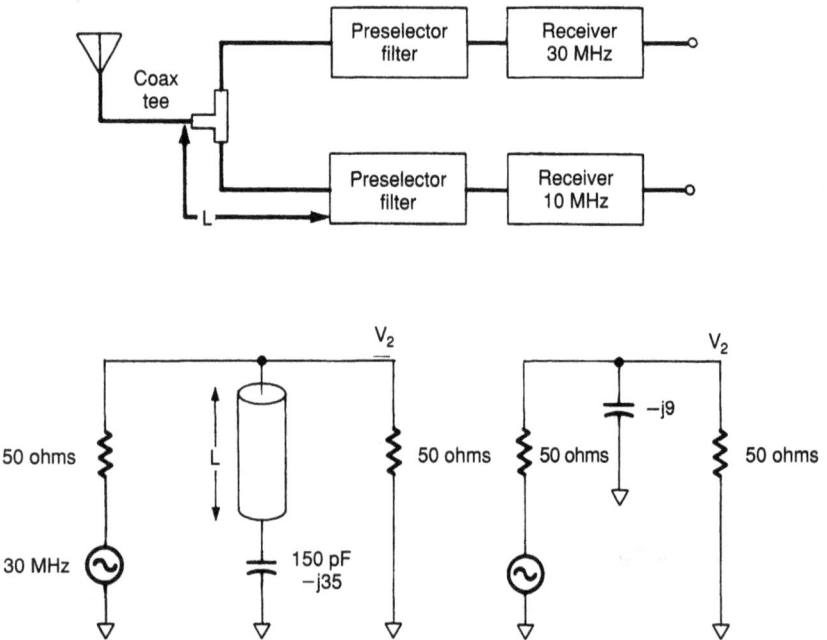

Figure 9.17 Worst-case multicoupling analysis for two units.

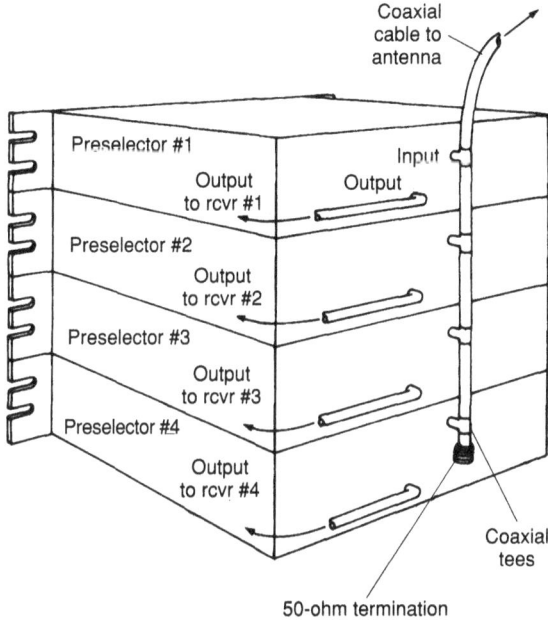

Figure 9.18 High-level multicoupling with preselectors.

ing the 50-ohm input to the second preselector, which sees a 10-dB loss compared to having the antenna unshared.

This type of analysis leads to the rule of thumb that if the input impedance off-frequency is capacitive, the coax length to the tee connection must be as short as possible. If the input impedance is always inductive off-frequency, a small coax length may improve the performance, but a long coax will cause bad degradation at some frequency.

If many units are connected together, a useful starting point is a daisy chain of tee connections, all with short coax lengths, and a resistive termination at the end (Figure 9.18). The best termination may not be the coax impedance, and the best location may not be at the end of the chain, since the system is full of mismatched impedances. This model can be varied to study the effects of coax lengths, termination impedance, and of inserting reactances between the units.

References

1. Anatol I. Zverev, *Handbook of Filter Synthesis* (New York: John Wiley & Sons, 1967).

2. W. H. Hayward, *Introduction to Radio Frequency Design*, chaps. 2–3 (Englewood Cliffs, NJ: Prentice-Hall, 1982).

3. James K. Hardy, *High Frequency Circuit Design*, Chapter 2 (Reston, VA: Reston Publishing, 1978).

4. ITT, *Reference Data for Radio Engineers*, 6th ed., pp. 24-28–24-30 (Indianapolis: Howard W. Sams, 1975).

5. Anatol I. Zverev, *Handbook of Filter Synthesis*, Chapter 9 (New York, John Wiley & Sons, 1967).

6. ITT, *Reference Data for Radio Engineers*, 6th ed., pp. 4-28–4-31 (Indianapolis: Howard W. Sams, 1975).

7. ITT, *Reference Data for Radio Engineers*, 6th ed., pp. 8–25, Figure 28, 9-2 (Indianapolis: Howard W. Sams, 1975).

8. William E. Sabin, "Designing Narrow Band-Pass Filters with a BASIC Program," *QST* LXVII (May 1983): 23–29.

9. Seymour B. Cohn, "Direct Coupled Resonator Filters," *Proc. of IRE* 45 (February 1957): 187–196.

10. Seymour B. Cohn, "Dissipation Loss in Multiple-Coupled Resonator Filters," *Proc. of IRE* 47 (August 1959): 1342–1348.

11. Jesse J. Taub, "Design of Minimum Loss Bandpass Filters," *Microwave J.* 6 (November 1963): 67–76.

12. R. W. Carroll and D. B. Hallock, "Application of the Minimum-Loss Filter to Optimum Receiver Input Design," Collins Radio Co. Working Paper WP-8521 (August 1965).

13. R. W. Carroll, "Synthesis of Optimal Receiver Preselectors," Collins Radio Co. Working Paper WP-8522 (August 1965).

14. Robert L. Sleven, "Pseudo-exact Bandpass Filter Design," *Microwaves*, 11–12 (August 1968–July 1969) (appeared in eight parts).

15. Herman J. Blinchikoff and Anatol I. Zverev, *Filtering in the Time and Frequency Domains*, sec. 6.3 (New York: John Wiley & Sons, 1976).

16. H. T. Friis, "Noise Figures of Radio Receivers," *Proc. of IRE* 32 (July 1944): 419–422. (Note: Equation 15 on p. 421 is garbled and should read: $F^{ab} = F_a + (F_b - 1)/G_a$.)

17. W. E. Sabin, "A BASIC Approach to Calculating Cascaded Intercept Points and Noise Figure," *QST* LXV (October 1981): 21–24.

18. W. H. Hayward, *Introduction to Radio Frequency Design*, Chapter 6 (Englewood Cliffs, NJ: Prentice-Hall, 1982).

19. M. Dishal, "Alignment and Adjustment of Synchronously Tuned Multiple-Resonant-Circuit Filters," *Proc. of IRE*, vol. 63, November 1951, pp. 1448–1455.

20. W. H. Hayward, *Introduction to Radio Frequency Design*, sec. 3.4 (Englewood Cliffs, NJ: Prentice-Hall, 1982).

10

Synthesizers for SSB

Donald E. Phillips
William R. Weaverling

This chapter will deal with frequency synthesizers appropriate for medium- and high-performance HF SSB radios. Space limitations preclude a thorough treatise on synthesizer design, many aspects of which are covered in the literature (see references 1–5, for example). We will concentrate on tuning speed, phase noise, spurious output signals, and the generation of small frequency steps needed in the HF band. The most important phaselock loop (PLL) configurations will be discussed; newer, all-digital techniques for producing small steps will be introduced, and the methods of combining these with traditional loops will be shown.

The method of analysis is, we believe, unique. Most of the literature on synthesizers presents either a detailed mathematical analysis of single loops or a discussion of multiloop synthesizers that have already been designed. Here, we will concentrate on the important middle ground, where it is assumed that the individual building blocks exist to build a loop and the designer must understand in detail the overall loop properties and the way loops relate to each other. Block diagrams will be used extensively; the blocks that make up a loop will be assigned terminal characteristics and analyzed. Computer analysis will be extensively used. Graphic results will be shown for each configuration, and it will become clear why certain arrangements are preferred and what level of complexity is required to produce the desired output.

10.1 Receiver-Synthesizer Relationship

Overall frequency scheme

Synthesizer design requires coordination with the receiver or transmitter frequency schemes, which should be created together to optimize important trade-offs in the choice of all frequencies. Figure 10.1 shows a receiver scheme that will be used as a design example throughout this chapter. This design is typical of a modern, high-performance upconverter type like those discussed in Chapter 4. The first IF at $f_1 = 109.35$ MHz is commonly used and is high enough to reduce the first-mixer spurious responses but low enough that crystal filters at f_1 are available. The second IF at $f_2 = 10.7$ MHz gives the designer a wide choice of available crystal filters, and f_3 at 455 kHz makes a wide selection of mechanical, ceramic, or crystal filters usable.

The synthesizer is required to produce frequencies f_4, f_5, f_6, and the 455-kHz product detector injection f_7 with certain specifications on the level, frequency stability, purity, and tuning time. If the receiver interference rejection specifications require a narrow filter for f_1, then f_4 must provide all the frequency variation with f_5 fixed. If a wider IF at f_1 can be permitted, the synthesizer design can provide coarse variations in f_4 and fine variations in f_5; here we assume all of the variations must be in f_4.

Most SSB synthesizers use a PLL to produce f_4, for its built-in filtering, economy, flexibility, and producibility rather than direct analog synthesis. This chapter initially features all PLLs, but later includes other digital techniques.

Mixer injection choices

The choice of high- or low-side receiver first-mixer injection involves both receiver and synthesizer design trade-offs that need to be made before considering the synthesizer in more detail. Assuming that a PLL will be used to provide f_4 (Figure 10.1), the choice will depend upon the frequency range of the VCO, the divider and frequency control capabilities, the phase noise allowed for f_4, and the spurious levels allowed in the first receiver mixer.

The frequency ratio of the VCO is less for high-side injection. Lower VCO tuning ratios mean easier VCO design, assuming that the frequency does

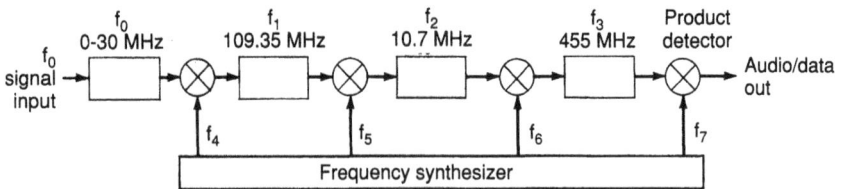

Figure 10.1 Relationship of the receiver to the synthesizer.

not go too high for the available circuit components. High-side injection with a 1.27 ratio (139.35/109.35) is preferred when only this factor is considered, although low-side injection with a 1.38 ratio is also possible. Ratios become difficult in the 1.5 to 1.75 range and impossible above about 2.0. To cover higher ratios, it is necessary to band switch either whole oscillators or the frequency-determining components of one oscillator.

Phase noise originating in the VCO is usually lower at the high-frequency end of any given VCO tuning range, partly because of the higher varactor diode Q at higher tuning voltages. Low-side injection would provide lowest VCO noise near 109.35 MHz (receive frequencies near 0 Hz), while high-side injection would provide lower VCO noise near 139.35 MHz (receive frequencies near 30 MHz). However, the lower tuning ratio required with high-side injection might counteract this effect since the varactor diode tuning voltage would not have to go as low.

The divider upper frequency limit and the range and direction of counts available may also help determine whether f_4 is above or below 109.35 MHz. For high-side injection using a PLL, the divide ratio N will increase as receive frequency increases, while for low-side injection N will decrease. With microprocessor control, either choice may be acceptable. The need to control N with mechanical switches or to retrofit the synthesizer into older equipment may influence the decision, since dividers using both variable-modulus prescalers and counters are easier to implement as downcounters, with N increasing as channel frequency increases.

In the first receiver mixer, high-side injected difference mixing typically results in lower levels of spurious responses (see Chapter 4). When very low VCO phase noise at low receiver input frequencies is needed, and the first receiver-mixer spurious responses are low enough, low-side injection would be the proper choice. For the design example in this chapter we have chosen high-side injection with f_4 ranging from 109.35 to 139.35 MHz.

Synthesizer configurations

Single-sideband synthesizers may use one or many PLLs, depending upon the required noise, spurious signals, and settling time requirements. Small frequency increments are needed for SSB (100 Hz or less), which result in a correspondingly low reference frequency for a single PLL. A loop with such a low reference would have a bandwidth of 5 Hz or less, and would not reject noise, mechanical vibration, magnetic field, and power supply ripple components in the AF range. Also, SSB is more sensitive to phase noise than conventional AM; multiple-loop methods must therefore be used.

Section 10.2 describes loops with improved performance, which generate coarse (100-kHz) steps in the "output" loop providing f_4 (the variable injection), and with fine frequency steps mixed into the loop from a separate source. Section 10.3 describes the highest-performance two-loop schemes for generating coarse steps. Section 10.4 then describes methods of generating fine frequency increments, including special techniques to obtain wider band with small frequency steps in single or dual loops.

10.2 Generating Coarse Frequency Steps in the Output Loop

The output loops providing f_4 in these examples will use a 100-kHz reference to allow the loop bandwidth to cover audio frequencies. Fine frequency steps are supplied through a loop mixer. The basic components (Figure 10.2) will be discussed in sufficient detail for analytical purposes, and the same principles will then apply to later designs described in this chapter.

Basic PLL design

The VCO tuning curve (frequency versus control voltage V_c) is basic to loop design. A VCO and its tuning curve are shown in Figure 10.3. The 109.35- to 139.35-MHz operating range is tuned with a V_c ranging from 5 to 13 Vdc (with a maximum VCO range of 105 to 142 MHz and a maximum V_c range of 3.0 to 13.5 Vdc). The slope at any point is designated as K_v (in megahertz/volt), which decreases with increasing frequency. It is especially important to know the extremes of K_v, as well as the extreme frequency limits past the desired operating range. The latter are important because under transient conditions when the loop is tuning, V_c could go above 13 Vdc or below 5 Vdc. This could create problems if, for example, the loop divider ceased operation above 150 MHz or the VCO stopped running below $V_c = 2$ Vdc. Also, it is necessary to know the tuning curve over the temperature range to be encountered. K_v plays a key role in many loop characteristics and must be a stable, repeatable parameter.

The PLL shown in Figure 10.2 has a mixer between the VCO and the variable divider. The mixer allows introduction of fine frequency increments

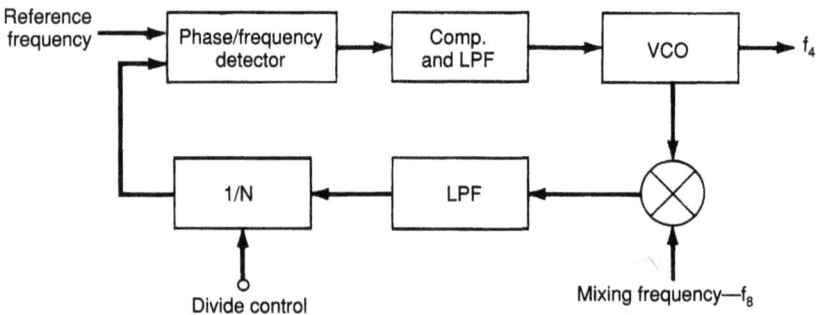

Figure 10.2 Basic elements of a PLL to product f_4.

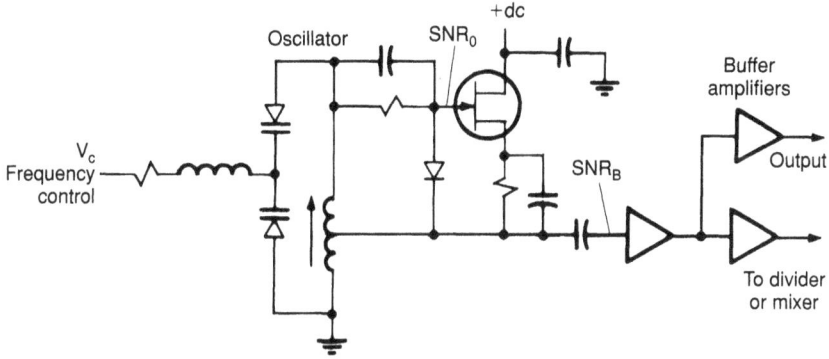

A

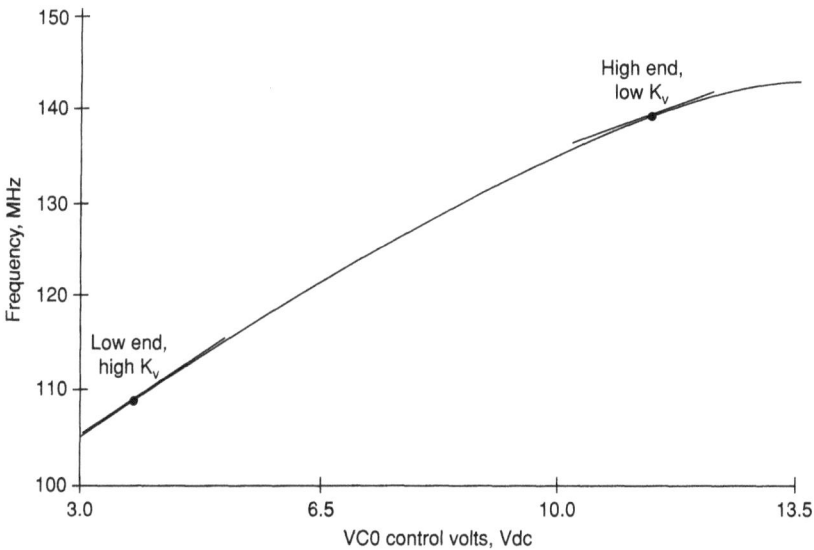

B

Figure 10.3 Field-effect transistor VCO: (A) circuit; (B) tuning curve.

through the mixing frequency f_8, and can reduce the divider input frequency and count ratio. A lowpass filter at the mixer output reduces interference from the mixer input signals. An IF amplifier would be used, if necessary, to drive the divider.

The frequency divider divides the VCO or mixer output frequency by some integer value N, resulting in f_D, which is applied to one of the phase detector inputs. The divider usually consists of a digital programmable counter, with a variable-modulus prescaler to extend the upper frequency limit.

The phase detector produces a voltage related to the phase difference between the reference input f_{R1} and the divider output f_D. The digital phase detectors discussed in this chapter develop a variable duty-ratio rectangular-wave error signal composed of both ac (v_c) and dc (V_c) components. Phase frequency detectors are usually used to ensure rapid locking from an unlocked condition.

Two types of phase/frequency detectors are shown in Figure 10-4. Because of some confusion in the literature about phase detector terminology (such as "edge-triggered," etc.), we will adopt the following functional definitions, regardless of the hardware used. A *phase/frequency detector* (discriminator) is any phase detector having a strong off-frequency sense, to ensure rapid capture. All phase detectors used in this chapter are of the phase/frequency type.

A *proportional phase detector* has a phase range of 2π radians, as shown in Figure 10-4 [6, 7]. The phase difference between the inputs (horizontal axis) causes the output rectangular wave to vary its duty ratio. This wave is filtered and becomes the VCO control voltage (vertical axis), so that the phase difference is roughly proportional to the VCO frequency. The slope of the phase-sensitive region is K_p (in volts/radian). If this phase difference exceeds the 2π range, causing a loss-of-lock (LOL) condition, the additional frequency detection provides a strong error signal to move the VCO in the proper direction until phase lock occurs. Frequency detection in the proportional detector is shown by the horizontal lines at either limit, instead of a repeating sawtooth function. If the detector output swing is insufficient for the required V_c range, it may be amplified by a switching level shifter which has less noise than an analog op-amp.

A proportional detector is intended for use in a type 1 PLL, which means that there is only one integrator (the VCO frequency-to-phase conversion) in

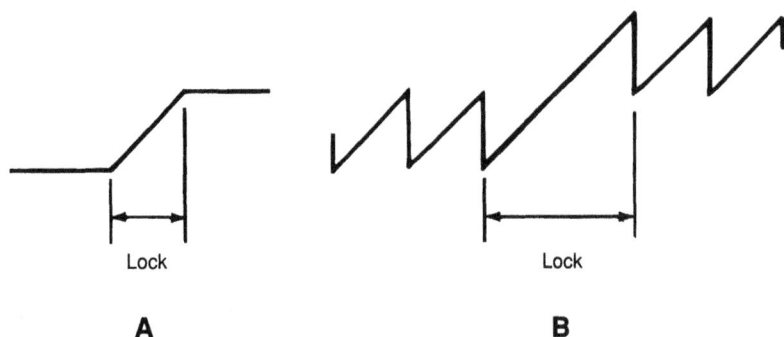

Figure 10.4 Phase/frequency detector average output voltage versus phase difference: (A) proportional 2π lock range; (B) differential 4π lock range.

the open-loop transfer function. When used without lead-lag compensation, the proportional detector results in the highest tuning speed and minimum overshoot, but this requires careful loop gain adjustment, with prepositioning of the VCO control voltage if necessary.

A *differential phase/frequency detector* has a phase range of 4π radians, as shown in Figure 10.4B [8]. The phase difference results in a voltage that varies on either side of some zero-phase center voltage. Frequency detection occurs because of the net dc shift in the sawtooth wave. An ac component exists when the phase difference is nonzero, but nearly disappears at the center operating point. This reduces the reference frequency filtering requirements. A differential phase/frequency detector is intended for use with a compensated integrator circuit, resulting in a type 2 PLL.

The type 2 loop has higher gain at lower frequencies because of the added integrator, which tends to reduce noise at these frequencies, to reduce posttuning drift due to component instabilities, and to tune more rapidly than a lead-lag-compensated type 1 loop. But it will tune more slowly than an uncompensated type 1 loop and may have more noise at some Fourier frequencies (Δf from carrier) because of the use of an op-amp integrator.

Type 2 loops will generally be used in our examples because of the reduced reference filtering requirements and the general availability of differential phase detectors as integrated circuits. Other types of phase detectors or combinations of types have been used, but the two discussed here are found most often in SSB synthesizers.

The lowpass filter plays a crucial role in loop stability, dynamics, and spectral purity of f_4. Its purpose is to reduce the loop reference frequency component, its harmonics, and high-frequency random noise on the VCO control voltage without introducing enough phase shift to cause loop instability or poor dynamic characteristics. The term *lowpass filter* used here does not include any compensating or integrator networks, which, of course, may have some additional filtering effect. Lowpass filter types include passive and active RC filters, which generally have the largest phase shift, high-order elliptic LC filters with less phase shift, and sample-and-hold (S/H) types having the least phase shift. Sample-and-hold filters are most practical at reference frequencies below 5 kHz, and can be used with either proportional or differential phase/frequency detectors. Active RC filters using wideband op-amps are practical up to 100 kHz, where the attenuation requirement is not too severe. High-order elliptic LC filters are most useful at 100 kHz and above, where the inductance values are smaller.

Loop compensation circuits (Figure 10.5) are also important. A type 1 uncompensated loop is partially shown in Figure 10.5A. A switching limiter amplifier is used to provide the required V_c range. If a type 1 loop is unstable or underdamped because of insufficient phase margin, it may be stabilized with an RC lead-lag network (Figure 10.5B). A type 2 loop, by

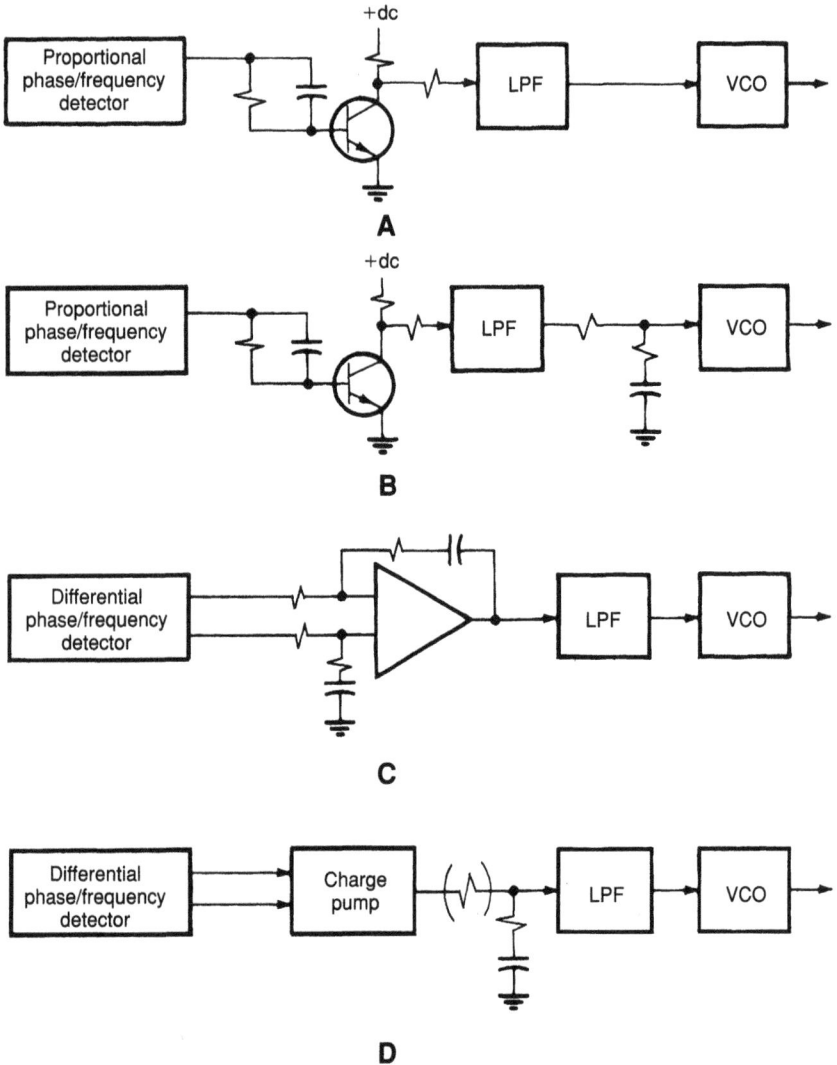

Figure 10.5 Compensation circuits: (A) uncompensated loop; (B) lead-lag network; (C) dual-input integrator; (D) charge-pump integrator.

definition, requires an integrator compensated for stability (Figure 10.5C). A *charge pump* (Figure 10.5D) combines the two phase detector outputs digitally to feed a single-ended network whose dynamic characteristics may differ from that shown in Figure 10.5C in the amount of lag (or delay) contributed to the loop, depending upon whether the charge pump has a voltage or current output. Linearity in the zero-phase center region is more difficult to achieve with digital combining. The op-amp integrator is used in our type 2 loop examples to ensure greater linearity through zero phase, and consequently better control of loop dynamics.

Direct- and reverse-count mixer loops with variable dividers

As shown in Figure 10.6 and Table 10.1, the VCO output is combined with a mixing frequency that lowers the divider input frequency (or loop IF), resulting in a lower division ratio N compared to a loop with no mixer. The lower N ratio reduces reference noise multiplication, and a wide loop bandwidth helps suppress VCO noise. The 100-kHz reference results in 100-kHz steps. Finer steps enter through the mixing frequency.

If the mixing frequency is lower than the VCO frequency (low-side injection), the VCO frequency rises with increasing N, resulting in a direct-count loop. If it is higher than the VCO frequency (high-side injection), the VCO frequency is lowered by raising N, resulting in a reverse-count loop.

The usefulness of the reverse-count design lies in its ability to minimize loop bandwidth changes as the loop is tuned. Referring again to the VCO

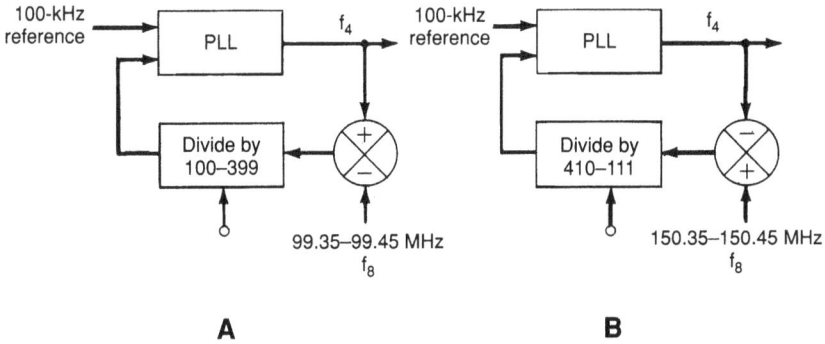

Figure 10.6 Loops with mixers and 100-kHz reference signals: (A) direct count; (B) reverse count.

TABLE 10.1 Loop Parameters for Figure 10.6

Parameters	Direct count	Reverse count
Reference frequency, kHz	100	100
VCO frequency f_4 range, MHz	109.35–139.35	109.35–139.35
Mixing frequency f_8 range, MHz	99.35–99.45	150.35–150.45
N division range	100–399	410–111
Loop IF range, MHz	10.0–39.9	41.0–11.1
Phase-frequency detector	Differential	Differential
Loop type	2	2
LPF, 45° at, kHz	10	10
Integrator lead corner, kHz	1	2.5
Integrator gain	0.5	0.7

characteristic, Figure 10.3B, the slope of the VCO curve (K_v) decreases as frequency increases. Since loop bandwidth is proportional to K_v/N, the bandwidth varies less in the reverse-count loop because K_v and N vary in the same direction. With proper VCO design and choice of IF tuning ratio (that is, maximum-to-minimum ratio of N), it is possible to achieve nearly constant bandwidth over a wide VCO range, which then allows a wider stable bandwidth, as will be shown in the following paragraphs. This is especially important for wideband, fast-hop synthesizers where every possible microsecond must be removed from the tuning time. Other schemes are available to optimize bandwidth over a wide tuning range, but these usually involve switching integrator component values and can add considerable complexity and more noise to the loop. Further comparisons between these loops will be made in the following analysis.

Frequency domain analysis

The next step after the VCO is the lowpass filter design. A Cauer (elliptic) lowpass filter is often used to minimize phase shift within the loop bandwidth, with a large suppression of reference frequency sidebands, and can be easily built with small RF inductors and capacitors, with the first notch at 100 kHz and another notch at 200 kHz to help attenuate the first two harmonics. When the lowpass filter phase characteristics are known, the open-loop response can be obtained and adjusted for stability and eventually for transient performance.

The open-loop responses of the direct- and reverse-count loops are shown in Figures 10.7A and 10.7B (dashed lines). In the direct-count loop, the open-loop bandwidth (at 0-dB gain) varies considerably from the high to low ends of the VCO range, because K_v and N change in opposite directions, gain being proportional to K_v/N. The phase rises because of the integrator lead compensation, and then falls off rapidly because of the lowpass filter (total loop phase shift is shown including the 180° inversion for negative feedback). The phase margin (phase at 0-dB gain) falls within the peak of phase. The compensation parameters were chosen for transient performance, to be described later. The smaller variations in the reverse-count loop are quite noticeable.

The open-loop response is the product of all the functions around the loop, including the phase detector gain K_p (in volts/radian), the VCO gain K_v (in hertz/volt), the divide function $1/N$, the frequency-to-phase conversion $1/s$ (the La Place integration), the lowpass filter, and compensation, if used. The only time delay is the logic propagation delay, which is negligible in this case. Transport lag is not inherent in a PLL [9, 10], although it is often assumed in some charge pumps (Figure 10.5C) and S/H filters. In these cases, an exact response is more accurate than an assumed delay [1, sec 5.3].

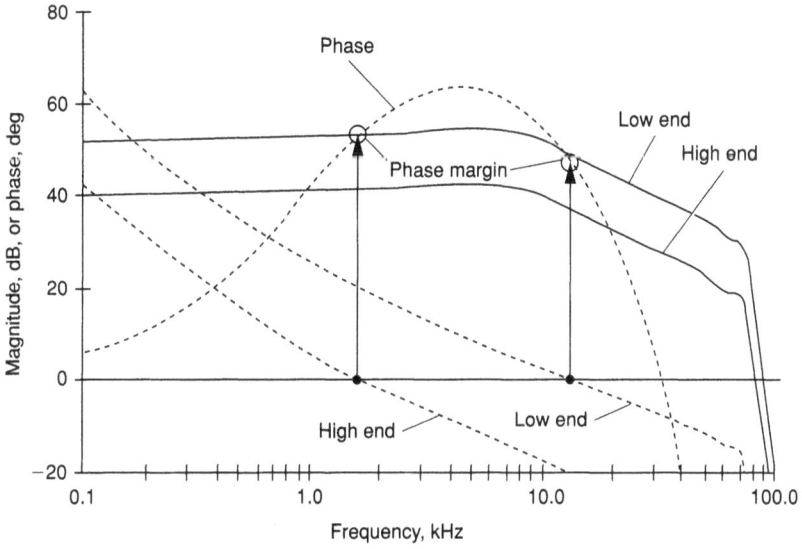

A

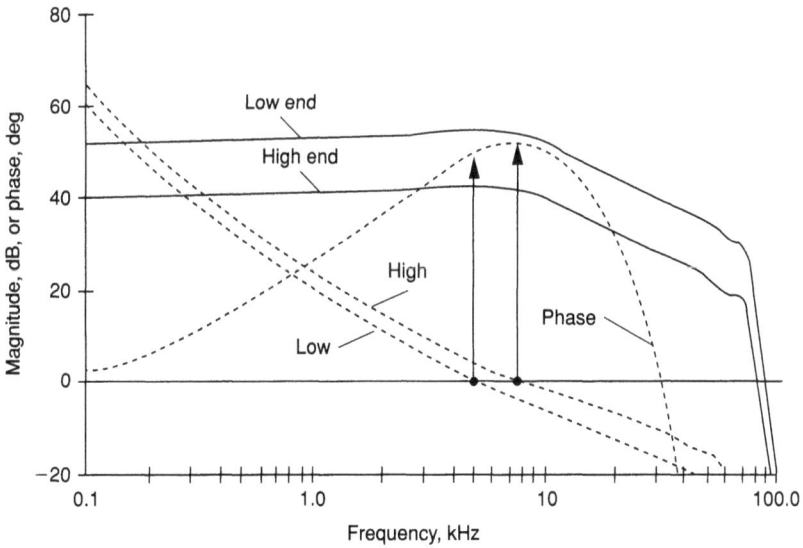

B

Figure 10.7 Loop frequency responses for Figure 10.6: (A) direct count; (B) reverse count. Solid line—closed-loop response; dashed line—open-loop response.

The closed-loop responses in Figure 10.7 also show that the reverse-count loop has less bandwidth variation, actually overcompensated with the high end of the band having a slightly wider bandwidth. The sharp cutoff of the Cauer lowpass filter is seen at 70 kHz.

Time domain analysis

The settling time variations in the two approaches are even more dramatic (Figure 10.8). Tuning is from 109.35 to 139.35 MHz and back to 109.35 MHz in two periods of 3 ms each. The direct-loop compensation parameters were optimized for similar settling times in both directions.

Transient phase error (Figure 10.8B) is the phase difference between the slewing VCO and a steady-state signal with the same phase as the final value of the VCO phase. For clarity, plotting is omitted for very large phase errors. Phase error is easily measured by mixing the VCO output with a phase-stable signal generator, both synchronized to the same frequency standard. The computer analysis includes nonlinear VCO gain, op-amp slew rate and voltage limiting, and phase detector characteristics. Not included are the effects of posttuning drift caused by some types of capacitors and varactor diodes having charge penetration effects. Varactor diodes may also cause posttuning drift because of thermal variations as their dissipation changes with frequency.

The direct-count loop takes almost 2 ms to reach 1 radian of error, while the reverse-count loop only takes 0.7 ms.

Instantaneous frequency error (Figure 10.9) is computed versus time. The downward points are transitions through zero frequency error due to the slightly underdamped loop. Although an instantaneous frequency settling time is often specified, it is more difficult to measure, being the derivative of the phase error measurement. Where rapid settling time is required, as in frequency-hopping systems, the phase-settling time specification is usually more appropriate. Since a rapidly changing frequency has a transient spread to its spectrum, the signal is useful when the phase is stable, even before the instantaneous frequency reaches the system steady-state bandwidth. Since this frequency-error analysis does not represent a physical reality, it should not be used as a firm design goal, which could result in costly overdesign, but rather as a most useful graphical analytical tool for optimizing loop settling time as long as the final evaluation is done on a phase-settling basis [11].

The curves are accurate enough to speed up the early stages in the design process and give the designer added insight about the interrelationships of the basic loop parameters.

The effect of compensation on transient performance is not what one would expect. Its usual purpose is to lower the loop bandwidth for stability or noise reasons (discussed under noise analysis). Lagging, for either type

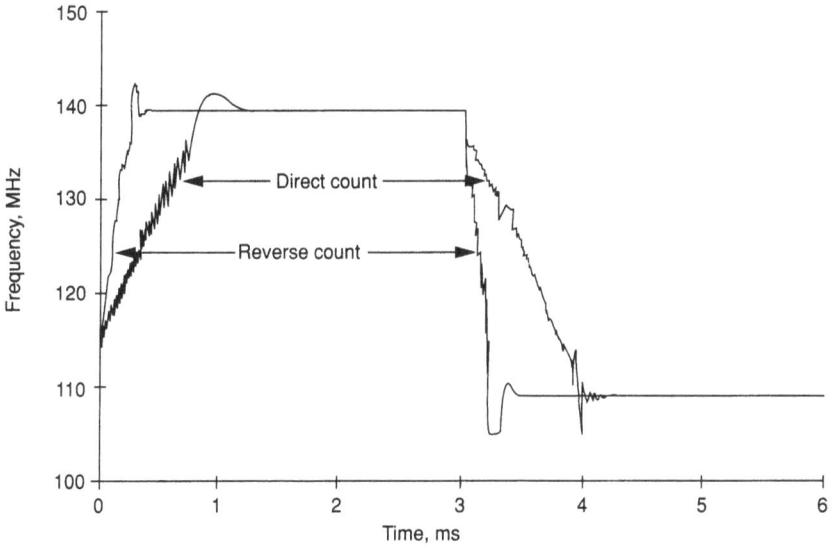

A

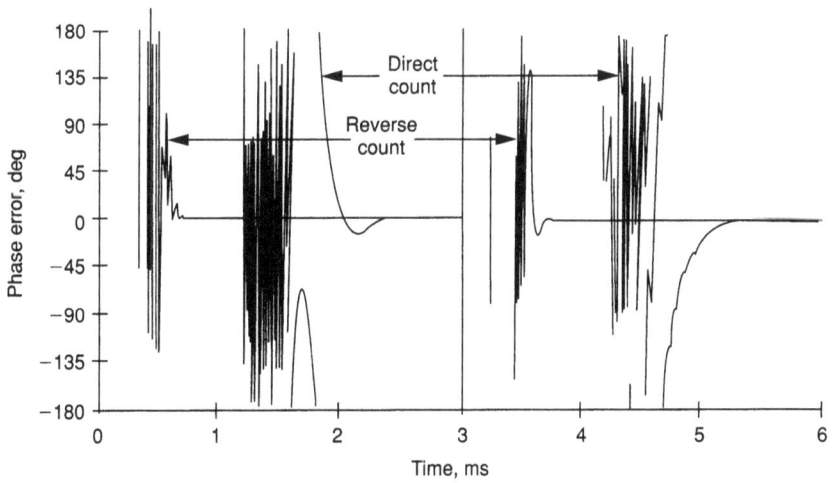

B

Figure 10.8 Transient responses for Figure 10.6: (A) absolute frequency; (B) phase error. Switching occurs at 0 and 3 ms.

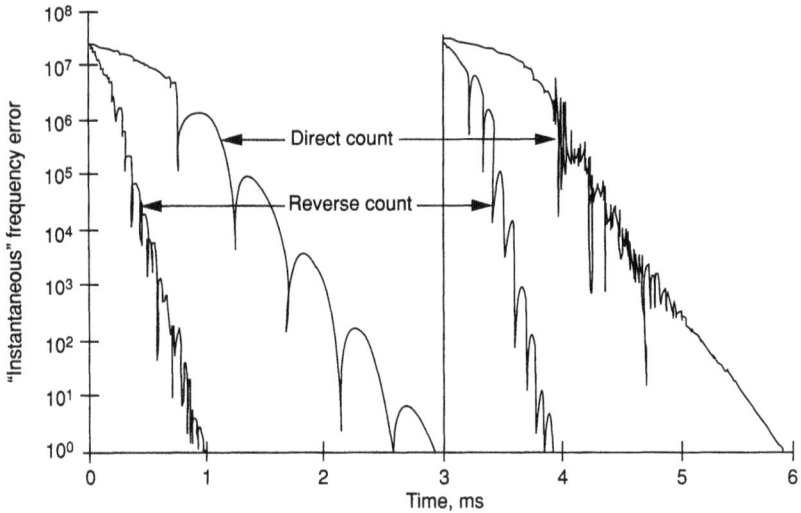

Figure 10.9 "Instantaneous" frequency error for Figure 10.6.

1 or type 2 loops, not only reduces the bandwidth with a corresponding increase in settling time, but also causes the loop to lose lock for large-frequency jumps, which further increases the settling time. Therefore, the total settling time is increased approximately as the square of the bandwidth reduction. Prepositioning of the VCO control voltage results in faster settling, more because of the gain reduction without lagging than the prepositioning itself.

In our example, frequency overshoot of the VCO driving the first receiver mixer can degrade the receiver performance whenever a large step is taken in a downward direction, especially in a frequency-hopping system. If the injection frequency passes through the receiver first IF, it leaks through the mixer with enough power to excite the receiver IF filter, creating a large momentary signal and possibly upsetting the AGC system. If the signal frequency hopped from 30 to 2 MHz the first receiver mixer injection frequency would change from 139.35 to 111.35 MHz, just 2 MHz short of the IF at 109.35 MHz. Compensated loops inherently overshoot unless extremely overdamped. In fact, a type 2 second-order loop with a damping factor of 1 allows a 6% overshoot, which in the above case would be just short of reaching the IF. Since most loops designed for speed are somewhat underdamped and the lowpass filter phase shift tends to increase the overshoot, this can be a significant design consideration. Even uncompensated type 1 loops must be carefully designed to control overshoot.

Figure 10.10 shows an enlarged overshoot area for the reverse-count loop. The original design overshoots past the receiver IF, hitting it twice (curve A). This occurs even for smaller frequency jumps (curve B). Narrowing the bandwidth and increasing the damping of the loop (curve C) reduces the overshoot at a considerable increase in settling time.

Prepositioning the VCO control voltage can greatly reduce the overshoot while maintaining faster settling (curve D). This requires precise voltage control. Prepositioning after the lowpass filter results in even less overshoot, but some filtering may be necessary to prevent high-frequency noise from entering the VCO via the prepositioning circuit.

Another solution is windowing or blanking (between the oscillator and the receiver first IF), which may also be used for hopping spectrum control. However, very high attenuation is required to eliminate the effect of overshoot.

Loop noise sources

Noise levels in this chapter are normalized to a 1-Hz bandwidth. The various noise sources in a PLL are defined as shown in Figure 10.11, with numerical examples for the reverse-count loop (Figure 10.6B). "Fourier frequency" is frequency-offset from the carrier.

Reference oscillator noise. The 5-MHz frequency standard will be assumed to have a typical noise level of –140 dBc, which is divided to 100 kHz. Theoretically the noise would be reduced by $20 \log_{50}$ (35 dB) to –174 dBc, but CMOS dividers have a higher output noise level at 100 kHz, at least –160 dBc.

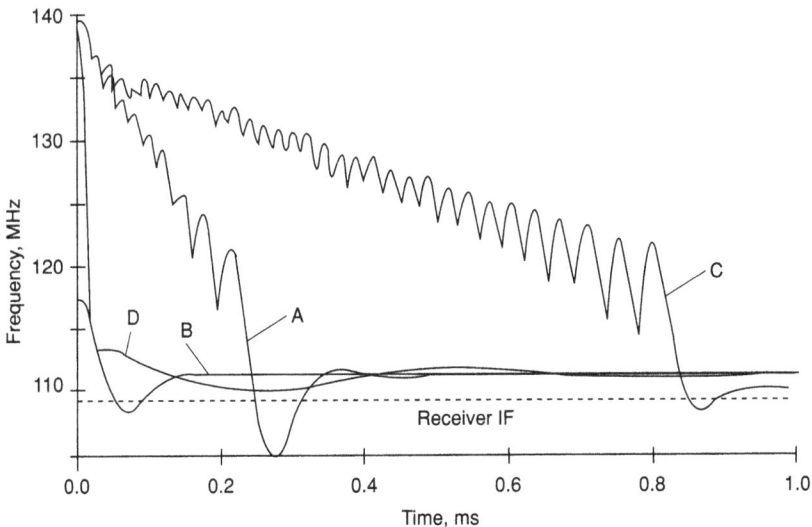

Figure 10.10 Synthesizer overshoot across receiver IF.

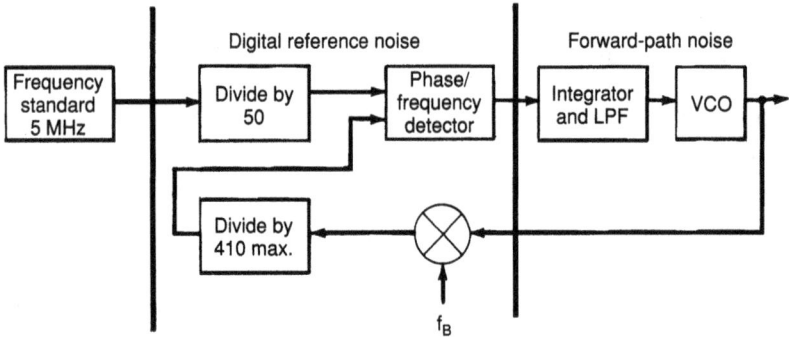

Figure 10.11 Phase-lock loop noise sources.

Other reference noises. Other sources combine with the reference to make the equivalent reference noise to the loop, which includes the reference oscillator and divider, mixer, IF amplifier, loop divider, reference divider, and phase detector output. The closed-loop effect on all reference noise is that of a lowpass filter. The loop multiplies this by 52 dB to –108 dBc at the low end of the VCO range (Figure 10.12 [solid line]). After the 5-kHz loop bandwidth, the loop lowpass effect rolls the noise off gradually until at 70 kHz the sharp-cutoff lowpass filter takes over.

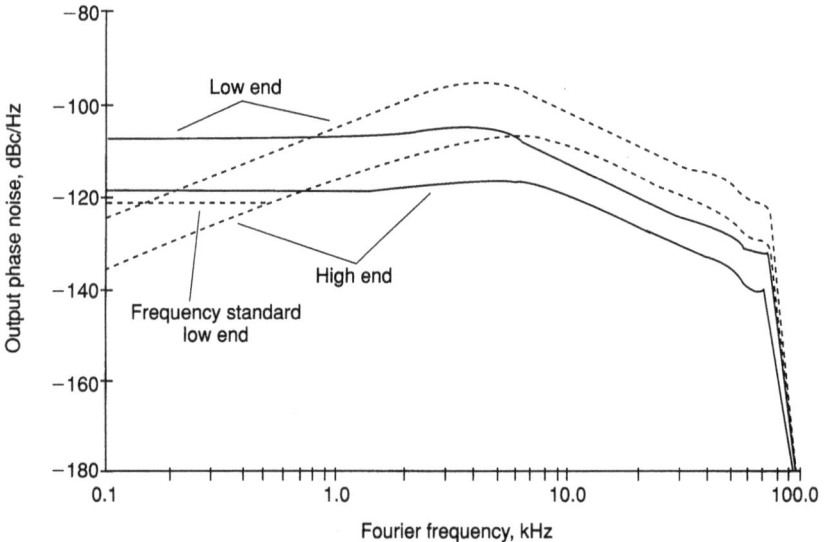

Figure 10.12 Noise sources (except the VCO) for loop in Figure 10.6B. Solid line—digital reference noise; dashed line—forward-path noise before LPF and before the integrator.

Forward-path noise. Voltage sources from the phase detector output that are sent to the VCO add to the VCO phase noise itself. The loop response is basically highpass for voltage noise modulating the VCO creating FM sidebands. However, phase noise has a 6 dB/octave slope compared to FM, so the loop has a lowpass effect for that phase noise output that is caused by noise voltage on the VCO control line.

A type 2 loop integrator creates an additional highpass effect to forward-path noise voltages after the integrator, so that the low-frequency noise has greater attenuation. An op-amp after the integrator has a spectrum as shown in Figure 10.12 (dashed lines). A total op-amp noise of 20 nV was assumed, consisting of 4 nV from a 5534 op-amp, plus resistor and current noise. Note that the noise of a single op-amp is noticeably higher than the digital reference noise, which clearly indicates the need for lower noise op-amps or linearized high-voltage charge pumps.

The open-loop VCO noise is plotted in Figure 10.13 (solid lines). The open-loop VCO noise is predominantly phase noise close to the carrier, falling off with increasing Δf until, near 1 MHz, it reaches the uncorrelated additive noise shelf of the broadband amplifiers. The following equation (derived from reference 12) is useful to describe the basic form of VCO noise-to-signal ratio (even though it omits shot noise and varactor diode parametric effects):

$$\text{VCO NSR} = 10 \log \left\{ \text{NSR}_b + \text{NSR}_o \left[\left(\frac{f_o}{2Q\Delta f} \right)^2 + \frac{f_a}{\Delta f} \right] \right\} \text{dBc} \qquad (10.1)$$

where Q = loaded Q (including all circuit losses, but no Q multiplication)
NSR_b = noise-to-signal power ratio at the first buffer amplifier input (Figure 10.3A)
NSR_o = noise-to-signal power ratio at oscillator device input (Figure 10.3A)
f_o = oscillator frequency
Δf = frequency separation (Fourier frequency) from f_o
f_a = upper frequency corner of extra 3 dB/octave slope (typically about 1000 Hz)

Note that the signal-to-noise ratios are inverted to simplify adding the two noise powers. Above $\Delta f = f_o/2Q$ (700 kHz in Figure 10.13, assuming equal signal-to-noise ratios for the buffer and oscillator), the noise flattens out to a shelf of −162 dBc/Hz, which depends on the buffer amplifier noise figure and the signal power at its input. The active-device noise figure in the VCO depends upon its input source impedance, which varies rapidly on either side of the carrier. Therefore, it is difficult to optimize oscillator operating conditions to minimize the noise figure as is done in low-noise amplifier designs. At frequencies lower than noise shelf corner, the sidebands become

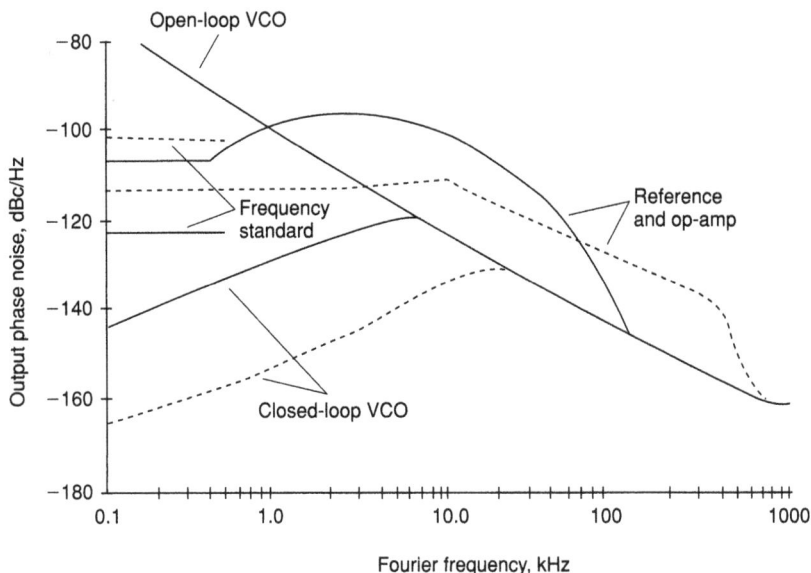

Figure 10.13 VCO noise and total noise in loops of Figures 10.6B and 10.15. Solid line—100-kHz reference; dashed line—1-MHz reference.

correlated phase noise with a slope of –20 dB/decade. The slope increases to –30 dB/decade below a corner frequency f_a because of additional low-frequency noise sources [13, Chapter 5]. f_a is shown at a typical value of 1 kHz. Additional noise sources with steeper slopes at very low frequencies are not shown here.

Parametric effects in the varactor tuning diodes cause a noise level that is greater than their effect on tank Q would predict. This varactor noise and tuned circuit Q are the major factors in determining phase noise. There is a compromise between the noise shelf and close-in noise, since more power drawn from the tank reduces the loaded Q. To achieve a high carrier-to-noise ratio in the VCO, the buffer must have a low noise figure and a large signal-handling ability, and the oscillator should have a high level of available power in its LC tank.

The closed-loop VCO noise is also shown in Figure 10.13, where the high-pass filtering effect of the loop decreases noise at lower frequencies so that it would be about level with a type 1 loop. The type 2 loop integrator causes an additional 6 dB/octave drop at some of the low-frequency range. Above the loop bandwidth, the VCO noise is unaffected. The solid line is for a 100-kHz reference and the dashed line is for a 1-MHz reference.

Total closed-loop noise is also shown in Figure 10.13. The lines marked "Ref & op-amp" show their addition to the VCO noise, which is the total noise except for the lowpass filter rolloff at high frequencies. For the 100-

khz reference, the solid lines below 700 Hz show that the multiplied digital reference noise predominates, with the op-amp noise strongest between 1 and 50 kHz and the VCO taking over above 100 kHz. With a 1-MHz reference, the dashed lines indicate a lower reference noise because of the lower multiplication ratio, extending past 7 kHz, where the op-amp noise takes over, but still less than with the lower reference frequency. Eventually, the VCO noise takes over as described before.

This example illustrates that a low-noise VCO does not guarantee a low-noise output over much of the Δf range. Even more important are the digital reference noise, the size of the loop multiplication ratio N, and the op-amp noise. Any additional noise from the loop-mixing frequency must be considered if it is higher than the closed-loop output noise shown. This loop could be improved from a noise standpoint by reducing reference noise sources and op-amp noise, and changing the overall frequency scheme to lower N.

If the tuning speed can be degraded, noise levels can be reduced by narrowing the bandwidth to 1500 Hz, where the op-amp noise would peak at –101 dBc and roll off above 1500 Hz. The VCO contribution would move up at 1500 Hz, but the VCO and reference noise combined would still be less than the op-amp contribution. The large noise bulge to about 24 dB above the VCO curve in the 5-kHz bandwidth design would be cut down to about 8 dB above the VCO curve with the narrower 1500-Hz bandwidth.

A type 1 loop with a low-noise switching amplifier and passive lowpass filter (as in Figure 10.5A) can reduce the forward-path noise to almost the reference level, while allowing even faster tuning time than the example shown.

Loop mixer spurious signals

While a loop mixer has the advantages of lowering N and providing a convenient means of combining loops, the biggest disadvantage is the presence of unwanted discrete spurious signals in the mixer output. Since the VCO, which provides the LO input to the mixer, must vary over a 30-MHz range for this example, the ratio of mixer input frequencies varies enough to allow spurious signals as low as fifth order to cross over the IF output; seventh- and ninth-order crossovers are also present.

The fifth-order spurious signal crossovers at the mixer output are illustrated in Figure 10.14A with the synthesizer IF at 33.1 and 33.2 MHz and the mixing frequency varied from 150.35 to 150.45 MHz.

This corresponds to the receiver bands 23.1 to 23.2 and 23.2 to 23.3 MHz. Because the synthesizer IF moves in 0.1-MHz steps, the $3f_8 - 2f_4$ fifth-order spurious signal mixer output never falls right on the IF signal; it ranges from 50 to 150 kHz above when the IF is at 33.1 MHz to 250 to 150 kHz below when the IF is at 33.2 MHz. However, these spurious signals beat just as ef-

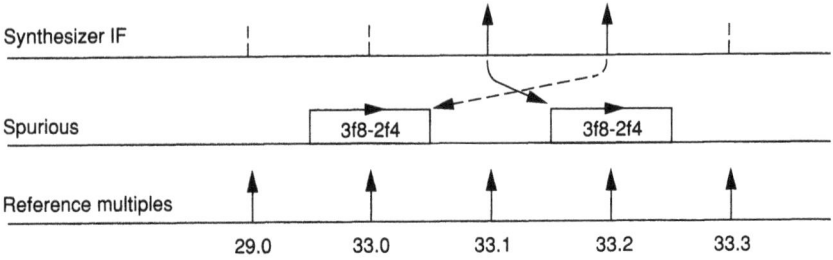

A

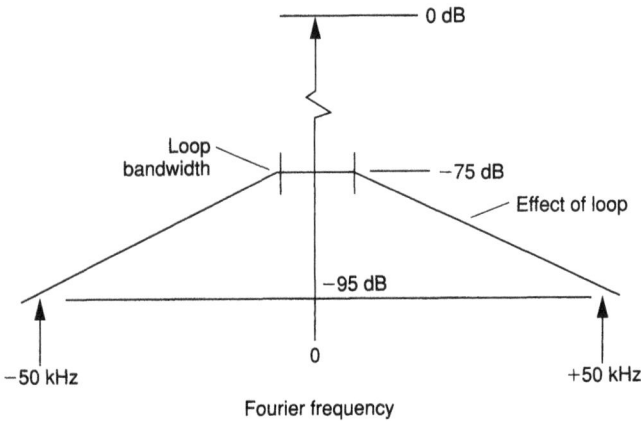

B

Figure 10.14 Mixer spurious signals in loop of Figure 10.6B: (A) spurious beating with reference harmonics; (B) loop filtering effect on spurious signals.

fectively with other harmonics of the reference, as shown. This occurs in the digital divider mechanism itself [1, pp. 15–23].

Typical fifth-order spurious signal suppression for a properly terminated doubly balanced mixer with a –10-dBm signal at the RF port is 69 dB. A digital divider converts the discrete spurious signal into two FM sidebands, 6 dB lower. (A single sideband represents combined AM and FM, and the AM sidebands are removed by clipping.) The resulting spurious signal at the divider output is $-69 - 6$ dBc/N, or –125 dBc, the beat note ranging from 50 kHz to 0 as the spurious signal crosses over any reference harmonics. Inside the loop bandwidth (Figure 10.14B), the loop multiplies the spurious signal levels by N (20 log 331 = 50 dB) to –75 dBc, the same degree to which they were lowered by the divider, except for the 6-dB SSB-to-DSB conversion loss due to the divider. Outside the loop bandwidth, the spurious signals are attenuated by the lowpass effect of the loop.

Other spurious production mechanisms exist, such as leakage from either end of the divider to the other, leakage around the IF filter, and leakage between loops, both conducted and radiated. The degree to which the spurious signals are actually seen in the synthesizer output depends on the close-in phase noise performance of the whole synthesizer. In the design example above, –75-dBc spurious signals at ±5-kHz separation due to the output loop mixer would be below the phase noise of most synthesizers in a 3-kHz noise bandwidth.

If low-side receiver mixer injection (below 109.35 MHz) had been chosen, the resulting synthesizer mixer loop with moderate values of N would produce a third-order crossover because of the higher variation in the ratio of mixer input frequencies (lower frequency divided by the higher). In a typical mixer this third-order spurious signal can be held to about –35 dB below the IF output, resulting in a possible –41-dBc spurious signal at the synthesizer output, allowing for mixer compression and converting to DSB. In this example, with a 109.35-MHz first IF, the spurious signal crossover would occur at 5.325 MHz in the HF band. To achieve lower spurious signal levels, the output loop can be used for a purpose other than producing frequency increments, as described in section 10.3.

10.3 Generating Coarse Frequency Steps with Combining Loops

Loop configuration

Figure 10.15 shows a unity gain ($N = 1$) combining a first loop mixed with a second divider loop generating 1-MHz increments. The finer frequency steps are provided via the combining loop's high variable reference from another circuit in the synthesizer. The problem of low-order loop mixer spurious output signals has been eliminated because as loop 2 steps from 105 to 134 MHz, the loop 1 VCO also moves upward, remaining above the loop 2 output by the value of the loop 1 reference. This keeps the ratio of the two mixer inputs nearly constant at approximately 105/109.35 = 0.96, a value where lower-order spurious signal outputs are not troublesome. A combining loop sometimes has an N greater than 1, to alleviate tracking problems, but with a low fixed ratio.

The price paid for this excellent mixer performance is complexity. If the loop 1 output f_4 ever gets below the mixer image frequency (f_2 – variable reference), loop 1 will latch up on this wrong mixer sideband. To prevent this, the loop 1 VCO is coarsely tracked by the control voltage from loop 2, which keeps loop 1 in a safe range. The two VCOs must have similar tuning characteristics so that f_4 and f_8 keep the correct relationship across their respective bands. An alternative is to generate the tracking voltage with a D/A converter and curve shaper driven by the loop 2 frequency control

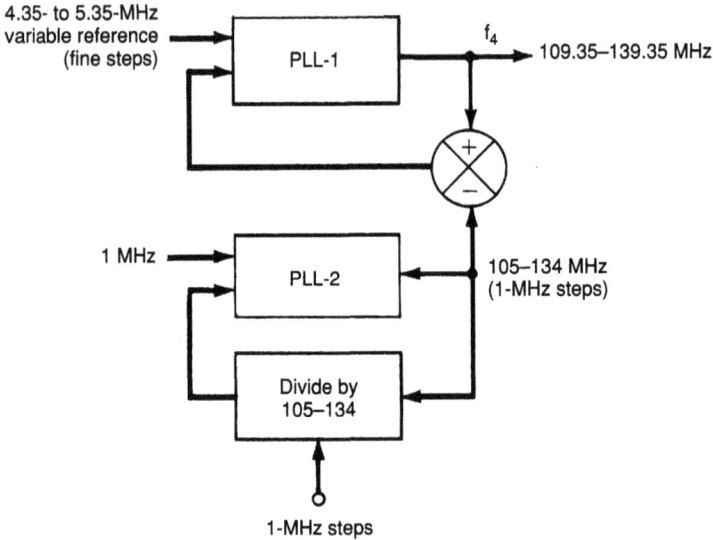

Figure 10.15 Two-loop approach with improved mixer and noise performance.

lines; a programmable read-only memory (PROM) and D/A combination can also be used.

Loop analysis

The loop 2 bandwidth is held narrower than that for loop 1, and will control the transient performance of the combination, which is an order of magnitude faster than the loops with 100-kHz reference described in section 10.2. The Bode plots are similar, at correspondingly higher frequencies. The frequency overshoot problem also exists in the coarse-step loop. However, noise and spurious signal levels do not change proportionally, and will be described in more detail.

Chapter 17, "Software for SSB," describes a computer program that was designed to expedite the detailed design of synthesizer PLLs. The reader is encouraged to become familiar with this versatile program.

Noise analysis

In the two-loop combination shown in Figure 10.15, the loop 1 output noise level will depend primarily on loop 2 because of its higher multiplication of reference noise.

The loop 2 digital reference noise, shown in Figure 10.16, is lower than that in Figure 10.12 for the 100-kHz reference, both derived from the same 5-MHz frequency standard. This implies that broadband noise in digital dividers is not reduced in proportion to the division ratio, perhaps because of the sampling effect [1, pp. 75–81]. Measurements with HCMOS logic, operating from low-noise power sources and good RF construction techniques, indicate that the digital reference noise of –160 dBc at 100 kHz rises to only –157 dBc if divided down to only 1 MHz, and to –145 dBc at 5 MHz. Direct measurement is difficult, and the stated values have mostly been inferred by reducing noise from all other sources and knowing the values of loop parameters accurately. New measuring methods should lead to improved analysis and design techniques in the next few years [14].

The forward-path (op-amp) noise before the lowpass filter is greatly reduced by the wider loop bandwidth. However, the 5-MHz frequency standard typical noise of –140 dBc in this scheme will result in an output noise of –102 dBc, which is higher than the –114-dBc digital reference noise (all levels referred to the VCO output frequency at the high end of the band). Therefore, it is now helpful to use a lower noise frequency standard or a crystal filter after the frequency standard to achieve the overall performance shown in Figure 10.13 (dashed lines).

Thus, the choice of reference frequency has a great effect on synthesizer noise. The narrowband slow-tuning synthesizer with large N piles up noise near the carrier, but uses the narrow bandwidth and lowpass filter to roll off the noise before it contaminates adjacent signals. Lower audio frequencies are contaminated with a high noise level, and the loop is very susceptible to external disturbances such as vibration, magnetic fields, and power supply

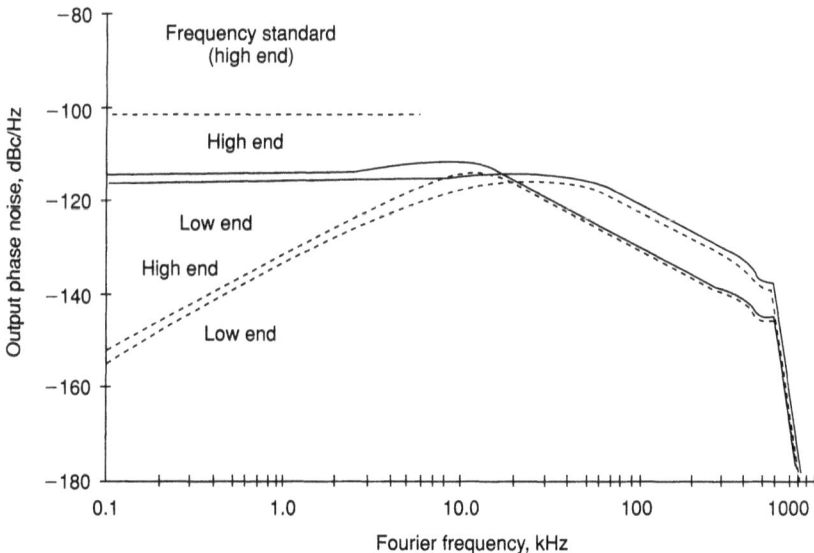

Figure 10.16 Noise sources (except the VCO) for loop 2 in Figure. 10.15. Solid line—digital reference noise; dashed line—forward-path noise before LPF.

ripple. The wideband, low-N synthesizer reduces noise and vibration effects and tunes fast, but requires clean reference sources and low-noise hardware for its benefits to be fully realized.

10.4 Generating Fine Frequency Increments

PLL methods

The generation of fine frequency increments involves the application of the same principles that have been discussed in connection with coarse-increment loops. Reference frequencies must be high to provide loop bandwidths wide enough to meet the required tuning time specification. Division ratios (N) must be held low to keep multiplied digital reference and frequency standard noise low enough to meet the required phase-noise specification within the loop bandwidth of the loop (or loops) that will translate the fine increments to the output.

For small increments such as 1 to 10 Hz, it is necessary to generate steps at higher frequencies and larger increments, which are then divided to obtain the smaller steps; this also reduces the noise contribution from the fine-increment generator. The VCO and divider frequency ranges are critical design considerations. $N = 1$ (or at least low fixed-N) combining loops are frequently required to raise the output of the fine-increment section to a frequency that will interface with the coarse loops. These principles are best illustrated by the following two examples that interface with the designs shown in Figures 10.6B and 10.15, respectively.

Generating fine increments with two loops. In Figure 10.17, loop 3 generates a 250- to 300-MHz signal with 5-kHz steps; this reference is high enough to provide a 250-Hz loop bandwidth and tuning time (end to end) of 20 ms. Though N_3 maximum is 60,000, the output of loop 3 is divided by 500, for a net multiplication of only 60,000/500 = 120. Because the signal at the divide-by-500 output is translated (not multiplied) to the synthesizer output, noise contributions by loop 3 are low. The range of the loop 3 VCO is 50 MHz in order to produce the 100-kHz range required by the loop in Figure 10.6B. The sole function of loop 2 is to translate the 500- to 600-kHz signal to 51.7 to 51.8 MHz, to be used in the final translation to 150.35 to 150.45 MHz.

Also illustrated is the use of fixed frequencies 51.2 MHz (eighth harmonic of the 6.4-MHz frequency standard) and 98.65 MHz, to be used as the second receiver injection (Figure 10.1). Making maximum use of such available signals to minimize complexity is a major and challenging aspect of multiloop synthesizer design. Loop 2 also demonstrates a potential problem with $N = 1$ translating loops. The 51.7- to 51.8-MHz VCO must be stable enough over the required operating temperature range, on an

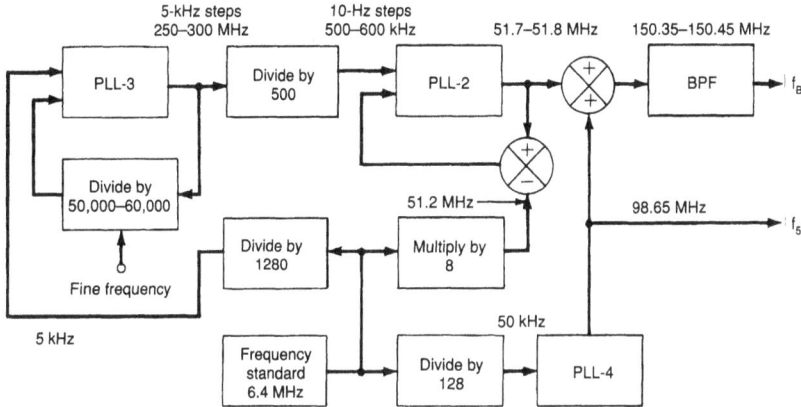

Figure 10.17 Loops to provide 10-Hz steps to the coarse loop in Figure 10.6B and the second receiver injection in Figure 10.1.

open-loop basis, so as not to drift more than 500 kHz below 51.2 MHz, or the loop will latch up past the mixer image frequency. Thus, there is a practical limit on how small the reference frequency can be relative to the VCO frequency; the limit is 0.5 to 1% for a temperature-compensated VCO. Therefore, it may not be possible to make the translation in one loop from the low-frequency range where the finest steps are generated to the required frequency (500 to 600 kHz and 150.35 to 150.45 MHz, in the example). Another reason might be a lack of the required fixed-frequency signals of sufficient stability and spectral purity.

Generating fine increments with three loops. Figure 10.18 shows a method of generating the 4.35- to 5.35-MHz signal with 10-Hz steps to interface with the two coarse-loop approach of Figure 10.15. Loop 3 acts as an $N = 1$ combining loop for signals from loops 4 and 5. Loop 5 generates 10-kHz steps across the 4.26- to 5.25-MHz range, which is translated by loop 3. The net worst-case multiplication of 100-kHz reference noise is only $N_{5, max}/10 = 525/10 = 52.5$ or $20 \log 52.5 = 34$ dB. Loop 5 must track the loop 3 VCO to prevent loop 3 from latching up past the image when loop 5 makes its largest steps. This tracking voltage could also be provided by an outside source, such as a microprocessor, that also controls the division ratios.

The finest 10-Hz steps are provided by loop 4, where the VCO produces 10-kHz steps across the 90- to 100-MHz range. This is divided by 1000 to produce 10-Hz steps over the 90- to 100-kHz range, which acts as a reference for loop 3, where it is converted to the desired 4.35- to 5.35-MHz range. The net worst-case multiplication of 10-kHz reference noise is only 10,000/1000 or 20 dB. Unity gain loop 3 and loop 1 (in Figure 10.15) provide no additional multiplication.

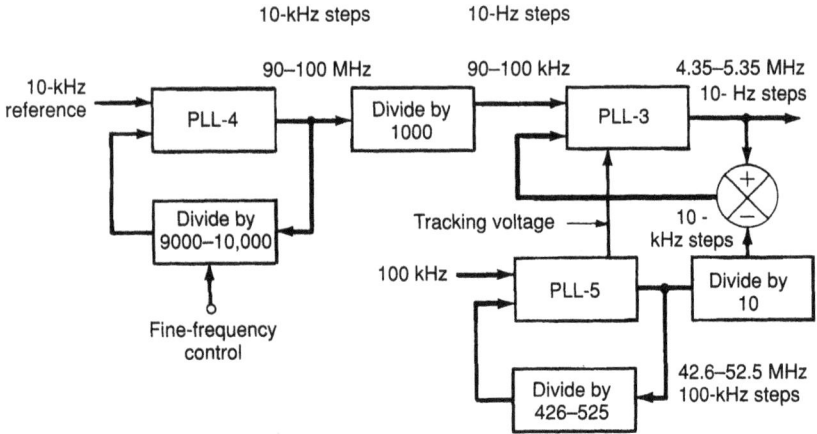

Figure 10.18 Loops to provide 10-Hz steps to the coarse loop in Figure 10.15.

The settling time can be estimated for loops in these frequency ranges as 50 to 200 reference periods (depending on loop type). For loop 4 (the slowest in this scheme), which has a 10-kHz reference, this would be 5 to 20 ms, which is adequate for dial-tuned SSB radios. The division by 1000 will reduce this settling time by a small amount, but not by 1000 because of the exponential characteristic of loop settling. Loops 3 and 5 would be about 10 times faster because of the higher references.

The *difference-loop scheme* shown in Figure 10.19 [15] uses two loops with references that differ by the smallest desired increment, in this case 100 Hz. Loop 2 tunes in 10-kHz steps, with the output f_8 always above f_4. For steps in f_4 of 10 kHz, only N_2 changes, tuning the 114.3- to 144.3-MHz range in 3000 10-kHz steps. The loop 1 VCO is coarsely tracked by the loop 2 tuning voltage, and its divider N_1 is changed at the same time as N_2 for 100-Hz steps. The difference in the frequency changes of the two loops appears at f_4. Since the smallest changes in N are the same for both dividers, $\Delta f_4 = \Delta N \times 10\ \text{kHz} - \Delta N \times 9.9\ \text{kHz} = \Delta N \times 100\ \text{Hz}$. N_1 increments from 500 to 599, while N_2 increments 99 counts. At the next 100-Hz increment, N_1 recycles by 99 back to 500, and N_2 decreases by 98, and the cycle repeats.

This economical scheme has been used effectively for SSB synthesizers, in spite of the higher phase noise close to the carrier caused by large values of N_2 multiplying the 10-kHz digital reference noise, the slower tuning time of 20 ms, and tuning transients on the 100-Hz increments. The latter results from the fact that generation of each 100-Hz step requires tuning both loops simultaneously, with the settling of loop 1 dependent on the settling of loop 2; therefore, the two opposing transients cannot cancel completely, causing large transients for even the smallest steps. The resulting audio noise can be muted if necessary while tuning. This scheme is at its best in low-cost, channelized HF radios.

The examples shown in Figures 10.17 through 10.19 illustrate the complexity of generating small steps at even modest high frequencies using PLLs. Multiloop VHF technology is required, just as in the coarse-increment sections of the synthesizer. As medium-scale integration (MSI) and LSI circuit technologies have advanced in the mid-1990s, digital methods have become available to generate small frequency increments up to at least a few megahertz. These methods are examined in the next sections.

Direct digital synthesis

Direct digital synthesis (DDS), in contrast to the previously described indirect synthesis techniques, combines rapid phase-continuous frequency switching with closely spaced frequency increments. Notwithstanding its problems of comparatively high spurious frequency content and low output frequency range, the design has many applications when combined with wideband PLLs. The circuits are highly digital in nature, which leads to production economies.

A simplified block diagram for a direct digital synthesizer is shown in Figure 10.20. A clocked phase accumulator creates a sawtooth wave in dig-

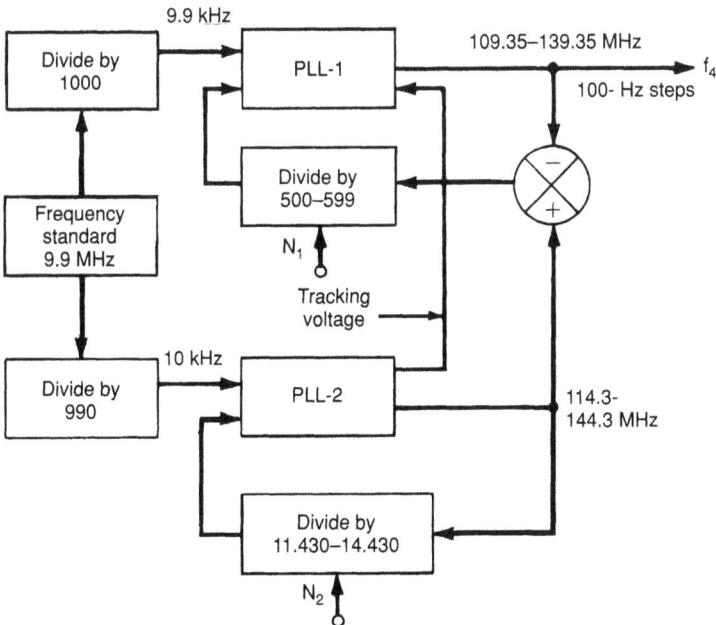

Figure 10.19 Difference-loop synthesizer.

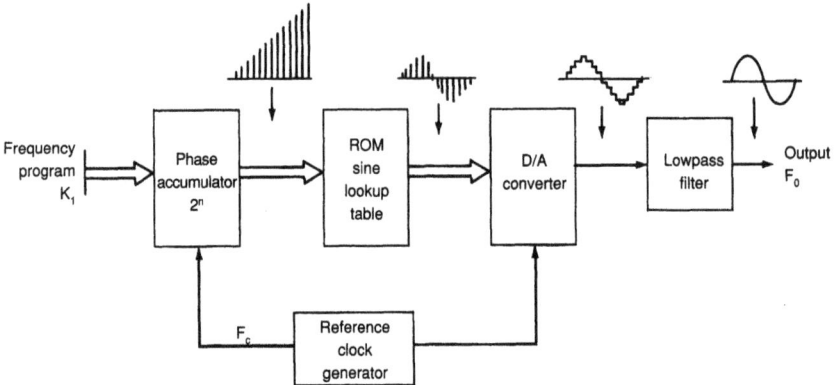

Figure 10.20 Direct digital synthesizer with sine lookup table.

ital (numeric) form. The accumulator most significant bit (MSB) is an actual rectangular wave, which can be used directly. Alternatively, the digital sawtooth can be used to address a ROM that stores sine wave sample values that can be converted to a sine wave voltage by a D/A converter and lowpass filter [3, 16]. It should be apparent that arbitrary waveforms may be generated by appropriate storage in the ROM, and that the accumulator itself can produce other waveforms, such as triangular.

The accumulator output frequency $F_o = K_1 F_b$, where K_1 is the binary frequency control word, and F_b (base frequency) $= F_c/2^n$ (F_c is the accumulator clock frequency, and n is the number of binary stages in the accumulator). Decimal and other schemes have been used [17], but binary accumulators are the simplest and most efficient, and can be effectively controlled if the radio has a microprocessor.

The accumulator output (MSB or carry) has an average frequency equal to the desired F_o, and is spurious-free if K_1 is an integer power of 2. Noninteger powers of 2 have zero crossings that ideally would have to occur between clocks. Since there can be no output except on a clock, the outputs will then occur on the nearest clock, resulting in a spurious modulation as the output phase moves with the clock for awhile, then jumps ahead or behind one clock period to catch up. The ratio of spurious amplitude to accumulator output is approximately equal to the output-to-clock frequency ratio.

The sine ROM and D/A converter greatly reduce these spurious signals by creating a sine wave with zero crossings that can occur between clocks at the desired times for each output frequency. These correct zero crossings are detected when the sine wave is finally converted to a rectangular wave again. The spurious signal levels in this approach depend upon the number of bits in the sine table input and output and glitches in the D/A converter.

Efforts to predict the quantization error levels using statistical methods usually result in levels that are much higher than measured. Fairly good correlation seems to result from a Fourier transformation of the complete sine table, which for a 10-bit table predicts spurious levels of –72 dBc. Typical spurious levels of -65 dBc have been achieved with a low-glitch D/A converter, up to 1- or 2-MHz output, and a clock frequency of 8 to 10 MHz.

A synthesizer using a DDS for fine increments is shown in Figure 10.21 where loop 1 combines 1-MHz steps from loop 2 with the fine increments Δf_{min} 10 on the loop 1 reference. Loop 3 raises the DDS frequency to the 10.35- to 11.35-MHz range where loop 1 will track properly with loop 2. The two divide-by-10 circuits at the loop 1 reference lower the phase detector operating frequency to the 1.035- to 1.135-MHz range, while maintaining the 10.35- to 11.35-MHz range at the mixer output. Thus the net division from the DDS to the output f_4 is unity, and the smallest output frequency increments are those of the DDS.

The 10-MHz clock for the DDS requires a BCD or some decimal form of accumulator. If the more usual and efficient binary accumulator is used, a clock generator PLL would be required (between the frequency standard and the DDS) to provide a power-of-2 clock, such as 10.48576 MHz for 10-Hz increments.

A DDS output can also be mixed up to a higher frequency and used with the loops shown in Figure 10.6.

Fractional division

Like the DDS, fractional division [1, pp. 196–202; 18] deals with noninteger relationships and corresponding spurious problems. The difference is in the

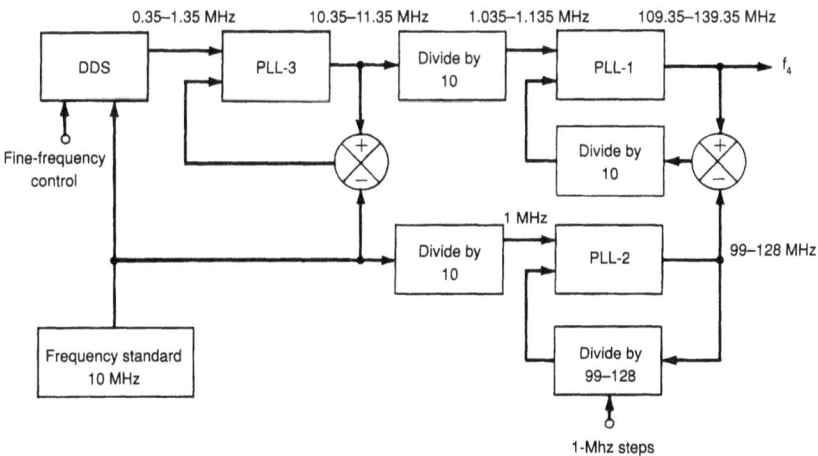

Figure 10.21 Synthesizer with fine steps produced by a DDS.

relationship between the frequency increment and the input and output frequencies. The DDS is intended for a fixed input frequency, with equal frequency increments in the output. A fractional divider has a constant output frequency with equal frequency increments in the input. This presents a difficulty in using a sine ROM, as the accumulator size would then have to vary with the division ratio.

In order to use a constant-size accumulator, most fractional dividers use some form of analog bridge to balance out the spurious components. Figure 10.22 shows how a fractional divider could be used in a single-loop synthesizer. In spite of the 100-kHz reference, very small frequency increments are possible, such as 1 Hz, but at the price of a wide spectrum of spurious signals.

The fractional part M of the total divide ratio programs an accumulator that alters the main divider count N at intervals to obtain the correct average total count in the loop. The accumulator also creates a spurious modulation error signal, which is converted to an analog signal of polarity opposite to the modulation on the desired signal and is combined at the phase detector output. The degree of spurious signal reduction is related to the degree of balance, affected by component variations with temperature and aging. Uncorrected spurious signals at the phase detector would be approximately $1/N$, but multiplied by N at the VCO output, so the output spurious signal depends primarily on the degree of balance.

Since more than a 20- or 30-dB reduction is difficult to maintain with wide temperature variations and other tolerances, fractional division is usu-

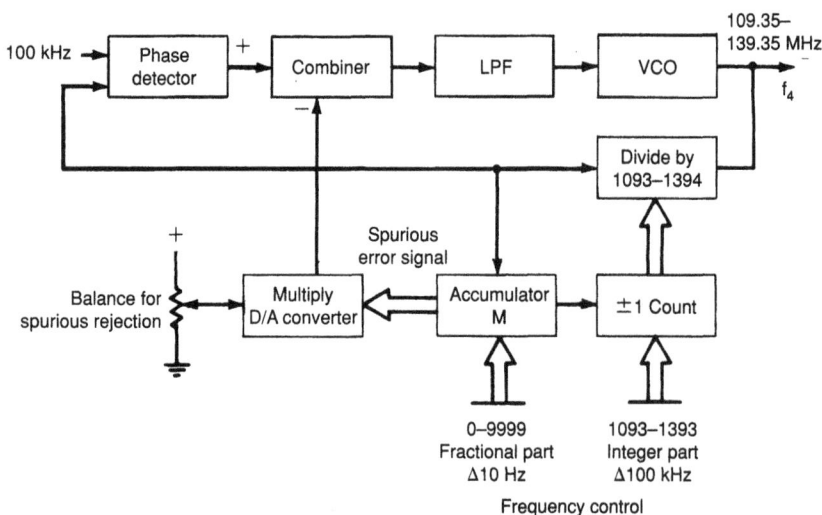

Figure 10.22 Single-loop synthesizer with fractional division.

ally used with a very narrow loop bandwidth to filter the spurious signals, and much of the advantage of a high reference frequency is lost.

References

1. W. F. Egan, *Frequency Synthesis by Phase Lock* (New York: John Wiley & Sons, 1981).

2. U. L. Rohde, *Digital PLL Frequency Synthesizers*, Figure 2-12 (Englewood Cliffs, NJ: Prentice-Hall, 1983).

3. J. Tierney, "Digital Frequency Synthesizers," in Gorski-Popiel (ed.), *Frequency Synthesis: Techniques & Applications*, pp. 121–149 (Piscataway, NJ: IEEE Press, 1975).

4. F. M. Gardner, *Phaselock Techniques*, 2d ed. (New York: John Wiley & Sons, 1979).

5. J. Noordanus, "Frequency Synthesizers—A Survey of Techniques," *IEEE Trans. Commun. Tech.* COM-17, no. 2 (April 1969): 257–271.

6. J. J. Andrea, Phase Locked Loop with Digitalized Frequency and Phase Discriminator. U.S. Patent 3,431,509 (March 4, 1969) (assigned to Rockwell International).

7. R. C. Debloois and N. E. Hogue, Digital Frequency-Phase Discriminator Circuit, U.S. Patent 3,866,133 (February 11, 1975) (assigned to Rockwell International).

8. J. M. Laune, Digital Frequency and/or Phase Detector Charge Pump, U.S. Patent 3,714,463 (January 30, 1973) (assigned to Motorola).

9. D. M. Mitchell, "Pulsewidth Modulator Phase Shift," *IEEE Trans. Aerospace Electron Syst.* AES-16, no. 3 (May 1980): 272–278.

10. W. E. Sabin, "Phase Relations in the Frequency Divider and Phase Detector of a Locked Digital Frequency Synthesizer," Rockwell-Collins Working Paper WP3881 (1974).

11. M. A. Caloyannides, "An Analytic Evaluation of Synthesizer-Switching Performance Loss," Rockwell-Collins Working Paper WP81-3005 (1981).

12. E. A. Janning, "A Low Noise Oscillator," Master's thesis, University of Cincinnati, Cincinnati, Ohio (1967).

13. W. P. Robins, *Phase Noise in Signal Sources*, Chapter 5, IEE Telecommunications Series 9 (London: Peter Peregrinus, 1982).

14. Hewlett-Packard, Spectrum Analyzer System model HP3047A.

15. R. D. Tollefson, Frequency Synthesizer, U.S. Patent 3,588,732 (June 28, 1971) (assigned to Rockwell International).

16. W. J. Melvin, Digitalized Tone Generator, U.S. Patent 3,597,599 (August 3, 1971) (assigned to Rockwell International).

17. L. B. Jackson, Digital Frequency Synthesizer, U.S. Patent 3,735,269 (May 22, 1976) (assigned to Rockland Systems Corp.).

18. G. C. Gillette, "The Digiphase Synthesizer," *Freq. Tech.* 7 (August 1969): 25–29.

11

Frequency Standards for SSB[1]

Marvin E. Frerking

The application of precision frequency control has played an important role in establishing SSB as a reliable means of radio communications. The crystal oscillator thus provides one of the vital links required for the implementation of SSB. As will be seen subsequently, the crystal oscillators used for SSB are often required to maintain a higher degree of accuracy than those used in many other applications.

The material covered in this chapter deals primarily with crystal oscillators in the frequency stability range required for SSB equipment operating at HF or VHF bands. The design of higher-precision frequency standards that might be used for microwave SSB links is beyond the scope of this book, and the reader is referred to the references at the end of this chapter. Thus, both temperature-compensated crystal oscillators and crystal ovens of the type used to achieve a frequency stability in the ppm range are addressed. Since the quartz crystal itself plays such a vital role in determining the oscillator performance, a section is included discussing many of the characteristics of the crystal itself. A brief sketch is also included presenting a historical review of frequency control in SSB equipment.

[1]Much of the material in this chapter is taken from M. E. Frerking, *Crystal Oscillator Design and Temperature Compensation* (New York: Van Nostrand, 1978), and appears by courtesy of the Van Nostrand Company.

11.1 Historical Notes on Frequency Stability

Historically, one of the most effective frequency control schemes for general receiver coverage has been to use a crystal oscillator in combination with a variable LC oscillator. The LC oscillator, often in the 1- to 3-MHz range, had a tuning range of a few hundred kilohertz to perhaps 1 MHz. The permeability tuned oscillator (PTO), developed by Hunter, Hodgin, Mifflin, and others at Collins Radio, served this function well and was used for over 30 years in the manufacture of communications receivers and transceivers. The PTO uses a Colpitts oscillator in which the inductor is tuned by means of a ferrite core on a special lead screw with a linearization cam. As a result, the PTO has a very linear tuning range and also possesses an inherent frequency stability of approximately ±200 ppm. The PTO has now been displaced by frequency synthesizers in modern communications equipment. One of the early frequency synthesizer designs using discrete tuning steps controlled by a tuning knob was used in the Collins 651S receiver. This receiver tuned in 100-Hz increments. As frequency synthesizer hardware continued to improve, frequency synthesizers with 10- and even 1-Hz steps came into use. These designs give the impression of continuous tuning.

The crystal oscillators used for SSB communications vary considerably in frequency stability depending on the application. For military applications requiring an accuracy of about 1 ppm, crystal ovens were first used. These units were relatively large, consumed considerable amounts of power, and unfortunately required a warm-up time to stabilize. The introduction of the varactor tuning diode during the late 1950s, however, soon made temperature compensation of crystal oscillators practical and eliminated the warm-up time. Recent improvements in crystal ovens have reduced the warm-up time to about 2 min, and a few designs have allowed the use of an oven at a somewhat lower cost than a temperature-compensated crystal oscillator (TCXO).

11.2 Crystal Oscillator Design

The design of precision crystal oscillators is somewhat unique in that not only must the circuit oscillate, but it must also possess a high degree of frequency stability as the environment changes and over long periods of time. A considerable portion of this chapter is, therefore, devoted to showing how to achieve a high degree of frequency stability from the oscillator.

Basically a crystal oscillator can be thought of as an amplifier and a feedback network. The crystal is placed in the feedback network in a position where it has a large effect on the phase shift. When the oscillator is turned on, the amplitude of oscillation builds up at a frequency where the phase shift around the loop is a multiple of 360° and the loop gain is greater than unity. The amplitude continues to increase until saturation or limiting effects reduce the loop gain to unity. If a frequency does not exist where the phase-shift requirement is satisfied and the loop gain is greater than unity, oscillation will not take place. A crystal oscillator is unique in that the impedance of the crystal changes so rapidly with frequency that all the other components in the oscillator can be considered to be of constant reactance, that reactance being calculated at the nominal frequency of the crystal. Variations in the oscillator components do, of course, have a secondary effect on the phase shift, and the frequency of oscillation then adjusts itself so that the resulting impedance change of the crystal exactly compensates for the phase change caused by the oscillator component.

The application of these principles combined with the circuits listed in this chapter and in the references at its end allows the design and understanding of many of the frequency standards used in SSB equipment.

Crystal oscillators have been designed from a few kilohertz to several hundred megahertz, with frequency stabilities ranging from several hundred parts per million to parts in 10^{10}. The frequency region from 3 to 5 MHz is very desirable for the design of temperature-compensated crystal oscillators, although other frequencies can be used, and many reference oscillators have been built in this range for frequency synthesizers.

An AT-cut quartz resonator is often used for SSB equipment. This resonator, if optimized for stability in a particular temperature range, can have a frequency stability of about ±1 ppm over a limited temperature range of 0 to 50°C and about ±15 to ±20 ppm over –55 to +105°C. Improved frequency stability is normally obtained by the use of temperature compensation or temperature control in a crystal oven.

Quartz crystals

The quartz crystal itself plays such an important role in the design of precision crystal oscillators that the designer should become familiar with the fundamentals of the crystal prior to undertaking an oscillator design. This section, therefore, summarizes many of the important characteristics of quartz crystals. For more information, the reader is referred to the many excellent discussions in the literature, especially references 1 through 3.

A quartz crystal possesses an extremely high Q, it has a very stable resonant frequency, and is also small in size and available at reasonable cost. A quartz crystal utilizes the piezoelectric properties of quartz. If a stress is applied to the crystal in a certain direction, an electric field appears in a perpendicular direction. The converse is also true, so that if an electric field is applied across the crystal, a small mechanical deformation will result. In a quartz crystal resonator a thin slab of quartz, called the *crystal blank*, is placed between two electrodes. An alternating voltage applied to these electrodes causes the crystal to vibrate at the frequency of the applied voltage. If this frequency approaches a natural resonance in the blank, the amplitude of the deflection becomes relatively large. The mechanical vibration causes charges to be induced in the electrodes and greatly influences the impedance observed between the electrodes.

The electrical equivalent circuit of the crystal is shown in Figure 11.1. The inductor is associated with the mass of the crystal, C_1 is associated with the stiffness of the quartz, and R_1 is associated with the loss in the quartz as well as in the mounting structure. C_0 is associated with the quartz dielectric material between the two electrodes, but also includes the stray capacitance of the crystal mounting structure. C_0 is often in the 3- to 5-pF range, while C_1 may be on the order of 0.01 pF for a fundamental-mode AT-cut resonator. A list of typical equivalent circuit values is given in Table 11.1.

The equivalent circuit shown in Figure 11.1A can be simplified to that shown in Figure 11.1B at any specific frequency. A reactance-versus-frequency plot of the equivalent circuit is given in Figure 11.2. The portion circled on this plot is expanded in Figure 11.3. The corresponding equivalent resistance is shown in Figure 11.4.

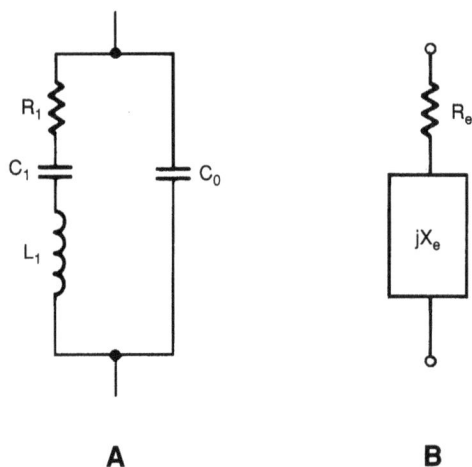

Figure 11.1 (A) Approximate equivalent circuit of a quartz crystal. (B) Impedance representation of a quartz crystal.

A   **B**

TABLE 11.1 Typical Crystal Parameter Values

Frequency, MHz	Cut	Overtone	C_0, pF	C_1 pF	L_1	R_1, Ω
2	AT	1	4	0.012	520 mH	100
3	AT	1	3.2	0.011	256 mH	20
5	AT	1	5.0	0.020	51 mH	7
10	AT	3	5.0	0.0025	102 mH	20
30	AT	3	6	0.0026	11 mH	20
4	SC	3	3.3	0.0002	7.9 H	300
10	SC	3	3.5	0.00027	940 mH	60

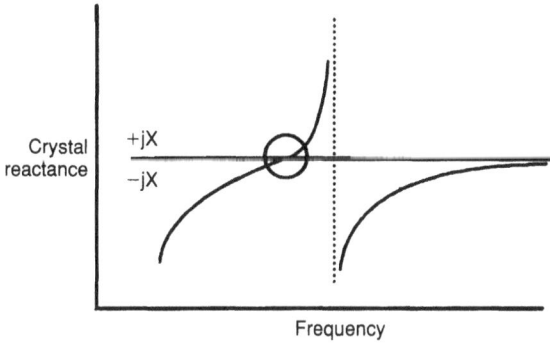

Figure 11.2 Reactance versus frequency for a quartz crystal.

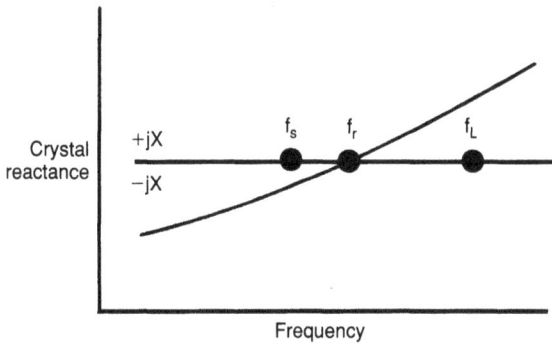

Figure 11.3 Expanded portion of crystal reactance near resonant region.

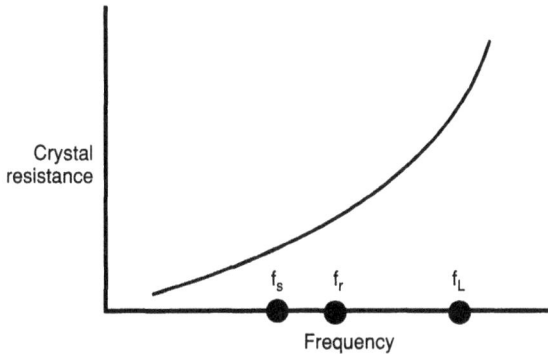

Figure 11.4 Equivalent resistance of crystal near resonant region.

Several frequencies are marked on these figures. The first of these is f_s. This is the frequency where the motional arm consisting of R_1, L_1, and C_1 is series resonant and is given by

$$f_s = \frac{1}{2\pi\sqrt{L_1 C_1}} \tag{11.1}$$

The second point, f_r, is where the crystal is purely resistive. It is different from f_s because of the presence of C_0, and for practical purposes can be considered to be equal to f_s. The third point, f_L, is the frequency at which the crystal is resonant with an external capacitor C_L. If Δf is the frequency shift between f_L and f_s, it can be shown [1] that

$$\frac{\Delta f}{f_s} = \frac{C_1}{2(C_0 + C_L)} \tag{11.2}$$

where Δf = frequency shift $f_L - f_s$
C_1 = motional capacitance, pF
C_0 = crystal holder capacitance, pF
C_L = external load capacitance, pF

In the equivalent circuit shown in Fig 11.1B, jX_e has a value equal to, but opposite in sign to, the reactance of C_L at a frequency f_L. The antiresonant frequency of the crystal (the peak in Figure 11.2) is the frequency where the motional arm is resonant with C_0. This can be found by setting $C_L = 0$ in Equation 11.2 and setting $\Delta f = f_a - f_s$. It is given by

$$f_a = f_s\left(1 + \frac{C_1}{2C_0}\right) \tag{11.3}$$

The equivalent resistance R_e of the crystal in the region between series and antiresonance is given by

$$R_e = R_1\left(\frac{C_L + C_0}{C_L}\right)^2 \tag{11.4}$$

provided $|X_{c0}[C_L/(C_L + C_0)]| >> R_1$ is true, where $X_{c0} = -1/2\pi f_s C_0$.

It is often quite convenient to calculate the reactance of the crystal as a function of frequency using Equation 11.2. This will be discussed further in section 11.4.

The crystal is normally operated between series and antiresonance so that the reactance of the crystal is either zero or inductive.

The properties of the crystal may be varied considerably by the angle at which the blank is cut from the raw quartz and by the mode of vibration. This topic is primarily a concern of the crystal manufacturer and will not be discussed in detail here. (An excellent treatment of crystal resonator design is given in reference 3.) But several properties of crystal resonators are of concern to the designer of crystal oscillators and will be discussed in the following paragraphs.

Load capacitance

From Figure 11.3 it can be seen that the frequency of the crystal will vary to some extent depending on the reactance that the crystal must present to the circuit. It is important that the crystal be plated to frequency at the load reactance value at which it will be used in the oscillator circuit. Several load conditions have become standard and are nearly always used. Among these are 20, 30, and 32 pF. These crystals must be used in parallel resonant oscillators to operate on frequency. Another common load condition is series resonance, where the crystal is purely resistive and has a value R_1. Crystals of this type must be used in series resonant oscillators if they are to operate at the specified nominal frequency.

Resistance

The resistance of a crystal is specified at the rated load capacitance, and usually does not differ greatly from the series resistance R_1. It may vary considerably from unit to unit, however, and it is important to ensure that the oscillator will function properly with a crystal that has a resistance as large as the specification allows. The maximum allowable resistance for a given crystal type may vary from about 40 ohms for VHF crystals to over 500 kohms for audio frequency crystals.

Rated drive level

Drive level refers to the power dissipated in the crystal. The drive level specification should be reasonably duplicated in the oscillator.[2] The drive level ratings vary from 5 μW below 100 kHz to about 10 MW in the 1- to 20-MHz region for fundamental crystals. Precision crystals normally operate at considerably lower drive levels in the 10- to 50-μW region. This leads to greater stability because the frequency effect of drive level is reduced at low drive levels, as is the aging rate. The frequency effect of drive level is less predominant in SC-cut crystals than in AT-cut crystals.

---

[2]At the time of this writing, consideration is being given to specifying the drive current rather than the power dissipation since this more nearly determines the amplitude of displacement.

Frequency stability

The frequency stability of a quartz crystal is limited primarily by its temperature coefficient and the aging rate. Common temperature specifications are ±0.0025% from −55 to +105°C for wide-temperature-range units or ±0.001% over a 10°C range for oven crystals. The frequency-versus-temperature characteristics of a given crystal are determined by the angle at which the crystal is cut from the raw quartz bar, and typically the manufacturer can cut a given angle plus or minus a few minutes of arc. Tighter tolerances are available by selection. Figure 11.5 shows the frequency-temperature-angle characteristics of typical AT-cut crystals. The points of zero slope are called the lower and upper turning point temperatures.

The aging rate of a crystal is caused primarily by a gradual mass transfer to or from the crystal blank and by stress relaxation. The cleaning of the unit and the seal of the holder are thus extremely important. Aging rates in the order of parts in 10^9 to a few parts in 10^8 per week are typical for precision oscillators used in SSB equipment. This requires considerable care, and glass or cold-weld enclosures are often used.

Other factors

A number of other factors are also of concern to a crystal oscillator designer and the reader is referred especially to references 1 through 3 for a more detailed treatment. Among these are the finishing tolerance, Q, stiffness, spurious modes, pin-to-pin capacitance, vibration susceptibility, and others.

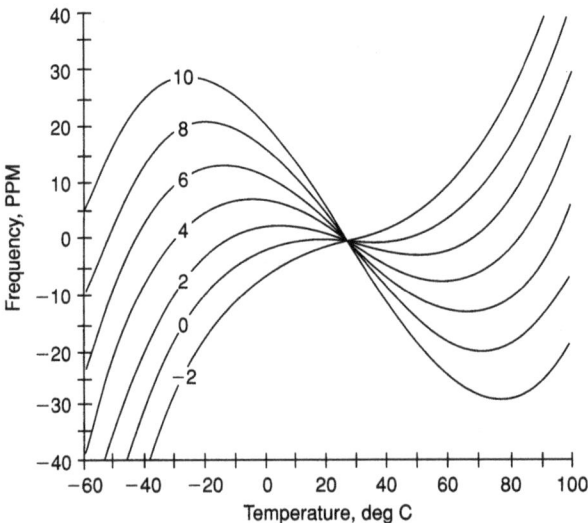

Figure 11.5 Frequency-temperature-angle characteristics of AT-type quartz resonators.

It should also be noted that quartz crystals can be used on odd mechanical overtones, notably the third and fifth. If these are used, it is often necessary to use a trap in the oscillator to ensure oscillation on the desired mode. If this is not done, it is necessary to ensure that the resistance of the fundamental mode and lower overtones is considerably higher than that of the desired mode.

Oscillator circuits

A large number of crystal oscillator circuits have been used in various applications, and it is beyond the scope of this chapter to discuss even the majority of these circuits. A very comprehensive treatment can be found in references 1, 2, and 4. Since a large number of oscillators for SSB have used the Pierce or Colpitts circuits, they will be discussed in more detail here.

These circuits have the same RF equivalent circuit but with the ac ground at different points. The ac ground in the Colpitts is at the collector, while in the Pierce it is at the emitter. Practically, since the biasing resistors and the stray capacitance fall across different elements, the circuits may perform considerably differently. Both circuits can be made to perform well over the 1- to 20-MHz frequency range; however, the Pierce circuit generally operates better at very low or very high frequencies. The Pierce circuit has been used in the frequency range extending beyond 60 MHz; however, above the 30- to 60-MHz range the grounded-base (Butler) circuit may be more appropriate to use.

The Pierce circuit is generally easier to design than the Colpitts. It is less prone to spurious oscillations and squegging and in some cases possesses superior stability. A significant disadvantage is the fact that the pin of the crystal can be grounded. The Colpitts oscillator, on the other hand, does allow one side of the crystal to be grounded. Also, since the collector may be tied directly to the supply voltage, it can be operated from lower supply voltages without using an inductor to feed the collector.

A number of crystal oscillators have also been designed using logic gates and IC oscillators. The performance of these circuits varies greatly and generally tends to be inferior to that obtained using a well-designed transistor stage. For this reason, crystal oscillators used in SSB equipment often use a transistor oscillator. Gate oscillators are treated in references 1, 2, and 4. Gate oscillators having the best frequency stability often use a single gate with a π network containing the crystal.

Pierce oscillator. A schematic diagram of the Pierce oscillator is shown in Figure 11.6. The conditions for oscillation are satisfied in the following manner. The basic phase-shift network is composed of C_1, C_2, and the crystal, which looks inductive. The capacitors are normally large enough to effectively swamp out the transistor input and output impedances. Under

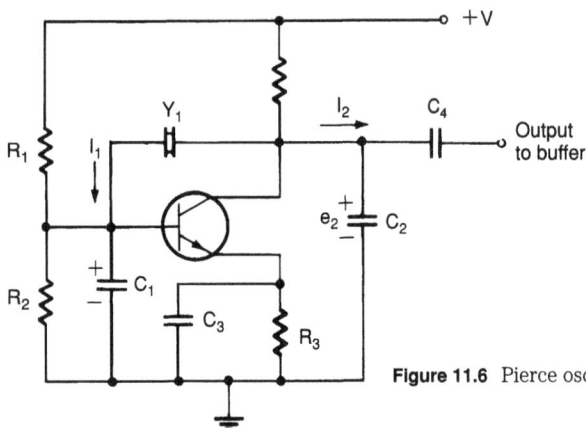

Figure 11.6 Pierce oscillator schematic diagram.

these conditions, if the crystal resistance is not too large, the following explanation is applicable. The frequency of oscillation adjusts itself so that the crystal is inductive and resonant with the series combination of C_1 and C_2. Because the inductive reactance of the crystal is larger than the capacitive reactance of C_1, the current I_1 lags the voltage e_2 by 90°. The voltage e_1, being developed across C_1, lags I_1 by 90°, making it 180° behind e_2. Now, since C_2 is resonant with the resulting reactance from C_1 and the crystal, the collector of the transistor looks into a resistive load. This being the case, the collector voltage e_2 is in phase with the current generator in the transistor, which is 180° out of phase with the base voltage e_1. The total phase shift around the loop is, therefore, 360°. If either of the capacitors C_1 or C_2 changes slightly with temperature, the frequency of oscillation shifts so that the change in crystal reactance restores the phase balance. When the circuit is turned on, if the conditions for oscillation exist, the amplitude builds up until saturation takes place and the loop gain is reduced to unity. Depending on how the transistor is biased, saturation may take place by the base-to-emitter junction being cut off during part of the RF cycle or by the base-to-collector junction becoming forward biased during part of the RF cycle. Generally the former results in better stability while the latter gives better amplitude control. Limiting effects are difficult to analyze mathematically; however, a good treatment of them is presented in reference 2.

A small-signal analysis can be used to determine the conditions necessary for the onset of oscillation. Reference 1 presents a treatment based on the y parameters of the transistor. It is shown that for oscillation to take place

$$g_{fe}X_1X_2 \geq R_e + K_1 \text{ (gain equation)} \tag{11.5}$$

$$X_1 + X_2 + X_e = 0 + K_2 \text{ (phase equation)} \tag{11.6}$$

where g_{fe} = real part of the transistor forward transfer admittance, sometimes referred to as the *transconductance*

$$X_1 = \frac{-1}{\omega C_1}$$

$$X_2 = \frac{-1}{\omega C_2}$$

R_e = effective crystal resistance

X_e = crystal reactance

K_1 and K_2 are corrective terms that produce only secondary effects if the previous assumptions are fulfilled (see reference 1, Equations 7-13 and 7-14).

It can be shown that to a first approximation, loading of the output manifests itself as an effective increase of the crystal resistance, given by $\Delta R_e = X_2^2/R_L$, where R_L is the load resistance. The loading due to the bias resistors results in a $\Delta R_e = X_1^2 R_e$, where R_e is the equivalent resistance of R_1 and R_2 in parallel. To a first approximation, at frequencies well below the cutoff, the transconductance is given by

$$g_{fe} = 0.04 I_e \text{ mhos} \tag{11.7}$$

where I_e is the emitter current (in milliamperes) (mhos are also known as Siemens, abbreviated S).

It is normally desirable to design an oscillator so that the gain is two to three times greater than that required for oscillation with a maximum resistance crystal to allow for production variations. The oscillator is normally trimmed to frequency by placing a variable reactive such as a varactor or trimmer capacitor in series with the crystal.

If the Pierce oscillator is used with overtone crystals, it is necessary to prevent oscillation on the fundamental mode. This can be done by placing an inductor in parallel with C_1 or C_2 with appropriate dc blocking. The resulting combination should be resonant below the desired overtone, but above the next lower odd overtone. Thus at the lower overtone the reactance of the parallel circuit changes sign and becomes inductive. As can be seen from Equation 11.5, X_1 and X_2 must be of like sign for oscillation to take place.

Colpitts oscillator. The Colpitts oscillator has been widely used because of its circuit simplicity and excellent frequency stability. The ac equivalent circuit is the same as that for the Pierce oscillator, but with the ac ground point moved from the emitter to the collector. Equations 11.5 through 11.7 can be used to predict oscillation, provided the loading assumptions are accounted for. Because the loading and biasing configuration is different than for the Pierce oscillator and because stray capacitance on the base falls across the crystal, Colpitts circuits may behave considerably differently. The base biasing resistors also effectively shunt the crystal, and they must be made as large as possible. Using an FET allows the use of a very large bias resistor, which is an advantage particularly at low frequencies.

A schematic diagram of a Colpitts crystal oscillator is shown in Figure 11.7. This circuit can be thought of as an emitter follower driving a capacitive tapped tank circuit. The frequency of the oscillator adjusts itself so that the crystal is inductive and resonant with the series combination of C_1 and C_2. The Colpitts oscillator is more prone to squegging than the Pierce circuit and it may be necessary to choose C_2 and R_3 to prevent this effect. If the Colpitts oscillator is used to operate with the crystal on its third overtone, an inductor must be placed across C_1 or C_2 to produce a resonant frequency between the fundamental and the third overtone. Appropriate dc blocking must, of course, be provided.

11.3 Temperature Control

The frequency stability of a crystal oscillator is primarily determined by the quartz crystal. A graph showing the temperature characteristics of the crystal for several angles of cut is shown in Figure 11.5. Temperature effects on other oscillator components result in additional frequency changes of a few ppm over a wide temperature range for a well-designed circuit. Since the requirements of SSB are often in the 1-ppm range, a considerable improvement in performance is required. Temperature control or temperature compensation are the primary means by which the frequency stability can be improved. Temperature control to achieve a frequency stability in the ppm range is relatively straightforward and requires less test labor than temper-

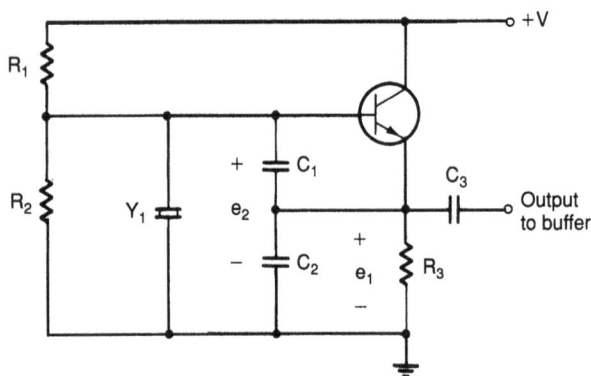

Figure 11.7 Colpitts oscillator circuit.

ature compensation. However, temperature control has the following disadvantages:

1. A warm-up stabilization period is required after turn-on.

2. The power consumption is higher than for temperature compensation because of thermal loss.

3. The size of the units may be larger than that required for temperature compensation because of the amount of thermal insulation required.

4. The reliability may be reduced because of thermal cycling if the unit is frequently turned on and off.

5. The aging rate is increased because of the higher operating temperature of the crystal.

In some applications these characteristics may be of little concern. For example, in a fixed-station installation, power consumption and size may be secondary to cost.

Temperature-controlled crystal oscillators can be built with extremely high frequency stability in the region of parts in 10^8 to parts in 10^9 over a wide temperature range. Oscillators of this type are beyond the scope of this chapter because the requirements of SSB are generally much less severe.

A variety of crystal ovens can be used to achieve frequency stabilities from a few ppm to a fraction of a part per million. In the 1- to 5-ppm range a number of commercial ovens are available that can be placed over the crystal or into which the crystal can be inserted. The warm-up time is usually not critical in these applications and may be in the 5- to 15-min range. Often, only the crystal is in the oven, with the other oscillator components exposed to the ambient temperature. A higher-performance oscillator with a frequency stability of a few parts in 10^7 can be achieved by placing the oscillator stage itself along with the crystal in the oven. The oven in some cases may consist of a metallic shell around the components to be temperature-controlled or in some cases may be a small block of metal to which the crystal and oscillator components are mounted. The heater may be a power transistor that is attached to the oven. The warm-up time may vary considerably with the size of the oven and the amount of warm-up power used.

With considerable power, in the 10- and 30-W region, and a small oven, the stabilization time can be reduced to the 2- to 3-min region even for a turn-on temperature of –55°C.

The crystal characteristics over the temperature range can be described by a cubic equation of the form

$$\frac{\Delta f}{f} = A_1\,(T - T_0) + A_2\,(T - T_0)^2 + A_3\,(T - T_0)^3 \tag{11.8}$$

where T_0 is an arbitrarily chosen reference temperature, usually selected to be in the 20 to 30°C range and A_1, A_2, and A_3 are constants for a given angle of cut and reference temperature T_0. For an AT-cut crystal and $T_0 = 25°C$, $A_1 = -5.08 \times 10^6 \Delta\theta$, $A_2 = -0.45 \times 10^9$, and $A_3 = 108.6 \times 10^{12}$. Here $\Delta\theta$ is the angle (in degrees) from the angle producing a zero slope at 25°C (see reference 5). Figure 11.5 shows these characteristics for AT-cut crystals.

It is desirable to operate the crystal oven at the upper turning-point temperature where the crystal has a zero slope or as near as possible to the turning point. In a precision oven, the temperature is adjusted to match the turning-point temperature of the particular crystal used. For a frequency standard of lesser stability, an error of several degrees may be tolerated. It can be shown from Equation 11.8 that the crystal slope as a function of distance from the turning-point temperature is given by

$$S = 2A_2(T - T_p) + 3A_3[(T^2 - T_p^2) - 2T_0(T - T_p)] \qquad (11.9)$$

where S = slope of curve, ppm/°C
$\quad\ T$ = temperature, °C
$\quad\ T_p$ = turning-point temperature, °C

For an AT-cut crystal with an 85°C turning point, the slope is about 0.2 ppm/°C if the oven is 5°C off the turning point. An oven must then hold the crystal temperature to within ±0.5°C to achieve a ±0.1-ppm frequency stability.

The schematic diagram of a typical oven temperature control circuit is shown in Figure 11.8. The heating element is the power transistor Q_2. This transistor is attached directly to the oven block or to the crystal. The tem-

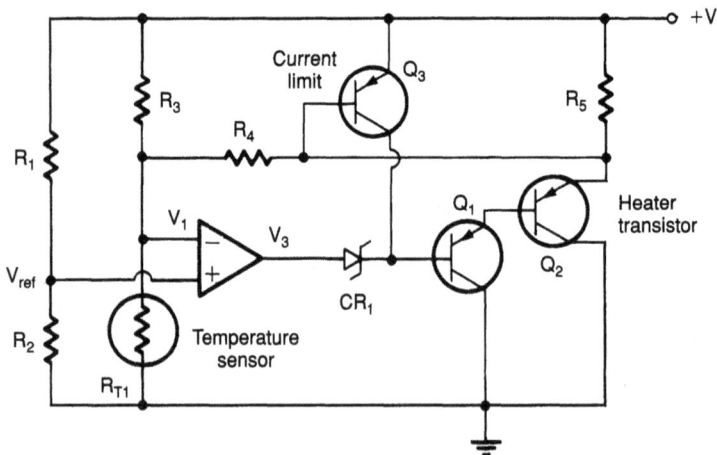

Figure 11.8 Schematic diagram of temperature control circuit.

perature sensor is a thermistor R_{T1}, which is also attached to the oven and very near the heater transistor. A significant problem in crystal ovens is that of temperature gradients. Crystals are quite sensitive to minute temperature differences between the pins and between the pins and the crystal case. When the crystal is operated at its turning-point temperature, the frequency-versus-temperature characteristic of the cubic curve can be completely masked by temperature gradients. Therefore, it is good practice to provide a metal block or enclosure to equalize the temperature. This becomes increasingly important in high-precision oscillators in the parts per 10^8 region where two heaters are often used, one on each side of the oven.

The circuit shown in Figure 11.8 works in the following manner. When the oven is cold, at turn-on, the thermistor RT_1 has a high resistance causing V_1 to exceed V_{ref}. This causes the output of the operational amplifier V_3 to go low, turning on Q_1 and Q_2. The power dissipated by Q_2 heats the oven, and when it reaches the set temperature determined by V_{ref}, the thermistor resistance decreases to the point where V_1 approaches V_{ref}. At that point the voltage V_3 increases, which reduces the power in Q_2. The oven stabilizes when the heater power is equal to the thermal loss at the operating temperature. The resistor R_4 is used to provide negative feedback which limits the gain. If the gain is too high, thermal oscillations will occur because the heater and/or oven becomes too hot before the temperature sensor cuts down the power. If the gain is too low, inadequate temperature control will result over the ambient temperature range. The transistor Q_3 in connection with R_5 can be used to limit the current during warm-up to a safe value.

The steady-state power is a function of the thermal loss, which consists primarily of the heat flow through the insulation and the loss in the wires going into the oven. The approximate heat loss for a crystal oven with foam insulation can be calculated using the formula

$$P = \frac{\lambda A \Delta T}{L} \tag{11.10}$$

on each of the six surfaces and for each of the wires going into the oven. Here

λ = thermal conductivity of the wire or insulation
A = cross-sectional area for volume under consideration
L = length of the thermal path or wire length
ΔT = difference between oven and ambient temperature

Depending on the size of the oven, ambient temperature, and insulation used, a thermal loss in the region of 1 to 5 W may be expected.

11.4 Temperature Compensation

Advantages

The frequency stability of a crystal oscillator can be greatly improved by temperature compensation, and the resulting stability is adequate for many SSB applications. The resulting crystal oscillators, with an accuracy in the ±1-ppm range, are fairly inexpensive, operate on low power, and stabilize very quickly after turn-on. A temperature-compensated crystal oscillator can be packaged in 2 or 3 in.^{3} (33 to 50 cm^3) using conventional components. TCXOs also are often packaged using thick film or hybrid techniques. Sizes vary; however, some units of modest stability are available in a volume of 0.3 in.^{3} (5 cm^3) or less.

Methods of temperature compensation

The frequency-versus-temperature curves for AT-cut quartz crystals are shown in Figure 11.5 for various angles of cut. The crystal is not the only contributor to frequency changes, and the other oscillator components affect the frequency, but to a much smaller degree. The overall uncompensated frequency stability of the oscillator is approximately the same as that shown in the crystal curves. To temperature compensate a crystal oscillator, such as the circuit shown in Figure 11.6, a varactor is normally placed in series with the crystal Y_1, where it has a maximum effect on the frequency. Direct current blocking capacitors are used so that a bias voltage can be applied to the varactor through large-value resistors which have no effect on the RF signal. The varactor bias voltage applied at any given temperature is the value required to pull the crystal by exactly the amount, but in the opposite direction, that it has drifted in temperature. If the varactor voltage is carefully developed as a function of temperature, a resulting frequency error of parts in 10^7 can be obtained.

Many methods have been devised to develop the required varactor bias voltage as a function of temperature. The majority of these methods fall into three categories: analog resistor thermistor compensations, digital temperature compensation, and microcomputer temperature compensation. A detailed treatment of temperature compensation is beyond the scope of this book, and the reader is referred to reference 1 for more detailed information. A brief summary of the methods in common use is given here to provide a reasonable understanding of how they work.

Analog temperature compensation. The method of analog temperature compensation was developed during the late 1950s by D. G. Newell and G. R. Hykes at Collins Radio for use over a wide temperature range [6]. The method uses a network consisting of several thermistors and resistors to develop a voltage curve that is the mirror image of the crystal curve. A considerable number of network variations have been used to provide advantages in one situation or another. The network shown in Figure 11.9 is fairly typical and has been used on over 50,000 TCXOs covering the temperature range from −55 to +75°C.

The thermistor resistance values are chosen so that the network achieves a reasonable degree of independence of adjustment between the cold temperature region and the room temperature region, and also between the hot temperature region and the rest of the curve. It is not difficult to show that the network transfer function is given by

$$(11.11)$$

$$\frac{V_2}{V_1} = \frac{RT_3\,(R_1 + RT_1)\,(R_2 + RT_2)}{(R_1 + RT_1)\,(R_2 + RT_2)\,(R_3 + RT_3) + R_2 RT_2[(R_1 + RT_1) + (R_3 + RT_3)]}$$

where the thermistor resistances are approximately given by

$$RT = RT\,(T_0)\,\exp\left[B\!\left(\frac{1}{T} - \frac{1}{T_0}\right)\right] \qquad (11.12)$$

Here $RT(T_0)$ is the nominal thermistor resistance at the reference temperature, T_0 usually taken to be 298 K (25°C), B is the thermistor exponential temperature coefficient, and T is the temperature (in Kelvin).

For precision temperature compensation, the thermistor values are usually taken from stored tables derived from measured values. It is not surprising that the computer has been used to aid in the initial selection of the resistors to minimize the number of temperature runs required.

A considerable latitude exists in the computer program and in the accuracy with which it predicts the resistance values. At one extreme, the required varactor voltage as well as the thermistor resistances are first measured at temperatures spaced 10 to 20°C apart. The computer then determines the

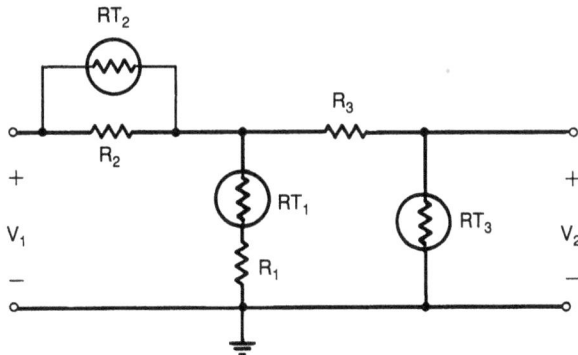

Figure 11.9 Analog temperature-compensation network.

resistance values giving the minimum error in some sense. An initial strategy may be desirable to calculate resistance values approximately to avoid local minima in the optimization program. One such strategy is to assume initial values and using Equation 11.11 solve for R_2 to obtain the desired voltage V_2 at the coldest temperature, for R_1 to obtain it at room temperature, and for R_3 to obtain it at the highest temperature. This procedure can be iterated until exact values of V_2 are obtained at each of the three temperatures with one set of resistances. An optimization program such as the steepest-descent algorithm can then be used to improve the values using additional temperature points. More elaborate networks than the one shown in Figure 11.9 can be effectively handled with computer programs.

One of the most difficult problems, however, seems to be finding a global minimum for the resistor configuration. TCXOs designed with the aid of computer programs have been produced with frequency stabilities in the order of a few parts in 10^7.

Another program strategy is to measure only the required varactor voltages over the temperature range and predict the thermistor resistances from prestored tables. This procedure requires less data taking, but sacrifices some accuracy in the process.

A still easier process is simply to use the constants for the particular crystal being used, from manufacturers' data, and predict the varactor voltage. This can be done by using the constants A_1, A_2, and A_3 in Equation 11.8 to calculate $\Delta f/f$ at a given temperature T. The value of $\Delta f/f$ can be used in connection with Equation 11.2 and the nominal load capacitance C_L to find the reactance change necessary to pull the crystal back on frequency. The reactance change is then used in the varactor equation to find the required voltage. The varactor capacitance is represented by an equation of the form

$$C = \frac{K}{(V + V_0)^n}$$

where K, V_0, and n are determined for the particular varactor used. Obviously this procedure results in additional errors, but if carefully done, a frequency stability of several ppm can be obtained. Experimental tweaking can then be done to improve the frequency stability. At the time of this writing most TCXOs are manufactured using analog networks in which the components are adjusted with the aid of computer programs, with the amount of final tweaking dependent on the frequency stability required. The difficulty of compensation increases considerably beyond ± 1 ppm.

Digital temperature compensation. Digital techniques are quite amenable to temperature compensation and eliminate much of the skill required to determine the component values in an analog network. At the present time an IC designed for temperature compensation is not available and the cost of the MSI parts exceeds the component cost of the analog net-work. Digital compensation can easily be automated, however, which re-duces the labor cost. The block diagram of a digital compensation circuit is shown in Figure 11.10. In this circuit the temperature sensor develops a voltage that approximates a linear function of temperature. A tempera-ture-dependent current source or a poorly biased transistor are both possible choices. This voltage is digitized by the A/D converter. The digi-tal word is then used as the address to a PROM which has been pro-grammed to contain the required correction voltage at every temperature. The output of the PROM is then converted to an analog voltage by the D/A converter and applied to the varactor. A detailed de-scription of a digitally compensated crystal oscillator (DCXO) is given in reference 1, along with an analysis of the digital word size required.

In practice it is desirable to use enough bits so that the frequency errors caused by the digital system are small compared with the required fre-quency tolerance. This allows most of the tolerance to be used for tem-perature tracking and for the frequency retrace characteristics of the crystal.

The PROM can be programmed by a test station which reads the address and simultaneously provides a tracking or searching voltage that deter-mines the proper word to apply to the D/A converter. In practice, it is not necessary to check every temperature during compensation, and the test station can interpolate between points 5 to 10°C apart.

Since digital temperature compensation is compatible with IC technol-ogy, TCXOs can be built using dedicated digital chips without adding undue amounts of hardware. It should also be noted that as IC memories become larger, the frequency jumps between memory locations can be made small enough for most voice applications. The digital TCXO with automatic pro-gramming of the memory during manufacture may significantly reduce the labor associated with analog temperature compensation.

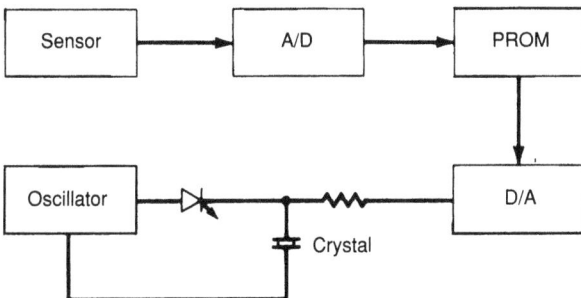

Figure 11.10 Digital temperature-compensation circuit.

Microcomputer temperature compensation. Another technique that can be used for temperature compensation is microcomputer temperature compensation. A block diagram is shown in Figure 11.11. This circuit appears to be more complex than the DCXO in Figure 11.10; however, the microcomputer offers several advantages.

One of the major advantages is that an *RC* oscillator such as an astable multivibrator can be used for the temperature sensor if the resistor is a thermistor. The microcomputer can then measure the period of oscillation to determine the temperature in digital form. A second advantage is the convenience of using eight-bit memory words. The microprocessor can then unpack the memory as required to achieve 9- or 10-bit words if desired. The microcomputer can also perform interpolation, so that a smaller number of memory words are required. Another advantage is that the flexibility of the microcomputer, particularly if it includes input compare and output capture registers, can be used to simplify the A/D and D/A conversion process; for example, an *RC* lowpass filter can be used with a port on the processor to form a high-accuracy D/A converter.

Microcomputers with (on-chip) electrically erasable and programmable read-only memory (EEPROM) as well as nonvolatile RAM make this technique attractive compared to the DCXO. As in the case of the DCXO, it is not necessary to provide compensation data at every temperature, and the test station can perform interpolation from a relatively small number of temperature points.

Basically, the microcomputer goes through a never-ending loop in which it reads the temperature, searches the table for the two closest data points, interpolates between them, and outputs the calculated voltages. The use of microcomputers holds the promise of ultimately producing improved accuracy over the analog compensated crystal oscillator. This results because sufficient bits can be used so that nearly all the frequency error is due to retrace errors in the crystal and the oscillator reactive elements.

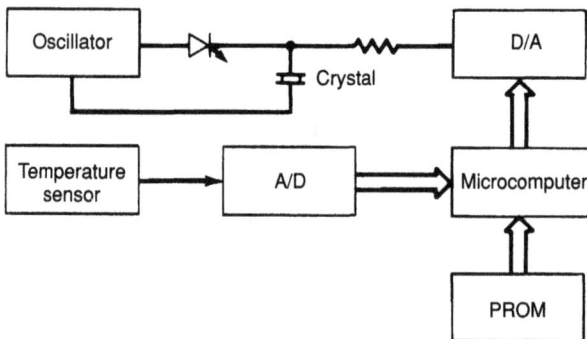

Figure 11.11 Microprocessor temperature-compensated crystal oscillator.

High-precision temperature compensated crystal oscillator. When the frequency accuracy required exceeds the region of 0.4 to 0.5 ppm, it may be desirable to use a temperature compensation system in which the reference crystal is not actually pulled to frequency, but the output frequency is synthesized in some way from the reference crystal whose temperature characteristics are stored in memory. A stress compensated (SC-cut) crystal can then be used. The SC-cut crystals, particularly overtone units, are generally too stiff to pull to frequency over the operating temperature range.

One system that has been investigated uses self temperature sensing in the reference crystal. This is accomplished by simultaneously exciting two modes in the same crystal that have different temperature coefficients. In one system, designated the MCXO, the fundamental and third overtone are simultaneously excited (see references 7 and 8). Another system uses the C-mode as the reference and the B-mode as the temperature sensor in an SC-cut crystal [7].

The self-sensing schemes generally require considerable complexity in the dual-mode oscillator. The fundamental third overtone system senses temperature by mixing the third harmonic of the fundamental with the third overtone oscillator. The resulting beat frequency is used as the temperature indicator. Unfortunately, since the fundamental and third modes have only moderately different temperature coefficients, any hysteresis in either mode limits the accuracy of the temperature compensation. Accuracies as high as a few parts in 10^8 have been reported in the laboratory with great care.

The second self-sensing scheme uses the B-mode of the SC-cut crystal as the temperature sensor. The B-mode is approximately 10% higher in frequency than the C-mode. The beat frequency is sometimes used as the temperature sensor signal. Unfortunately, for many SC-cut crystal designs, the B-mode is not well behaved and may undergo activity dips within the temperature range of interest.

Other schemes using the SC-cut crystal with some other form of temperature sensor have also been used; however, differences in the thermal time constant of the sensor and the reference crystal must be carefully matched and may severely limit the performance under temperature transient conditions.

References

1. M. E. Frerking, *Crystal Oscillator Design and Temperature Compensation* (New York: Van Nostrand Reinhold, 1978).

2. B. Parzen and A. Ballato, *Design of Crystal and Other Harmonic Oscillators* (New York: John Wiley & Sons, 1983).

3. V. E. Bottom, *Introduction to Quartz Crystal Unit Design* (New York: Van Nostrand Reinhold, 1982).

4. R. J. Matthys, *Crystal Oscillator Circuits* (New York: John Wiley & Sons, 1983).

5. A. Ballato, "Doubly Rotated Thickness Mode Plate Vibrators," in W. P. Mason and R. N. Thurston (eds.), *Physical Acoustics Principles and Methods*, vol. 3, Chapter 5, pp. 115–181 (New York: Academic, 1977).

6. G. R. Hykes and D. G. Newell, "A Temperature Compensated Frequency Standard," in *Proc. 15th Annual Frequency Control Symposium*, pp. 297–317 (Springfield, VA: National Technical Information Service, 1961).

7. A. Benjaminson and B. Rose, "Performance Tests on an MCXO Combining ASIC and Hybrid Construction," *Proc. Forty-Fifth Annual Symposium on Frequency Control* (1991): 393–397.

8. R.L. Filler, "Frequency-Temperature Considerations For Digital Temperature Compensation," *Proc. Forty-Fifth Annual Symposium on Frequency Control* (1991): 398–403.

Suggested additional reading

1. I. V. Abramson, "Two-Mode Quartz Resonators for Digital Temperature Compensated Quartz Oscillators," *Proc. 1992 IEEE Frequency Control Symposium* (1992): 442–447.

2. H. Kawashima and K. Sunaga, "Temperature Compensated Crystal Oscillator Employing New Shape GT Cut Quartz Crystal Resonator," *Proc. Forty-Fifth Annual Symposium on Frequency Control* (1991): 410–417.

3. M. Pavsek, D. Belaric, U. Kunaver, and M. Hrovat, "Crystal Oscillator with Temperature Compensation Realized in Thick Film Technology," *Hybrid Circuits*, No. 29, September 1992.

4. Y.L. Vorokhovsky, and B.G. Drakhilz, "High-Stability Quartz Oscillators on Internally Heated Quartz Resonators with AT and SC Cuts," *Proc. Forty-Fifth Annual Symposium on Frequency Control* (1991): 447–459.

5. M. Watanabe, Y. Sakuta, and Y. Sekine, "Digital TCXO Using Delta Modulation," *Proc. Forty-Fifth Annual Symposium on Frequency Control* (1991): 405–409.

12

Solid-State Power Amplifiers

Roderick K. Blocksome

Over the years, advances in transistor technology have produced devices capable of considerable RF power in the HF range. These modern devices make possible today's compact, yet powerful, RF power amplifiers (PAs).

This chapter will consider the design of linear PAs for the 1.6- to 30-MHz range at power levels up to 1000 W. Only linear operation will be covered, as it is commonly used in most SSB transmitters. It also forms the basis of the more sophisticated applications covered in Chapter 13. The normal HF range, usually considered as 3.0 and 30 MHz, has been extended over recent years to the top of the AM broadcast band at 1600 kHz. This is a result of increased military and maritime usage, coupled with advances in solid-state technology.

Transistors with ever-increasing power dissipation capabilities have been developed, making high-power solid-state linear amplifiers practical. As few as eight output transistors are used in a Rockwell 1000-W (average) PA. Such amplifiers have replaced vacuum tube amplifiers at lower power levels and seriously challenge them at the > 1000-W level as component costs decrease, linearity improves and higher-power devices become available.

12.1 A Typical Solid-State HF PA

Figure 12.1 shows a block diagram of a typical solid-state HF PA. A high-power system is shown in order to illustrate the combining of multiple output modules and their ancillary control systems. Multiple modules are usually required for output powers greater than 250 to 500 W. Depending on the types of devices used, the maximum ambient temperature, and the altitude of op-

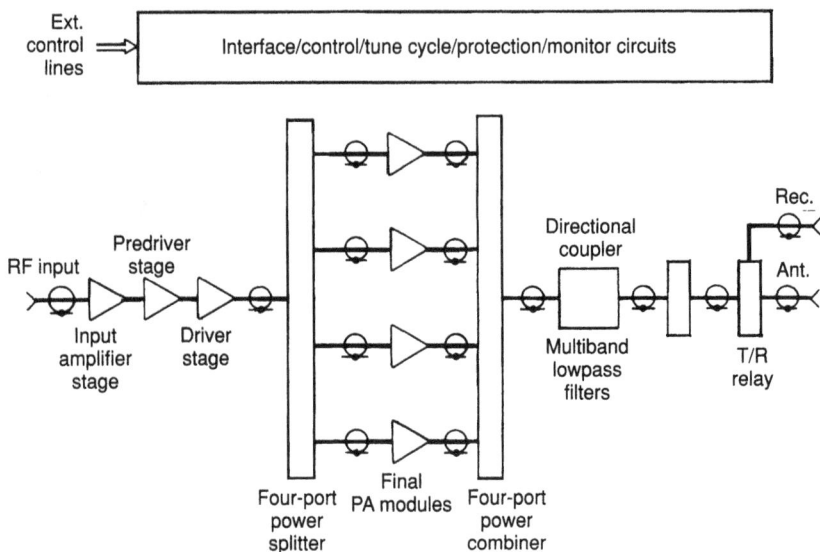

Figure 12.1 Block diagram of a typical solid-state power amplifier.

eration, it is presently feasible to produce 1000 W from a single amplifier module. There are several very important specifications that influence the output power obtainable from a single solid-state amplifier module or stage.

Power output ratings of PAs

A great deal of confusion and "specsmanship" is encountered in HF SSB transmitter output power specifications. The definitions and relationships of peak envelope power (PEP) and average RF output power have been explained very clearly in other publications. However, the amount of output power delivered into an SWR load by a tube-type PA versus a solid-state PA has been the subject of some confusion in recent years. A brief clarification follows.

Any PA must deal with two basic kinds of load. One is the 50-ohm resistive load used for the majority of laboratory performance tests and measurements, the other is the real-world antenna load that is bounded by a maximum specified SWR. Many problems arise when the power output into SWR loads is specified incorrectly or incompletely. This is due to a poor understanding of RF power relationships and how the PA output circuits behave when terminated with an SWR load. Recall the basic relationships between forward power, reflected power (the two quantities measured with a directional watt meter), and true power (the amount of "heating" power delivered into a load):

$$P_f - P_r = P_t \tag{12.1}$$

In the case of a 50-ohm dummy load, there is no reflected power and all the measured forward power is absorbed by the load. Forward power is equal to true power for only the special case of a matched load. Assume for the moment that an amplifier is capable of generating a constant forward power regardless of load SWR. It will deliver a decreasing amount of power to the load (true power) as the load SWR increases because reflected power increases as SWR increases. The true power absorbed by the load is the only power that has a chance to be useful. True power, factored by the antenna efficiency, is the amount that is radiated. True power is the quantity that we want to maximize in transmitter/antenna systems within the constraints of cost, weight, size, and prime power consumption.

Solid-state PAs are usually designed with broadband outputs and are limited by the maximum voltage, current, and dissipation ratings of the devices. They must therefore limit the maximum forward power that they can deliver into a given load impedance. Internal protection circuits monitor these parameters and control the drive to the final stages in order to provide the maximum amount of forward power without exceeding any of these ratings. In many designs, forward power is forced to decrease with increasing load SWR. For all possible load impedances, on a given SWR circle, there is a range of forward power allowed by the PA protection circuits.

A good rule of thumb in estimating forward power from a solid-state broadband PA operating class AB or B is that forward power is bounded by $1/S$ and $1/S^2$ where S is the SWR number (S:1). Typically the rated output power is held constant by the internal control circuits for SWR values up to 1.3 or sometimes as high as 1.5. At this point forward power starts to turn down as SWR increases. Typical broadband HF antennas present load impedance values with SWR values as high as 3.0. In such cases the forward power could be as low as 1/9 of the value for a dummy load. The true power is 25% less than the 1/9 value due to the reflected power of a 3:1 SWR (25% of forward power is reflected by a 3:1 SWR). Thus, the worst-case *true power* delivered into a 3:1 SWR load is approximately 11 dB below the power delivered into a 50-ohm dummy load by the same PA.

In contrast, vacuum tube PAs typically are designed with narrowband, tuned output tank circuits. This tank circuit is usually designed to transform the complex load impedance values anywhere on or within a given SWR circle (usually 3:1) to the desired resistive plate load impedance. The output tank tuning may be done manually by a radio operator, or automatically in more sophisticated equipment. The result is the same; the tube is always loaded with the proper resistive load impedance. The tube never sees a load SWR even though the PA output is terminated in an SWR load. Normal practice is to design the amplifier with enough margin to absorb the slight differences in tank circuit efficiency due to tuning into different load impedance values. When this is done, the tube PA, with its tuned output network, can deliver a *constant true power* into the load at any impedance within the matching range of the tank circuit. This means that forward power is actually higher with a load SWR than a 50-ohm load. The tuned tube PA delivers *full true power* into the 3:1 load SWR, so it has an 11-dB higher signal than the equivalent solid-state PA in the previous example. Figure 12.2 illustrates the differences between true power and forward power in a typical 1-kW tube PA and a typical 1-kW solid-state PA.

Two conclusions can be drawn from the above examples. First, true power is the quantity that must be considered in system designs and controlled by specifications to meet system requirements. Second, if the frequency agility of a broadband PA and the true power capability of a tuned PA are required, then the required amount of solid-state PA is about 10 times greater than a tube PA with a tuned tank circuit. Clearly, future frequency-agile HF transmitter systems require the development of new broadband HF antennas with very low SWR or development of extremely fast-tuning, high-power, impedance matching networks to deliver power to antennas within reasonable cost, size, weight, and prime power consumption limits.

The factors determining maximum output power fall into two categories: device ratings and thermal considerations. The manufacturer's safe operating ratings of voltage breakdown and maximum current must not be exceeded under any circumstances. High-power RF transistors, unlike vacuum tubes, are very unforgiving of any transient condition which exceeds their electrical ratings. These ratings are usually adequately described in the device data sheet. Less well described are the subtleties of thermal limits.

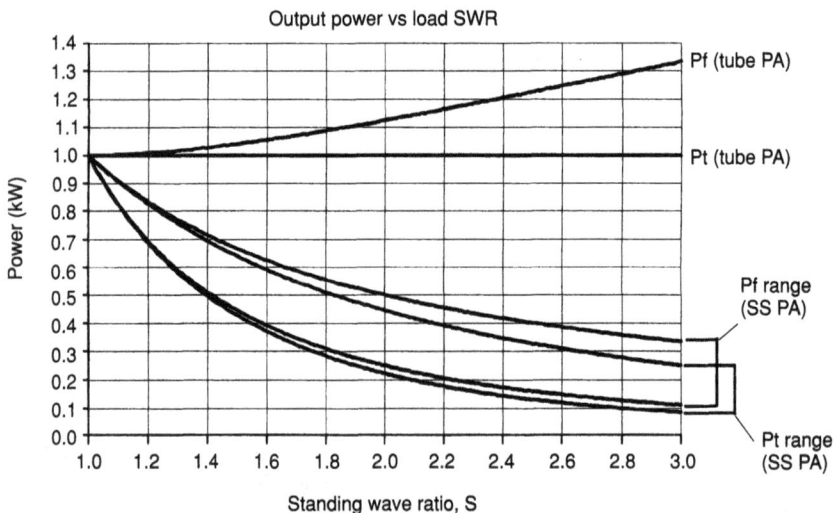

Figure 12.2 Comparison of true power and forward power between a 1-kW tube PA and a 1-kW solid-state PA.

Data sheet specifications are usually limited to specifying the *average* thermal resistance θ_{jc} of the junction to case and the maximum junction temperature of the device chip. Many devices specify 200°C as the maximum rated junction temperature. However, 160°C maximum should be used for reliable design.

θ_{jc} is a difficult number to obtain accurately because it is influenced by several variables inherent in the transistor manufacturing process [1]. The determination of θ_{jc} requires measuring the semiconductor junction and case temperatures during actual RF operation. Conventional thermocouples can be used to measure the case temperature, but a scanning infrared microscope must be used to measure junction temperatures on the die. Each junction temperature on the die is recorded and the average obtained. Therein lies a potential problem for the design engineer. The transistor data sheets generally do not specify the maximum (or minimum) junction temperature, only the average. Yet the variation of junction temperatures across a single die is often considerable because of bonding voids under the die, effects of reactive loads (voltage standing wave ratio or VSWR), and variations in emitter ballasting. The hottest junction is stressed the most and therefore leads to an early failure of devices operated at or near the specified maximum limits. A conservative derating factor must be applied to the device thermal ratings to ensure a reliable design. The entire cooling system design must be an important and early task in designing solid-state PAs.

Intermodulation distortion

Intermodulation distortion is a very important PA specification because it directly affects performance of the HF communication link and interference to adjacent HF channels. Intermodulation distortion products are formed when any device or component acts on the desired signal in a nonlinear fashion.

Generally, narrowband, tube-type HF amplifiers are capable of lower levels of IMD than current solid-state amplifiers. Many solid-state amplifiers on the market today exhibit relatively poor IMD because of a greater priority placed on RF output power at the expense of linearity. Aside from this, the better IMD performance of tube amplifiers can be attributed to the longer history of design improvements of RF power grid tubes for linear service and the relative ease of single- or multistage RF feedback circuits in tuned output tube circuits. Refer to Chapter 14 for a thorough discussion.

There are two ways to specify IMD performance: the two-tone and the noise loading test. The noise loading test more accurately simulates the conditions encountered in high-power fixed-station PAs transmitting multi-tone data. This technique and the mathematical derivation of IMD products for both methods are discussed in Chapter 14.

A really good test to closely simulate IMD performance of an SSB voice transmitter does not exist. The standard two-tone test is the one most readily made that provides reproducible results. This requires injection of two independent (not phase-related) equal-amplitude RF signals into the PA RF input. The frequency of the signals should be within the range of the typical SSB exciter baseband output, that is, 300 to 2700 Hz. Typically, tones of 800 and 1800 Hz are used, giving a tone separation of 1000 Hz. The test is then performed at several RF frequencies throughout the operating range of the PA under test. The amplifier output is observed on a good-quality spectrum analyzer, the amplitudes of all spectra being noted in relationship to the two injected tones. The amplitude of each input tone must be adjusted to produce an output power exactly 6 dB down from the rated PEP of the PA. This ensures that the combined two tones drive the PA exactly to the PEP rating. The odd-order distortion products will lie on either side of the input tones, each spaced in frequency from each other (and the input tones) by the separation of the original input tones.

Herein lies an opportunity for playing the game of "specsmanship." Should the level of IMD products be specified in relation to PEP or the level of the two injected tones? If the former is used (as is nearly always the case in amateur products and many commercial amplifiers), a figure 6 dB larger results. However, a strong case may be made to reference the IMD level to each of the two equal desired tones. Since we are measuring IMD products in the frequency domain, they ought to be specified against a reference level also measured in the frequency domain. Figure 12.3 illustrates a typical PA output spectrum when driven by a two-tone test

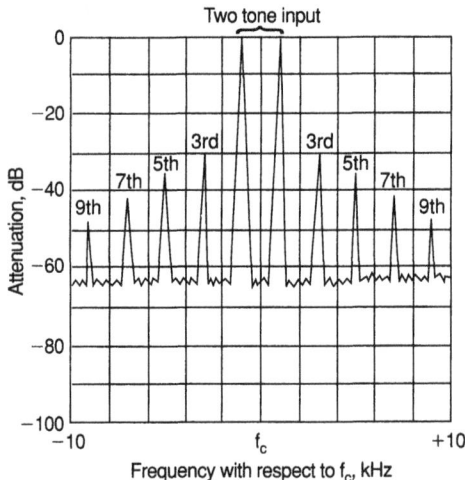

Figure 12.3 Typical spectral output of a PA driven with two-tone test signal.

signal. The amplifier output in Figure 12.3 exhibits IMD levels "not less than 30 dB below one of two equal tones." This is a typical specification for good-quality, commercial, high-power HF PAs. Typical amateur-grade solid-state PA specifications range from –30 dB to as poor as –20 dB below one of two equal tones. More expensive military solid-state PA specifications range from –26 to –40 dB.

How do these two-tone test results relate to the real-world problems of error-free data transmissions and minimum adjacent channel interference to other HF spectrum users? Consider a 1000-W PA producing third-order IMD levels of –24 dB. The unwanted distortion products are being transmitted in an adjacent channel at a power level of 1 W! This seems small until one considers the fickle characteristics of HF propagation which enable amateurs to communicate worldwide with a few tens of milliwatts!

Another, more effective, method of evaluating the effects of various IMD levels may be accomplished by operating an SSB voice transmitter for a long period of time while recording and storing the peak spectra of the output signal. The result is a power density display of in-band as well as out-of-band signals. This test was applied to three different grades of SSB equipment, each exhibiting markedly different IMD levels on the standard two-tone tests. Figure 12.4 shows the results for comparison. The power density curve of the in-band signal display is typical for a male voice into a particular type of microphone. The results shown in Figure 12.4 graphically indicate the need for low-IMD designs in HF linear PAs.

Frequency range

The frequency range of an HF wideband solid-state PA is usually 1.6 to 30 MHz, with some applications requiring only a 2.0-MHz lower limit. The frequency range specification affects the design primarily in the area of wideband RF transformers and decoupling networks on the dc line. A range of 1.6 to 30 MHz spans more than four octaves, requiring transformers and networks of more than casual design. Special ferrite cores have been developed to handle wide bandwidths and power levels of 1000 W or more. Transmission line transformers are preferred over conventional wirewound transformers for their wide bandwidth and low insertion loss.

Harmonics

The harmonic levels specified for most HF SSB communications range from –46 to –80 dB below the fundamental. The radiation of unwanted harmonic signals and the interference that these signals cause are of primary concern. Only a class A amplifier can produce harmonic levels this low without output filters. However, it has very low efficiency—typically 25 down to 10%, depending upon allowable harmonic levels.

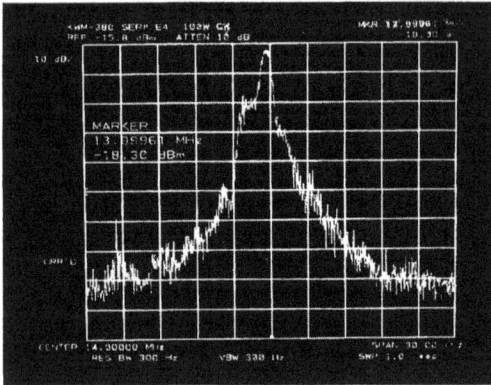

Figure 12.4A Single-sideband signal comparisons. KWM-380 transceiver. IMD: –24 dB below one of two tones; vertical: 10 dB/div; horizontal: 3 kHz/div; center: 14.0 MHz.

A

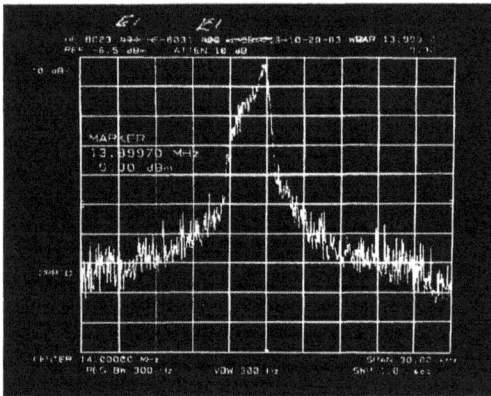

Figure 12.4B Single-sideband signal comparisons. HF-8023/8031, 1-kW PA/PS. IMD: –32 dB below one of two tones; vertical: 10 dB/div; horizontal: 3 kHz/div; center: 14.0 MHz.

B

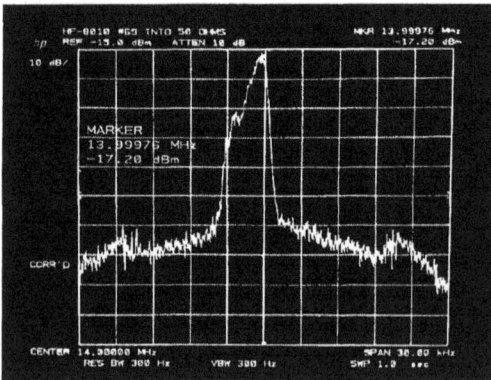

Figure 12.4C Single-sideband signal comparisons. HF-8010 exciter. IMD: –50 dB below one of two tones; vertical: 10 dB/div; horizontal: 3 kHz/div; center: 14.0 MHz.

C

Class AB or B operation is commonly used for linear amplification. Typical collector/drain efficiencies are 40 to 55%, but with significant harmonic output. In a push-pull stage, even-order (second, fourth, sixth, etc.) harmonics cancel to a degree (depending on circuit balance and device matching) while odd-order harmonics are significantly higher. The third harmonic is typically only 10 to 13 dB below the fundamental. A band-switched lowpass filter can be used to improve the PA harmonic output levels.

Hum and noise

The hum and noise specification of modern SSB transmitters is usually 40 to 55 dB below the rated PEP output power. Hum on the signal refers to modulation of the signal by ripple products from the power supply rectifier and filter circuits. Noise can be caused by any other circuit, such as voltage regulators, that put a random low-level modulation on the PA output. Solid-state PAs offer only about 3 to 12 dB of ripple rejection, depending upon whether or not negative feedback is employed. The remaining ripple rejection must be accomplished in the power supply filter.

Applications

The solid-state linear PA finds application in several broad categories which, due to prime power considerations, ends up affecting the power level of the amplifier. Amateur radio applications often use 13.8-Vdc devices to allow operation directly from vehicular batteries. This low collector voltage limits the practical output power to around 150 W for a push-pull transistor pair. Higher power may be obtained by combining multiple modules; however, total current drain on the automotive battery becomes very high.

Amplifiers for military vehicular applications standardize on 28 Vdc. Correspondingly higher powers can be obtained from a single push-pull pair of RF transistors. Radio frequency power transistors for 28-Vdc operation are available in a wide range of power and frequency capabilities. As a result, these devices are often used for fixed-station or transportable applications. However, for higher power (up to 1 or 2 kW), higher-voltage devices are desirable for two reasons:

1. The higher collector voltage allows reasonable values (a few ohms or greater) of collector/drain load impedances. This allows efficient wideband transformer design and implementation to couple the RF to a reasonable load impedance such as 50 ohms.

2. A minimum number of modules (push-pull transistor pairs) are required to be combined to achieve high power by operating each module at the highest practical output power.

Recent developments in FET PAs

Bipolar and FET transistors operating at 50 Vdc are available for designs that produce 1000 W or more of RF output power. There are RF power FETs that operate from 100-Vdc supplies. However, their thermal and current limitations do not provide a significant increase in power output compared to 50-V PA designs.

In recent years, RF power FETs for HF SSB applications have continued to show significant improvements. Bipolar junction transistors for HF SSB, being a more mature technology, have not enjoyed the same improvements in contrast to FETs. Advancements in FET technology have been primarily in the development of large, rugged die to withstand circuit stresses, higher gain-bandwidth products, and packages containing multiple die. A brief review of these two types of devices in HF PA circuits will be given here. Reference 2 is an excellent source for more details on FET and BJT HF amplifiers.

Unlike BJTs, high-power HF FETs can be successfully paralleled. Careful layout design is required to provide equal and symmetrical path lengths for each device. Simple, low-current, bias circuits and closer matching of dc and RF gain characteristics are advantages that FETs enjoy over BJTs. Thus, parallel connected FETs can be made to share drive power and the output load more equally than BJTs.

Field-effect transistors have a high but constant value of input capacitance (Ciss) and an input resistance in the hundreds of megohms. Very simple, low-current, gate bias circuits can be used. Adjustable, regulated voltage sources are all that is required. However, the RF input circuit must be designed to drive the high input capacitance. This is accomplished by either loading the FET input with a low-value shunt resistor, employing negative feedback, or a combination of both. The more devices that are paralleled, the more gate capacitance, and the lower the value of load resistance needed to maintain a low input SWR over the bandwidth of the amplifier.

An input impedance lower than 50 ohms usually results when the input gate capacitance is adequately swamped. A wideband transformer can be used to match the impedance back to the desired 50-ohm input. Achieving an adequate impedance match to the amplifier input over a wide bandwidth is more easily accomplished with FETs than with BJTs. The swamped gate capacitance presents a well-behaved impedance over wide bandwidths compared to the large, low-value impedance excursions of a high-power BJT. The FET PA designer has considerable freedom in setting the input impedance and at the same time achieving a low input SWR simultaneously with adequate gain over the PA bandwidth.

Two FETs in push-pull have the advantage of a gate-to-gate balanced input impedance two times higher than the unbalanced gate-to-ground input impedance of a single device in the same type circuit. For higher power in a single stage, each side of the push-pull circuit can consist of two or more

FETs in parallel. With proper swamping and/or feedback, the input can be matched to 50 ohms unbalanced with a wideband balun transformer. The disadvantage of this scheme is the requirement for equal and symmetrical current paths for each device. A planar layout, as commonly used on a printed circuit board layout, requires some compromise in these design criteria. The main problem is that the common impedance through the ground plane from the FET source leads on each side of the push-pull pair is not the same for all devices. The result is unequal drive and load sharing which will cause the gain bandwidth product, efficiency, and output power to be lower than expected. In recent years two new packaging techniques have been developed to improve the performance in these areas.

One new family of FET devices have multiple die mounted in proximity to each other inside an especially large flange package. The gates and drains are connected in parallel with wire bonds and their sources are connected directly to the case flange to achieve an extremely low source impedance to ground. Motorola's MRF-153 300-W and MRF-154 600-W FETs are examples that contain two and four (respectively) parallel-connected die. A single push-pull pair of MRF-154 FETs can achieve 1 kW in a single stage HF PA. However, the PA designer now has to design and implement high-power wideband RF balun transformers that must work at very low impedance values. Also, the design must have a more effective cooling system, because a large amount of heat is generated in two very small areas rather than spread out over the larger area that would be available if individually packaged FETs were used in an equivalent parallel/push-pull PA circuit. Because of the very high input capacitance of these large parallel die FET devices, it is very difficult to design a PA with a bandwidth from 1.6 MHz to the VHF region. All interconnection of components must consider the amount of stray inductance involved and its effect on the low impedance values of the circuits. Innovative circuit layout and construction techniques to solve these interconnect problems are required for a successful design.

A second packaging innovation improves gain-bandwidth performance by mounting two FET die in close proximity with only their sources internally connected directly to the flange. This provides an extremely low source-to-source common impedance, which is critical to wide bandwidths in a push-push circuit. Individual gate and drain leads are connected to each of the two die in the package. This packaging technique was first introduced by CTC and later refined by Motorola. It is often called a "Gemini" package due to the two closely matched FET die packaged for optimum wide-bandwidth push-pull PA designs. Motorola's MRF-151G is an example of this technique. Bandwidths of 1.6 to 200 MHz are achievable at 150 W with only one of these devices.

Wideband transformer design and construction along with circuit interconnections are critical in this type of amplifier, but for a different reason.

Instead of extremely low impedances at 1-kW power levels, the design here must be adequate for moderate power well into the VHF region.

12.2 Design Considerations of an Amplifier Stage

Collector/drain load impedance matching [3, 4]

Solid-state PAs with large amounts of power gain often require several stages of amplification. The first stages must have a high degree of linearity in order to achieve low IMD of the overall PA. Class A is often chosen for the driver stages for its high linearity. Since driver amplifiers are usually operated at comparatively low power, the low efficiency of class A can be tolerated. The class A amplifier may be either single-ended or push-pull. Wideband load impedance determination for the single-ended class A stage will be examined. The push-pull configuration is essentially the same as two single-ended stages operating 180° out of phase and combined in the output transformer.

A class A BJT amplifier is illustrated in Figure 12.5. The bias is selected to achieve the desired quiescent collector current within the device dissipation ratings and yet allow sufficient current "swing" to achieve the desired RF power output. The device will operate as a current source as long as the collector voltage does not swing into the saturation or cutoff regions of the transistor.

The amount of power that may be obtained from a class A stage may be found, starting with the fact that the current in the load R_L is $I_{RL} = V_{RL}/R_L$. Since the voltage cannot swing into either saturation or cutoff, the peak RF voltage across the load resistor must be equal to or less than V_{CC}. This assumes that the transistor quiescent bias point has been selected such that RF collector voltage peaks are symmetrical with respect to saturation and cutoff voltages.

The RF output power is

$$P_o = \frac{\left(V_{R,\,\text{peak}}/\sqrt{2}\,\right)^2}{R_L} = \frac{V_{R,\,\text{peak}}^2}{2R_L} \tag{12.2}$$

but

$$V_{R,\,\text{peak}} \leq V_{CC}$$

Therefore $P_o \leq V^2{}_{CC}/2R_L$). In practice $V_{R,\,\text{peak}}$ cannot equal V_{CC} because of the saturation voltage $V_{CE,\,\text{sat}}$ of BJTs and the "on-resistance" of FETs.

The load resistance for maximum class A output power (for BJTs) is then given by

$$R_L \leq \frac{\left(V_{CC} - V_{CE,\,\text{sat}}\right)^2}{2P_o} \tag{12.3}$$

A

B

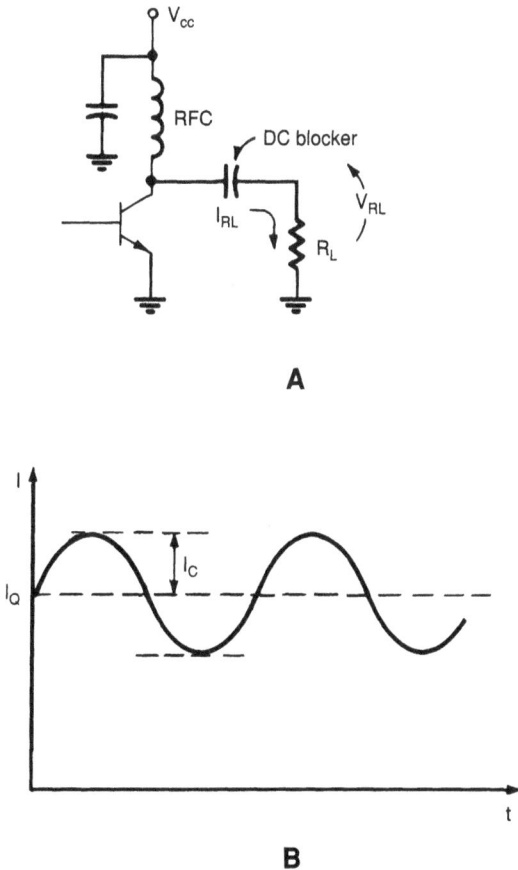

Figure 12.5 Class A BJT amplifier: (A) single-ended; (B) collector current.

The collector dissipation for class A is the difference between input dc power P_i and the RF output power P_o. The important point is that the device dissipation is at its maximum with *no* RF drive and at its minimum when maximum RF power is delivered to the load.

The basic design problem for a class A collector/drain circuit becomes

1. A device must be chosen that has sufficient dissipation rating to deliver the required RF output power.

2. The device must have sufficiently high maximum current and voltage ratings.

3. Steps 1 and 2 must be reevaluated in the light of any required wideband impedance transformer requirements of the class A stage output and the actual RF load impedance.

Design consideration 3 has a large influence on the overall design. Transmission line transformers offer superior performance in terms of bandwidth and low loss. Simple transmission line transformers only provide certain impedance transformation ratios such as 1:1, 1:4, 1:9, 1:16, and so on. However, the equal-delay transmission line transformer (described later) can provide even wider bandwidths at a larger number of impedance transformation ratios with only slightly more complexity. To illustrate the design relationship between device ratings, class A power output, and transmission line wideband impedance transformers, Table 12.1 is offered. The example assumes the RF power is delivered to a 50-ohm load and a nominal value of $V_{CE, sat}$ is included.

Class B (or class AB) operation offers high-efficiency RF linear amplification. It may be used single-ended with a tuned (narrowband) impedance matching network. More often it is found in a push-pull configuration with wideband impedance transformers on the input and output.

Figure 12.6 shows a push-pull Class B circuit and the collector current waveform. Each device delivers current (and also the *entire* power output $2P_0$, where P_0 is the average power per transistor) to the load during *one-half* of the RF sine wave. When one transistor is conducting through one-half the primary winding, the other transistor is turned off. The voltage on the turned-off collector then rises to a value of about $2V_{CC}$ due to auto-transformer action in the primary winding. The peak collector RF voltage $V_{c, peak}$ should not be allowed to exceed the supply voltage V_{CC} in either the positive or negative direction. This means that each transistor must be rated to supply the peak current $V_{c, peak}/R$) when conducting and withstand a peak instantaneous voltage of at least $2V_{CC}$ when it is turned off. The average (over a full cycle) RF output power from each device is

$$P_o \leq \frac{\left(\dfrac{V_c, \text{peak}}{\sqrt{2}} \right) \cdot \dfrac{1}{2}}{R} \qquad (12.4)$$

where R is the collector-to-ground load resistance. Note that R is 1/4 of the collector-to-collector load impedance.

TABLE 12.1 Maximum Class A Power for Common Loads and Voltages

Impedance ratio	1:1	1:4	1:9	1:16		
Load impedance	50	12.5	5.56	3.125	V_{CC}	$V_{CE,sat}$
RF output power	1.3 W	5 W	12 W	21 W	12.5 V	1.0 V
RF output power	1.6 W	6 W	14 W	25 W	13.5 V	1.0 V
RF output power	7.0 W	28 W	63 W	112 W	28.0 V	1.5 V
RF output power	23 W	90 W	203 W	361 W	50.0 V	2.5 V

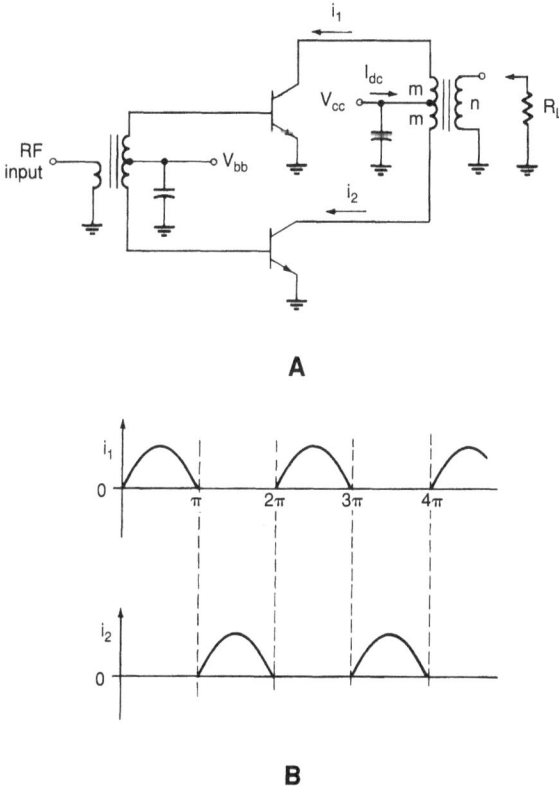

Figure 12.6 Class B amplifier: (A) push-pull; (B) collector current waveforms.

Considering that $V_{c,\,peak} \le (V_{CC} - V_{CE,\,sat})$, Equation 12.4 becomes

$$P_o \le \frac{\left(V_{CC} - V_{CE,\,sat}\right)^2}{4R} \qquad (12.5)$$

The amplifier load impedance R_L is reflected to each collector by the turns ratio squared of the output transformer as shown in Figure 12.7. Since

$$R = \left(\frac{m}{n}\right)^2 R_L \qquad (12.6)$$

then the full-cycle average output power for each transistor is

$$P_o \le \frac{\left(V_{CC} - V_{CE,\,sat}\right)^2}{4\left(\dfrac{m}{n}\right)^2 R_L} \qquad (12.7)$$

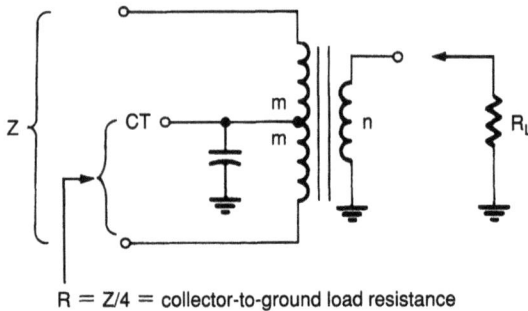

R = Z/4 = collector-to-ground load resistance

Turns ratio: $\dfrac{n}{2m}$

Impedance ratio: $\dfrac{n^2}{(2m)^2} = \dfrac{n^2}{4m^2}$

$$Z = \dfrac{R_L}{n^2/4m^2} = \dfrac{4m^2 R_L}{n^2}$$

Figure 12.7 Wideband impedance-matching transformer.

and the total power delivered to R_L by the push-pull pair is $2P_0$. Table 12.2 shows the actual collector load impedance for various transformer ratios with a secondary load of 50 ohms.

There are two basic types of wideband impedance matching transformers used in modern solid-state PAs [5, 6]. They are commonly referred to as wire-wound types and transmission line types. The term wire-wound refers to the common type of transformer in which the power is transferred from one winding to another by a magnetic flux linking the two windings. The transmission line type refers to transformers constructed of sections of transmission lines interconnected to give the desired impedance transformations. Magnetic core materials are also used, but their function is to suppress common mode or "nontransmission line" currents.

TABLE 12.2 Transistor Load Impedances for Various Common Transformers

Turns ratio	Impedance ratio	Transformer model values	R_L, ohms	Collector-collector load impedance Z, ohms	Transistor collector load impedance R, ohms
1:1	1:1	$n = 2, m = 1$	50	50	12.5
1:2	1:4	$n = 4, m = 1$	50	12.5	3.125
1:3	1:9	$n = 6, m = 1$	50	5.556	1.389
1:4	1:16	$n = 8, m = 1$	50	3.125	0.781

TABLE 12.3 Wideband Transformer Configurations

Source	Load
Balanced, floating	Balanced, floating
Balanced, center-tap grounded	Balanced, floating
Unbalanced (single-ended)	Balanced, floating
Balanced, floating	Balanced, center-tap grounded
Balanced, center-tap grounded	Balanced, center-tap grounded
Unbalanced (single-ended)	Balanced, center-tap grounded
Balanced, floating	Unbalanced (single-ended)
Balanced, center-tap grounded	Unbalanced (single-ended)
Unbalanced (single-ended)	Unbalanced (single-ended)

The wire-wound transformer can provide a large number of impedance transformation ratios [7]. There are no inherent restrictions on the primary-to-secondary turns ratio that one may employ. The bandwidth that one obtains for a given turns ratio is another matter. In general it is not as great as a comparable transmission line transformer. The wire-wound transformer may also be a conventional autotransformer.

The wire-wound transformer with separate primary and secondary windings is easily connected to all combinations of source and load configurations. Table 12.3 lists all possible combinations of source and load configurations for a wideband transformer.

Figure 12.8 illustrates two common construction techniques for wire-wound wideband RF transformers. The balun core in Figure 12.8A does not have to be a single block of ferrite (as shown); it may also be constructed of two separate ferrite tubes or sleeves. A number of toroidal cores can be stacked to make a ferrite sleeve also. The type of construction shown in Figure 12.8A has been proven in numerous designs and is especially useful for matching to impedances as low as 1 ohm on the secondary and has the capability of a good symmetrical center tap on the secondary. Copper tubes are inserted in the core to form the low-impedance secondary. Teflon-insulated wire wound through the copper tubes (and thus also the ferrite core) forms the primary. Tin-plated copper braid may also be successfully used in lieu of the copper tubes and copper-clad circuit board end pieces.

The toroidal core wire-wound transformer shown in Figure 12.8B illustrates another construction technique. Although the windings are shown physically separate, the winding could also be wound bifilar (two wires wound side by side), trifilar, quadrifilar, or even multifilar. Various transformation ratios are made by series connecting two or more of the windings. The multifilar winding technique provides close coupling between the windings and improves the high-frequency performance of the transformer.

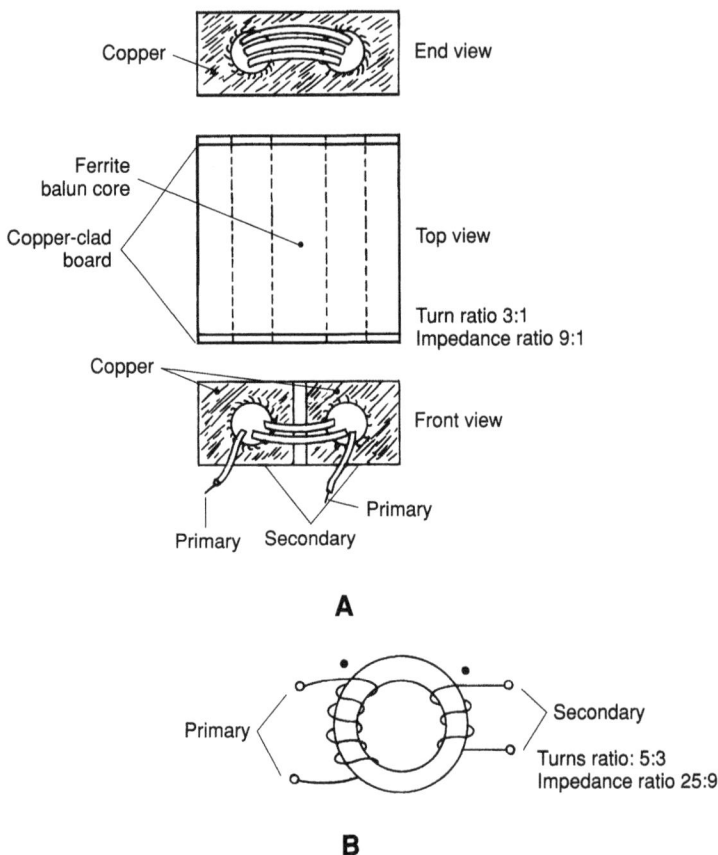

Figure 12.8 Two examples of wire-wound wideband RF transformers: (A) balun core construction; (B) toroidal core construction technique.

If a bifilar winding is placed on a core, the source connected across the two ends at the start of the winding, and the load connected across the opposite end of the two wires, a transmission line balun transformer is formed. The term bifilar implies that the two wires lie parallel to each other; however, the two wires may be twisted to achieve the same effect. The desired characteristic impedance Z_0 of such transmission lines is determined by the following factors:

1. Wire gauge

2. Type and thickness of insulation on the wire

3. Distance between wires (for parallel or bifilar)

4. Number of twists per length (for twisted pair)

Four or more wires may be twisted together, then half of the wires paralleled to achieve a lower transmission line characteristic impedance.

The 1:1 transmission line balun transformer is basic to understanding more complex transmission line transformer designs [8]. Figure 12.9A shows a typical balun using a twisted-pair wire transmission line wound on a toroidal core. The transformer is depicted as a physical implementation, a schematic representation, and an equivalent circuit to fully illustrate its operation. In this case a single-ended (unbalanced) source is connected via the balun to a balanced and floating load R_L.

The current which magnetizes the core must flow in a path that does not cause an imbalance in the signal current and thus upset the balanced output voltages. The magnetizing current i_m [9] flows through both windings and the load.

Figure 12.9B illustrates the same circuit except the balanced load has a grounded center. It is no longer floating, and the magnetizing current no longer flows equally in the windings and the load. The result is a current imbalance in the load. This situation can be corrected by adding a third or tertiary winding to the transformer as shown in Figure 12.9C. The tertiary winding now provides a path for the magnetizing current around the load. This type of transformer is made by winding the core trifilar, or with three wires twisted. At first glance, the tertiary winding would indicate a wire-wound transformer. Only when the source and load are connected as in Figure 12.9C can one identify it as a transmission line balun (with tertiary winding).

The twisted-pair or bifilar transmission line winding may also be constructed of coaxial transmission line. Coaxial cables with characteristic impedances of 25, 50, 62, 75, and 95 ohms are readily available. Coaxial cables with other characteristic impedances can be obtained by special order and at premium prices. Lower-impedance lines may be obtained by paralleling two or more coaxial lines and winding them on a core. For example, two 50-ohm coaxial cables of equal length and connected in parallel will provide a 25-ohm line. Similarly, paralleling four 50-ohm lines results in an equivalent 12.5-ohm line.

The 4:1 transmission line transformer shown in Figure 12.10 illustrates the practical details of interconnecting a ferrite-loaded transmission line to achieve unbalanced-to-unbalanced operation. Suppose the source must be loaded with 50 ohms. The transformer steps down the 200-ohm load impedance R_L by using a 100-ohm ferrite-loaded coaxial cable. The transformer may be used in reverse to step a 12.5-ohm load up to 50 ohms by using a ferrite-loaded 25-ohm coaxial cable. Other, more standard, coaxial cable impedances may be used if some degradation in bandwidth is acceptable.

The rule of thumb for selecting ferrite cores for transmission line transformers is to use a sufficient combination of core size, permeability, and winding turns to achieve a winding reactance at the lowest frequency of operation equal to five times the transformer load impedance. Using more than a factor

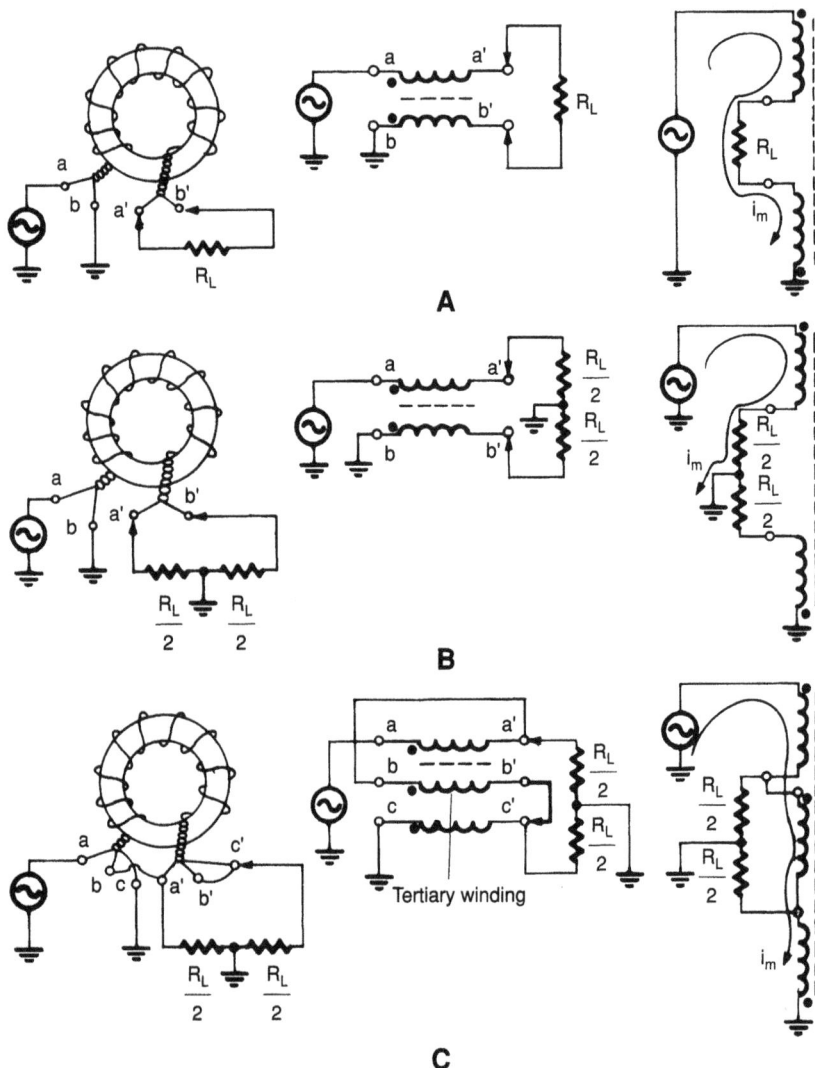

Figure 12.9 1:1 transmission line balun transformers with balanced load: (A) floating; (B) center-grounded; (C) center-grounded with tertiary winding.

of five will do no harm except it will tend to make the winding go series-resonant at a lower frequency and thus limit the highest frequency for which the transformer is useful. In general, the best results involve using a high-permeability ferrite with low-loss, sufficiently high cross-sectional area, and as few winding turns as possible. Some examples of suitable core materials are Stackpole's 7D, Ceramic Magnetics' CMD 5005, and Fair-Rite's No. 43.

The high-frequency response also degrades when the length of coaxial cable in the Figure 12.10 transformer approaches one-quarter of a wavelength. As long as the line is electrically short, there is negligible phase difference between the signal at the line input and its output, where the output shield is connected to the input center conductor.

An important subclass of transmission line transformer, called an equal-delay transformer, eliminates the phase difference and extends the upper frequency limit. The equal-delay transformer was first investigated by G. Guanella in 1944 [10] and later by W. A. Lewis [11] at Collins Radio in 1965. Lewis' results were never published outside the company. Figure 12.11A illustrates the metamorphosis of an ordinary 4:1 transmission line trans-

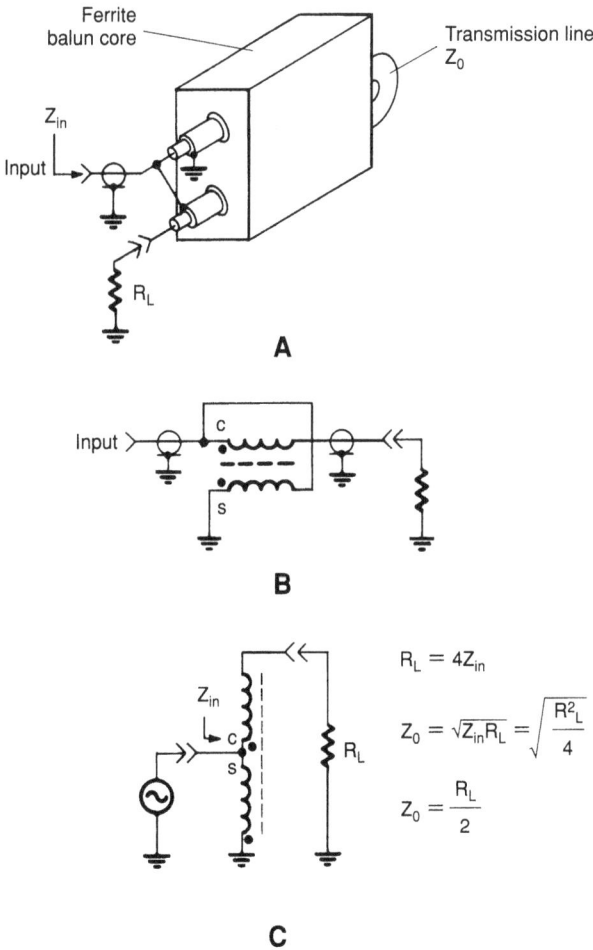

$$R_L = 4Z_{in}$$

$$Z_0 = \sqrt{Z_{in}R_L} = \sqrt{\frac{R_L^2}{4}}$$

$$Z_0 = \frac{R_L}{2}$$

Figure 12.10 Transmission line transformer (4:1 unbalanced-to-unbalanced): (A) pictorial view; (B) schematic; (C) equivalent circuit.

former, Figure 12.11B that of a "stretched-out" version, and Figure 12.11C that of the equal-delay transformer. The interconnecting wire from shield to opposite end center conductor is replaced with a transmission line equal in length (and thus equal in phase delay) to the original transformer transmission line.

The equal-delay transformer's physical configuration lends itself nicely to good layout techniques because its input and output terminals are on opposite sides of the transformer. Additional impedance transformation ratios are possible by adding to the equal-delay transformer as shown in Figure 12.12. An intuitive way to look at the equal-delay transformer is to recognize that the two transmission lines are connected in parallel at the load and in series at the input.

The "top" line must be ferrite-loaded to provide a high impedance to common-mode shield currents, in effect to isolate the input side of the shield from its grounded output side. In like manner three and four transmission lines may be connected to give impedance transformation ratios of 9:1 and 16:1, respectively. If the second line in a two-line (4:1) transformer requires one unit of ferrite for adequate isolation at the lowest frequency of operation, the third line requires two units and the fourth line requires three units to provide adequate common-mode isolation to the higher voltages on the input shields.

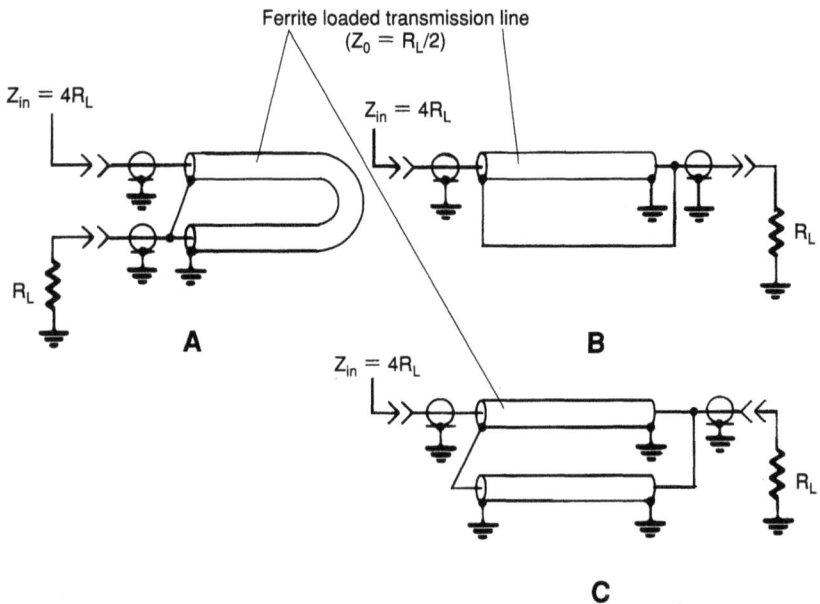

Figure 12.11 Derivation of the equal-delay transformer from the standard 4:1 transmission line transformer.

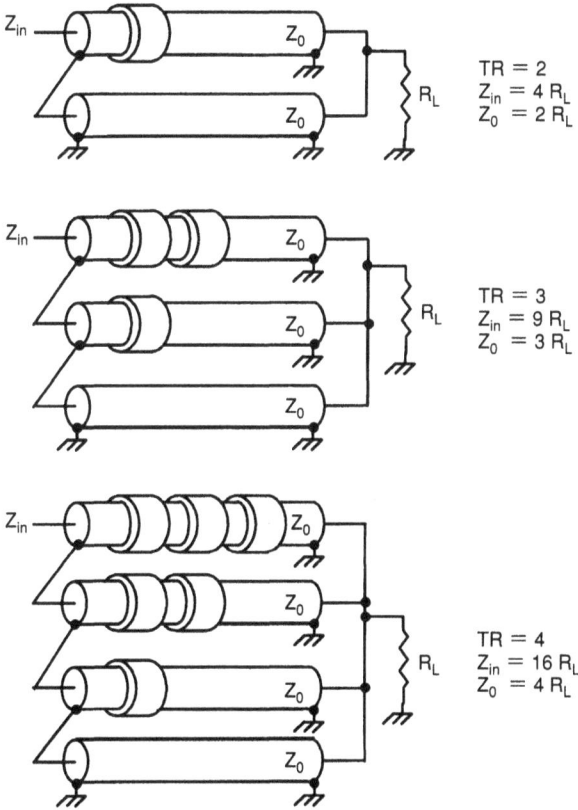

TR = 2
$Z_{in} = 4 R_L$
$Z_0 = 2 R_L$

TR = 3
$Z_{in} = 9 R_L$
$Z_0 = 3 R_L$

TR = 4
$Z_{in} = 16 R_L$
$Z_0 = 4 R_L$

Figure 12.12 Equal-delay transformer configurations.

Noninteger ratio equal-delay transmission line transformers

Equal-delay transmission line transformers are not restricted to integer transformation ratios. Transformation ratios of any rational number (m/n) are possible. The m/n transformation ratios are accomplished by using combinations of series and parallel connections at each end of the transformer. This class of transformers was investigated at Rockwell in 1985. McClure at RCA conducted a similar investigation predating the one at Rockwell and published the results in an excellent article [12]. A brief description of how to design these transformers will be presented.

General design rules to design any equal-delay transmission line transformer are as follows:

1. The turns ratio (TR) can be any rational number, that is, the quotient of any two integers.

2. The quotient of any two integers may be broken down by a continued fraction expansion. The first term is the number of lines with their inputs series connected. The next denominator is the number of line inputs parallel connected. The third denominator (if there is one) is the number of line inputs series connected and so on until the expansion is complete.

3. The transmission line output connections must be opposite the input line connection. If the input is series connected then the output must be parallel connected.

4. The total number of transmission lines in the transformer may be determined by adding the first term and all of the denominators in the fractional expansion of the turns ratio number.

5. The Z_0 of the transmission lines should be equal to the turns ratio multiplied by the output load impedance. If the transformer is employed in a circuit in which the exact design ratio is not used, then determine Z_0 from the square root of the product of the input and load resistance. This will give slightly better bandwidth.

6. The impedance looking into the input of the transformer is the product of the turns ratio squared and the load resistance.

For example: An equal-delay transformer with a 10/7 turns ratio is required. A continued fraction expansion yields

$$\text{TR} = \frac{10}{7} = 1 + \frac{3}{7} = 1 + \frac{1}{\dfrac{7}{3}} = \mathbf{1} + \cfrac{1}{\mathbf{2} + \cfrac{1}{\mathbf{3}}} \tag{12.8}$$

Referring to the bold numbers in the fractional expansion of 10/7, we can determine how many transmission lines are required $(1 + 2 + 3 = 6)$ and the interconnections required at the input side of the transformer. In this case we would have one line series connected, two lines parallel connected, and three lines series connected. The connections between groups of lines follows the same alternating pattern, and it is the same (series or parallel) as the preceding group of lines are interconnected. Figure 12.13 is a schematic of the resulting transformer. The connections are drawn in the same hierarchy as the fractional expression of the turns ratio.

The number of transformers that can be constructed with up to four transmission lines are shown in Table 12.4. The number of transformation ratios are twice those listed since each transformer may be connected in reverse to obtain the inverse transformation ratio. Obviously, more transformers are possible with more than four lines using the design guidelines above.

Another very useful transformer is the 4:1 balun illustrated in Figure 12.14. It may be constructed on either balun cores, as shown, or on toroidal

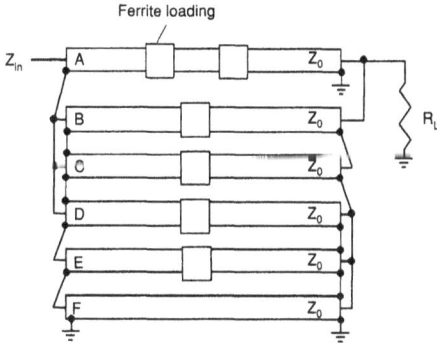

Ferrite loading

$$TR = \frac{10}{7} = 1 + \cfrac{1}{2 + \cfrac{1}{3}} = 1.4286$$

$$Z_{in} = \left(\frac{10}{7}\right)^2 R_L$$

$$Z_0 = \left(\frac{10}{7}\right) R_L$$

Connections on the input, corresponding to the fractional expansion of 10/7 are:

One line(s) (A) series-connected as a group and the group series-connected to two lines (B & C), parallel-connected as a group and the group parallel-connected to three lines (D, E, & F), series connected as a group.

Figure 12.13 Example of a 10:7 turns ratio equal-delay transmission line transformer.

TABLE 12.4 Transmission Line Transformer Ratios

No. of lines	Voltage/current transformation ratio	Impedance transformation ratio
1	1	1
2	2	4
3	3	9
3	3/2 = 1.5000	9/4 = 2.2500
4	4	16
4	4/3 = 1.3333	16/9 = 1.7778
4	5/3 = 1.6667	25/9 = 2.7778
4	5/2 = 2.5000	25/4 = 6.2500

cores. This transformer is a proven, compact method of coupling push-pull transistor collectors to a 50-ohm unbalanced load.

Shunt capacitance is required to compensate the transformer leakage reactance in order to obtain the maximum bandwidth from a given transformer

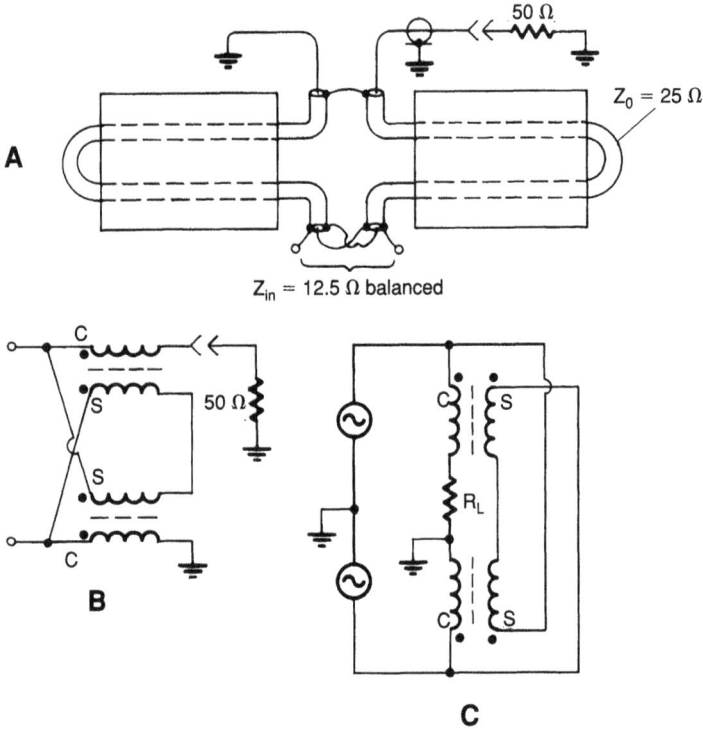

Figure 12.14 4:1 transmission line balun transformer: (A) physical construction; (B) schematic; (C) equivalent circuit.

design. The exact value often cannot be determined until the transformer is mounted to the final version of the printed circuit card. A clever method of evaluating transformer designs and leakage inductance compensation is to build two identical transformers and operate them back to back. Insertion loss and impedance data can then be taken with fairly accurate results. Construction of low-impedance loads for transformer testing usually gives poorer results the lower the impedance.

Base/gate load impedance matching

The techniques of wideband matching into the push-pull bases of BJTs are different from those used on FET gates. The BJT base presents a quite low impedance that must be transformed to 50 ohms at a reasonable VSWR over the frequency range of the amplifier. A low-input VSWR is desired to present a good load impedance for the driver stage. The RF power-FET input impedance is highly capacitive. It is possible to swamp this capacitance with a low value of resistance to obtain a good input VSWR and a flat frequency response from the FET amplifier.

Figure 12.15 is a typical BJT base circuit showing the circuit elements to consider in designing a wideband matching circuit. There is an element of "art" in designing such circuits because of the very low impedances involved and the uncertainty of the base input impedance changes under RF drive and various possible VSWR loads on the collectors. A recommended practical design technique is presented.

Start with a wire-wound transformer using a 1:16 impedance ratio. There should be no input compensation network, transformer leakage compensation capacitors, or base RLC networks connected at this point. Apply normal operating collector voltage and bias, and then sweep the frequency range. Obtain the input impedance data versus frequency and plot on a Smith chart. Increase the base bias to several higher values, taking care not to exceed the transistor current or dissipation ratings. Record and plot the impedance data for each bias setting.

Next, apply RF at normal operating bias and drive the amplifier to full power. With a quality directional coupler in the RF input path, measure and record the VSWR versus frequency. Compare this VSWR plot with the impedance plots taken at various bias settings. Choose the bias setting whose impedance plots show VSWRs most nearly like the VSWR plots from the RF test. This bias setting will then be used in the remainder of the tests to evaluate the additional base circuit impedance determining elements. Once the circuit values have been tentatively assigned, a final check of input VSWR under drive must be made. The use of the impedance data under higher than normal bias greatly speeds up this iterative design process.

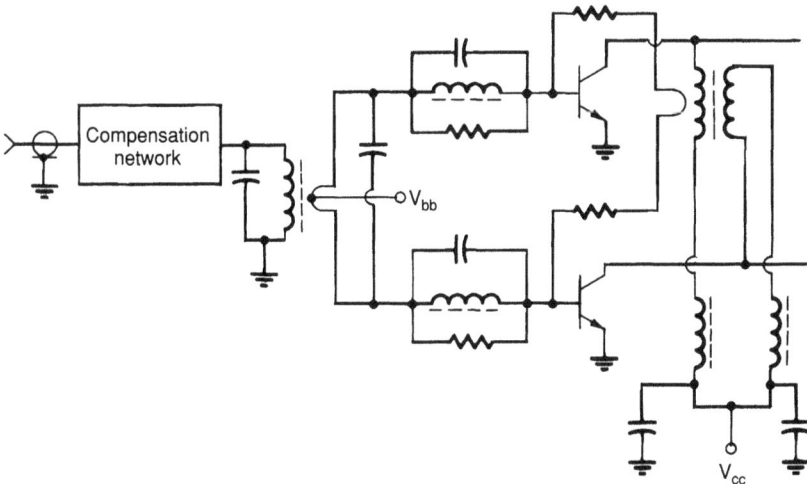

Figure 12.15 Typical BJT base impedance-matching circuit with negative feedback.

The series base RLC circuit should be investigated next. This circuit helps "swamp" the base input impedance and also prevents parametric oscillations. The type and value of these components can be first selected by considering the impedance data plots. The values are generally not critical, provided the parts are mounted with very short leads. Lap soldering of these components is recommended. The inductance is simply a ferrite bead on bus wire. If the inductor is not used, the bias should be fed to the bases via chokes rather than through the transformer center tap. Otherwise the series resistors will degrade the bias regulation and adversely affect IMD performance.

Then look at compensating the input transformer leakage inductance by adding shunt capacitance on the secondary and/or the primary. This will have the largest effect on the high frequencies.

The next step is to add the negative feedback circuit, if one is to be used. This circuit affects not only the amplifier input impedance but amplifier stability, IMD, gain variation, and to some extent collector efficiency. At this point it is wise to go back and evaluate the values previously selected to obtain further input VSWR improvement. The turns ratio of the transformer may have to be changed in this process.

The input compensation network may be required to achieve improved input VSWR and also to flatten the gain variations encountered in high-power BJT amplifiers. A good computer-aided design (CAD) program will be useful in designing this network, but the cut-and-try method will work if one does not have access to CAD programs.

Once the circuit values have been refined, it is wise to go back and investigate the amplifier for any signs of instability. Keeping the networks as simple as possible and at low impedances will minimize the amount of frustration one must endure to stabilize an amplifier design.

Compared with the design of BJT amplifiers, that for an FET input wideband circuit is more straightforward. Figure 12.16 gives three examples of commonly used circuits. The FET input is mostly capacitive. Swamping this capacitance with a low-value resistor is an effective way to flatten the gain and to choose the secondary impedance for the wideband input transformer. Negative feedback, either coupled through dc blocking capacitors from the drain or transformer-coupled as illustrated on the BJT amplifier, will lower the input impedance and must be accounted for in the circuit design. The wide gain-bandwidth products of RF power FETs coupled with the swamping resistor give a very good input VSWR with comparatively simple circuits.

Direct current feed methods, decoupling

The method of feeding the collector/drain dc voltage and decoupling the RF for a particular design is determined after careful laboratory experimentation. Figure 12.17 illustrates four common circuit topologies. Each has its merits and demerits. Component values and choke/transformer windings

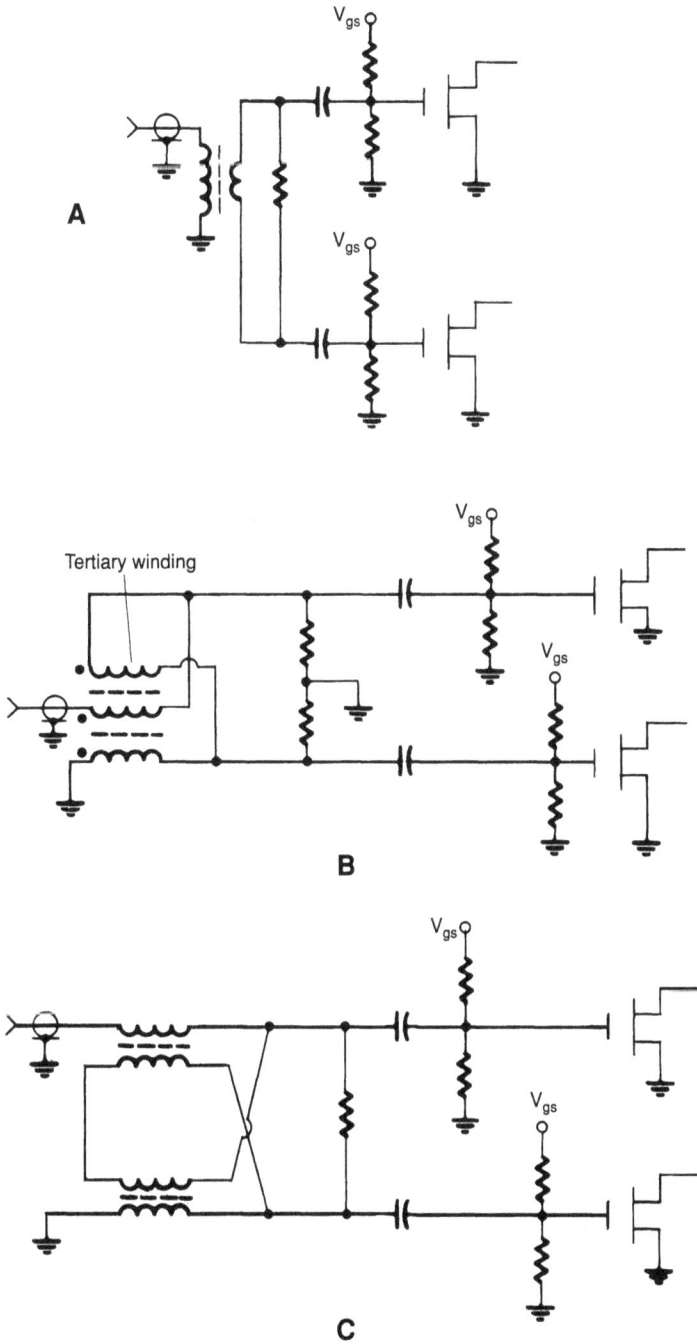

Figure 12.16 Typical FET gate impedance matching circuits: (A) wire-wound transformer; (B) transmission line 1:1 balun; (C) transmission line 4:1 balun.

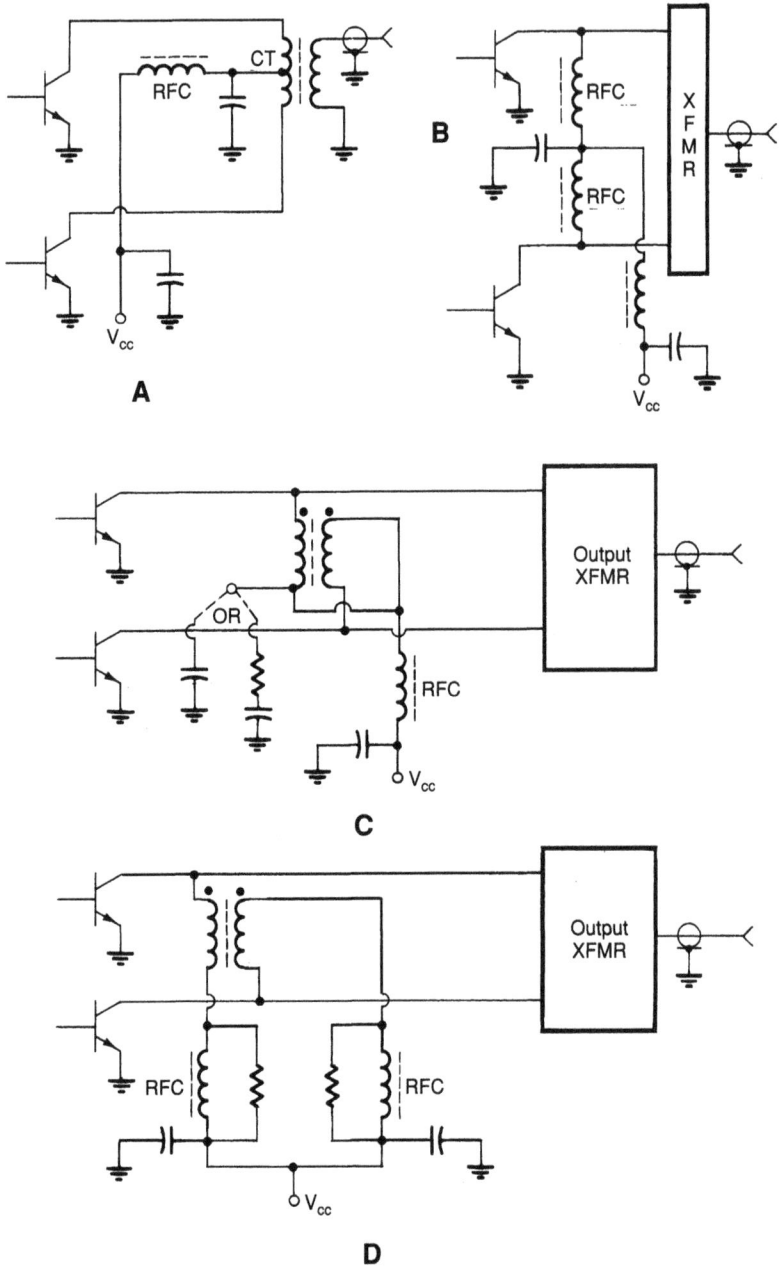

Figure 12.17 Collector dc voltage feed methods and RF coupling; (A) center-tap transformer; (B) split choke; (C) transformer, center-bypassed; (D) transformer, split center.

and core materials are the variables with which the designer works to optimize the operation of a transistorized push-pull wideband amplifier. The amplifier parameters usually affected by the dc feed circuit design are

1. IMD
2. Collector/drain efficiency
3. Even-order harmonic output
4. Amplifier stability

These parameters vary with frequency. A change in the dc feed circuit that improves a particular parameter at low frequencies will possibly either degrade it at high frequencies or degrade another parameter. It is important to evaluate all amplifier parameters over the operating frequency range when optimizing the topology and component values in the dc feed circuit.

We now present a few general remarks about each circuit topology. The center-tap feed on the wire-wound output transformer is useful for lower-powered (less than 100-W) PAs and those operating with collector-to-collector load impedances of a few ohms or less. The problem to watch for with this type of feed is dc saturation of the output transformer core, which will ruin the transformer action.

The split choke feed shown in Figure 12.17B is used for medium power levels and where the collector-to-collector impedances are more than a few ohms. The chokes are constructed of wire, wound on a ferrite core, to give a low-frequency reactance at least five times the collector load impedance.

Another technique is shown in Figure 12.17C and uses an RF transformer feed. This circuit, together with an output balun transformer, forms a 180° hybrid combiner. The bridging resistor could be connected from the dc feed point to ground. The bridging resistor, if used in this application, must have a dc blocking capacitor to prevent dissipating dc power. The resistor will dissipate any differences in phase (from 180°) and/or amplitude of the two transistor collector waveforms.

Simply providing an RF ground at the hybrid center with an RF bypass capacitor will provide adequate results in most designs. However, at high frequencies, IMD and collector efficiency can be improved by modifying the circuit to match that shown in Figure 12.17D. This technique provides "light" bypassing at high frequencies while the series chokes allow "heavier" bypassing at lower frequencies. Depending on the choke design, damping resistors may be required to stabilize the amplifier at low frequencies. The RF current-handling capability of the bypass capacitors should be checked when a final dc feed design is chosen. Some designs use high-current, low-loss chip capacitors.

Additional decoupling chokes and bypass capacitors may be required to ensure that the dc lines back to the power supplies are free of RF. Failure to adequately filter the RF from the various power supply voltage feeds to a PA is the cause of many stability problems. The decoupling must be effective at frequencies outside the operating range where the transistors have appreciable gain. Decouple the low-frequency and medium-frequency range for BJT amplifiers and also the VHF range for FET amplifiers.

The base-bias feed techniques for BJT amplifiers follow basically the same topologies as the collector feeds. The transformer feed is not commonly used. However, the center-tapped transformer feed shown in Figure 12.18A is often used with good results because the secondary load impedance presented by the bases is generally quite low. The split choke tech-

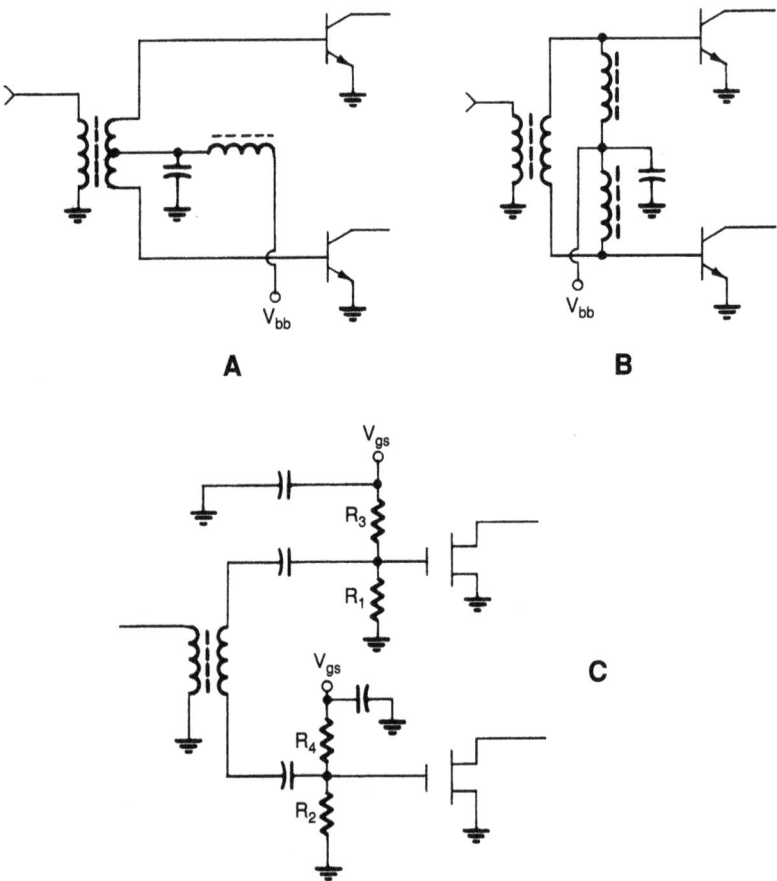

Figure 12.18 Base/gate bias feed methods: (A) center-stopped transformer; (B) split-choke bias feed; (C) separate bias feeds for FETs.

nique shown in Figure 12.18B or individual RF chokes feeding each base also work well but require more parts and circuit board area.

Figure 12.18C is a typical gate bias feed circuit for FET amplifiers. Separate adjustable bias voltages are shown for each FET. This is usually required for FETs (but not in matched BJTs) in order to obtain balanced linear amplification. Radio frequency power FETs generally have quite high transconductances, and simple low-current bias supplies allow individual bias adjustments. The gates draw only leakage current, so R_1 and R_2 are provided to load the bias supply. R_3 and R_4 are higher values and provide sufficient RF decoupling. Typical values are 10,000 ohms for R_3 and R_4 and 1000 ohms for R_1 and R_2.

Printed circuit design

Solid-state amplifier designs are usually implemented on a printed circuit (PC) board for ease of assembly and maintenance, reproducible phase and gain characteristics, and the ability to use low-impedance microstrip transmission line interconnects.

There are several layout techniques that the designer should consider to achieve top performance from a wideband solid-state PA. One of the most important is to use a symmetrical layout of the RF devices and the major components in the RF path. The interconnecting paths to or from each transistor in a push-pull amplifier must be equal in length. The circuit traces to each transistor should be symmetrical about a center line between the two push-pull transistors.

The bottom side of the PC card must be a solid ground plane with as few breaks as necessary for circuit traces and clearance around through-holes. The goal is to achieve a low-impedance ground plane with all points on the surface equipotential and as close to ground as possible when the high currents from the PA are returned to ground. This will help prevent unwanted inter-circuit coupling through ground loops. This bottom ground plane must have a good electrical connection to the transistor heat sink for the same reason.

Figure 12.19 is an example of a 280-W PA module. The RF input and output are designed for 50 ohms to facilitate module testing and troubleshooting with standard RF test equipment. The PA module contains a pair of high-power bipolar transistors of unique low thermal resistance design. Each transistor die is attached to a special cone slug which in turn is mounted in a high-efficiency heat sink. The transistors are operated push-pull class AB, into a 12.5-ohm balanced load line. A 4:1 (impedance ratio) transmission line balun transformer couples the collectors to the RF output 50-ohm load. The output module also contains a low-impedance bias regulator, RF gain compensation network, and current and temperature monitoring circuits. Simple, effective circuits were designed to reduce complexity and enhance overall reliability of the module.

Figure 12.19 Power amplifier module showing cone transistor design.

RF grounding in low-impedance circuits [13]

Ground areas on the top side of the PC board must also be connected to the bottom-side ground plane with low-impedance paths. Low impedance in the context here refers to inductive reactances in the milliohm range, since the circuit impedances are as low as a few ohms. These low-impedance ground plane interconnections are especially critical in the area where high currents are returned to the ground. The emitter or source leads and grounds in the input or output impedance networks and transformers are areas of critical grounding.

Some practical techniques to accomplish low-impedance grounding are

1. Solder a thin copper strap around the outside edges of the PC card, connecting top-layer ground areas to the bottom-side ground plane.

2. Solder a thin copper strap around the inside edges of the transistor cutout holes to connect the emitter/source lead grounds to the bottom ground plane.

3. Use several large (approximately 0.062- to 0.125-in. [1.5- to 3.0-mm] diameter) plated through-holes in areas of high ground return currents to connect the top ground to the bottom ground plane. This technique can be used in lieu of steps 1 and 2 when working at frequencies below 30 MHz and results in adequate grounding at a low cost, whereas for optimum performance at VHF frequencies the more expensive techniques of steps 1 and 2 have to be employed. Examples of grounding holes may be seen near each emitter lead in Figure 12.19.

4. In lieu of plated through-holes, eyelets or hollow brass rivets may be used to accomplish the same quality of grounding as in step 3. Do not

rely on mechanical compression; solder the rivets or eyelets on both surfaces but do not fill the eyelet hole with solder.

Avoid circuit layouts that force a low-level or small-signal circuit ground return to share the same path as a high-current circuit ground [14]. This type of ground loop will upset the operation of the low-level circuit. Typical circuits that are susceptible include the bias control and regulation and the PA current meter analog. The ground return for the collector/drain dc supply should attach to the PA module ground plane at a point that represents a nearly identical path shape and length to each emitter/source. Some deviation from this ideal is acceptable if there is a large ground area surrounding the emitter/source leads and the dc ground return connection.

Parasitic oscillation [15]

The elimination or suppression of parasitic oscillations (sometimes referred to as spurious oscillations) is one of the toughest tasks the solid-state PA designer faces. There are several modes that are peculiar to the solid-state PA because of the nature of the devices and the very low impedances at which power is coupled into and out of them. To further complicate things, a cure for one mode of parasitic oscillation may induce or aggravate another. The subject is quite involved, so only a cursory treatment will be given here. Experience and the scientific method in the laboratory are the most valuable tools a PA designer can apply to solving parasitic oscillation problems. The problem is compounded if the amplifier is to be built in production quantities because variations due to component tolerances must be considered.

Briefly, the basic test setup for searching out parasitic oscillations in a solid-state PA is to

1. Terminate the RF output in a power attenuator (rather than a dummy load). Insert sufficient attenuation to reduce the level to that required for a spectrum analyzer. This will give a flat response over a very wide band of frequencies. The range of measurement must include frequencies from LF through VHF. The analyzer must be used with both wide- and narrowband frequency sweeps.

2. Monitor one or both collector/drain RF waveforms with a high-frequency oscilloscope. This will help spot transient or unusual spurious oscillations that may otherwise be overlooked with the spectrum analyzer.

There are three recommended steps or guidelines to follow in the search for spurious oscillations (assuming none has been found in earlier testing):

1. Operate the amplifier without the lowpass filter (if one is normally used) in series with the RF output. Drive the amplifier at various power levels and slowly sweep the signal generator through the frequency range of

the amplifier. Look for abnormal or abrupt changes in the collector/drain waveforms and/or spurious spectra on the analyzer. Correct anything found before proceeding.

2. Install the lowpass filter and repeat step 1 for each filter band.

3. Operate the amplifier into various load impedances at several VSWRs and at many frequencies spaced throughout the HF range. This task quickly becomes a very lengthy and tedious process. To do a thorough search, one should check four to eight impedances per VSWR circle, two to four VSWR circles, and all at four to eight frequencies. This works out to anywhere from 32 to 256 combinations of load impedance and frequency. More tests may be prudent depending on the types of parasitic oscillations found. This test may also be used to evaluate the dissipation protection circuit performance at the same time.

Generating the various load impedances can become a time-consuming problem in itself. Three methods of generating mismatched loads are

1. Construct a tunable L network with a series inductor and shunt capacitor, both variable. Insert the network between the attenuator load and the PA. The input and output of the L network must be reversible to obtain impedances in all regions of the VSWR circle. With the aid of an impedance measuring instrument, the L network transforms the 50-ohm attenuator load to the desired load impedance. This method is very slow and time-consuming if a thorough search is to be conducted.

2. Similarly, "spot loads" may be constructed by shunting the 50-ohm attenuator load with capacitors or inductors. Various lengths of coaxial cable may be cut to move the spot loads around to other impedance points on the VSWR circle. This method is somewhat faster than the one described in step 1, but still limited in flexibility.

3. A continuously variable length of transmission line is used at VHF and UHF to rotate an impedance around the VSWR circle. A similar device [16] may be constructed for VSWR testing of HF amplifiers by a system of coaxial relays that switch in or out a series connection of binarily related lengths of transmission line. Thus the VHF/UHF "trombone" line can be approximated by the binary step size of the shortest line section. Impedances of various VSWRs may be connected to one end of the binary stepped transmission line, and a large number of impedances around that particular VSWR circle [17] may be obtained by energizing the proper relays. Unterminated power attenuators and parallel combinations of 50-ohm loads are ideal for this application.

The most common parasitic oscillation problem encountered in BJT solid-state PAs is at frequencies of 1 MHz and below. The BJT gain is quite high at these frequencies, and resonances in the collector dc feed decoupling network and/or the base-bias decoupling network often leads to this type of instability. The gain rolloff of HF power BJTs at VHF contributes to stability in this region. The RF power FET, however, has significant gain well into the VHF region. Therefore, it is likely to encounter VHF spurious oscillations. Common circuits involved are the input and output wideband matching transformers and the drain dc decoupling networks.

Three of the most common modes of parasitic oscillations are

1. Feedback via unwanted coupling in the amplifier circuit and/or in conjunction with the device internal feedback capacitance

2. Parametric oscillations caused by "pumping" a nonlinear reactance in the device by the drive signal

3. Base-bias circuit instability

Feedback oscillations may be readily identified with a spectrum analyzer. They may be either self-sustaining or driven oscillations. Some driven oscillations may require RF drive to get them started and then remain self-sustaining if the RF drive is removed. The cure for them is to locate the reactances involved and change values, add damping elements, add swamping elements, or trying alternative circuit topologies.

Parametric oscillations are almost always driven oscillations, and are only present during RF drive. They are usually a subharmonic of the RF drive frequency. Resistive loading or swamping in the external circuit is usually an effective cure. This type of oscillation is more prevalent in BJT amplifiers operated on low collector voltages.

Base-bias instabilities are similar to parametric oscillations except they are not subharmonics. This type of oscillation is sometimes due to base-bias regulator circuit interaction with the envelope of the RF drive modulation. Do not use excessive gain bandwidth in the bias regulator circuit. Ferrite beads and resistive damping along with adequate decoupling from audio through VHF on the bias circuit are effective techniques. Pay particular attention to the diode temperature compensation circuit frequently used on the bias regulator. It must be thermally coupled to the RF devices and is therefore susceptible to RF coupling.

Distortion reduction

Several design considerations for low-distortion, class A, AB, or B, solid-state wideband amplifiers are briefly discussed. They are meant to be an aid to the amplifier designer in the quest for adequate linearity for the intended application. High-linearity requirements are addressed in Chapter 13.

The most obvious first approach to good linearity is a low-distortion device that is operating at optimum voltages and currents and is free from instabilities. Generally the high-performance BJT devices are no better or worse than comparable FET devices regarding IMD performance in a well-designed circuit. The FET outperforms the BJT in the levels of the higher-order IMD products [18]. The higher-order products decrease in level at a faster rate as one examines higher and higher-order products in an RF FET rather than a BJT.

A well-regulated collector/drain dc supply is essential to obtaining good linearity. It allows optimizing the collector-to-collector (or drain) load resistance for good efficiency and controlled current swing without going into saturation. The collector-to-collector (or drain) load impedance must be maintained as close to resistive as possible. This in turn requires optimizing the output wideband transformer(s), combiner (if used), and lowpass filter passband VSWRs to as low values as possible.

The base-/gate-bias regulator must be absolutely stable and free from RF or envelope modulation. The bias regulator for a BJT amplifier must be capable of supplying the peak current required from the lowest-gain devices anticipated in production and at the highest RF power level. Careful attention to the bias supply performance is necessary for the best linearity that the devices are capable of.

Noise figure

The congestion of the HF spectrum places ever-increasing demands on the purity of the transmitted signals. This requirement is particularly severe when several transmitters and receivers must operate simultaneously in close proximity. The transmitted broadband noise from a solid-state PA is of particular concern because the gain bandwidth product is so large. Generally, the exciter synthesizer noise shelf will mask all but the noisiest PA. However, rapid improvements in synthesizer design and the use of automatically tuned postselectors between the exciter output and the PA input have placed more emphasis on low-noise-figure PA designs.

Normally noise figure measurement equipment is used only on receiver front ends and preamplifiers. The high-power capability of the PA will easily damage an automatic noise figure instrument. Therefore, the PA broadband noise output level is measured with a spectrum analyzer at a convenient IF bandwidth and is then converted to dBW/Hz. Some analyzers may require a correction factor when converting to dBW/Hz (consult the instruction manual). The small-signal gain of the PA must also be measured and the figure subtracted from the output noise level. This will give the noise level (in dBW) referenced to the amplifier input. A perfectly noiseless amplifier would be -204 dBW/Hz ($10 \log kTB$). The difference in the measured PA noise, referenced to its input, and -204 dBW/Hz is the noise figure of the PA. This number can readily be used in system performance calculations for collocated receivers and transmitters.

A noise figure of 8 to 15 dB is not difficult to obtain with the RF power FETs available. The noise figure for cascade amplifier stages is given by

$$\text{NF} = 10 \log \left(F_1 + \frac{F_2 - 1}{G_1} + \frac{F_3 - 1}{G_1 G_2} + \ldots + \frac{F_n - 1}{G_1 G_2 \ldots + G_{n-1}} \right) \text{dB} \quad (12.9)$$

where G = numerical gain
$\quad F$ = noise factor
$\quad$ NF = noise figure
$\quad n$ = amplifier stage designation (that is, 1 = first stage)

The goal is to distribute the gain between the cascaded amplifier stages such that the first stage sets the overall amplifier noise figure. Extraordinary low-noise circuit design must only be done on one relatively low-power stage. A good rule of thumb is to make the gain of an earlier stage equal to the noise figure (in decibels) of the succeeding stage plus 10 dB in order to make negligible the succeeding stage's contribution to the overall noise figure of the amplifier chain.

Source impedance of HF linear PAs

Occasionally, the source impedance of an HF linear PA becomes a design parameter of interest. One instance is during the design of output lowpass filters where the filter parameters depend upon the terminations at both the input and the output. The issue also arises when the PA designer wants, for some reason, to provide a load impedance that is a conjugate match to the source impedance.

In most designs, the source impedance of the HF linear PA is usually coincidentally determined by other, more important design requirements such as linearity, gain, efficiency, and harmonic output. There are some PA applications however, in which the source impedance is an important part of the design specifications. These requirements usually arise in sophisticated HF systems with special requirements.

The source impedance of broadband linear PAs can be established by the amount of negative feedback. Another technique is the use of quadrature combiners in the PA output circuit. If the source impedance of the amplifier is important at harmonic frequencies, then the feedback circuits and/or quadrature combiners must be effective over a much wider bandwidth than just the operating range of the amplifier.

An excellent article on a measurement technique for determining the source impedance of a linear HF amplifier may be found in reference 19.

12.3 Miscellaneous PA Circuits and Functions

T/R relay

Simplex operation (transmitting and receiving on the same frequency) often requires the use of a single antenna for both transmit and receive. An antenna switching relay is usually used to perform this function. Often it is required to be a part of the PA and is commonly called a transmit/receive or T/R relay.

Several specifications affecting the T/R relay design will be discussed. The relay must be capable of withstanding the maximum RF voltage encountered. This voltage is the peak RF value calculated for the highest VSWR expected, with allowance for RF transient power levels likely to occur during the ALC attack time. These same factors must be used to find the maximum expected current through the T/R relay.

The amount of receiver isolation provided by the T/R relay is important in high-power transmitters to avoid damage to the receiver or preselector input circuits. Usually 36 dB or more isolation is required. This value subjects the receiver front end to 0.25 W from a 1-kW transmitter. The receiver isolation usually degrades as the frequency of operation increases. Achieving high receiver isolation in a practical design requires a second relay to switch the receiver either to ground or to a 50-ohm termination or to an open circuit. The latter provides an additional interruption of the receiver input to the main T/R relay. The exact physical layout will determine which of the three techniques will ultimately provide the best isolation. Of course, extremely short direct interconnecting leads are required to optimize isolation and to maintain a low VSWR on the ports of the T/R relay.

The time required to switch the transmitter from transmit to receive to transmit is important in break-in CW and certain types of simplex error-correcting data communications systems. The T/R time requirements for these modes is 10 ms or less. The total T/R time includes relay contact bounce and any other transmitter (or receiver) delays from transmit command to RF output (or receive command to RF detection). Small, fast, vacuum relays can accomplish switching times in the low milliseconds, but the isolation is not as good as larger, slower relays because their contact spacing is very close. Over the life of the relay, the operation will slow down and contact bounce time may increase.

The ultimate expected life of the relay is critical when operation in the above modes is extensive. It is not unusual to find that the expected lifetime is less than 6 months of continuous duty in these modes. For this reason, a great deal of research has been conducted to develop a practical, high-powered HF solid-state T/R switch. Positive intrinsic negative diodes [20] have been successfully used at lower (100 W) power levels and raise frequencies. Positive intrinsic negative diode T/R switches at power levels of 1 kW and above, and at frequencies down to 1.6 MHz, generally exhibit higher than desired insertion loss and require higher dc power to operate. As better devices are developed with longer carrier lifetimes and lower "on-resistance" the solid-state T/R switch will replace the T/R relay.

In the past, vacuum tubes were used as electronic T/R switches with very fast switching times. But their circuit complexity, obsolescence, and the special voltages required have prevented them from becoming a practical solution.

Table 12.5 gives a comparison of the various T/R relay techniques.

Directional couplers for wattmeter applications

There are basically three types of directional coupler circuits that find practical application as wideband (over four octaves) wattmeter circuits. The inductive loop coupler [21] is a common technique. It is used in BIRD Electronic Corporation products to measure forward and reflected power on a transmission line section. A second technique [22], developed by Warren Bruene of the Collins Radio Company in the 1950s, is a transformer-coupled bridge circuit. The third technique uses cross-coupled transformers and has the advantage of requiring no RF balancing adjustments. Each circuit will be discussed in detail.

The inductive loop coupler is shown in Figure 12.20. It is a series resistor and a small loop inductively coupled to the center line of a coaxial line section by mutual inductance M and at the same time capacitively coupled by capacitance C. The current in the line and voltage across the line are represented by I and E, respectively, as shown in Figure 12.20. The resistor and the capacitance of the loop to line form a voltage divider. The value of capacitive reactance (over the frequency range of interest) must be much larger than the resistor value to minimize errors due to phase shift. The value of mutual inductance M is chosen to equal the product of the characteristic impedance of the line, $Z_0 \times C \times R$. The coupler output voltage e is the sum of two components: e_1 from the division of E by R and C and e_2 from induction of the current I, as shown in Figure 12-21. Anywhere along the line the voltage E is the vector sum of the forward voltage E_f and the reflected voltage E_r. Likewise, the line current I is the vector sum of the for-

TABLE 12.5 T/R Relay Techniques

	Open-frame relay	Coaxial relay	Vacuum relay	PIN diode	Electron tube
Cost	Low	Medium	High	High	Medium
Speed	Slowest	Slow	Medium	Fast	Fast
Complexity	Low	Low	Medium	High	High
Receiver isolation	Low	Medium	High	*	*
Transmitter loss	Low	Low	Low	Higher	Higher

*Depends upon complexity of specific design.

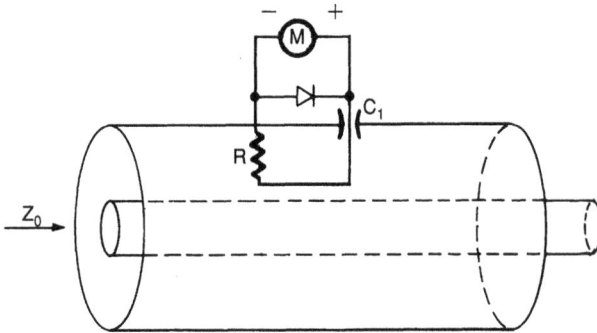

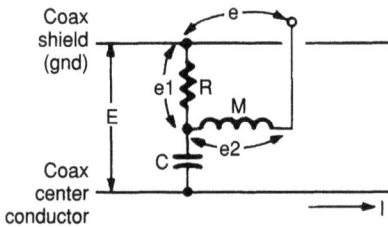

E = RF voltage across transmission line

I = RF current in transmission line

M = Mutual inductance between loop and center conductor

C = Capacitance between loop and center conductor

R = Resistor inserted in series with loop

• Coupling loop dimensions must be a small fraction of a wavelength.

• Resistance R must be very small compared to the capacitive reactance X_c.

• $CR = M/Z_0$ required for good directivity.

Figure 12.20 Inductive loop directional coupler.

ward current I_f and the reflected current I_r. The forward current is defined as a positive direction since it flows from the source to the load. If the load reflects energy, it causes a current to flow in the opposite direction and is given a negative sign.

By simply reversing the sense of the mutual coupling inductance, either the forward or the reflected current may be sampled. The output voltage e is directly proportional to frequency. If the output is terminated in a capacitor (C_1) of 500 to 2000 pF a wideband response is obtained. The output may then be rectified, filtered, and used to drive a sensitive microammeter calibrated to display power in watts.

The transformer-coupled wattmeter operates on the same principles as the inductive loop. A toroidal transformer is used to obtain a larger sample of the line current. Two voltage samples are taken via capacitive dividers. This allows displaying both forward and reflected power at the same time without the need to physically rotate the inductive loop coupling probe described previously. Figure 12.22 shows a schematic representative of this type of directional-coupler wattmeter. Note that a Faraday shield is used to prevent unwanted capacitive coupling between the line and the toroidal windings. The basic difference between this circuit and the previous one centers on the inductive loop coupler's need for precision parts to control the mutual coupling and a precision series resistance. Typically, a very sensitive and expensive microammeter is required to display the power. The transformer-coupled wattmeter operates with larger current and voltage samples and therefore does not require a sensitive output meter. It is, however, a more complex circuit. Depending on the care taken in physical layout, the latter circuit is capable of very wide bandwidths. The low-frequency

$e = e_1 + e_2$

$e_1 = \dfrac{RE}{X_c}$ for $R \ll X_c$

$e_2 = \pm IX_M$

$e = REj\omega C \pm Ij\omega M$ $\quad \begin{cases} + \text{ for current in forward direction} \\ - \text{ for current in reverse direction} \end{cases}$

$CR = M/Z_0$ $\quad$ chosen for directivity

$e = j\omega \left(\dfrac{EM}{Z_0} \pm IM \right)$

$e = j\omega M \left(\dfrac{E}{Z_0} \pm I \right)$

The voltage across the transmission line (at any point) is the sum of the forward component E_f and the reflected component E_r.

$E = E_f + E_r$

The current in the transmission line (at any point) is the sum of the forward component I_f and the reflection component I_r.

$I = I_f - I_r$ ($-$ because reflected current travels in the opposite direction)

$I = \dfrac{E_f}{Z_0} - \dfrac{E_r}{Z_0}$

$e_f = j\omega M \left(\dfrac{E_f + E_r}{Z_0} + \dfrac{E_f - E_r}{Z_0} \right)$ forward direction

$e_r = j\omega M \left(\dfrac{E_f + E_r}{Z_0} - \dfrac{E_f - E_r}{Z_0} \right)$ reflected direction

$e_f = \dfrac{j\omega M}{Z_0} (2E_f)$ forward voltage

$e_r = \dfrac{j\omega M}{Z_0} (2E_r)$ reflected voltage

Figure 12.21   Inductive loop directional coupler equations.

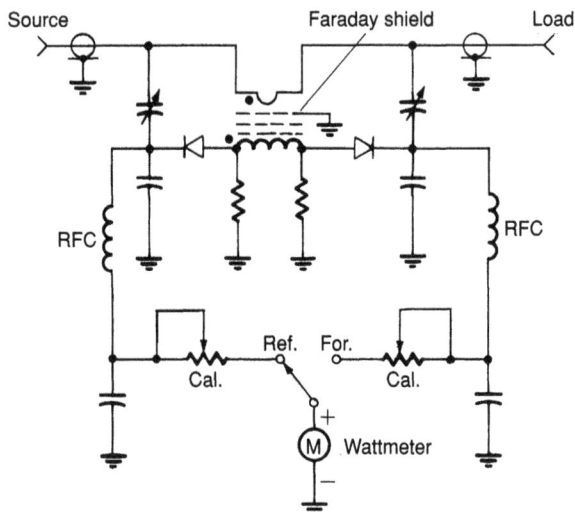

Figure 12.22 Transformer-coupled directional coupler.

response is limited when the toroidal transformer winding reactance lowers to values comparable with those for the load resistance, resulting in significant phase-shift errors. The high-frequency response is limited by the series resonance of the winding and, additionally, by any inductive reactance in the load resistors.

The third type of directional coupler consists of two cross-coupled toroidal transformers and suitable RF detector circuits, as shown in Figure 12.23. The amount of coupling and the input impedance depend upon the turns ratio of the transformers. A small turns ratio gives close coupling but introduces significant VSWR at the input port. It also increases the loss from input to output ports.

The input-to-output coupling (loss) is given by

$$S_{oi} = -20 \log \left(1 + \frac{1}{2N^2} \right) \qquad (12.10a)$$

The input-to-forward port coupling is given by

$$S_{fi} = -20 \log \left(\frac{1 + (1/2N^2)}{1/N} \right) \qquad (12.10b)$$

The input return loss is given by

$$S_{ii} = -20 \log \left(\frac{1}{2N^2 + 1} \right) \qquad (12.10c)$$

Using these formulas, a directional coupler may be designed for any desired amount of coupling. The forward and reflected coupled ports are terminated in good low-reactance dummy loads of a value equal to the characteristic impedance of the system. In most cases this is 50 ohms. The rms RF voltage measured across the forward port, squared and divided by 50 ohms, will be equal to the forward RF power applied to the input port decreased by the amount of coupling. In like manner, the RF voltage at the reflected port is related to the RF power entering the output port, that is, reflected from the load connected to the output port.

The RF voltage on the forward and reflected ports may be rectified and filtered to drive a dc meter with a power scale indicating forward and reflected power (in watts). Figure 12.23 shows a typical design for a 1000-W solid-state PA. If $N = 32$ turns (for a one-turn primary), the coupling is 30.11 dB. The power coupled to the forward port will be approximately 1 W. This gives a reasonable dissipation level for the terminations and good drive level to the wattmeter. The return loss calculates as –66.23 dB, and the insertion loss calculates as 0.004 dB.

Practical construction of this type of wattmeter circuit can result in very wideband responses. The interconnecting leads of the toroidal transformers should be very short and direct. A Faraday shield is recommended between the primary turn and the secondary windings of the transformers. A symmetrical layout will ensure accuracy at various power levels and VSWRs.

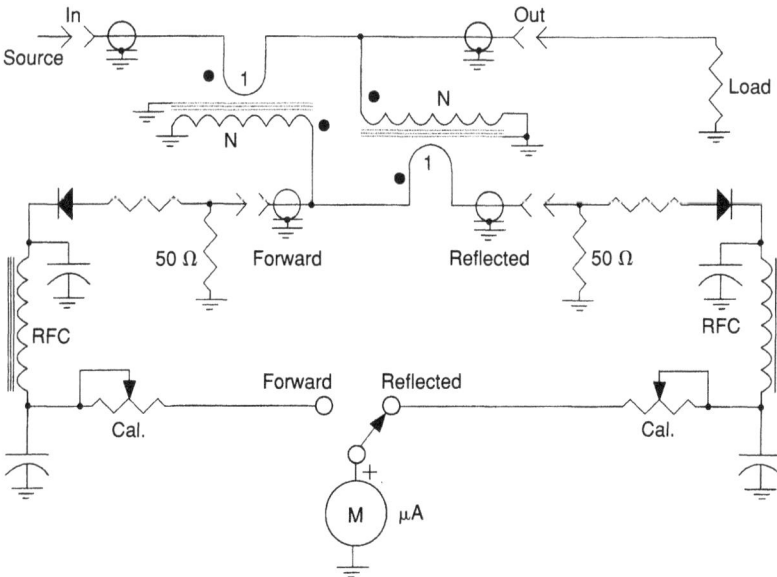

Figure 12.23 Cross-coupled transformer directional coupler.

Teflon insulation on the core and/or the secondary wire is necessary to prevent corona or RF breakdown of the secondary winding through the core. The core material must be rated for the maximum anticipated flux density over the frequency range and load VSWRs. In general, it has been found that ferrite materials are superior to powdered iron at the 1-kW level in the HF band. Practical designs have been accomplished at the 1-kW power level over a bandwidth of 1 to 100 MHz.

Lowpass filters for solid-state PAs

Most solid-state PAs require some type of RF output filtering to suppress harmonics. The filter could either be bandpass or lowpass. For a given level of harmonic attenuation, the lowpass configuration contains fewer elements and is thus less costly. Other considerations such as protection against unwanted RF coupled from collocated broadband transmitters or transmitter broadband noise attenuation may dictate bandpass filtering.

Invulnerable filters

The typical PA output lowpass filter is designed as a reflective type. Within the passband, the PA sees a low SWR when the filter output is properly terminated. In the stopband, the PA sees a very high SWR load. This high reflection coefficient reflects harmonic energy from the filter input, preventing the harmonic energy from reaching the output port. The reflected harmonic energy modifies the transistor collector/drain waveform from what it would be if the PA were terminated in a wideband resistive load (that is, no lowpass filter). This in turn changes the amplifier efficiency. It can also affect linearity performance, depending on the type and amount of negative feedback employed.

Another type of filter, called an invulnerable filter [23], circumvents the above problems. The invulnerable filter, shown in Figure 12.24, is essentially a highpass filter and a lowpass filter with the same corner frequency. The inputs of the two filters are connected in parallel while the outputs are separate. The lowpass output connects to the antenna while the highpass output is terminated into a 50-ohm load. The load must be of sufficient power rating to dissipate all of the harmonic energy. The PA output is filtered, yet it "sees" a low SWR at all frequencies, including the harmonics. Its performance is very consistent across its operating frequency range. The disadvantage is that the invulnerable filter is about twice as complex as the reflective filter.

There is much excellent literature on RF filter design to which the reader is directed (see references 24 through 27). The practical aspects and design considerations of implementing RF lowpass reflective filtering on a solid-state PA will be discussed here.

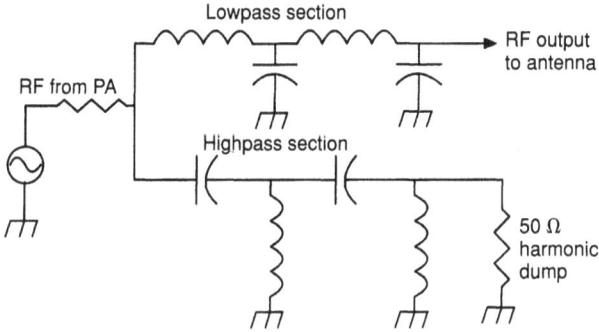

Figure 12.24 Typical invulnerable filter circuit topology.

The first step in filter design for a solid-state PA is to decide how many individual filters are required to cover the frequency range of the PA, given the
required harmonic output suppression. Typical class AB or B push-pull PAs
will generate a second harmonic that is at least 20 dB below the fundamental
if the push-pull pair of transistors have reasonably close matched gain characteristics. The third harmonic can be as high as 9 dB down. Higher-order
odd and even harmonics will fall off in amplitude from these respective values. A typical PA output spectrum is shown in Figure 12.25. Harmonics that
fall in the transistor gain rolloff region and above will be at a considerably
lower level. The designer can take advantage of this characteristic to reduce
the complexity of the higher-band filters. This will also reduce the insertion
loss and cost of the filter for these higher-frequency bands.

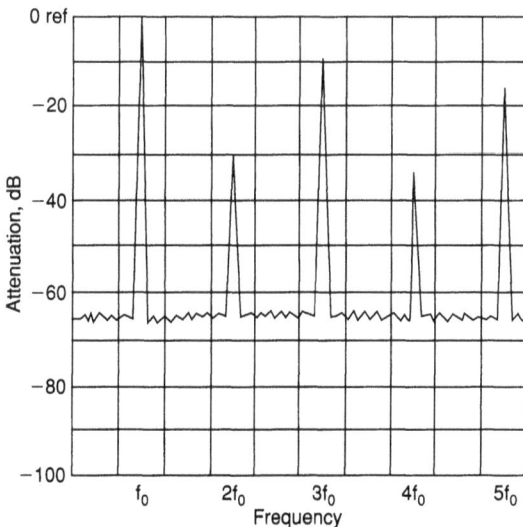

Figure 12.25 Typical unfiltered power amplifier harmonic
spectrum.

The next important practical consideration in choosing a lowpass filter design concerns the effect of load impedance on harmonic generation in the solid-state PA. The harmonic level before filtering determines how much stopband attenuation the output filter must have to suppress the harmonics below the design specification. A common design pitfall is to first operate the amplifier without a filter into a 50-ohm load and carefully measure the harmonic levels across the operating frequency range. A filter design is selected with a minimum stopband attenuation equal to the difference between the measured harmonics and the harmonic specification. The filter is then built, aligned, and tested with the amplifier. The resulting output harmonic levels are not what was predicted from the "filterless" data plus the filter's attenuation measured on a network analyzer. What went wrong?

The answer lies in the fact that most cataloged filter designs (for a 50-ohm load) are either for a 50-ohm source or a high-impedance source. The PA's source impedance is rarely close to either of these values. Thus the filter stopband response could vary as much as 6 dB from the predicted catalog value compared with its response when driven with the actual source impedance of the PA. The RF waveform on the transistor collectors/drains depends upon whether the PA is terminated in a 50-ohm load or a harmonic filter. A dummy load on a filterless PA provides a 50-ohm load to all harmonics as well as the fundamental. Insert the lowpass filter and now only the fundamental is terminated in 50 ohms. The filter input impedance presents a highly reactive termination to the harmonic energy. The harmonics see a high VSWR at the filter input and are reflected back to the transistors, modifying the collector RF waveform. This phenomenon is also responsible for the different collector/drain efficiencies encountered in measurements made with and without RF output filters.

An iterative process is required to find the optimum number of filter bands and the lowest filter complexity. For example, given a certain harmonic output specification for the transmitter, is the lowest-cost design (for the lowpass filter bank) accomplished with numerous filter bands covering small segments of the frequency range or with few filter bands covering large segments? The first approach requires a simple filter design (few components) since the ratio of the lowest harmonic frequency to the highest operating frequency is relatively large. The second requires a more complex design but fewer bands. The limiting factor is that the filter operating frequency range cannot exceed (or even approach, in practical designs) one octave without allowing the second harmonic of the lowest operating frequency to fall in the passband.

The design trade-off becomes a problem of the number of bands (filters) versus filter complexity for the lowest cost that provides adequate harmonic attenuation above the cutoff frequency. This allows a wider operating bandwidth for a particular filter. Figure 12.26 illustrates the problem.

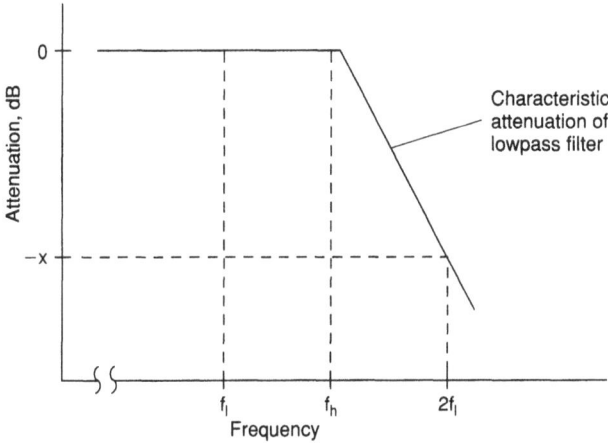

Figure 12.26 Lowpass filter bandwidth. x dB = minimum attenuation for adequate suppression of lowest second harmonic $(2f_l)$; f_l = lowest operating frequency of filter band; f_h = highest operating frequency; $f_h - f_l$ = filter operating bandwidth.

The operating bandwidth ratio, if kept constant, will allow using the same normalized filter design for all bands. The band ratio may be found by

$$\left(\frac{f_h}{f_l}\right)^{1/n} = \text{band ratio} \tag{12.11}$$

where n = number of desired bands
  f_h = upper frequency limit of the PA
  f_l = lower frequency limit of the PA

This band ratio, coupled with filter input harmonic levels, is then used in selecting a normalized elliptic filter design that provides sufficient attenuation at the harmonics of the lowest operating frequency. If a wide margin of attenuation is found to exist, the filter order (complexity) may be reduced, or the number of bands n may be reduced to better fit the filter to the requirements. Several iterations of this process will eventually reveal the most economical solution to this problem.

Typical filter designs have operating ranges of approximately one-half octave. The most cost-effective design solution is a careful weighing of the above factors to achieve the specified harmonic output level or better with the fewest number of capacitors and inductors in the bank of lowpass filters. The Cauer-Chebyshev or elliptic-function filters are commonly used because they have sharp rolloff characteristics. But they also have ripples in the stopband attenuation that prevents increasing attenuation with frequency.

There are three main types of components in a filter bank for a solid-state PA: switches, capacitors, and inductors. The proper filter must be connected to the amplifier RF output and the external load or antenna. The switching arrangement must be capable of selecting the proper filter and maintaining the correct characteristic impedance, usually 50 ohms, through the system. Several approaches may be used. In general they involve rotary switches with either very short wire leads or coaxial leads, miniature relays and microstrip transmission lines [28], or a tree of coaxial relays and interconnecting coaxial cables.

The capacitors used in a lowpass filter must be rated for the maximum RF current expected in the design. Computer analysis of the proposed design is almost mandatory to determine the worst-case voltages and currents for each filter component. High-Q components are required to keep insertion loss low and to prevent excessive component heat. Attempts to parallel low-current, low-cost leaded capacitors usually are not successful. However, modern low-cost ceramic chip capacitors can usually be paralleled with reasonable circuit layout to ensure equal symmetrical current paths.

The coils used in filters for 1 kW and above are usually air-wound, although successful filters have been made using toroidal core inductances at the 1-kW level. The coil Q is important for achieving as low insertion loss as possible. The length-to-diameter ratio is key to optimizing the Q for a given inductance. If the coils must be shielded, the dimensions of the shielding relative to the coil diameter are important considerations. Coupling between coils in different sections of the same filter or between adjacent filter bands will cause unwanted responses in the filter passband characteristics. This unwanted coupling can be minimized or eliminated by using a layout scheme that places the solenoid axis of any one coil orthogonal to any two adjacent coils within a filter band. Judicious interleaving of the locations of HF-band filters and LF-band filters can help eliminate the effects of coil coupling between adjacent filters. Shields between bands and/or between filter sections are the most effective but can have a detrimental effect on coil Q. Using silver-plated wire may have some slight advantage in reducing losses (in the higher-frequency bands) but will certainly increase the filter costs. Enameled copper wire or Teflon-insulated tin-plated bus wire is suitable for coils that must be close-wound, while tin- or silver-plated bus wire may be used for space-wound coils. Final alignment of the filters is accomplished by pushing or pulling the coil turns slightly while watching the filter characteristics (input impedance, passband, and stopband attenuation) on an automatic network analyzer. This process can become very tedious without the aid of the automatic network analyzer. If one uses close-tolerance parts and is not too fussy with the filter results, it is possible to dispense with the coil alignment and still achieve adequate results for many applications.

At power levels below 1 kW, the use of toroidal coils becomes practical. Coil currents are lower and therefore toroid cores of reasonable physical size can be used. The main advantage is the self-shielding property of this type of inductor. The various inductors may be physically close (but not too close) to each other and still have negligible magnetic coupling between coils. Elimination of the shielding and the separation that is necessary with air-wound coils allows construction of small, compact lowpass filter assemblies for PAs of a few hundred watts or less. The disadvantage is the limited range of inductance adjustment possible by expanding or compressing the turns spacing. This can usually be overcome with the wide variety of core sizes and ferrite permeabilities that are available.

Many lowpass filter assemblies are constructed using printed circuit boards. The common fiberglass-epoxy (G-10) boards are adequate up to power levels of a few hundred watts. At higher power levels, a board with a lower dielectric loss is usually required. A Teflon-filled board is the usual choice. G-10 will overheat at high power levels wherever sufficient current flows in the capacitors formed from component pads and the ground plane on the bottom of the board. If the lowpass filter assembly is constructed on a single large board, some type of mechanical stiffening may be required for the Teflon boards, since they are considerably more flexible than G-10 and the component solder joints may deteriorate in time from mechanical stress if the Teflon board is allowed to flex.

The entire bank of lowpass filters can be assembled on one large circuit board to facilitate the switching method. However, a case may be made for mounting each filter band on an individual circuit board. It complicates the interconnect and switching arrangement but offers the advantage of a lower-cost replaceable subassembly in the event of a capacitor failure, which can easily damage the circuit board.

The lowpass filter bank should be arranged for maximum physical separation between the input and the filter outputs. This is necessary to achieve the best stopband attenuation that the filter is capable of. This technique should also be considered in conjunction with controlling the filter currents (from shunt elements) that flow in the ground plane. The ground return current from the input RF should not intersect with the ground return current path for the output RF. Shielding of the input and output sections of the filter may also be necessary in some applications in order to achieve maximum stopband attenuation.

Figure 12.27 shows an example of a lowpass filter assembly for a 1-kW PA. It consists of eight individual filters to cover the 1.6- to 30-MHz frequency range. The filter band ranges are: 1.6 to 2.3, 2.3 to 3.4, 3.4 to 4.9, 4.9 to 7.0, 7.0 to 10.0, 10.0 to 14.5, 14.5 to 21.0, and 21.0 to 30.0 MHz. The proper filter is automatically selected and switched in series with the RF output. This particular design required a seven-pole elliptic function lowpass filter to ensure harmonic suppression at least 55 dB below the fundamental. The elliptic function filter offers the advantages of a steep attenuation characteristic at cutoff and low attenuation ripple in the pass-

Figure 12.27 Lowpass filter assembly.

band. The filter was constructed with high-Q porcelain capacitors and air-wound silver-plated coils for the HF bands and close-wound enameled wire for the LF bands. Each coil was carefully designed for maximum Q within the constraints of the necessary shielding requirements. The resulting filter has a measured maximum passband attenuation of 0.17 dB and suppresses harmonics to 60 dB or better below the fundamental.

The filters are switched in and out of the RF output line by a unique patented (U.S. Patent 4,349,799) transmission line switch. The filter is constructed on a large Teflon-glass PC board. A solenoid-activated switch contact is located at the ends of each filter section. A 50-ohm printed microstrip transmission line runs along the filter inputs and outputs and carries the other half of the switch contacts. There is a transmission line stub in parallel with any selected filter except the one located at the far end of the line. The reactance of this stub is known and can be compensated by the values of the input and output filter elements. If the filters are arranged so that the longest stub is in parallel with the lowest-frequency filter, progressing to shorter stubs as the filter frequency increases, the stubs' effect on filter element values will be negligible. This technique of filter switching is fast and reliable.

12.4 Power-Combining Techniques

Multiple output modules

Multiple output modules may be combined to achieve higher power levels than are possible from a push-pull pair of transistors. A push-pull pair is commonly termed a module, but a module subassembly may in fact have two or even four pairs of transistors whose outputs are combined to a single 50-ohm output connector. Module subassembly outputs may be further combined to produce a single high-power output. High-level module combiners usually operate with 50-ohm impedance ports to facilitate testing and troubleshooting the equipment with common test equipment. However, the combining that takes place within a module of multiple pairs of transistors may be done at any impedance level. This impedance is often chosen in conjunction with other considerations such as matching to the collector/drain load impedance with easily achievable transformation ratios. These types of combiners will be discussed in detail in this section.

Power dividers and combiners [6, 29, 30]

A power divider will always be found on the input side of the modules where a combiner is used. In the discussion that follows, the combiners described can always be used in reverse to perform power dividing. It is imperative that the same type of divider be used on the input side of the amplifiers as the combiner used on the outputs. A 180° divider on the inputs and an in-phase combiner on the outputs will spell disaster.

A wideband power combiner must perform the following basic functions:

1. Provide isolation (minimum coupling) between the input ports
2. Provide low insertion loss over the required bandwidth
3. Provide a low VSWR load at the input ports over the required bandwidth

There are three basic types of combiners:

1. In-phase combiners (two or more ports)
2. Hybrid or 180° combiners (two ports)
3. Quadrature or 90° combiners (two ports)

All three types will be described here and practical implementations will be presented. The following definitions apply:

- R_L: output load resistance
- R_B: bridging resistor
- Z_0: transmission lines characteristic impedance
- Z_{in}: input impedance (with output port terminated)
- S: shield connection of coaxial cable
- C: center connection of coaxial cable

In-phase combiners operate with two or more inputs of equal phase and amplitude to combine into a single output. There are two basic topologies for in-phase combiners, examples of which are shown in Figures 12.28 and 12.29. The differences are in the number and configurations of the ferrite cores which must be used and the value of the bridging resistor. The type I configuration uses a single balun core or toroidal core and a bridging resistor equal to four times the output load. The type II combiner must use two separate cores, which may be either straight tubular or toroidal. The bridging resistor is equal to the output load.

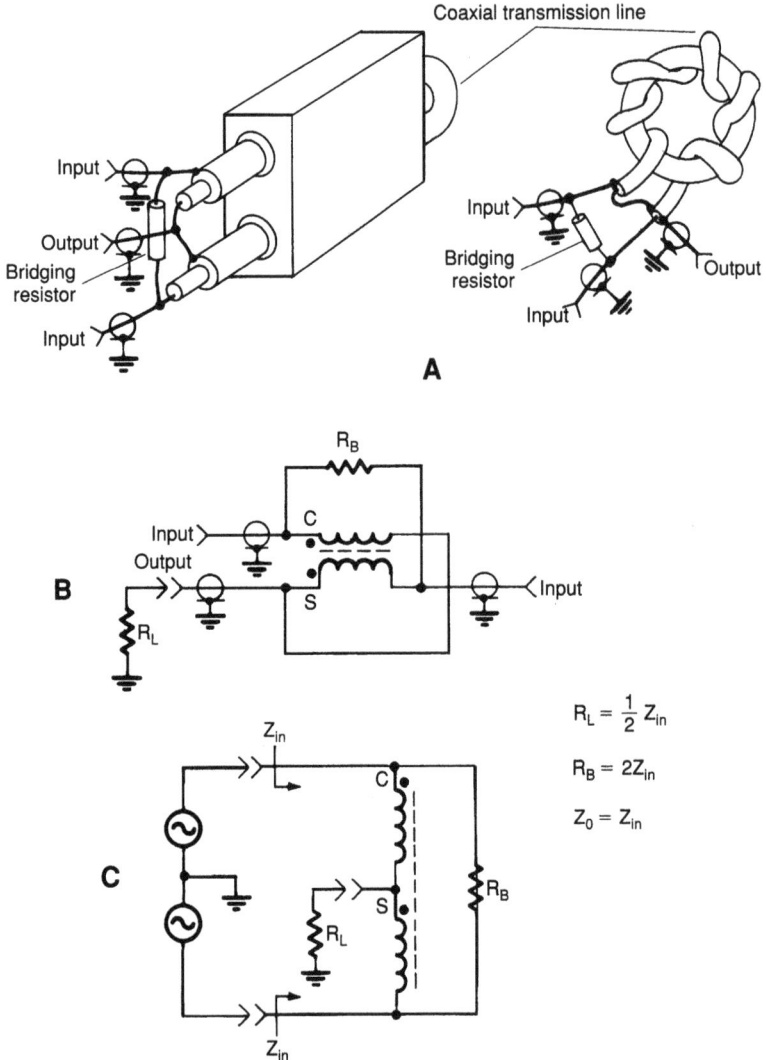

$$R_L = \frac{1}{2} Z_{in}$$

$$R_B = 2Z_{in}$$

$$Z_0 = Z_{in}$$

Figure 12.28 Type I in-phase two-port combiner: (A) pictorial view, balun and toroidal cores; (B) schematic; (C) equivalent circuit.

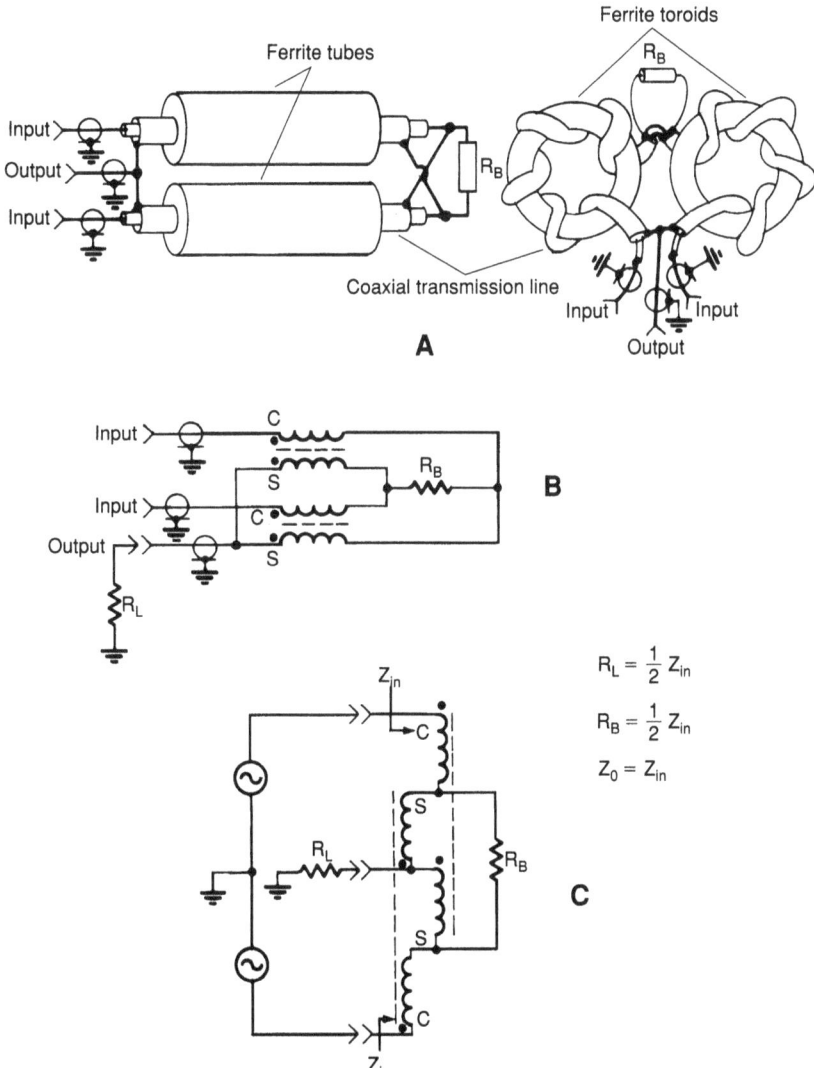

Figure 12.29 Type II in-phase two-port combiner: (A) pictorial view, tubular and toroidal cores; (B) schematic; (C) equivalent circuit.

Physical layout considerations as well as practical values of the transmission line impedance Z_0 and the bridging resistor R_B will determine which type of combiner to choose for a particular design. A comparison of input impedance and port-to-port isolation between a typical type I and type II combiner yields interesting results, as shown in Figure 12.30. Note that an expanded Smith chart is used—the outside rim is the 1.67:1 VSWR circle. Both combiners were constructed with a single turn of 50-ohm

A

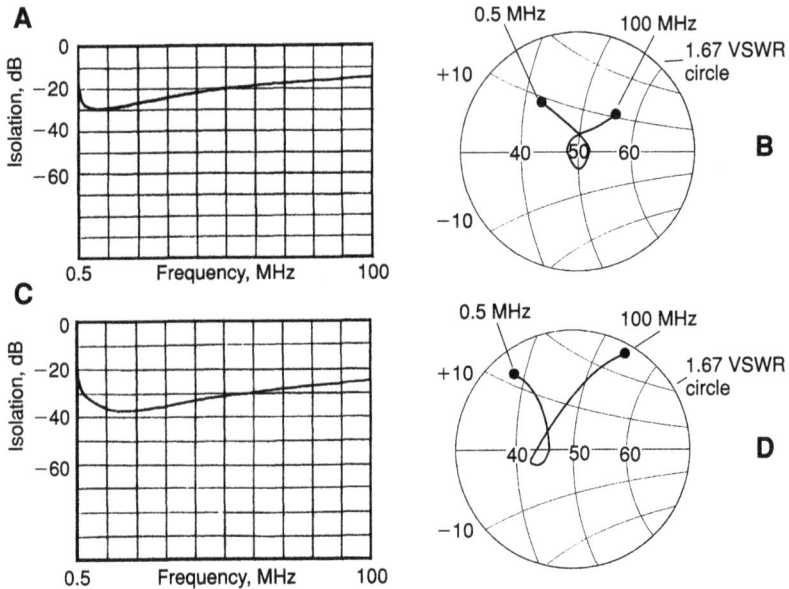

B

C

D

Figure 12.30 Comparison of isolation and input impedance for typical types I and II in-phase combiners: (A) port-to-port isolation, type I in-phase combiner; (B) input imped-ance, type I in-phase combiner; (C) port-to-port isolation, type II in-phase combiner; (D) input impedance, type II in-phase combiner.

coaxial cable in the cores. Core material was identical: Stackpole 7D. The test data indicates better port-to-port isolation with a type II combiner, but better input VSWR with a type I combiner.

The combiner output load impedance is usually transformed to another desired value such as 50 ohms. This is easily accomplished by one of the wideband transformers described in section 12.2. Usually the output impedance transformer is physically integrated into the combiner assembly so that the odd impedance value can be handled easily. Strip-line techniques are often required to interconnect the wideband transformer and the combiner.

Theoretically any number of inputs may be combined with an in-phase combiner, but a practical limit is reached when the output impedance becomes too low to allow efficient wideband transformation back to the desired load impedance. An example of a type II four-port in-phase combiner is given in Figure 12.31.

Four-port combiners may also be implemented by cascading two-port combiners. This technique is illustrated in Figure 12.32 for both types of two-port combiners.

The in-phase combiners all use a floating bridging resistor. This may be difficult to implement, especially in combiners handling high power. A wideband balun transformer allows the use of a single-ended or unbalanced load. This balun can also transform the balanced impedance to 50 ohms. Standard coaxial dummy loads connected to the combiner with coaxial cable can then be used as bridging resistors.

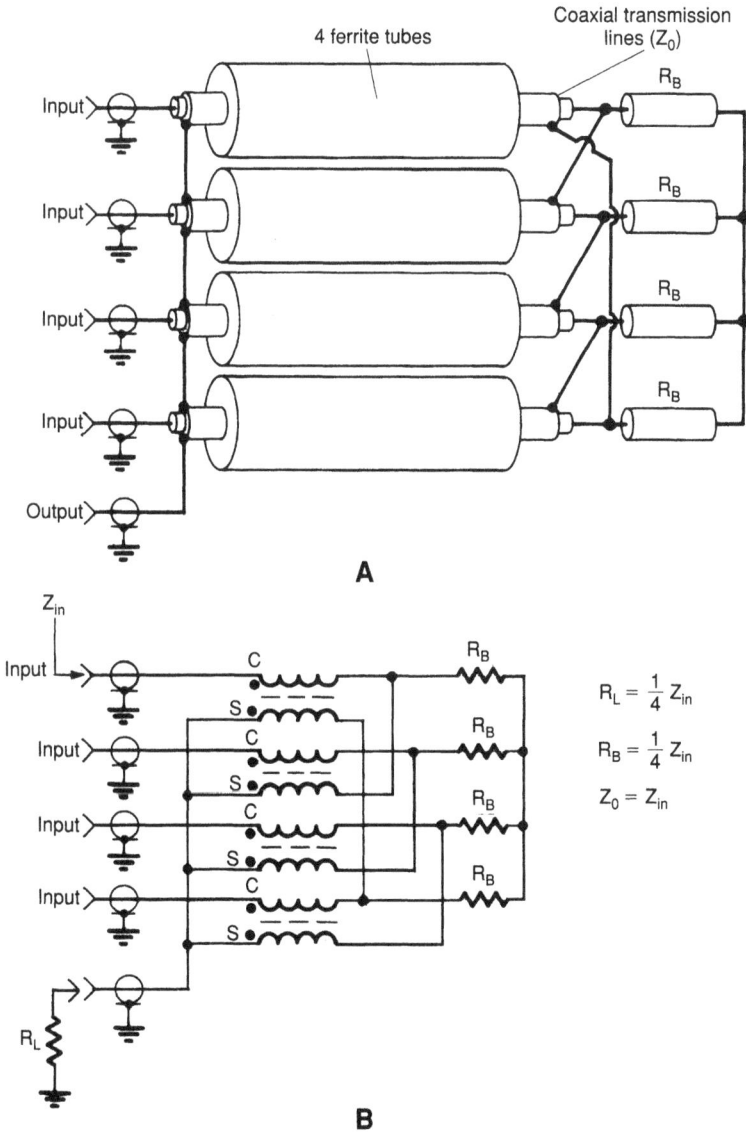

$$R_L = \frac{1}{4} Z_{in}$$

$$R_B = \frac{1}{4} Z_{in}$$

$$Z_0 = Z_{in}$$

Figure 12.31 Type II in-phase four-port combiner: (A) pictorial view; (B) schematic representation.

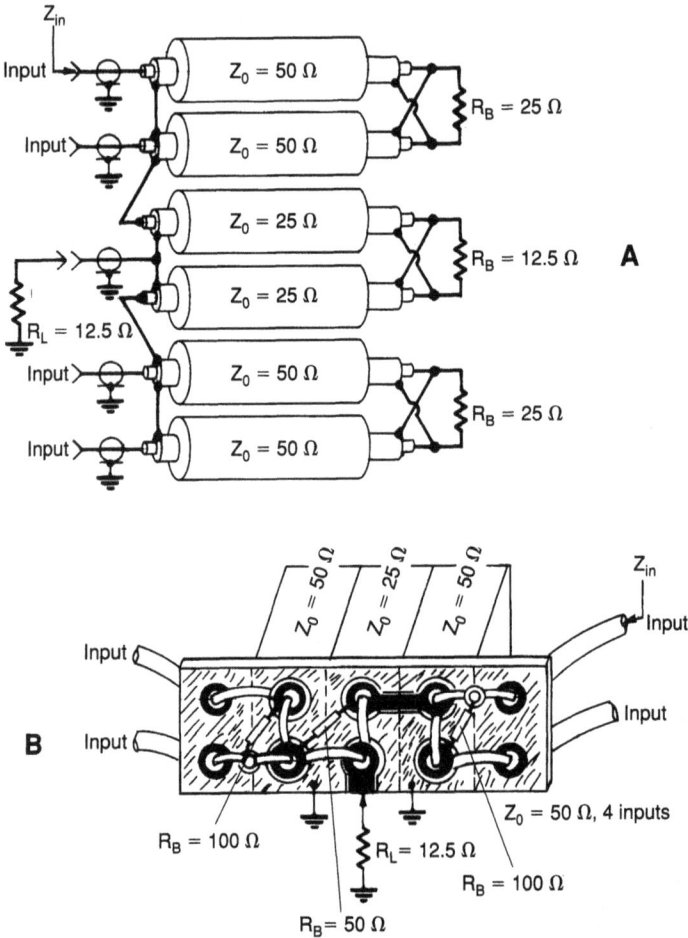

Figure 12.32 Four-port combiners implemented with two-port in-phase hybrids: (A) type II combiner; (B) type I combiner.

If the roles of the bridging resistor and the load are interchanged, the result is a 180° hybrid combiner. The two input signals must be 180° out of phase and of equal amplitude. The output is balanced to ground unless the usual balun is used. Examples of type I and type II 180° hybrid combiners with output baluns are shown in Figures 12.33 and 12.34.

Many unique combiner designs are possible by using various combinations of basic combiner types and balun transformers. Figures 12.35 and 12.36 are examples of a four-port combiner each using two type I in-phase combiners (cores A and F) and two parallel connected type II 180° hybrid combiners (cores D and C) with a 4:1 balun transformer (cores B and E) to couple the combined output to a 50-ohm load. Connecting two 180° hybrids in parallel avoids the need to use 25-ohm coaxial cable and provides the extra core material to handle the higher RF power.

Quadrature combiners

The quadrature combiner with a bandwidth covering the HF band (or wider) is implemented by two allpass networks, each in series with the inputs of an in-phase or 180° combiner as shown in Figure 12.37. The two allpass networks are designed to produce a constant phase difference of 90°

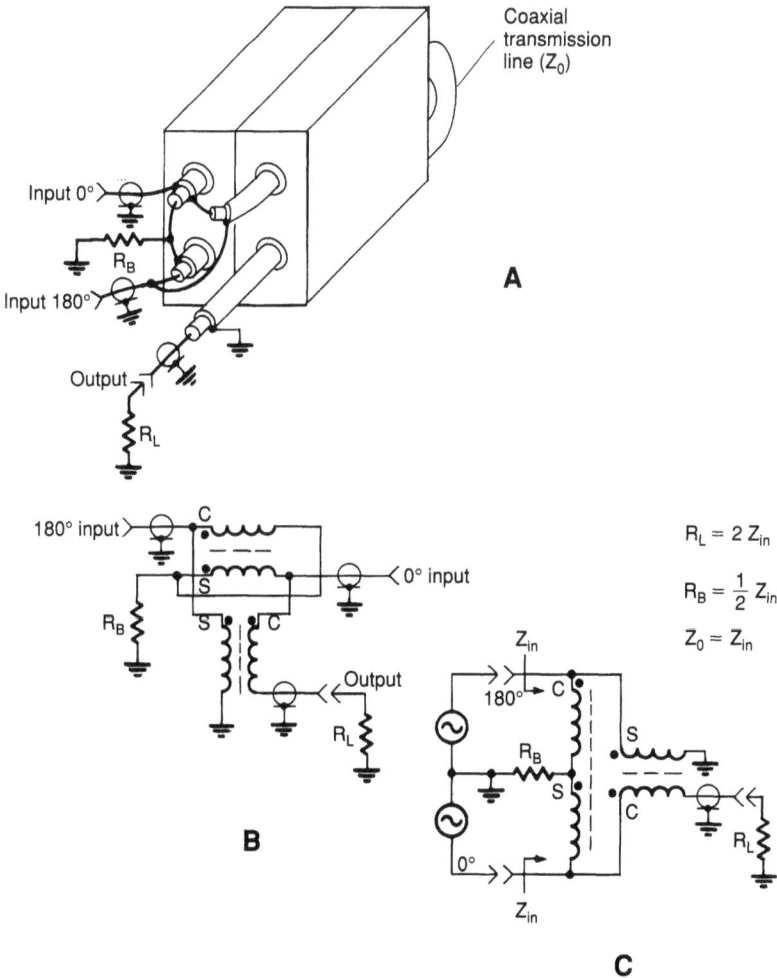

$R_L = 2\,Z_{in}$

$R_B = \dfrac{1}{2}\,Z_{in}$

$Z_0 = Z_{in}$

Figure 12.33 Type I 180° hybrid combiner: (A) pictorial view, balun core; (B) schematic; (C) equivalent circuit.

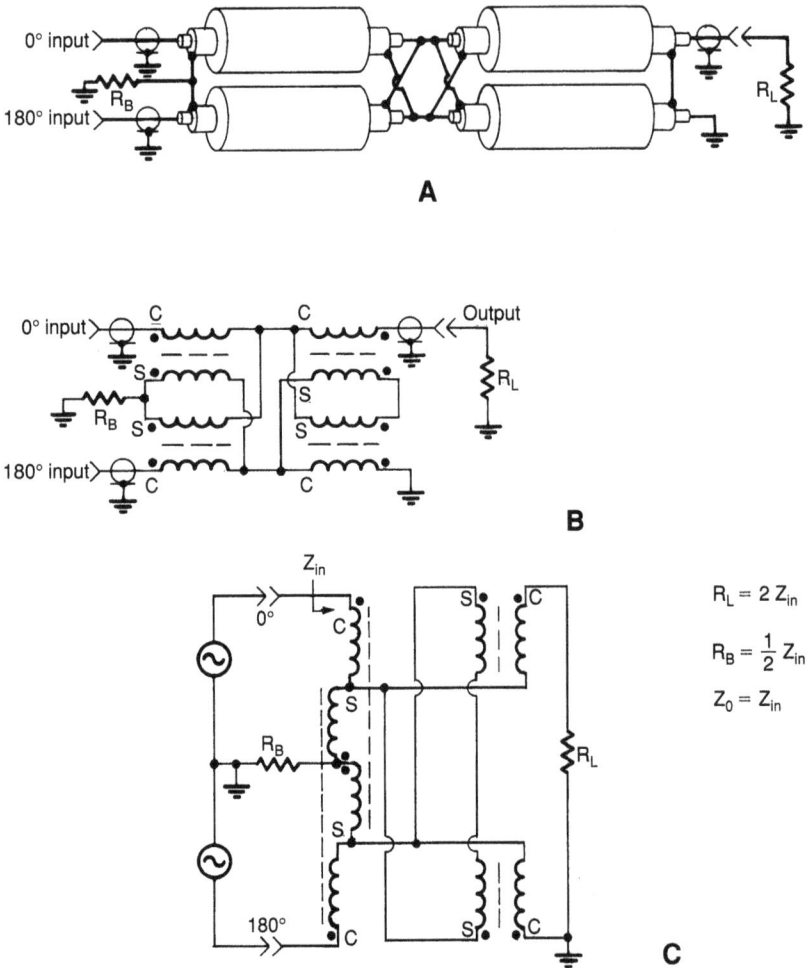

Figure 12.34 Type II 180° hybrid combiner: (A) pictorial view, tubular cores; (B) schematic; (C) equivalent circuit.

at their outputs across the bandwidth of interest. The absolute phase shift from input to output of the allpass networks is not important as long as the phase difference between the two outputs remains very close to 90°. A typical circuit topology is shown in Figure 12.38. Refer to references 31, 32, 33, and 34 for practical details on designing allpass networks.

Are there any inherent advantages or disadvantages to using one combiner topology over another? Basically there is no difference between inphase and 180° combining as far as their effect on overall PA performance. There are physical implementation differences that will dictate choosing one or the other for a given design. Several differences exist between these two and the 90° quadrature combiner which will be briefly described.

The major disadvantage of the quadrature combiner at HF is the need for the two allpass networks and the requirement for accurate phase tracking between the two filters. This is not insurmountable technically, but adds complexity, cost, and size penalties to the overall PA.

Advantages to quadrature combining (depending upon PA design requirements) are reduced power supply current requirements, cancellation of certain odd-order reverse IMD products, and improved output power for SWR loads.

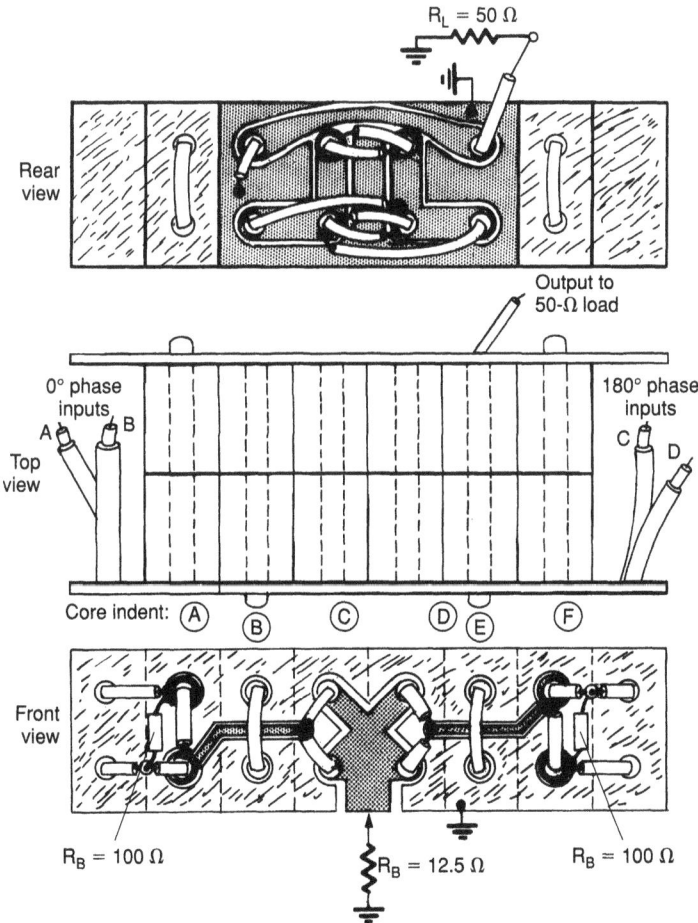

Figure 12.35 Four-port, two-stage combiner using types I and II hybrids, pictorial. Note: All transmission lines Z_0 = 50 ohms.

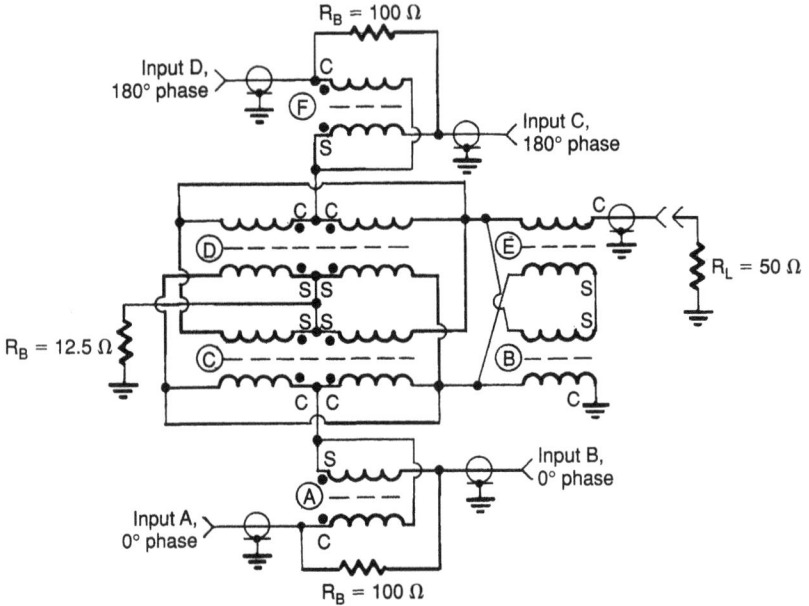

Figure 12.36 Four-port, two-stage combiner using types I and II hybrids, schematic. Inputs A and B combine in-phase, as do inputs C and D. The A/B output and the C/D output combine in two parallel connected 180° hybrids. Two in parallel avoids 25-ohm Z_0 coaxial cable. 180° combiner gives balanced 12.5-ohm load impedance; ideal for transformation with a 4:1 balun to 50 ohms. Circled letters A, B, C, etc. are the core identifications per Figure 12.35.

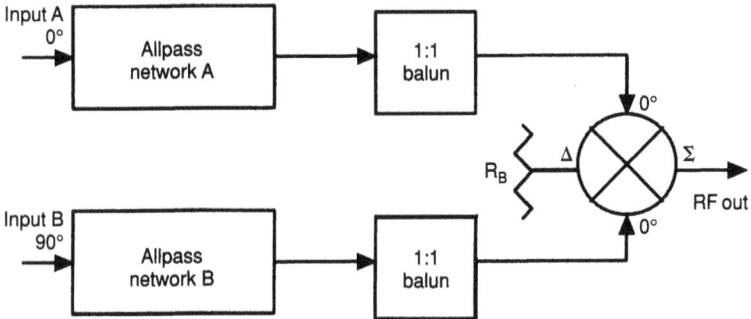

Figure 12.37 Wideband HF quadrature combiner.

The reduction in power supply current may be realized by considering the example of two modules, quadrature combined, compared with two modules combined with either an in-phase or 180° combiner. In the latter case, there is a load impedance on any given SWR circle that will cause a maximum drain current, compared to all other load impedance values on the same SWR circle. Both modules will experience the same load impedance and thus the same maximum current. But if the modules are combined in quadrature the phase shifts cause each module's load to be the complex conjugate of the opposite module's load impedance. Thus, if one module's load causes a current maximum, the other module's load will cause a current minimum and the total maximum dc that both modules require from the common power supply will be lower.

This same situation (that is, SWR loads) applies to total power output from the combined module. In the case where the load impedance is such that the power output is at a minimum for the worst-case load impedance on a given specified SWR circle, the quadrature combined module pair will deliver more power output than either in-phase or 180° combined modules. This is because both modules never see the same load impedance (for quadrature combined modules), whereas they do in the case of in-phase or 180° combined modules.

Reverse IMD (as opposed to forward IMD) is caused when a strong external signal is coupled into the output of a transmitter. The external signal mixes with the desired signal in the output devices and the distortion products are delivered to the PA load along with the desired signal. By carefully tracing the phase shifts that the fundamental and each harmonic of both the desired and the external undesired signal undergo, it can be seen that the odd-order IMD products are terminated in the bridging resistor of the

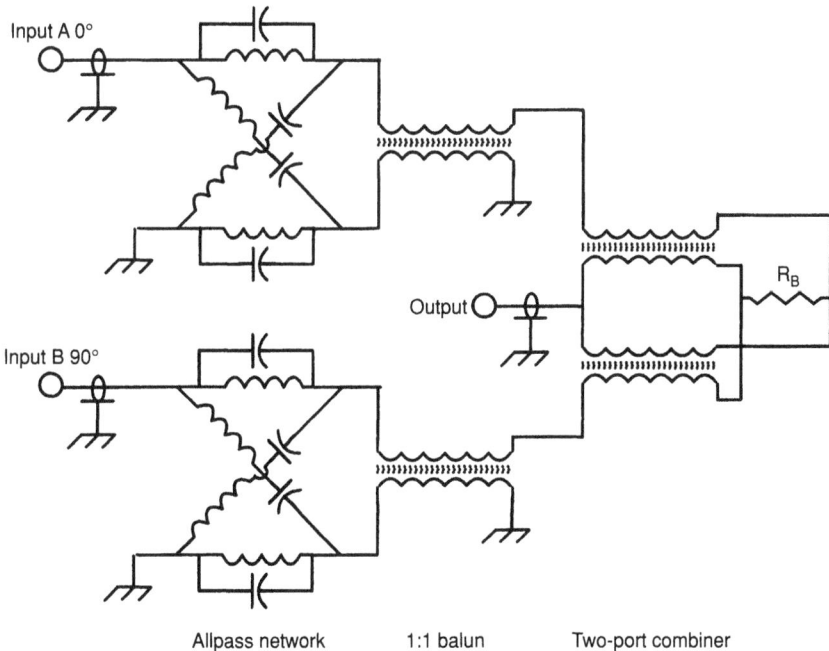

Figure 12.38 Typical HF quadrature combiner circuit topology. Single-section allpass lattice networks are shown with type II in-phase combiner.

quadrature combiner. This is true only if the allpass network phase-shift networks and the wideband transformers that make up the quadrature combiner hold their required phase and amplitude characteristics over a sufficiently wide bandwidth to include the odd-order harmonics.

For an amplifier operating at 30 MHz, this would require the quadrature combiner characteristics hold correctly to at least 150 MHz to include the fifth harmonic and 210 MHz to include the seventh harmonic. The amount of improvement depends greatly upon the tolerance to which the combiner input ports can maintain 90° phase differential and equal amplitude balance. It also depends upon how closely matched the source impedance of the two PA modules can be maintained. Allpass networks of more than casual design and layout are required for bandwidths of 1.6 to 210 MHz in order to realize these benefits. The 0° and 180° combiners do not provide any improvement for reverse IMD control.

Module gain-matching techniques

In order to effectively combine solid-state PA modules, each PA module must have the same power gain and phase shift from input to output [35]. Otherwise, some power will be dissipated in the bridging resistors of the power divider and combiner. Figure 12.39 can be used to determine the effect of power gain and/or phase-shift differences for various amounts of power lost in the bridging resistor power. Additional details with some useful graphs may be found in reference 36.

Phase-shift differences between identical modules are usually small for HF PAs, provided the modules are constructed on a PC board and use identical components (especially the wideband transformers). Another obvious precaution: always use identical lengths of transmission line between the divider and the module RF inputs and between the module RF outputs and the power combiner.

Gain matching of the modules is usually required since the relative gain between matched pairs of BJT transistors will vary considerably. Some type of "test-selected" attenuator normalizes the PA module gain to a given range. The attenuator is simply a "Tee" or "Pi" configuration resistive attenuator.

The value is chosen depending upon the measured module gain. Figure 12.40 gives the topology for the two types of attenuators and the formulas to calculate the values for any given impedance and attenuation. The minimum-loss L pad shown in Figure 12.40C is sometimes useful in conjunction with an attenuator. It provides impedance matching without the use of reactive elements, but with additional loss. If the load impedance is greater than the desired source impedance, the network may simply be operated in reverse. An attenuator at the module input will improve the VSWR variation across the frequency range. The power dissipation and/or value required for each resistor element will influence the configuration selected.

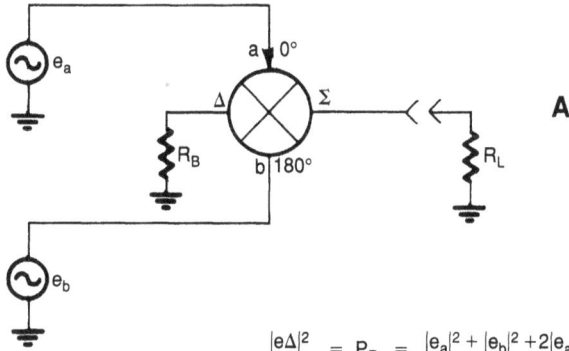

$$\frac{|e\Delta|^2}{R_B} = P_{R_B} = \frac{|e_a|^2 + |e_b|^2 + 2|e_a \ e_b|\cos \theta}{4R_B}$$

where θ = amount of phase difference between e_a and e_b
$R_B = R_a = R_b$ (input port resistance)

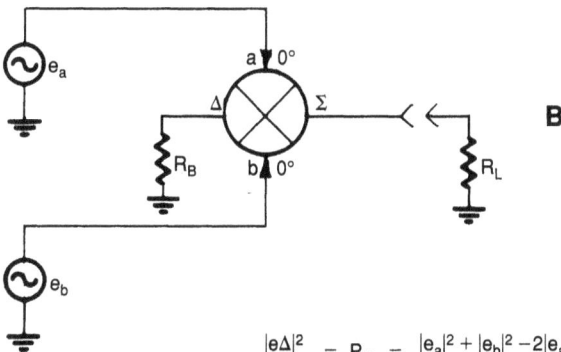

$$\frac{|e\Delta|^2}{R_B} = P_{R_B} = \frac{|e_a|^2 + |e_b|^2 - 2|e_a \ e_b|\cos \theta}{4R_B}$$

where θ = amount of phase difference between e_a and e_b
$R_B = R_a = R_b$ (input port resistance)

Figure 12.39 Calculations of combiner losses due to amplitude/phase imbalances: (A) 180° hybrid combiner; (B) in-phase combiner.

The PA module gain variation over the frequency range should track from module to module, otherwise the gain matching will not be valid everywhere. Bipolar transistor amplifiers are more likely to experience problems with this than FET amplifiers. Low-Q swamping networks are often used in the input circuitry to reduce the higher gain at low frequencies. A swamping network on the input also tends to narrow the gain differences between modules at low frequencies. Negative RF feedback around the PA stage will also flatten the gain response of BJT and FET amplifiers and narrow the gain differences between modules.

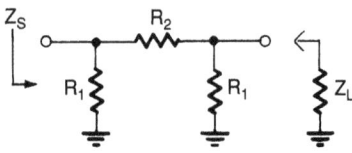

$$R_1 = Z \frac{10^{0.05A} + 1}{10^{0.05A} - 1}$$

$$R_2 = \frac{ZR_1(10^{0.05A} + 1)}{Z + R_1}$$

$$Z_S = Z_L$$

A

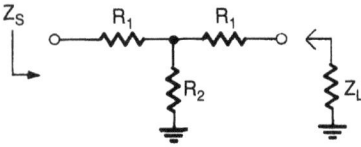

$$R_1 = Z \frac{10^{0.05A} - 1}{10^{0.05A} + 1}$$

$$R_2 = \frac{R_1 + Z}{10^{0.05A} - 1}$$

$$Z_S = Z_L$$

B

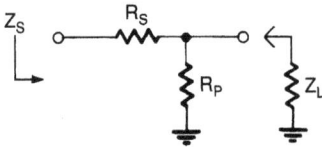

Given: $Z_S > Z_L$

$$R_S = \sqrt{Z_S (Z_S - Z_L)}$$

$$R_P = \frac{Z_S Z_L}{R_S}$$

$$A = 20 \log \left[\frac{(Z_L + R_P) \sqrt{Z_S Z_L}}{Z_L R_P} \right]$$

C

Figure 12.40 Attenuator designs: (A) pi; (B) tee; (C) minimum-loss L pad.
Note: A = loss (in decibels).

References

1. E. T. Rodriguez, "Model Semiconductor Thermal Designs Even with Scanty Vendor Data," *Electronic Design* 27 (February 15, 1979): 102–105.

2. Norm Dye and Helge Granberg, *Radio Frequency Transistors Principles and Practical Applications* (Stoneham, MA: Butterworth-Heinemann, 1993).

3. H. L. Krauss, C. W. Bostian, and F. H. Raab, *Solid State Radio Engineering*, Chapter 12 (New York: John Wiley & Sons, 1980).

4. J. Johnson, *Solid Circuits* (San Carlo, CA: Communication Transistor Co., 1973).

5. D. DeMaw, *Practical RF Design Manual* (Englewood Cliffs, NJ: Prentice-Hall, 1982).

6. R. K. Blocksome, "Practical Wideband RF Power Transformers, Combiners, and Splitters," *Proc. RF Technology Exp. 86*, Anaheim, CA (January 30–February 1, 1986): 207–227.

7. A. J. Burwasser, "Wideband Monofilar Autotransformers," part 1, *RF Design* 4 (January/February 1981): 38–44; part 2, *RF Design* 4 (March/April 1981): 20–29.

8. G. Badger, "A New Class of Coaxial-Line Transformers," part 1, *Ham Radio* 13 (February 1980): 12–18; part 2, *Ham Radio* 13 (March 1980): 18–29.

9. J. J. Nagle, "Testing Baluns," *Ham Radio* 16 (August 1983): 30–39.

10. G. Gaunella, "New Method of Impedance Matching in Radio Frequency Circuits," *Brown Boveri Review* 31 (September 1944): 327–329.

11. W. A. Lewis, "Low-Impedance Broadband Transformer Techniques in the HF and VHF Range," Collins Radio Working Paper WP-8088 (July 1, 1965).

12. Donald A. McClure, "Broadband Transmission Line Transformer Family Matches a Wide Range of Impedances," *RF Design* 17 (February 1994): 62–66.

13. H. O. Granberg, "Good RF Construction Practices and Techniques," *RF Design* 3 (September/October 1980): 51–59.

14. P. M. Rostek, "Avoid Wiring-Inductance Problems," *Electronic Design* 22 (December 6, 1974): 62–65.

15. N. O. Sokal, "Parasitic Oscillation in Solid-State RF Power Amplifiers," *RF Design* 3 (November/December 1980): 32–36.

16. R. K. Blocksome, "A Binary Stepped Transmission Line," *RF Design* 5 (July/August 1982): 22–29.

17. P. H. Smith, *Electronic Applications of the Smith Chart* (New York: McGraw-Hill, 1969).

18. H. O. Granberg, "Power MOSFETs Versus Bipolar Transistors," *RF Design* 4 (November/December 1981): 11–15.

19. Warren Bruene, "RF Power Amplifiers and the Conjugate Match," *QST* 75 (November 1991): 31–32, 35.

20. G. Hiller, "Design with PIN Diodes," part I, *RF Design* 2 (March/April 1979): 34–49; part II, *RF Design* 2 (May/June 1979): 40–47.

21. *Watt's New from Bird* 2, no. 2 (March-April 1965) (published by Bird Electronic Corp., Cleveland, OH).

22. W. B. Bruene, "An Inside Picture of Directional Wattmeters," *QST* 43 (April 1959): 24–28.

23. R. Weinrich and R. W. Carroll, "Absorptive Filter for Harmonics," *QST* (November 1968): 20–25.

24. W. H. Hayward, *Introduction to Radio Frequency Design* (Englewood Cliffs, NJ: Prentice-Hall, 1982).

25. P. R. Geffe, *Simplified Modern Filter Design* (New York: J. F. Rider, 1963).

26. R. Saal, *Der Entwurf von Filtern mit Hilfe des Kataloges Normierter Tiefpasse* (Backnang, West Germany: Telefunken GmbH, 1963).

27. A. I. Zverev, *Handbook of Filter Synthesis* (New York: John Wiley & Sons, 1967).

28. J. R. Fisk, "Microstrip Transmission Line," *Ham Radio* 11 (January 1978): 28–37.

29. H. O. Granberg, "Combine Power without Compromising Performance," *Electronic Design* 28 (July 19, 1980): 181–187.

30. A. J. Burwasser, "Taking the Magic out of the Magic Tee," *RF Design* 6 (May/June 1983): 44–60.

31. R. B. Wilds, "Formation of Passive Lumped Constant 90-degree Phase Difference Networks," *Ham Radio* (March 1979): 70–73.

32. Sidney Darlington, "Realization of a Constant Phase Difference," *Bell System Tech. J.* (January 1950): 4.

33. S. D. Bedrosian, "Normalized Design of 90-degree Phase Difference Networks," *Wireless Engineer* (March 1950): 72.

34. Thomas A. Keely, "Design of Constant Phase Difference Networks," *RF Design* 12 (April 1989): 32–42.

35. J. A. Benjamin, "Use Hybrid Junctions for More VHF Power," *Electronic Design* 16 (August 1, 1968): 54–59.

36. Roderick K. Blocksome, "Predicting RF Output Power from Combined Power Amplifier Modules," *RF Design* 11 (February 1988): 40–44.

13

Ultralow-Distortion Power Amplifiers

Edward G. Silagi

In applications where a high degree of linearity and low noise are required from the transmitter, external error correction may be added to improve performance. This chapter describes feedforward techniques for low transmitter noise and distortion.

13.1 Introduction

Linear PA output

Linear power amplifier (LPA) performance may be characterized by the contents of the amplifier output spectrum. The output of the LPA generally includes the desired signal plus a number of undesired signals. The undesired signals take the form of noise and distortion. Noise transmitted by the LPA includes amplified exciter noise and internally generated noise. It may be described by the equivalent amount of power existing in a 1-Hz bandwidth (dBW/Hz).

The distortion existing at the LPA output is usually described in terms of harmonic and IMD content and also spurious responses. Harmonic frequencies exist at integer multiples of the desired transmit frequency. Harmonic amplitudes may be measured by injecting a CW signal at the input of the LPA and observing the output at the harmonic frequencies. Intermodulation distortion is created by the mixing of two or more different frequency tones in the LPA. The IMD products exist at sum and difference frequencies of the mixing tones and their harmonics. They may be measured by injecting two equal amplitude tones of different frequency at the input of the LPA while observing the output with a spectrum analyzer. Spurious responses, or oscillations, occur at random frequencies and can exist while the LPA is being driven or undriven. All three types of distortion may be described by the number of decibels they are below the desired output (dBc—below either tone in the two-tone test case). A typical LPA output spectrum showing the desired transmit signal(s), noise, and distortion products is shown in Figure 13.1.

Current LPAs may be divided into two categories: vacuum tube and solid-state designs. Each type has advantages and drawbacks. Vacuum tube LPAs generally exhibit a high degree of linearity (IMD = –40 dBc). Because of the high output impedance of the vacuum tube, the matching networks are narrowband designs. This provides for filtering of the noise, harmonics, and IMD far away from the desired signal frequency. It also reduces the mixing of collocated transmitter signals in the tube output circuit ("backdoor" IMD).

The narrow bandwidth of the vacuum tube LPA also proves to be disadvantageous for certain modern requirements. It does not allow for simultaneous transmission of a number of signal frequencies if the frequency separation is wider than the bandwidth. Also, to prevent the desired signal from being jammed by outside sources, frequency-hopping schemes have been developed. Because of the hop rates, the vacuum tube LPA servo-tuned output network does not have time between hops to retune to each new frequency.

Solid-state LPA designs are inherently broadband. This results in part from the broadband low-impedance matching transformers used in the transistor output networks. The instantaneous bandwidth of a solid-state

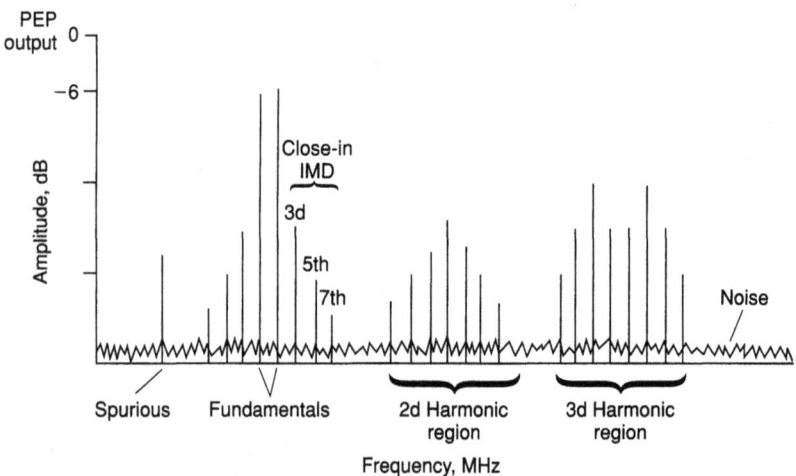

Figure 13.1 Typical PA output spectrum.

LPA is generally limited only by the half-octave filters used in the output path to reduce harmonic level.

The linearity provided by current solid-state LPA designs has not matched the vacuum tube designs described in Chapter 14. A push-pull transistor configuration may be used to reduce even-order harmonics, and quadrature hybrid combining may be used to reduce certain IMD products. Still existing are the close-in IMD products generated by the transistors. Typical IMD performance for a class AB solid-state LPA is –25 to –35 dBc. To increase the linearity beyond the limitations of the transistors of the solid-state LPA, external distortion reduction must be introduced.

Feedforward versus feedback

The most familiar form of external error correction used in many types of amplifier design is negative feedback. This concept was developed by H. S. Black in the 1930s. Several years earlier, Black [1] patented another method of error correction known as *feedforward*. Feedforward offers a number of advantages over feedback. The single fact that makes feedforward a more acceptable concept is that "it is not correcting for an error which has already occurred." That is, the correction is taking place in the same time frame that the error is occurring. Also, because there are no direct feedback paths, the correction concept is inherently stable over an infinite instantaneous bandwidth.

Early work by H. Seidel and others dealt with feedforward at VHF and microwave frequencies. The purpose of this chapter is to give the reader a working knowledge of feedforward as it applies to the HF range of 2 to 30 MHz. For detailed derivations and additional information beyond the scope of this book, the reader is directed to the references listed at the end of this chapter.

13.2 Feedforward Error Correction

To gain an understanding of the feedforward concept, a number of terms pertaining to the basic operation will be defined. During the explanation, the reader is referred to Figure 13.2.

The heart of the feedforward system is the main amplifier. Its purpose is to amplify the input signal to the desired output level. The output of the main amplifier contains the fundamental (desired) signal plus noise and distortion. As noted earlier, the noise and distortion are a combination of that already present on the input signal and of the products generated by the main amplifier. Feedforward will not correct for noise and distortion already existing on the input signal.

To accomplish the desired error correction, the process may be broken down into two "loops" (the term *loop* is used even though signal flow is

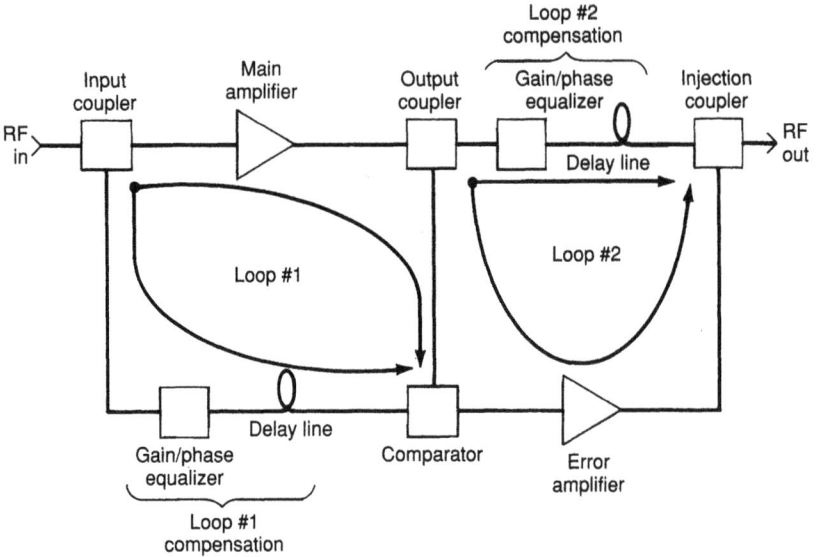

Figure 13.2 Feedforward amplifier block diagram.

not continuous in a single rotational direction). Loop 1 consists of the main amplifier, the input coupler, the output coupler, the loop 1 compensation, and the comparator. The purpose of the first loop is to obtain a sample of the distortion and noise added to the output signal by the main amplifier. This is accomplished by subtracting a sample of the "undistorted" input signal from a sample of the output signal. The input sample is obtained from the input coupler while the output sample is obtained from the output coupler. The subtraction takes place in the comparator. To make sure proper subtraction occurs, the input sample is adjusted in amplitude and phase by the loop 1 gain/phase equalizer and is delayed in time by the loop 1 delay line. The amount of delay is equal to the delay introduced by the main amplifier. The complexity of the gain/phase equalizer depends upon the characteristics of the main amplifier. The output from the comparator, in theory, consists only of a sample of the distortion and noise added by the main amplifier.

The distortion and noise output from the comparator is raised to a certain power level by the error amplifier. This amplified signal is then inverted and injected into the main output path such that it cancels the noise and distortion created by the main amplifier. The output signal is then an exact amplified replica of the input signal. To force proper time frame subtraction, the output signal is delayed by the loop 2 delay line. This delay is equal to the delay introduced by the error amplifier. The inversion of the correction signal may take place in either the error amplifier, the output coupler, or the injection coupler. This process of removing the noise and distortion from the main output is accomplished by loop 2. The second loop is therefore comprised of the output coupler, the loop 2 compensation, the injection coupler, the comparator, and the error amplifier.

13.3 System Component Requirements

Main amplifier

As stated previously, the primary purpose of the main amplifier is to raise the input signal to the desired output level. This output level may range from less than 1 W to a number of kilowatts, depending on the application. Regardless of the power level, the main amplifier of a feedforward system has a number of requirements unique to the application. These requirements are in the areas of gain, phase/delay response, IMD performance, and output impedance.

Gain. The gain requirement for the main amplifier is usually 5 to 10 dB greater than that needed by an equivalent amplifier in a nonfeedforward system. The increased gain is needed to make up for losses incurred in the input coupler, the output coupler, the loop 2 compensation, and the injection coupler. It is very desirable to have as flat a frequency response as possible to simplify the loop 1 compensation. The importance of flat response will become more apparent when loop null theory is discussed. As an example, a main amplifier with a gain of approximately +47 dB is required for a system with 100-mW input and 500-W output. This amplifier could be realized using three or four stages of solid-state devices.

Phase/delay response. The second area of consideration in designing the main amplifier portion of a feedforward system is the phase/delay response. It may be viewed from either a time-delay or a phase-shift perspective where the two are related by the equation

$$\text{Time delay} = \frac{\text{phase shift}}{360 \times \text{frequency}} \tag{13.1}$$

The unit for time delay is seconds and the unit for phase shift is degrees.
  The phase characteristic of a solid-state amplifier is determined by

- The transit delay of the solid-state devices
- The phase shift caused by the power splitters, combiners, and transformers used to interconnect the multiple-stage amplifier
- The phase shift resulting from dc coupling chokes and RF coupling capacitors

The desired phase response from the main amplifier is linear (constant time delay) with as small a delay as possible. As with the gain flatness, the phase flatness will determine the complexity of the loop 1 compensation network. An example of the phase response of a solid-state PA is shown in Figure 13.3. This response curve demonstrates the practical deviations from the ideal situation. On the phase response curve, the three areas of concern are the amount of delay, the phase linearity, and the zero-frequency phase intercept point.

The amount of delay is determined by the number and types of components existing in the signal path. The major contributors to the amplifier delay are the power transistors and the power-combining devices (splitters, combiners, and transformers). To obtain the required power output, either BJT or MOSFET type devices may be used. In the case of a feedforward amplifier, the MOSFET offers some distinct advantages over the BJT in terms of phase response. These advantages are

1. *Gain.* The power MOSFET generally exhibits more gain than its BJT counterpart. This allows amplifier realization with a minimum number of stages; the fewer the stages, the less delay there is.

2. *Delay.* The power MOSFET transistor shows less transit delay than the power BJT. Typical delay for a MOSFET is 5 ns; typical delay for a BJT is 10 ns.

3. *High input impedance.* The relativity high input impedance of the MOSFET along with its high gain allows for resistive pad input matching instead of transformer matching. The elimination of transformers reduces delay (typically 2 ns/transformer) and also maximizes phase linearity.

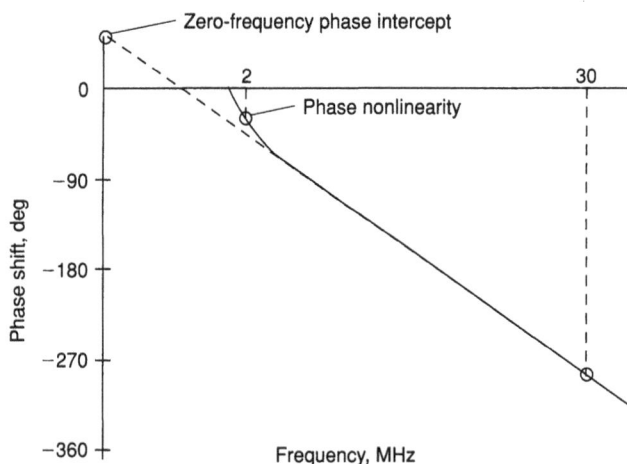

Figure 13.3 Solid-state PA phase response.

The phase linearity is defined by how well the phase response follows a straight-line path over frequency. This is determined by a number of factors. One is the phase linearity of the power transistor being used. Radio frequency power MOSFETs have an f_t that extends well above 100 MHz. This results in a linear phase response beyond the high end of the HF band. Some deviation, however, does occur in the 2- to 5-MHz region (as shown in Figure 13.3).

The low-frequency phase deviation is only partially due to the power transistor. A second cause of this deviation is passive component frequency response. The passive components include transformers, splitters, combiners, dc coupling chokes, and dc blocking capacitors. In the case of the inductive-type components, the deviation is due primarily to shunt inductance to ground. This can be minimized by winding the components on a ferrite material. Care must be taken, however, to prevent overheating of the ferrite at low frequencies.

From Equation 13.1 it can be deduced that the phase shift at zero frequency (dc) should be 0°. As shown in Figure 13.3, if a straight-line plot of the phase response is extrapolated to zero frequency, zero phase shift is not obtained. The amount of this zero-frequency phase shift is referred to as the *zero-frequency phase intercept point*. In solid-state amplifiers, this phase shift may be as great as 15 to 20°. The apparent nonzero phase intercept is caused by characteristics of both the power transistors and the passive components. If broadband operation of the feedforward system is required, the nonzero phase intercept and the phase nonlinearity of the main amplifier must be compensated. This can be done in the loop 1 compensation network. As with the gain flatness, the importance of the phase response will be seen in the discussion of null theory.

IMD performance and linearity. The linearity of the main amplifier in a feedforward system must be inherently good before the error correction is added. This refers primarily to low levels of IMD and harmonic products within the passband of the PA. Because filters limit the instantaneous bandwidth of the system, they can only be used for products outside the passband.

As stated earlier, typical IMD performance from a solid-state PA is –25 to –35 dBc. This is true for push-pull transistors biased for class AB operation and transmitting a PEP output of 150 to 200 W. For a feedforward system it is desirable to have a main amplifier IMD performance of –40 dBc or better. The reasons for this will be seen in the discussions of the error amplifier and the number of correction loops required. There are a number of ways to obtain optimum linearity from a solid-state PA before the feedforward is applied. These include

- Reduction of power output
- Negative feedback

- Bias adjustment
- Using optimum circuit topology
- Choosing the proper transistors

Solid-state devices exhibit inherent linearity limitations which are a function of power output, bias level, and frequency. Both BJT and MOSFET devices show good linearity (–35- to –40-dBc IMD) at low frequencies in the 2- to 5-MHz range. For higher frequencies, the performance is degraded for BJTs, the middle HF frequencies (5 to 20 MHz) show performance in the –25- to –35-dBc range, with some isolated frequencies showing poor IMD and/or efficiency. These "holes" are due in part to transformer resonances. Above 20 MHz (and up to 30 MHz) a phenomenon called *cancellation* may occur. With it, the third-order IMD actually decreases when power output is increased. The cancellation of the third-order products may be controlled somewhat by capacitive compensation of the output transformers and also by changing the bias level.

Power MOSFET devices show a more gradual IMD degradation with increasing frequency. This IMD degradation is also accompanied by decreased efficiency and decreased gain. The cancellation effect almost always seen with the BJT has been seen with some MOSFETs, but it occurs at a higher power level. The MOSFET does show some advantages over the BJT. New technology is pushing toward high drain-source breakdown voltages (greater than 100 V). This allows the use of increased supply voltage and higher-impedance output transformers (greater than the 2-ohm loading for low supply voltage, where series strays are difficult to overcome). The MOSFET also shows much lower-level high-order IMD (seventh order and greater) than the BJT.

Lowering the IMD levels to more than –40 dBc requires a reduction in output power and an increase in dc bias. For a pair of standard devices, the power output must be reduced from 150 W to the 30- to 70-W PEP range. Also, the dc bias level must be increased from low-current class AB (50 to 100 mA/device) to high-current class AB or class A mode (1 to 3 A/device). With the increased bias level, the efficiency of the PA drops from about 40% to about 15%. Thermal performance now becomes a limiting factor, and the type of heat removal (forced air, water, etc.) is an integral part of the amplifier design.

Negative feedback applied by resistance from drain to gate and source to ground improves IMD performance on low- and medium-PA stages (up to 30 W). The feedback also flattens gain and makes the input impedance of the transistors appear less reactive. This aids the impedance matching of the devices. For high-power stages, the feedback may help IMD at some frequencies and degrade it at others.

The methods described for reducing IMD also help to reduce harmonics generated by the PA. If class AB bias is used for increased efficiency, the push-pull configuration should be used to reduce the even-order harmonics. Typical high-power performance produces third harmonics in the –10- to –20-dBc range and gradually decreasing higher odd-order harmonics. The even-order harmonics are usually in the –30- to –40-dB range for a push-pull amplifier. This holds true as long as the two devices are closely matched in their gain characteristics.

PA output impedance. Because of second-loop problems (to be discussed) it is desirable to have the output impedance of the main amplifier equal to the characteristic impedance of the system (normally 50 ohms). There are three methods that can be used to accomplish this:

1. Design the main amplifier for excess gain and power output capabilities and place an attenuator after the amplifier to dominate the output impedance.

2. Design the amplifier final stage to have an output impedance equal to the desired value.

3. Use a quadrature (90°) combiner at the output of the main amplifier.

The output attenuator method is a "brute force" way of obtaining a 50-ohm output impedance. With it, the main amplifier must be designed to overcome the loss of the attenuator. As an example, a 10-W system must have a 100-W main amplifier if a 10-dB attenuator is used. For this example, the attenuator would dissipate 90 W. Although inefficient, the attenuator method is simple in design as long as power requirements are low. It also provides a constant load to the main amplifier regardless of system load (VSWR condition).

The second method that can be used to set the main amplifier output impedance is to design the transistor stages for a specific output impedance. This may be done by using (for FETs) drain-gate feedback and source degeneration. The values of resistance may be approximated by using dc analysis if the g_m of the FET is known. From the design, the input impedance, output impedance, and gain of the transistor stage will be set.

As with the output attenuator method, the feedback method has practical limitations that become more apparent at power levels greater than 10 to 20 W. Because the FET characteristics are frequency dependent, the dc approximation is less accurate as frequency increases. Some circuit tuning may be used as long as amplifier bandwidth is not substantially reduced. The FET characteristics are also dependent on the voltage and current applied to the device. In particular, g_m variation with drain current affects gain and matching of the device. This may be stabilized by using class A biasing for low- and medium-power applications (limited to 20 to 30 W because of the high power dissipation in the device). The source degeneration also limits maximum voltage swing at the output of the FET, which again limits maximum power output.

The third method that can be used to realize a 50-ohm output impedance is quadrature combining. This is used in most high-power applications because of the minimal additional requirements placed on the amplifier. It allows the main amplifier to be designed using standard modular techniques.

Quadrature hybrid combining. A quadrature hybrid is a four-port device with a unique relative phase relationship between the ports. When used as a splitter, a signal incident at the input is divided into two equal amplitude components that are different in phase by 90°. When used as a combiner, the two incoming signals must be equal in amplitude and 90° different in phase for the resultant to appear only at the output port. A PA containing quadrature combining is shown in Figure 13.4. This figure will also be used to describe why the quadrature combining results in an apparent 50-ohm output impedance. A wave traveling back to the PA is split by the quadrature combiner into equal magnitude components that have a 90° relative phase difference. The two signals reach the PA modules and are reflected back, fully or partially, depending on the output impedance of the module. At this point, one requirement for the modules becomes important. They must have identical output impedances so that the relative relationship of the signals does not change after reflection. The absolute impedance does not matter as long as the two are identical. This requirement should be easily satisfied since the output modules of a PA are usually identical in construction.

The reflected signals returned to the quadrature hybrid remain equal in amplitude and quadrature (90° difference) in phase. The 90°-shifted signal is shifted another 90° by the quadrature hybrid. The new phase difference

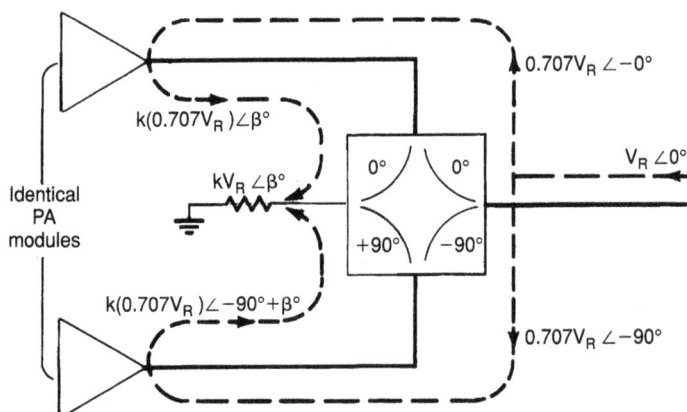

Figure 13.4 Quadrature hybrid combined PA.

results in destructive combining at the output port and constructive combining at the dump port (the resistively terminated fourth port of the combiner). Because the recombined signal is dissipated in the dump port, there is no reflected power returning to the source. By definition, zero reflected power implies a matched impedance condition at the output port of the quadrature combiner.

Amplifier modules combined using a quadrature hybrid must have their input signals 90° out of phase to obtain maximum power output. This may be accomplished by using a second quadrature hybrid as the power splitter preceding the modules. The splitter should exhibit the same amplitude and phase characteristics as the combiner. It may be scaled down, however, since its power-handling requirement is much less.

Quadrature hybrid realization. For operation in the HF band, the quadrature hybrid may be broken into two sections. The first is the power-combining network and the second the 90° phase difference network. The power-combining network requires two input ports, a summing port, and a difference port. It may be realized by standard HF power-combining techniques described in Chapter 12 as long as a resistive load is provided at the difference port.

The 90° phase difference network consists of two allpass filters that maintain a 90° phase difference over the required frequency range. The ability of the filters to keep their 90° difference is critical to the operation of the hybrid. As described in the previous section, the apparent 50-ohm output impedance of the PA depends upon the acquired phase difference of the signal which ultimately ends in the dump port. If, after passing through the phase-shift networks, the phase difference at the output port is not 180°, part of the signal travels back toward its source. This "reflected" portion of the signal may be characterized by calculating an effective VSWR at the amplifier output (see Table 13.1). The effective VSWR is now also dependent on the output impedances of the amplifier modules being combined.

The allpass filter is a two-port network that has an increasing phase shift as a function of frequency. It is designed to have a characteristic impedance at all frequencies and theoretically introduces no attenuation. For a quad-

**TABLE 13.1 Effective Output
VSWR versus Quadrature
Hybrid**

Phase deviation, deg	VSWR
0	1.00:1
1	1.05:1
2	1.09:1
5	1.25:1
10	1.57:1

rature hybrid application, two allpass networks that maintain a 90° phase difference over the given frequency band are required. The most common form of circuit topology for allpass realization is the lattice network. The first-order lattice network along with its normalized phase-response is illustrated in Figure 13.5. As shown, the lattice network is a balanced circuit configuration. To change it to an unbalanced circuit, a balun must be added to one side. The other side may then be treated as an unbalanced port. The maximum phase shift that may be obtained from a first-order network is 180°. To obtain higher-order filters, first- and/or second-order networks of the same characteristic impedance may be cascaded.

The realization of two allpass filters with 90° phase difference involves filter synthesis techniques. One method is to use a Chebyshev approximation of the phase difference (see reference 2). This approximation method will produce a phase-difference ripple. That is, the phase difference between the two networks is 90° plus or minus a ripple magnitude. By specifying the ripple magnitude and the bandwidth, the order of the allpass networks may be determined. Allpass filter order versus ripple for a 2- to 30-MHz quadrature hybrid is shown in Table 13.2.

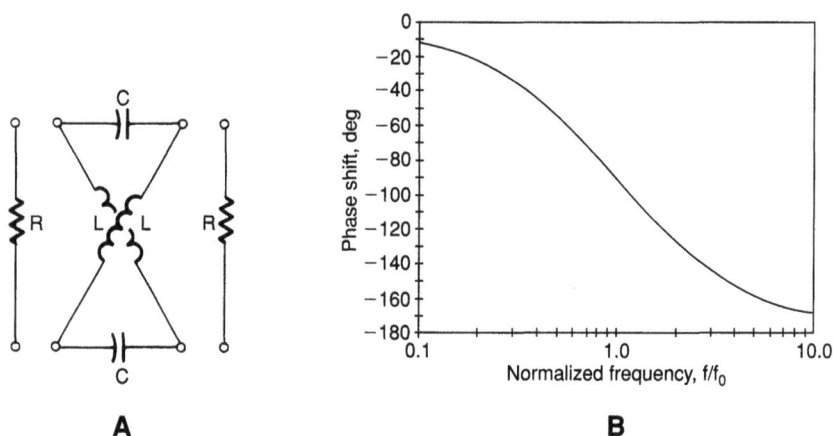

Figure 13.5 (A) First-order lattice allpass network; (B) first-order allpass filter phase response.

TABLE 13.2 Allpass Filter Order versus Phase Ripple for a 2- to 30-MHz Quadrature Hybrid

Ripple, deg	Combined filter order
0.5	6
1	5
2	4
10	3

Figure 13.6 shows an example of a quadrature hybrid design for use from 2 to 30 MHz. It is designed for 50-ohm characteristic impedance and a phase ripple of ±2°. The combined filter order is four. This means that each allpass filter is of order 2. They have been realized by cascading two first-order lattice networks per side.

Directional couplers

The directional coupler is a four-port device used for obtaining samples of forward and reflected power. In it, power incident at the input port is sampled at the forward port. Similarly, power incident at the output port is sampled at the reflected port. The amount of power obtained in each sample is specified by the coupling ratio. Ideally, no forward power will appear at the reflected port and no reflected power will appear at the forward port. For a directional coupler, a measure of this isolation is referred to as the *directivity*. Typical directivity in the HF band is 20 to 30 dB. The directivity is also a function of the terminating resistances at the four ports.

Figure 13.7 shows the schematic of a specific type of coupler called the *bidirectional coupler*. It is bidirectional in that power may be fed into any of the four ports with a sample being obtained at the corresponding coupling port. It is not truly symmetrical because one set of coupling ports (in-forward or out-reflected) does not invert the sample while the other does (180° phase shift). This feature is utilized to obtain correct polarity in the feedforward loops.

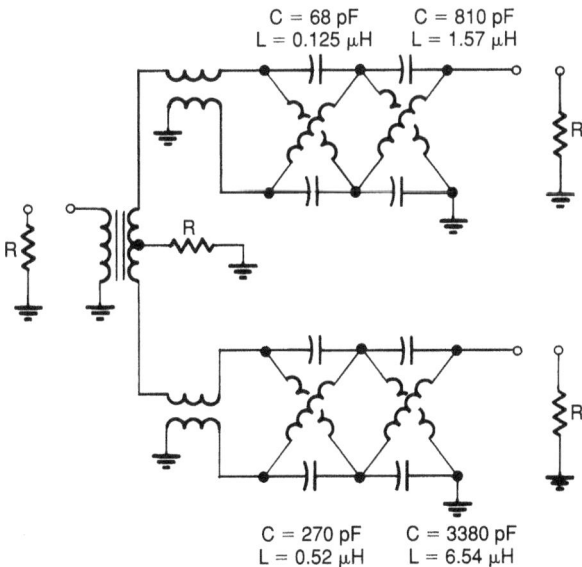

Figure 13.6 The 2- to 30-MHz quadrature hybrid.

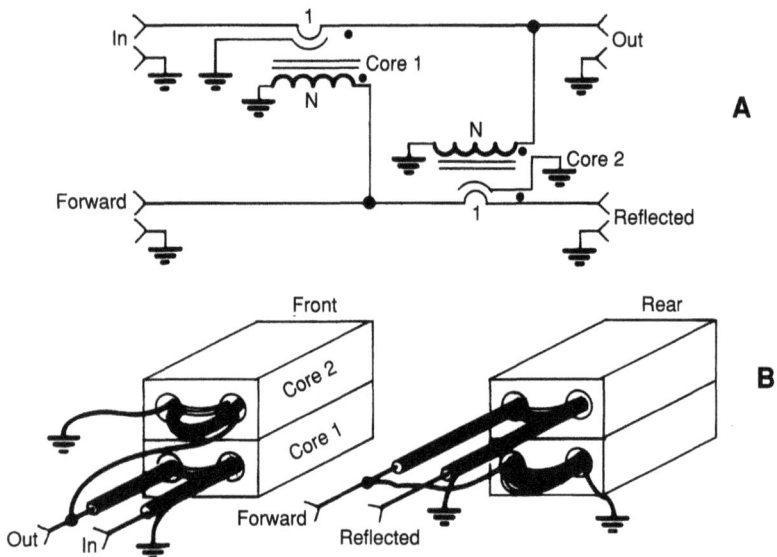

Figure 13.7 Bidirectional coupler: (A) schematic; (B) pictorial.

The bidirectional coupler may also be used to inject power into signal paths. Such a case is the injection coupler of the feedforward system. There the correction signal is injected into the main power path to cancel distortion. In a feedforward system, all the directional couplers may be constructed using similar techniques. The differences will be in the amount of coupling required and the power-handling capability. A review of the couplers required for the system is shown in Table 13.3.

The construction of a bidirectional coupler at HF frequencies uses cross-coupled transformers (as shown in Figure 13.7). The coupling ratio is determined by the turns ratio of the transformers. It may be approximately calculated using the following equation:

$$\text{Coupling ratio (dB)} = 20 \log (\text{turns ratio}) \qquad (3.2)$$

TABLE 13.3 Directional Couplers Required in a Feedforward System

Coupler	Purpose	Typical coupling ratio, dB	Power handling in 500-W system, W
Input	Obtains input sample	3–10	<1 for 30-dB system gain
Output	Obtains distorted output sample	20–30	500
Comparator	Removes input sample from output sample-output distortion remains	3–10	<1
Injection	Removes distortion from main output	6–10	500

The transformers are usually made with an N:1 turns ratio where the one turn is the main signal path. This minimizes delay (typically 1 to 2 ns) and series stray inductances that limit frequency response. The one turn may be a piece of coaxial transmission line with one side of the shield terminated to ground. This aids the high-frequency response and also acts as a Faraday shield to protect against high-voltage arcing (electromagnetic pulse [EMP], lightning, etc.) to the surrounding circuits.

To minimize shunt inductive effects, the transformers are usually wound on some magnetic core material (ferrite or powdered iron). High-permeability ferrite offers the greatest improvement in the low-frequency response (coupling flatness and phase linearity). The use of ferrite, however, does introduce some problems. Besides getting hot, ferrites create distortion at low frequencies from 2 to about 5 MHz. This must be taken into consideration in the paths where the feedforward will not cancel the coupler-created distortion (that is, in the injection coupler).

Error amplifier

The purpose of the error amplifier is to raise the amplitude of the error signal (main amplifier distortion and noise) obtained from the comparator. The amplified signal level is equal to the distortion amplitude at the main amplifier output raised by an amount equal to the coupling ratio of the injection coupler.

The error amplifier has a number of requirements:

1. *High linearity and low noise.* The success of the feedforward operation depends on three areas: the first loop null, the second loop null (see section 13.4), and the linearity of the error amplifier. Distortion products created by the error amplifier will be injected into the main output path along with the correction signal. Figure 13.8 shows this effect on the system performance. The error amplifier also controls the noise figure for the feedforward system, so it should be a low-noise design.

2. *Power output lower than main amplifier.* The power required from the error amplifier is determined by the distortion level of the main amplifier, the first loop null, and the coupling ratio of the injection coupler. It is typically 10 dB less than the main amplifier power output level.

3. *High-gain, flat-frequency response.* The gain required is typically 30 to 50 dB. This is needed to make up for output coupler, comparator, and injection coupler losses. Flat gain versus frequency is required to minimize loop 2 high-power amplitude compensation.

4. *Linear delay, short delay.* Linear delay is required to minimize loop 2 high-power phase compensation. Short delay is required to minimize the length of the loop 2 delay line.

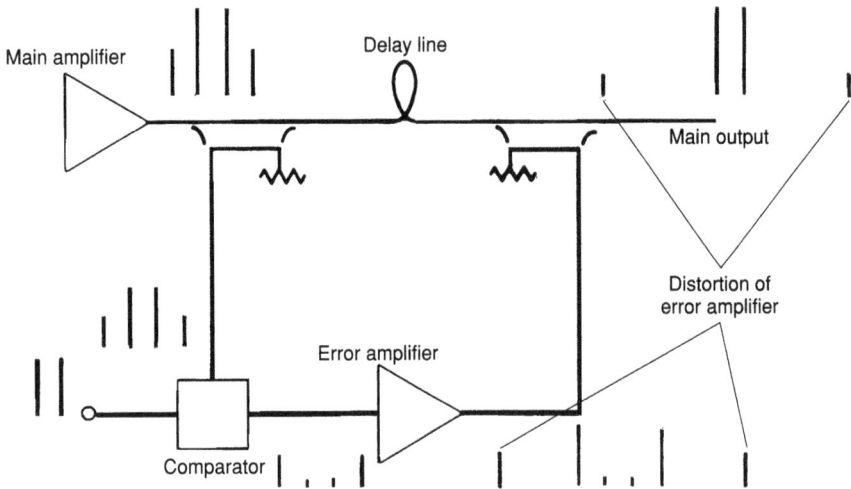

Figure 13.8 Error amplifier distortion at main output.

The error amplifier design includes a number of specialized circuit techniques to reduce noise, distortion, and delay. The first stage of the error amplifier may be a low-noise preamplifier utilizing transformer coupled lossless feedback for gain flattening and impedance matching. Low-power stages may include hybrid amplifier modules. These amplifiers exhibit 30 to 40 dB of gain with only 5 ns of delay. They are, however, limited in power output to less than 1 W for linear performance. A way of increasing power out of a stage with minimum additional delay is to parallel power transistors. Power FETs may be paralleled as long as some resistive isolation is provided between the gates to ensure stable operation. By eliminating an input and output transformer, up to 4 ns of delay may be saved. In the extreme case where ultralinear operation is required, the error amplifier may in itself be a feedforward system. This does, however, increase both its complexity and its delay.

The power output from the error amplifier is involved in a trade-off with the main output power. This is because of the operation of the injection coupler. When using a bidirectional coupler for injection, the error amplifier is fed into a coupled port with respect to the main output. The majority of the power from the error amplifier is dissipated in the dump resistor. This configuration is necessary to provide proper combining of the main and error outputs for distortion cancellation. By making the coupling ratio of the injection coupler small, the error amplifier output power is minimized at the expense of main power loss. As an example, if the injection coupler is changed from a 10-dB coupler to a 6-dB coupler, the error amplifier output and gain requirements are reduced by 4 dB. However, the in-out port loss (main output path through injection coupler) is increased from 0.5 to 1.23 dB. For a 500-W system, the main output loss would increase from 50 to 125 W.

Amplitude and phase/delay equalizing networks

Both loop 1 and loop 2 contain dual signal paths that must exhibit similar magnitude and phase characteristics. In each case, one side of the loop contains an amplifier that must be simulated in the opposite path. By choosing the proper coupling ratios for the directional couplers, the nominal amplitude of each path is equalized. The remaining deviations from nominal amplitude must still be corrected. Also, the delay and phase deviations must be compensated in the paths opposite the amplifiers. Therefore, the need exists for amplitude variation, delay, and phase equalizing networks to balance both loops.

Amplitude equalizers may be divided into two categories: frequency independent and frequency dependent. The frequency-independent amplitude equalizer has a fixed loss that exists for all frequencies. Because coupling ratios are limited by the integral values of turns in the transformers, only discrete values may be realized. The difference between the coupling ratio and the nominal amplifier gain can be eliminated by using a fixed-amplitude equalizer (Pi or Tee pad). If possible, the pad should be placed in a low-power path to minimize power dissipation.

Frequency-dependent (gain-sloping) networks are required if the gain of each amplifier is not flat over the frequency range of operation. In typical HF amplifiers, the gain has a tendency to decrease as frequency increases to correct for this, a number of gain-sloping circuit configurations may be used (see reference 3). The phase response of these circuits is not linear, however. Therefore, only small amounts of rolloff can be corrected without disrupting the phase balance significantly.

The phase and delay characteristics of an amplifier are closely related. For the purpose of discussing equalizer networks, the delay is considered to be the linear portion of the phase response (slope of the phase versus frequency plot). The phase error is then considered to be the deviation from constant delay.

As discussed in the earlier section, the phase deviation can be divided into two problems: the nonlinear and the nonzero phase intercept. The nonlinear phase intercept appears as a rolloff at the low-frequency end. The nonzero phase intercept may be characterized as a constant phase offset that is independent of frequency. Both problems can be corrected by using allpass phase networks cascaded to form the complex phase function. These networks are best designed by using approximation methods (for example, least squares) with the aid of a computer. If desired, more empirical methods of realizing the phase equalization may be used. The phase rolloff at the low-frequency end can be simulated by placing an inductor shunted to ground in the path opposite the amplifier. If the value is not too large, the phase will be corrected without disrupting the amplitude balance.

The nonzero phase intercept may be corrected by one of two methods. The first utilizes a quadrature hybrid and three hybrid junctions to create a constant phase-shift network. This method is described in reference 4. The second is to dynamically adjust the phase using a circuit such as the one in Figure 13.9. This circuit uses a quadrature hybrid, one or two voltage-variable attenuators, and a combiner to form an active phase-shift network. In it the input signal is split into quadrature (90° phase difference) components. The two components are then recombined, performing a vector summation. By adjusting the magnitude of one of the components with respect to the other, the phase of the resultant may be varied by as much as ±45°. The circuit does, however, introduce a loss to the incoming signal. The minimum loss may vary from 3 to 6 dB, depending on the phase setting. This loss may be flattened in the circuit by adjusting both attenuators simultaneously. In many feedforward applications, only 5 to 10° of phase-shift range is necessary to balance the loop. The loss, then, is kept close to 3 dB. In high-power systems, the active phase-shift network would only be used in loop 1 because of the loss introduced.

Besides providing a way of adjusting the phase, placing the network in the path opposite the main amplifier also compensates for the phase shift due to one of the quadrature hybrids in the main amplifier. A second quad-

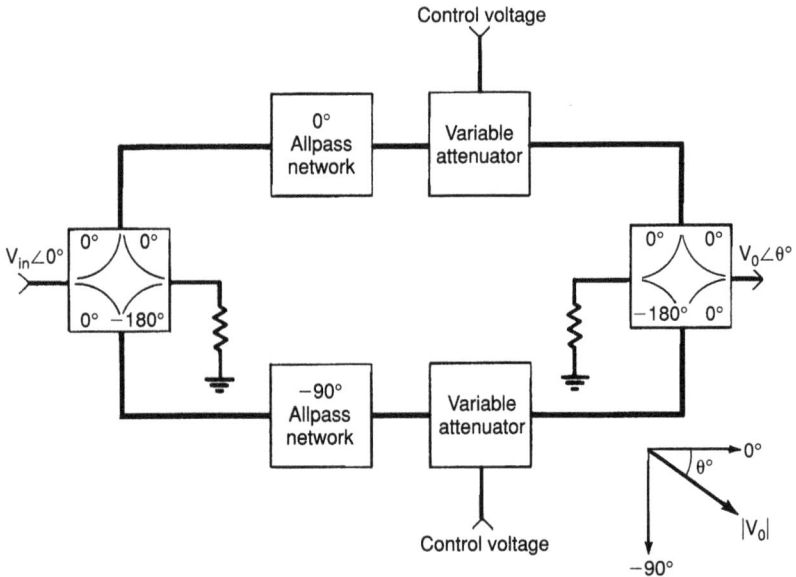

Figure 13.9 Active phase-shift network.

rature hybrid can be cascaded with the phase-shift network to compensate for the other quadrature hybrid in the main amplifier. Another advantage of the active network is that it can be used to obtain very accurate loop balance. It provides a means of maintaining balance over temperature and aging of the system. There are also some disadvantages in using the active network. The feedforward system is no longer instantaneous in its correction. Time must be allowed to adjust the balance. The control circuitry design now becomes a key factor to the success of the system. Finally, the loss introduced will degrade the noise performance of the system.

A delay line in each loop is required to compensate for the delays of the main amplifier and the error amplifier. In many applications, the delay line is realized by using a length of coaxial transmission line. Coaxial cable has a very flat delay characteristic in the HF range (approximately 1.44 ns-ft). The length required to simulate high-PA is 30 to 40 ft. In this case, the power lost in the coaxial cable can be substantial (0.7 dB).

13.4 Null Theory

Null depth

Feedforward operation depends upon the cancellation of signals at two points in the system. These two points are at the comparator and at the injection coupler. The degree to which this cancellation takes place is referred to as the *loop null*. The amount of null is specified in decibels and is referenced to the signal level present if no cancellation is occurring.

The first loop null is a measure of how well the input sample is removed from the sample of the main amplifier output spectrum. This "error signal" (output sample minus input sample) exists at the output of the comparator. The second loop null is a measure of how well the distortion and noise have been removed from the output of the main amplifier. This undistorted signal exists at the output of the injection coupler, which is also the output of the system.

Both the first and second loop nulls utilize the same cancellation mechanism. They use the sum and difference properties of a directional coupler to perform a vector addition on the incoming signals. The canceled signal energy is not destroyed but rather diverted to the termination resistor on the remaining (fourth) port of the coupler, as shown in Figure 13.10.

As stated throughout this chapter, the quality of the null depends upon the amount of amplitude and phase balance of the two opposing signals. Figure 13.11 shows the amount of null that can be expected as a function of both of these factors. As an example, for a 20-dB null, a 0.6-dB amplitude and a 4° phase balance would suffice. For a given null, amplitude and phase balance can be traded to a certain extent. However, as the null becomes deeper, the amount of available trade-off diminishes. Typical broadband

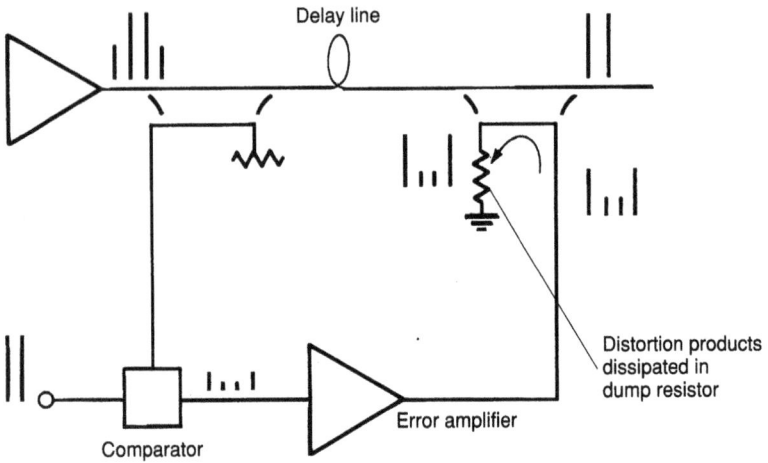

Figure 13.10 Distortion power dissipation.

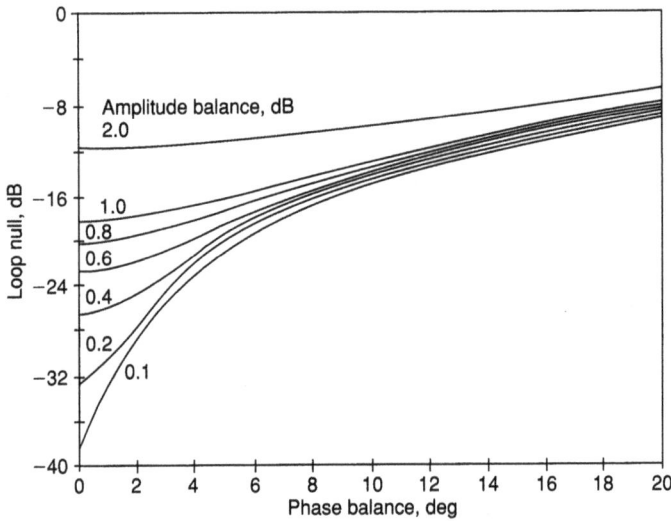

Figure 13.11 Loop null versus amplitude and phase balance.

nulls in an HF feedforward system are on the order of 20 to 25 dB. If a deeper null (30 to 40 dB) is required, the active network must be used.

Null implications

The first loop null has a direct effect on the power capability of the error amplifier, but only a secondary effect on the distortion reduction of the system. The PEP output from the error amplifier is determined by the number and magnitude of the signals at its input. The effect of the first loop null on this output varies, depending on the relative magnitude of the (nulled) fundamental signal with respect to the distortion product levels. If, after the first loop null, the fundamental signal level is still much greater than the distortion levels, the error amplifier output will be dominated by the fundamental power. If, on the other hand, the first loop null reduces the fundamental power to a level much less than the distortion, any additional improvement of the first loop null does not substantially reduce the error amplifier output.

The secondary effect of the first loop null on system linearity is a function only of error amplifier linearity. The fundamental power that is amplified by the error amplifier does not directly add distortion to the main output. However, distortion resulting from overdriving the error amplifier (because of an insufficient first loop null) will add directly to the main output at the injection coupler.

The second loop null is theoretically a measure of the distortion reduction from the main output. That is, a 20-dB second loop null implies that the distortion at the output has been reduced by 20 dB. The second loop null has only a secondary effect on fundamental power output. If significant fundamental power exists at the output of the error amplifier because of first loop null, part of it may add to or subtract from the main output at the injection coupler.

13.5 System Considerations

Noise figure

The noise figure of a feedforward system is determined by the noise figure of the error amplifier and the losses between the input to the system and the input to the error amplifier. With the assumption that both loop nulls are infinite, the noise figure may be calculated from the following equation:

$$\text{System noise figure (dB)} = \text{Error amplifier noise figure (dB)} \qquad (13.3)$$
$$+ \text{losses preceding error amplifier (dB)}$$

Perfect balance assumes that noise existing at the output of the main amplifier is totally canceled by the second loop. The remaining noise at the output of the system is then the noise at the output of the error amplifier reduced by the amount of coupling in the injection coupler. Referring this back to the input, the noise figure expression is derived. The losses preceding the error amplifier include input coupler, loop 1 compensation, and comparator losses. For this reason, it is advantageous to configure the couplers such that the input signal experiences minimal loss before reaching the error amplifier input (see Figure 13.12). This configuration, however, does increase the gain requirement for the main amplifier.

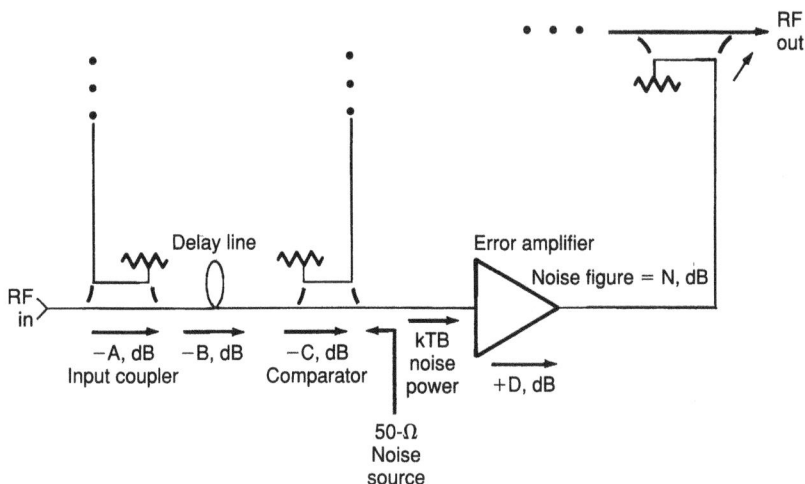

Gain = D−A−B−C−E, dB

Output noise = 10 log (kTB) +D +N −E, dBm

$$\text{Noise figure} = \frac{\text{Output noise}}{\text{Gain} \times \text{kTB}} = N +A +B +C, \text{dB}$$

Figure 13.12 Feedforward system noise figure.

Effects of VSWR and collocation operation

Operating the feedforward system into a VSWR load (load not equal to the characteristic impedance of the system) or in close proximity to another transmitter has similar effects on the system. In both cases, a signal is incident at the main output. In the case of the VSWR load, the signal is a partially reflected portion of the output from the main amplifier. In the case of a collocated transmitter, the signal is totally independent of the fundamental output of the system. Each case will be investigated separately.

VSWR. If the source impedance of the main amplifier is not matched to the characteristic impedance of the system (50 ohms), the returning signal from a VSWR load is rereflected by the main amplifier. A portion of this rereflected signal is coupled through the output coupler and comparator to the input of the error amplifier. The error amplifier can then be driven to the point of distortion (if not destruction). A second problem may be encountered even if the feedforward system is not being driven. If the VSWR load contains a reactive component, an oscillatory path may be established around loop 2. For these two problems, the solution is to design the main amplifier with a 50-ohm output impedance to prevent the rereflection of power (see reference 5 on quadrature hybrid combined amplifiers). A final problem resulting from the VSWR load exists because a portion of the rereflected power is coupled through the injection coupler to the output of the error amplifier. In this case, the error amplifier must be designed to dissipate this power without creating distortion or overheating.

Collocation operation. An interfering signal resulting from a collocated transmitter creates two problems for the feedforward system. The first is from the actual signal being reflected by the main amplifier, similar to the case of the VSWR load. This is handled in the same manner as the VSWR load. The second problem results from the mixing of the interfering signal with the desired signal in the main amplifier and error amplifier outputs. In the case of the main amplifier, the new IMD products will be reduced by feedforward as long as their frequencies fall within the correction bandwidth of the system. To eliminate products outside of the bandwidth, the system may be followed by a lowpass or bandpass filter. Feedforward will not correct for products generated in the error amplifier output. Again, the error amplifier must be designed to handle the reverse signal.

13.6 Working System Example

Figure 13.13 shows a schematic for a 150-W feedforward system. The main amplifier is a 165-W power MOSFET module with a nominal gain of +31 dB. By adding the gain-flattening networks, the gain variation was reduced from

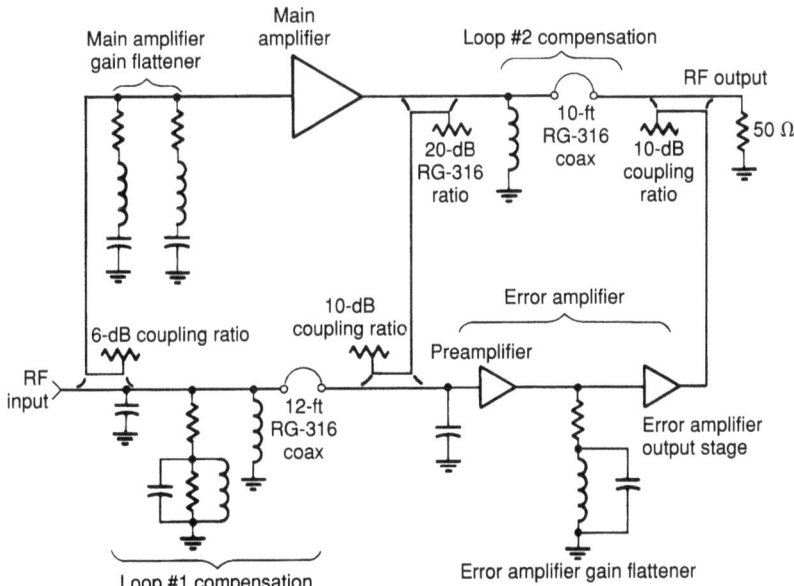

Figure 13.13 150-W feedforward system.

2 to 0.5 dB (nominal gain reduced to +27 dB), as shown in Figure 13.14A. The phase response of the main amplifier is shown in Figure 13.14B. The delay of the main amplifier is 17.2 ns with a 6° zero-frequency phase intercept and an additional nonlinearity of 14° at 2 MHz. The length of coaxial cable used to compensate for the main amplifier delay was approximately 12 ft (3.7 m). To compensate for the nonlinearity, a 6-µH coil was placed in the path opposite the main amplifier. Also placed in the opposite path was a gain-flattening network for the 2- to 5-MHz region and a coaxial "length-tweaking" capacitor (< 10 pF). The first loop null results are shown in Table 13.4.

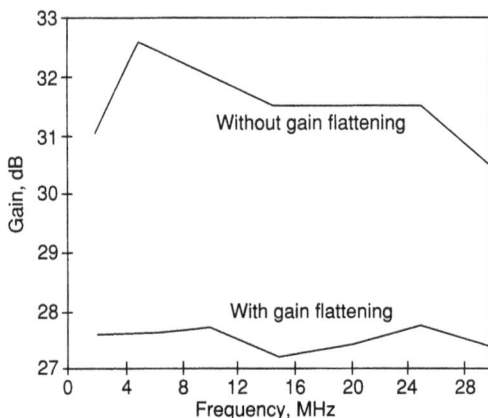

A

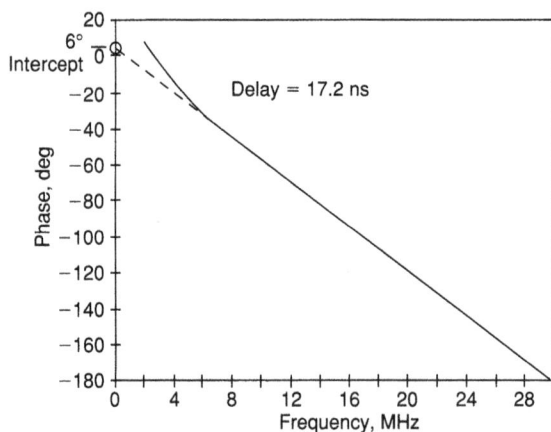

B

Figure 13.14 Main amplifier: (A) gain versus frequency; (B) phase response versus frequency.

TABLE 13.4 Performance of a 150-W, 2- to 30-MHz Feedforward System

			Third-order IMD products	
Frequency, MHz	First loop null, dB	System efficiency, c_0	Without feedforward, dBc	With feedforward, dBc
2	30	11	41	65
5	28	12	41	77
10	23	12	38	83
15	26	12	35	66
20	24	12	35	58
25	22	12	35	61
30	23	12	35	65

The error amplifier consisted of a parallel device amplifier driven by a two-stage MOSFET preamplifier. The gain of the error amplifier was approximately 40 dB and the delay was approximately 13.3 ns. Because of the lack of transformers, the frequency response of the error amplifier was much flatter than the main amplifier (0.2 dB). The compensation required for the second loop was one gain flattener, a 10-μH coil (for the phase nonlinearity), and 10 ft (3.0 m) of coaxial cable for the delay line.

The overall system gain was +21 dB, with a worst-case third-order IMD improvement of 23 dB. Higher-order products (fifths and sevenths) were also reduced by at least 20 dB along with a minimum of 15-dB in-band harmonic reduction. A summary of the system operation is shown in Table 13.4.

References

1. H. S. Black, Translating System, U.S. Patent 1,686,792 (October 9, 1928).

2. S. D. Bedrosian, "Normalized Design of 90 Degree Phase-Difference Networks," *IRE Trans. Circuit Theory* CT-9 (June 1960): 128–136.

3. Robert W. Landee, "Attenuators and Equalizers," sec. 7, in L. Giocoletto (ed.), *Electronics Designers Handbook* (New York: McGraw-Hill, 1977).

4. T. A. Harrington, Feed Forward Wideband Amplifier, U.S. Patent 4,348,642 (September 7, 1982).

5. T. A. Harrington, Feed Forward Amplifier with Enhanced Stability into Loads with High VSWR, U.S. Patent 4,352,072 (September 28, 1982).

14

High-Power Linear Amplifiers

Warren B. Bruene

This chapter describes transmitting tube circuits because tubes still provide the most practical and economical means of achieving the high performance necessary for multichannel (ISB) transmission at power levels above 1 kW (at present). The opening section provides a good discussion of IMD. This chapter will then show how to choose and compute tube-operating conditions. Tank circuit and coupling network requirements will be discussed, and neutralization and stabilization circuits will be shown. Finally, RF feedback circuits will be described as an effective means of improving IMD performance.

14.1 Intermodulation Distortion

High-power HF communication transmitters are typically designed to accommodate four 3-kHz voice channels in a 12-kHz band allocation. Each 3-kHz channel may carry voice or up to 16 RTTY or data tones. Intermodulation distortion can produce objectionable background "splatter" in voice channels in full-duplex operation. The IMD should be at least 40 dB below the voice signal level for a good-quality circuit. LINCOMPEX (Chapter 7) can be employed to reduce the apparent background noise by about 15 dB. This would practically eliminate any problem with IMD. LINCOMPEX achieves this advantage by compressing the SSB signal amplitude before transmission so that the weaker sounds are transmitted at a much higher level. The amount of compression used is transmitted to the receiver by the absolute frequency of a tone above the audio band. The receiver uses this compression information to change the receiver gain in an inverse manner to reduce the weak sound back to its original relative level. In so doing, it also reduces noise and interference (including IMD) by an equal amount. This greatly improves the apparent signal-to-noise ratio.

Low out-of-band emission is essential for achieving good spectrum utilization, or in other words, minimum interference to others occupying nearby channels. Figure 14.1 shows the out-of-band IM limits using the noise test signal specified in MIL-STD-188 for high-performance equipment.

Intermodulation distortion has been measured in the past by means of the two-tone test because it is the simplest signal that can be used to measure IM over an amplitude excursion from zero to rated PEP.

The two-tone test does not accurately represent the degree of amplifier linearity when the SSB signal consists of many tones or of speech, however. For example, consider a signal consisting of 16 equal amplitude but randomly phased tones. Furthermore, transmit the same signal in two voice channels to achieve frequency diversity. Let us assume 62.5 W/tone, so the total average power is $62.5 \times 32 = 2000$ W. The PEP would be $62.5(32)^2 = 64,000$ W if all tones (at RF) were in phase at some instant. Actually a composite SSB signal consisting of 12 or more tones exceeds five times the average power only about 0.7% of the time. The amplitude distribution (as a percent of time a given power is exceeded) is very close to that of a narrow band of white noise and therefore has a Rayleigh amplitude distribution. The curve is shown in Figure 14.2.

Clipping peaks above five times average power creates IMD, but the average IM level caused by this is over 40 dB below the signal. Power amplifiers that do not go into hard limiting until well above rated power will produce

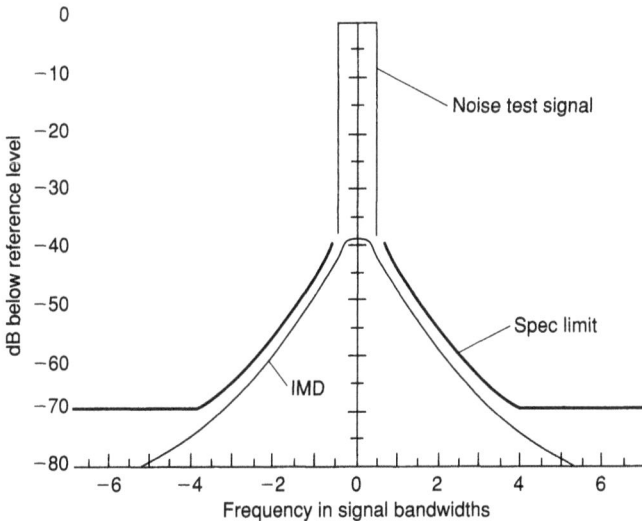

Figure 14.1 Power spectral density of IM products of a noise test signal and a representative specification limit.

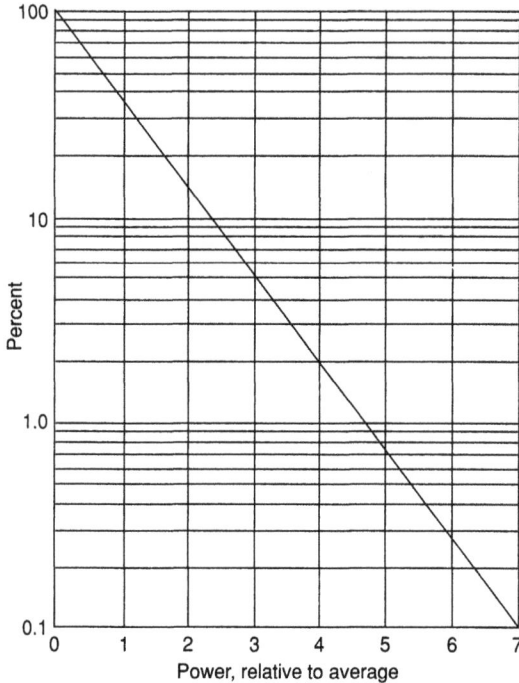

Figure 14.2 Amplitude distribution of a narrow band of white noise.

less peak clipping distortion than those with hard limiting just a little above rated peak power.

A noise test signal will drive the amplifier up to its peak power limit. Therefore, a noise test signal will include IM produced by peak clipping as well as from nonlinearity below rated PEP.

Another deficiency of the two-tone test is that the "third-order" products observed on a spectrum analyzer are actually the sum of the third and all higher odd-order components. Typically, the fifth-order component is out of phase with the third, which tends to produce distortion cancellation. This leaves the false impression that the IMD is better than it really is.

For a two-tone test signal passing through an amplifier tube

$$e_g = a_1 \cos \omega_1 t + a_2 \cos \omega_2 t \tag{14.1}$$

$$i_b = I_{BO} + k_1 e_g + k_2 e_g^2 + k_3 e_g^3 + k_4 e_g^4 + k_5 e_g^5 \tag{14.2}$$

The third-order components of concern are

$$(\tfrac{3}{4} k_3 a_1^2 a_2 + \tfrac{5}{4} k_5 a_1^4 a_2 + \tfrac{15}{8} k_5 a_1^2 a_2^3) \cos (2\omega_1 - \omega_2)t \tag{14.3}$$

Note that the first term in the coefficients is the part caused by third-order nonlinearity, while the other two terms are caused by fifth-order nonlinearity. Note also that the third-order terms vary as the cube of the two-tone signal level and that the fifth-order terms vary as the fifth power. This means that increasing the level of the two test tones by 1 dB each will cause the third-order components to increase by 3 dB and the fifth by 5 dB. The seventh and all higher odd-order nonlinearities also contribute to what is called the *third-order product*. They may be either in phase or out of phase with each other.

Perhaps of even more importance is that each combination of three tones of a multitone signal produces other third-order products that are 6 dB higher than the $(2f_1 - f_2)$ products. These other products for one combination of three frequencies are on frequencies of

$$f_1 + f_2 - f_3$$
$$f_1 - f_2 + f_3$$
$$-f_1 + f_2 + f_3$$

Other fifth-order products produced by each combination of five frequencies are on frequencies of

$$f_1 + f_2 + f_3 - f_4 - f_5$$
$$f_1 + f_2 - f_3 - f_4 + f_5$$
$$f_1 - f_2 - f_3 + f_4 + f_5$$
$$-f_1 - f_2 + f_3 + f_4 + f_5$$

One can easily see that a group of 16 data tones will produce literally hundreds of distortion products.

The vowel sounds of voice signals consist of many frequency components which are mostly harmonics of the fundamental frequency (about 100 Hz for a male voice). Noise is a better representation of voice signals than two tones.

At the time of this writing there is a considerable amount of activity directed toward the use of SSB for HF broadcasting. Power levels probably will range from 50- to 2000-kW PEP. A good-quality broadcast signal requires a wide dynamic range and a lower background noise and interference level. Intermodulation from a strong signal in a nearby channel could create a serious interference problem; therefore, IM of HF SSB broadcast transmitters needs to be kept as low as practical.

Adjacent channel IM caused by peak limiting in the PA can be greatly reduced by clipping the SSB signal envelope in the exciter and then filtering off the out-of-band IM before delivering it to the PA (see Chapter 7). This technique allows the use of a higher average signal level and hence an improved signal-to-noise ratio at the receiver. This is done at the expense of increased in-band IMD, but it offers a very advantageous trade-off when receiving conditions are not the best.

High-power transmitters are generally required to have better IMD performance than low-power transmitters designed to carry a single voice signal. Therefore, simply adding a grounded-grid linear to a low-performance transmitter is not a satisfactory solution.

The voltage phasors of the distortion products generated in successive stages in a linear amplifier add in amplitude. Therefore, two 35-dB cascaded amplifier stages might produce only a 29-dB IMD amplifier. See Chapter 4 for a discussion of distortion in cascaded amplifiers.

Most high-power linear amplifiers are designed to be tunable over a wide frequency range. The number of tuned circuits can be minimized by achieving high gain per stage. This can be realized with tetrode-type transmitting tubes, which have become very popular.

Transmitting tube manufacturers have learned how to design the geometry of the elements within the tube to achieve IMD levels in the –35- to –40-dB region. Radio frequency feedback can be employed around the driver and final stages to achieve transmitter IMD performance considerably better than –40 dB, depending upon the tube and operating conditions chosen.

14.2 Nature of an SSB Wave

An SSB wave may be considered (for amplifier design purposes) as a sine wave that changes relatively slowly in amplitude and in phase in a varying manner. Figure 14.3 illustrates an SSB wave. The envelope is determined by the audio signal components, while the number of cycles within the envelope is determined by the RF frequency. Typically, there are thousands of times as many RF cycles within an envelope as are illustrated.

Each cycle of the RF wave is slightly different from the preceding one, but the difference is very small. Other, more complex envelopes may be considered to have the same properties as far as the RF amplifier is concerned, even though the envelope may contain many frequency components.

Radio frequency amplifiers generally have tuned circuits in both their input and output circuits. One of the principal reasons for using tuned circuits is that they remove the shunting effect of the tube input and output capacitances as shown in Figure 14.4. These tube and stray circuit capacitances are absorbed into the tuned circuits so that the circuits present substantially a resistive impedance at radio frequencies. The second principal reason for using tuned circuits is rejection of unwanted frequencies.

For purposes of discussing tube operating conditions, it will be assumed that the signal consists of RF sine waves. Any harmonics of the signal voltage that may exist in the input or are generated by the tube will be considered to be removed by the tuned circuits unless stated otherwise. The tank circuit requirements to perform this function are discussed in section 14.7.

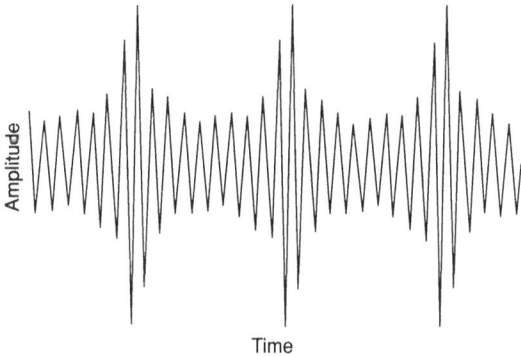

Figure 14.3 Representative SSB wave.

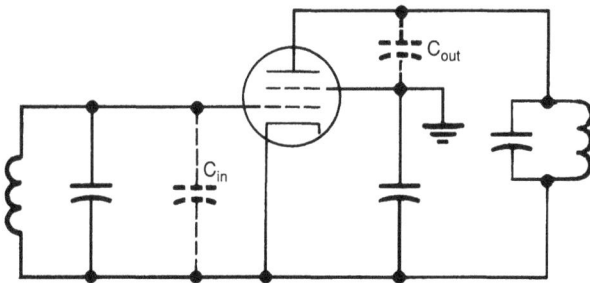

Figure 14.4 Tube and stray input and output capacitances are absorbed into tuned circuits to keep circuit impedances substantially resistive.

14.3 Classes of Operation

Radio frequency amplifiers are classified according to the angle of plate current flow, that is, according to the approximate number of degrees during one cycle of the RF wave:

- Class A: 360° or continuous plate current flow
- Class B: Approximately 180°
- Class C: Less than 180°
- Class AB: Between class A and class B

In addition, subscripts are used to indicate whether or not the tube is driven into the grid-current region. For example,

- AB_1 signifies class AB operation with no grid current
- AB_2 signifies class AB operation with grid current

Typically, class A operation is used in most small-signal applications, such as receivers, exciters, and low-level stages of PAs. Class A amplifiers are characterized by high gain, low distortion, and low efficiency.

Class AB operation is utilized in most linear amplifiers from a few watts upward. Gain is lower and distortion greater than for class A amplifiers, but the higher efficiency, smaller tube size, and lower cost become determining factors at higher power levels. Either AB_1 or AB_2 may be used as determined by the choice of operating condition selected. Tubes designed for low distortion are usually designed for class AB_1 operation. This substantially eliminates nonlinear grid current as a cause of distortion.

14.4   The Ideal Tube Transfer Curve

The main sources of distortion are the nonlinear characteristics of the RF PA tubes. One of the best ways to achieve low distortion is to avoid generating it in the first place. This is accomplished by the proper choice of tubes and their operating conditions.

Nonlinearity of plate current along the operating load line is the major source of distortion. Curves of plate current versus grid voltage along selected load lines are shown on the right-hand side in Figure 14.5.

An ideal zero-distortion plate current curve [1] for class AB_1 operation is illustrated in Figure 14.6. The plate current follows a square-law curve between grid voltages of –300 to –100 V. From –100 to 0 V the plate current continues in a straight line with the same slope. The zero-signal operating

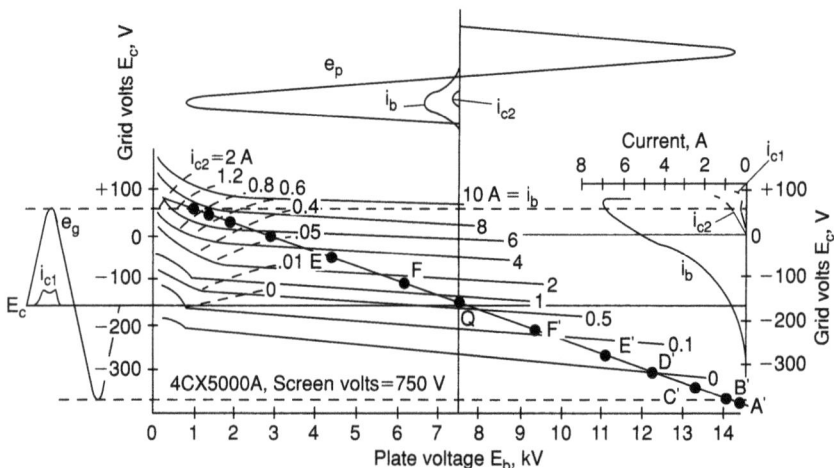

Figure 14.5 Load line (with 15° points marked) and associated relationships for 4CX5000A tetrode operating class AB_2 and delivering 12-kW FR power output. (*Basic tube curves courtesy Eimac Div. of Varian.*)

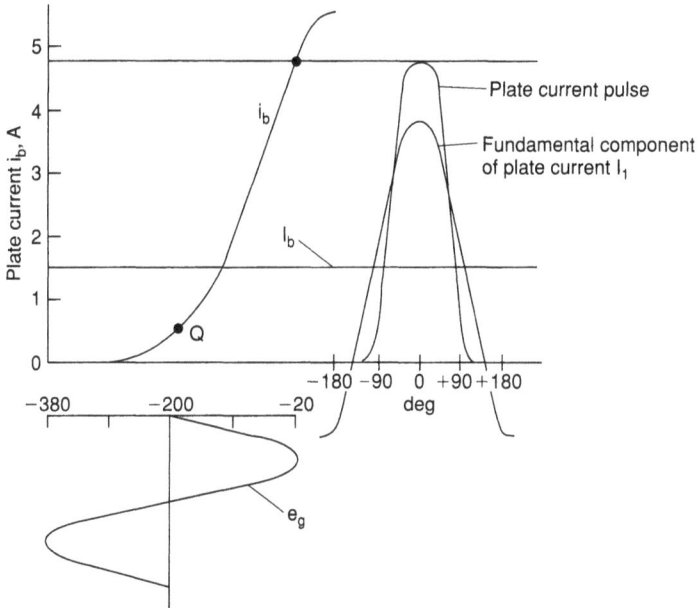

Figure 14.6 Ideal tube characteristic for class AB_1 operation suitable for 5-kW PEP output. i_b = 4.7-A peak plate current; I_b = 1.5-Adc plate current; Eb = 5000-Vdc plate voltage; e_p = 4500-V peak RF plate voltage; I_{b0} = 0.5-A zero-signal dc plate current; e_g = 180-V peak RF grid voltage; E_{c1} = –200-Vdc grid bias voltage; I_1 = 2.35-A peak fundamental component of plate current.

point Q is located midway (horizontally) between the end points of the square-law portion of the curve (–200 V in this example).

Small signals whose peak RF voltage is less than 100 V operate on the pure second-order curve and generate no odd-order IMD. When the peak grid voltage exceeds 100 V, the positive and negative peaks enter linear regions at the same time. When the slope of the linear region is correct, there is no change in the gain of the fundamental component and no IMD is produced by these larger signals either.

The current at point Q determines the zero-signal value of plate current, and when multiplied by the dc plate voltage determines the zero-signal plate dissipation.

The actual value of bias voltage is unimportant, but it must place the zero-signal plate current at the correct point Q on the curve. A sine-wave grid voltage and the resulting plate current pulse are shown, as well as the dc and fundamental component of this pulse. The difference between the actual pulse and the fundamental component represents even-order harmonics of plate current which should be returned to the cathode through a low-impedance path and should not be coupled to the load.

The optimum value of zero-signal plate current for tetrodes and pentodes depends upon the value of screen voltage as well as the basic tube characteristics. The optimum value of plate current varies approximately as the 3/2 power of the screen voltage. Lower screen voltages are desirable for this reason, but peak plate current is more limited. It is generally best to operate tubes at the screen voltage recommended by the tube manufacturer.

14.5 RF Power Output and IMD Ratings

The power output capability of SSB or ISB transmitters has been specified in several different ways:

1. *Single tone.* This continuous power rating is required if the transmitter is to be operated at full power with a single FSK TTY signal, for example. This rating is seldom applied to transmitters above the 10-kW level.

2. *Two tone.* Continuous operation with a two-tone test signal at rated PEP is often specified. This permits a reduction in the dc current rating of the high-voltage power supply.

3. *Noise power ratio (NPR).* Noise loading will load the high-voltage dc power supply the same as many data or TTY tones. The average dc plate current with voice modulation will be even less. This avoids power supply overdesign when the transmitter is not required to transmit a single-channel FSK signal (or equivalent).

Intermodulation distortion can also be specified in several ways. It is always associated with a power output rating such as "at least 40-dB S/D at any power level up to 10-kW PEP using a two-tone test signal." (S/D is the signal-to-distortion ratio.) This should be clarified to mean that the strongest IMD product as observed on a spectrum analyzer should always be at least 40 dB below the amplitude of either of the two equal-amplitude test tones. A specification of 40 dB at 10-kW PEP, for example, is not sufficient because the signal-to-distortion ratio may get worse when power is reduced if substantial distortion cancellation is present at rated power.

Sometimes one of the two third-order products of a two-tone test is larger than the other. This means that the PA has phase nonlinearity (or AM to PM conversion) as well as amplitude nonlinearity. Intermodulation distortion components produced by amplitude and phase nonlinearities add on one side and subtract on the other.

The notched-noise test provides the best measure of PA linearity for speech, multitone TTY, or data signals. Typically, the transmitter is driven to its noise power rating with a notched-noise test signal and the signal-to-distortion ratio measured. The noise IMD typically increases about four times as fast (in decibels) as the noise power output (near rated power output) as the input signal is increased. Thus for a 10% increase in power output (about 0.5 dB), the IMD increases about 2 dB, leaving a signal-to-distortion ratio decrease of about 1.5 dB.

There is no fixed relationship between the two-tone signal-to-distortion ratio and the notched-noise signal-to-distortion ratio. It is different for each PA design and operating condition. A transmitter with a 10-kW PEP rating and a two-tone signal-to-distortion ratio of 40 dB roughly corresponds to a 2.5-kW noise-power rating and an NPR measurement of 40 dB. There may be as many as a few decibels difference, however, which points to the value of the notched-noise test which is more representative of actual operation.

14.6 Analytical Methods of Estimating Tube Operating Conditions

Choosing the load line

Constant-current curves of a tube characteristic, as shown in Figure 14.7, are used because the load line is a simple straight line. This assumes that the grid and the plate voltages are dc plus pure sine waves and are in the proper phase relationship. The use of properly tuned high-Q tank circuits produces a condition that approaches this ideal.

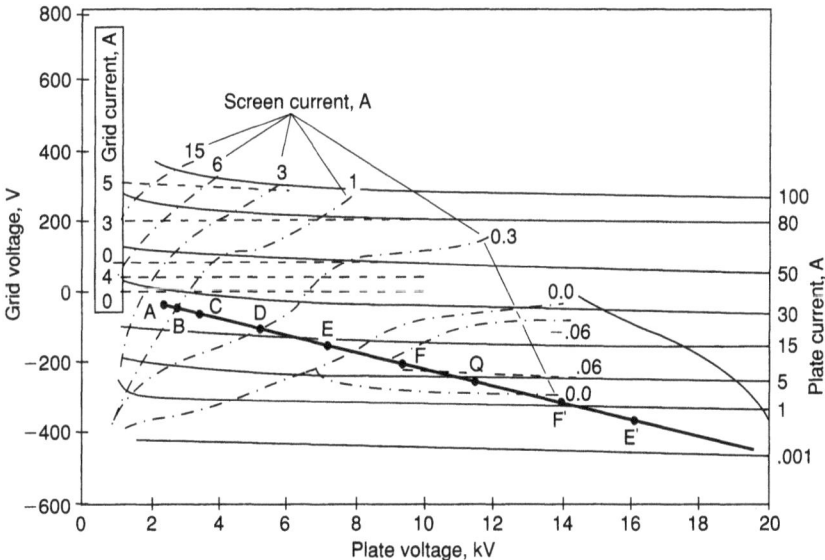

Figure 14.7 Load line (with 15° points marked) and associated waveforms for 4CW5000J operating class AB_1 with 60-kW PEP output. E_b = 11.5 kV, E_{c2} = 1500 V, E_{c1} = –260 V, I_{b0} = 8.5 A, e_p = 9100 V, i_p = 26 A, E_f (filament voltage) = 12 V, screen voltage = 1500 V. Solid line—plate current, A; dot-and-dash line—screen current, A; dashed line—grid current, A. *(Basic tube characteristics courtesy Eimac Div. of Varian.)*

Figure 14.7 illustrates a load line for class AB_1 operation on a tetrode-tube characteristic curve for a 50-kW amplifier. The grid voltage to plate current i_b transfer curve is also shown. The shape of the i_b-versus-e_g curve determines the amplifier distortion characteristic, as discussed in the preceding sections.

The plate current at the zero-signal operating point is chosen for the best compromise between zero-signal plate dissipation and low distortion.

The plate current flows in pulses for class AB operation. For large-signal class A operation, plate current may be quite nonlinear, but it never goes to zero. One function of the plate tank circuit is to provide a resistive load to the fundamental component of the plate current pulses and a very low impedance to all harmonic components. A loaded tank circuit Q of at least 5 is highly desirable.

The optimum load line for a desired amplifier operating condition is established by a trial-and-error process. A trial load line might be established as follows. The tube manufacturer's data is used to establish a value of dc plate voltage that should be suitable for the desired power output. A zero-signal plate current is then selected that will produce approximately two-thirds of rated plate dissipation. This establishes the trial zero-signal point Q on the load line. The formulas given below the next heading are used to assist in determining the coordinates of the load line end point A in Figure 14.7. The voltage for point A may be located where the constant-current curves depart from their linear region. This establishes a trial value for e_p. A value of peak plate current is then computed using the formula

$$i_b = \frac{4P_o}{e_p} \tag{14.4}$$

where P_o is the average power output. This establishes a trial value of i_b and locates point A on the tube characteristic curves. The approximate tube operating conditions for this load line are then computed using the formulas below. Additional trials will locate point A more precisely. Figure 14.8 illustrates the effect of moving points A and Q of the load line.

Formulas for estimating tube operating conditions [2]

Having selected a tentative load line, the following formulas can be used to estimate tube performance fairly accurately. The formulas are exact for a theoretical linear tube operating purely class B. Normal values of zero-signal plate current used for class AB operation have very little effect upon the accuracy of the following formulas at maximum signal conditions, however. The following formulas apply for a single-frequency signal:

DC plate current:

$$I_b = \frac{i_b}{\pi} \tag{14.5}$$

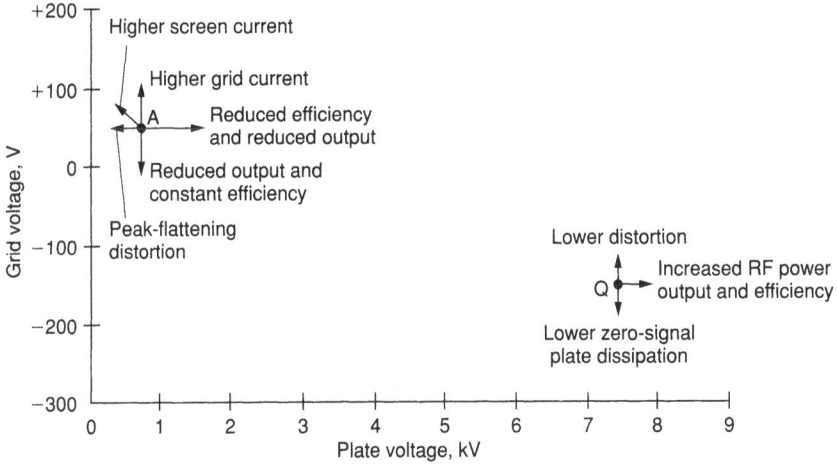

Figure 14.8 Effects of moving points A and Q on tube operation.

Watts of plate input:

$$P_i = \frac{i_b E_b}{\pi} \tag{14.6}$$

Watts of average RF output:

$$P_o = \frac{i_b e_p}{4} \tag{14.7}$$

Plate load resistance:

$$R_L = \frac{I_{ep}}{i_b} \tag{14.8}$$

Percent plate efficiency:

$$\text{Efficiency} = \frac{\pi e_p}{4 E_b}(100) \tag{14.9}$$

For a two-frequency SSB test signal the formulas are
DC plate current:

$$I_b = \frac{2 i_b}{\pi^2} \tag{14.10}$$

Watts of plate input:

$$P_i = \frac{2i_b E_b}{\pi^2} \tag{14.11}$$

Watts of average RF output:

$$P_o = \frac{i_b e_p}{8} \tag{14.12}$$

Watts of peak power output:

$$PEP = \frac{i_b e_p}{4} \tag{14.13}$$

Percent average plate efficiency:

$$\text{Efficiency} = \left(\frac{\pi}{4}\right)^2 \frac{e_p}{E_b}(100) \tag{14.14}$$

Figure 14.9 shows curves of plate efficiency and plate dissipation for both single-frequency and two-frequency signals as a function of signal level for a typical tube operating condition. It should be observed that plate efficiency is nearly a linear function of the ratio of maximum plate voltage swing to dc plate voltage.

The Chaffee analysis discussed in the following section will provide a more accurate computation of tube operating conditions.

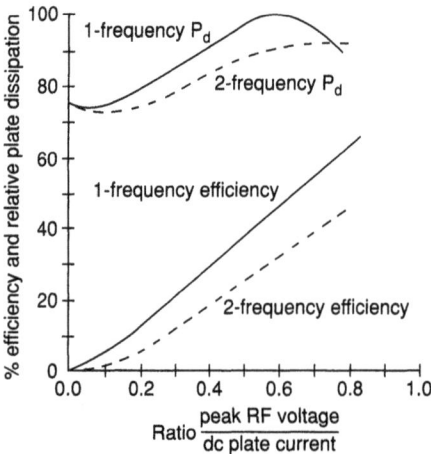

Figure 14.9 Efficiency and relative dissipation versus peak RF plate voltage for typical class AB operation.

Chaffee analysis

A method of calculating the tube operating condition that was originated by E. L. Chaffee [3] provides quite accurate results. A load line on a set of constant-current curves is selected as shown in Figure 14.7. At points on this load line $(A, B, C, \text{etc.})$ corresponding to each 15° along the RF cycle, values of plate current, grid current, and screen current are read. Mechanical aids [4, 5] have been devised to simplify determination of these points. Each point is taken to represent the average value of current over each 15° range centered on that point.

Substituting into the following equations provides calculations of the average dc values of current and the fundamental and harmonic RF components:

$$I_{av} = \tfrac{1}{12}\left(\frac{A}{2} + B + C + D + E + F + Q + F' + E' + D' + C' + B' + \frac{A'}{2} \right) \quad (14.15)$$

$$I_1 = \tfrac{1}{12}[(A - A') + 1.93\,(B - B') + 1.73\,(C - C') + 1.41\,(D - D') \quad (14.16)$$
$$+ (E - E') + 0.52\,(F - F')$$

$$I_2 = \tfrac{1}{12}[(A + A' + C + C' - E - E') + 1.93\,(B + B' - F - F') - Q] \quad (14.17)$$

The ac components calculated using the above equations are peak values, not rms values. The average I_{av} and fundamental I_1) components should be calculated for plate current. The fundamental component of grid current needs to be calculated for AB_2 operation also. Calculation of harmonics is generally not necessary.

These values are used in the following equations to complete the calculation of the tube operating conditions for the selected load line:

dc plate current:

$$I_{av} = I_b \quad (14.18)$$

Watts of input:

$$P_i = E_b I_{av,\,\text{plate}} \quad (14.19)$$

Watts of output:

$$P_o = \frac{I_{1,\,\text{plate}}e_p}{2} \quad (14.20)$$

Percent plate efficiency:

$$\text{Efficiency} = \frac{P_n}{P_t}\,(100) \quad (14.21)$$

Watts of plate dissipation:

$$P_p = P_i - P_o \tag{14.22}$$

dc control-grid current:

$$I_{av,\,grid} = I_{c1} \tag{14.23}$$

Watts of control-grid drive:

$$P_g = \frac{I_{1,\,grid}e_g}{2} \tag{14.24}$$

Drive consumed in bias supply:

$$P_c = I_{c1}E_{c1} \tag{14.25}$$

Watts of control-grid dissipation:

$$P_{g1} = P_g - P_c \tag{14.26}$$

dc screen-grid current:

$$I_{av,\,screen} = I_{c2} \tag{14.27}$$

Watts of screen-grid dissipation:

$$P_{g2} = I_{c2}E_{c2} \tag{14.28}$$

RF plate load resistance:

$$R_t = \frac{e_p}{I_{1,\,plate}} \tag{14.29}$$

The above equations can be programmed into a personal computer to greatly reduce the work of trying several different load lines.

It should be noted that all tubes do not have exactly the current shown in the tube data sheets. There are manufacturing tolerances and variations. Also, allowances should be made for tube aging. The tube manufacturer should be consulted regarding a suitable safety factor for peak plate current. Small, high-g_m tubes may vary $\pm 20\%$, whereas large tubes may be well within $\pm 10\%$ of published data.

The second harmonic component of plate current computed using Equation 14.17 will be very close to 7 dB below the fundamental component for class AB amplifiers. (It is 7.4 dB for a half sine wave.) The third harmonic will be small because it is produced by the same nonlinearity that generates IMD. Harmonics are considered to have current sources when computing the harmonic attenuation of the output network.

Effect of screen voltage

The effect of changing screen voltage by small amounts, such as 10 to 20%, can be estimated by applying multiplying factors to the set of curves. If the screen voltage is 10% higher than that on the curves, multiply the grid- and plate-voltage coordinates by 1.10 and multiply the values of the current lines by $(1.10)^{3/2}$.

In general, it is desirable to use as low a value of screen voltage as will permit desired output plus some reserve for variation in tubes from published data. Typically, lower screen voltage results in a little lower distortion for a given zero-signal plate current.

Operation with RF feedback injection

An RF feedback voltage is generally applied to a tube element other than the one that receives the excitation signal. This is done to minimize the reaction of the feedback voltage upon the excitation signal. Figure 14.10 shows a cathode-driven stage with the RF feedback voltage applied to the control grid. The feedback voltage must be in phase with the cathode drive voltage. Figure 14.11 shows a circuit with the excitation applied to the control grid and the RF feedback applied to the cathode. The feedback voltage is in phase with the excitation but, of course, is a little lower depending upon the amount of feedback used. For example, if 12 dB of feedback is used, the RF cathode voltage is three-fourths the value of RF grid excitation voltage.

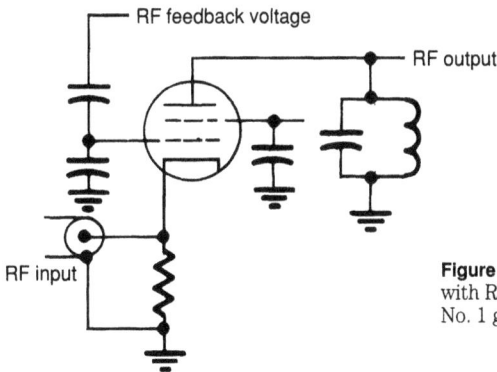

Figure 14.10 Cathode-driven stage with RF feedback applied to the No. 1 grid.

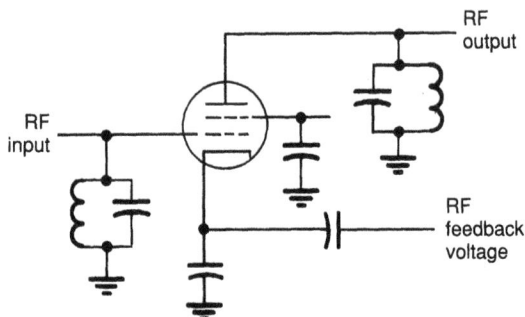

Figure 14.11   Grid-driven stage with RF feedback applied to the cathode.

When the tubes in the circuits shown in Figures 14.10 and 14.11 are operated class A with small signals, the operation is similar to that of a conventional class A amplifier with an applied signal equal to the difference between the excitation and the feedback voltage. There is an appreciable difference, however, when the RF voltage on the cathode is an appreciable fraction of the dc screen-grid voltage. In the circuit shown in Figure 14.10, the RF voltage on the cathode increases the cathode-to-screen grid voltage when the plate current is maximum, while in the circuit shown in Figure 14.11, it decreases it. In Figure 14.10, for example, a signal peak RF voltage of –40 V, a feedback peak RF voltage of –30 V, and a dc screen-grid voltage of 250 V result in an instantaneous cathode-to-screen voltage of 290 V when peak plate current flows, as compared with 220 V in Figure 14.11. For this reason the gain and power-output capabilities of a given tube are quite different in the two circuits. The choice between the two circuits is usually determined by whether or not a phase reversal of the RF signal is required.

14.7   Tank Circuits and Impedance-Matching Networks [6]

In contrast with solid-state amplifiers, high-power RF amplifiers generally use tuned circuits in both their input and output circuits. Amplifier tubes contain appreciable amounts of input and output capacitance, which presents an undesirable low-reactance shunting impedance. Tubes and transistors function best with a resistive load impedance at the signal frequency.

Tank-circuit functions

The following are the various functions that tank circuits are required to perform:

1. Absorb undesired capacitance across the input and output circuits, as previously mentioned.

2. Provide a flywheel effect to maintain substantially sine-wave voltages on the grids and plates of tubes operating class AB. The tank circuits absorb plate-current pulses and produce a nearly sine-wave voltage and current output. When tubes are driven into control-grid current, the grid tank circuit supplies the pulses of control-grid current while maintaining a nearly sine-wave voltage.

3. Provide a low-impedance return from both plate and control grid to cathode for the harmonics of plate and grid current. It is undesirable to allow significant amounts of harmonic voltage to appear superimposed upon the fundamental because loss of efficiency and increased harmonic output result. Requirements vary, but generally the harmonic voltage should not exceed 5% of the fundamental voltage. For class AB operation, the second-harmonic component of plate current is nearly one-half the value of the fundamental component, so to keep the second-harmonic voltage to 5% of the fundamental requires an impedance across the plate-to-cathode circuit at the second-harmonic frequency of one-tenth the value of the plate load impedance at fundamental frequency. The capacitors used for this purpose should provide a low impedance to harmonic currents. The Q of a tank circuit is often defined as the ratio of plate load resistance to the shunting capacitive reactance at fundamental frequency:

$$Q = \frac{R_L}{X_c} \qquad (14.30)$$

In the above example, the minimum Q required to prevent more than 5% second-harmonic voltage is 5. The value of R_L therefore places a limit on the maximum value of X_c.

4. Impedance matching is another function generally performed by tank circuits. The proper load resistances for most PA tubes are on the order of 500 to 5000 ohms. The final amplifier generally feeds a transmission line that couples the transmitter to the antenna. The transmission line input impedance may vary widely as a result of standing waves on the line. The transmission line characteristic impedance is usually 50 ohms unbalanced.

5. Harmonic attenuation is another important function. Licensing authorities typically require all harmonics of high-power transmitters to be 80 dB below the fundamental. The second-harmonic of a class AB PA tube plate current is only about 6 dB below the fundamental to start with, so at least 74 dB must be provided by the tuned circuits and filters.

6. Tank circuits are also often required to satisfy the needs of the neutralizing circuit. Basically this involves a means of obtaining a 180° phase shift at signal frequency.

7. Another practical requirement is circuit simplicity for low cost, simple tuning, and frequency changing.

Typical circuits

Tetrode tubes behave substantially like constant-current sources for their output harmonics. Therefore, the harmonic attenuation of the output network should be computed assuming a current source.

Figure 14.12 shows a good circuit upon which to base an output network design [7]. It consists of three parallel resonant circuits with inductive coupling between them. Loaded circuit Q's of 12, 10, and 4 will theoretically provide 76 dB of second-harmonic attenuation. The passband response is monotonic on both sides of the resonant frequency. The three resonators are tuned so that the phase delay between them is 90°. Top coupling, bottom coupling, or mutual coupling may be used.

Sufficient harmonic attenuation can be achieved by omitting the third resonator and raising the Q of the first two resonators to 20 as illustrated in Figure 14.13 [8]. Higher-loaded Q's mean higher circuit loss, but the simplicity sometimes is a good trade-off. Coupled variable inductors have been designed so that the coupling between them remains constant as the inductance is varied. It is possible to tune such a circuit with only three servos.

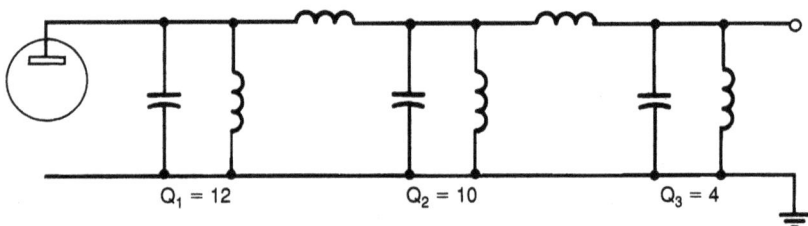

Figure 14.12 Output network with three inductively coupled resonators, which provides excellent harmonic attenuation and passband properties.

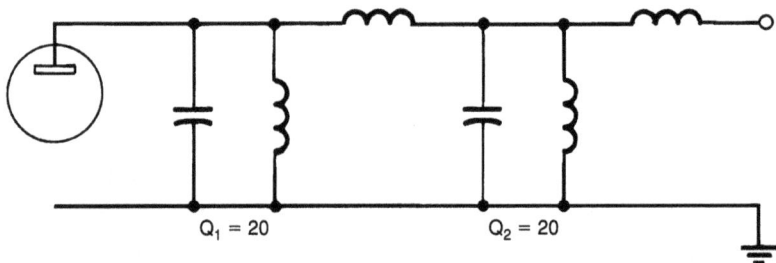

Figure 14.13 Output network with two inductively coupled resonators and combined L-network output coupling.

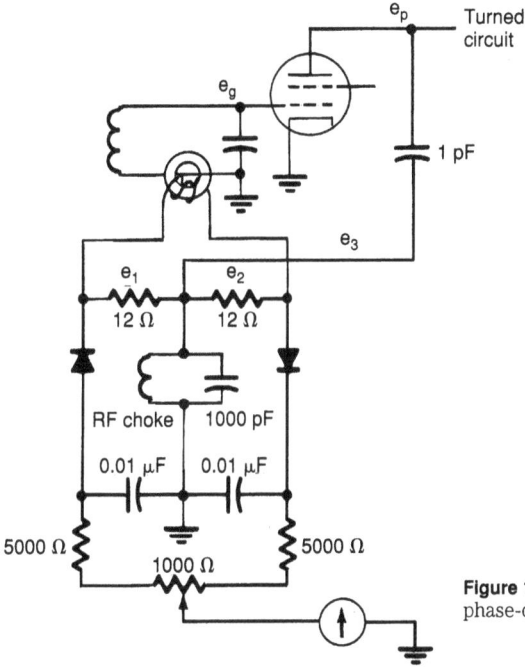

Figure 14.14 Radio frequency phase-detector circuit.

Methods of tuning

Tubes function best when their load impedance is resistive. When the circuit is not properly in tune, a reactive component appears in the load that causes an elliptical load line. An excessive amount of load reactance reduces efficiency and power output, increases dissipation, and changes the amplifier distortion characteristics. Generally, tank circuits should be tuned so that the phase angle of the tube load is within approximately ±5° of being resistive.

Since a plate load with a phase angle of 0° is the desired objective, it is logical that a phase detector be used for the tuning indicator. Figure 14.14 shows a typical phase-detector circuit using a zero-center meter as the resonance indicator. Two balanced voltages e_1 and e_2 are developed across the resistors in series with the secondary of the toroid coil.

The values of these resistors are low (10 to 100 ohms) compared with the inductive reactance of the secondary. The voltages e_1 and e_2 are independent of frequency because the voltage induced in the secondary, and the secondary reactance, are both proportional to frequency so their effects cancel. Voltages e_1 and e_2 are, therefore, proportional to the current in the grid tank coil. The current in the tank coil, which passes through the ferrite or powdered-iron-core toroid to form a one-turn primary, lags the grid voltage e_g by 90°. Voltages e_1 and e_2 are, therefore, ±90° from e_g. A sample e_3 of the RF plate voltage e_p is connected to the detector as shown. The output of one diode detector is proportional to the phasor sum of e_1 and e_3. The output of the other diode is proportional to the phasor sum of e_2 and e_3. When e_3 is phased 90° from e_1 and e_2, the detector outputs are equal and balanced, resulting in a zero meter reading. When the plate tank circuit is off resonance, e_p and hence e_3 have unequal phase angles with e_1 and e_2, so the unequal detector outputs cause a meter indication to one side of zero. Thus, it is only necessary to adjust the plate tuning for a zero meter reading to establish a resistive plate load for the tube. It should be noted that an off-resonance condition in the control-grid circuit does not create an error in the plate circuit resonance indication because the current in the control-grid tank coil lags e_g by 90° regardless of the resonance condition. The phase detector gives the proper indication regardless of the signal level, although it is less sensitive with smaller signals. Tuning does not require a constant-amplitude signal, and can even be adjusted on a speech signal, because the meter will kick to one side or the other if the circuit is slightly off resonance. In practice, it is found that more accurate tuning is easily accomplished with use of a phase detector than by other methods.

Automatically tuned transmitters use the phase-detector output to control a servomotor that positions the tuning elements to resonance when a signal is present.

Interstage coupling circuits

The coupling circuit between two small-signal amplifier tubes may be a simple parallel resonant circuit as shown in Figure 14.15. The load placed on the tube plate, R_L, is sometimes just the QX of the circuit, but usually a parallel shunting resistance is added (shown dashed in series with a blocking capacitor) to lower R_L. This is often done to keep R_L fairly constant over a given RF range and reduce the effects of stray feedback and loading.

Figure 14.16 shows a driver tube coupled to a PA tube. Impedance and voltage stepdown from plate to control grid is achieved by using a capacitive divider for the tank capacitance. A tuning range of two octaves can be covered in the HF range with a suitable variable inductor and fixed capacitors. A swamping resistance is placed across the circuit to stabilize the driver voltage gain.

In these two examples, the loading on the preceding stage is fixed by the transmitter design. This simplifies tuning, as no loading control is used. The trend in transmitter design is toward compact equipment designs with few controls and simple adjustments.

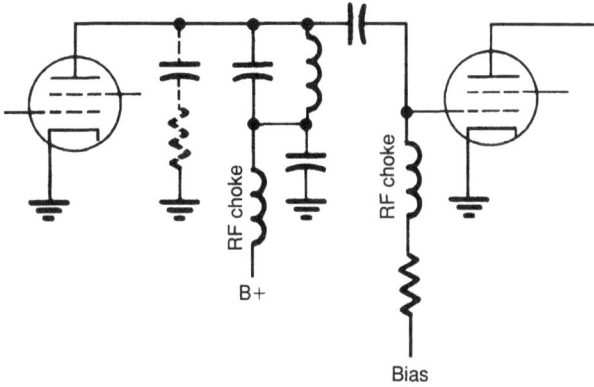

Figure 14.15 Capacitive coupling.

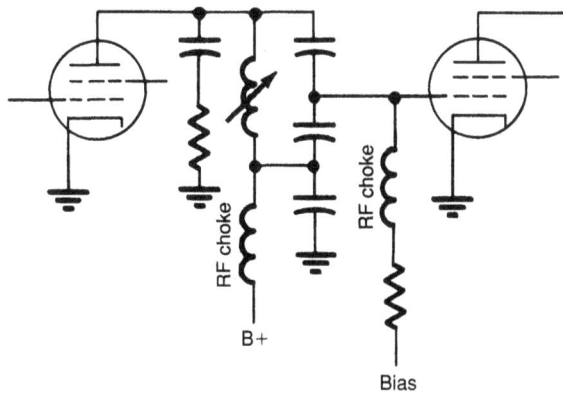

Figure 14.16 Impedance stepdown coupling circuit.

Power amplifier loading

Proper loading of a PA tube means that the load resistance presented to the tube plate is the value desired. Most output networks have loading as well as tuning controls so that a range of load impedances can be matched to R_L. Some means of proper loading is then necessary. Proper loading of linear amplifiers requires a specific relationship between two specific measurements. The only pair that has a linear relationship with varying signal level is e_g and e_p.

A practical means of using this principle for loading is shown in Figure 14.17. Radio frequency voltage detectors are connected to the grid and plate of the tube to sample the respective RF voltages. The detector outputs are adjusted and combined to feed a zero-center meter. When the cor-

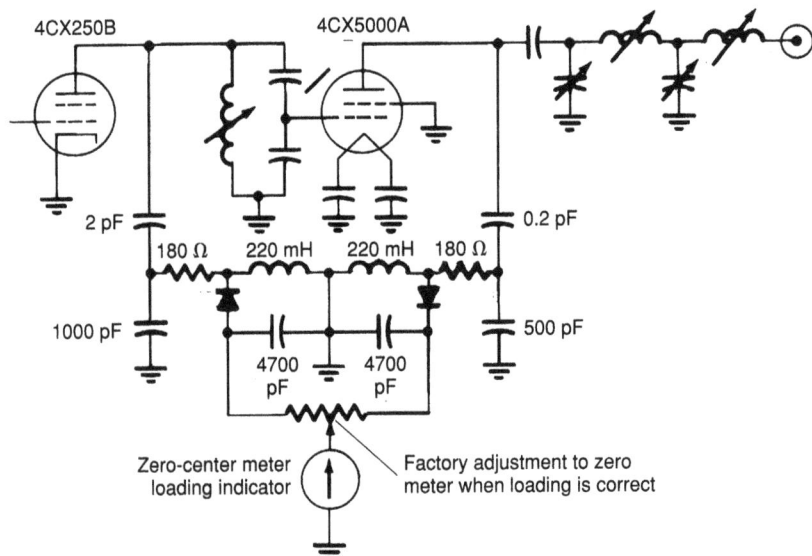

Figure 14.17 Loading indicator circuit.

rect ratio of e_p/e_g exists, the detector outputs are balanced and cause no meter deflection. This load indicator can indicate proper loading with nearly any signal of reasonable level. Therefore, the loading can be adjusted during signal transmission. Automatically tuned transmitters use this type of load sensing to provide a signal for the loading servomotor.

Low-frequency amplifiers

Linear amplifiers operating in the very low-frequency to low-frequency (VLF/LF) (14- to 300-kHz) region need special consideration because tank circuit components become very large and bandwidth is a crucial consideration.

The harmonic attenuation requirements of the output network can be greatly reduced by operating a pair of tubes in a balanced (push-pull) circuit. An output transformer is employed with tight coupling between the two halves of the primary. Balanced operation theoretically neutralizes all even-order harmonic components of plate current. Balancing adjustments can reduce the second harmonic to –40 dB or better.

The tube odd-order nonlinearities that produce IMD are the same ones that produce the third and higher odd-order harmonics. Therefore, choosing an operating condition with good linearity will also minimize odd-order harmonic output.

Radio frequency feedback (either current or voltage feedback) can be employed to reduce harmonics still further. This makes it possible to greatly reduce the harmonic attenuation requirements of the transmitter output network. It is usually necessary to control the phase delay between the PA tube plates and the antenna to achieve good passband symmetry. The phase delay from e_p to antenna current (in a series-resonated antenna) should be an integral multiple of 90° (including zero). The tube plate load variation is then minimum and symmetrical. A lowpass T-network is often used to increase the total phase delay of the entire PA output network, transmission line, and antenna coupling circuit to some multiple of 90°. The lowpass network provides some harmonic attenuation also.

14.8 Neutralization and Stabilization [2]

Effects of grid-to-plate capacitance

The effects of undesired coupling impedances, such as grid-to-plate capacitance C_{gp}, can be greatly reduced by means of circuits which balance out or neutralize these effects.

In a conventional tuned RF amplifier using a tetrode tube, the input resistance caused by unneutralized C_{gp} is

$$R = \frac{1}{2\pi f C_{gp}(A \sin \theta)} \tag{14.31}$$

where A = voltage amplification from grid to plate
$\quad \theta$ = phase angle of plate load

This resistance is in parallel with the grid-tank-coil equivalent-shunt resistance, driving source impedance, grid current loading, and any added swamping resistance.

This term can be either positive or negative, depending upon the phase angle or tuning of the tube plate circuit. When the plate circuit is tuned to the inductive (high-frequency) side of resonance, the term is negative. This means that energy is transferred from the plate to the grid circuit through C_{gp} resulting in positive feedback. When this negative resistance is lower than the equivalent positive resistance across the grid circuit, the amplifier will oscillate.

When the plate circuit is tuned to the capacitive (low-frequency) side of resonance, the term is positive and power is actually transferred from the grid to the plate circuit to produce added grid circuit loading.

This explains why grid current or RF grid voltage of an unneutralized amplifier swings considerably as the plate circuit is tuned through resonance. One important purpose of accurate neutralization, therefore, is to keep plate tuning from affecting the tube input resistance, which usually affects grid voltage and driver gain.

Unneutralized C_{gp} also causes a change in the effective grid-to-cathode capacitance by an amount $C_{gp}(1 + A \cos \theta)$. Tuning the plate circuit varies cos θ, and this reacts back on the grid-circuit resonance to some extent.

It should be noted that the effective values of input resistance and capacitance also depend upon amplifier gain A. In practice, the gain of linear amplifiers unavoidably varies a little with signal level. This causes slight variations in both input resistance and capacitance, which increases the amplifier distortion.

Accurate neutralization is, therefore, essential in nearly all high-gain linear amplifiers. Exceptions are those that have such a low impedance across the input circuit that input-impedance variations due to unneutralized C_{gp} are negligible.

Neutralizing circuits

Neutralization considerations have had a large influence on the evolution of RF PA design. In modern linear RF amplifiers, triodes are nearly always used in cathode-driven circuits, thus avoiding the need for neutralization. The grid is at RF ground potential and acts as a screen between the input and output circuits, as shown in Figure 14.18. As a result, the remaining plate-to-cathode capacitance C_{pk} is quite small. The tube-cathode input impedance is quite low, for example, 200 ohms or less, so the coupling from the plate to the input circuit through C_{pk} can generally be neglected.

Tetrode or pentode tubes are nearly always used in high-gain linear RF amplifier stages. The neutralizing circuit most widely used [9] is shown in Figure 14.19. It is sufficiently broadband that when properly designed it holds neutralization over the entire HF range.

The relationship for neutralization is

$$\frac{C_n}{C} = \frac{C_{gp}}{C_{gG}} \qquad (14.32)$$

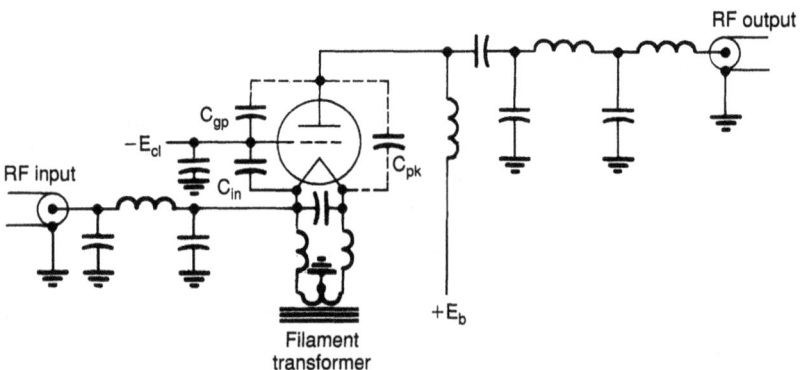

Figure 14.18 Cathode-driven amplifiers avoid need for neutralization.

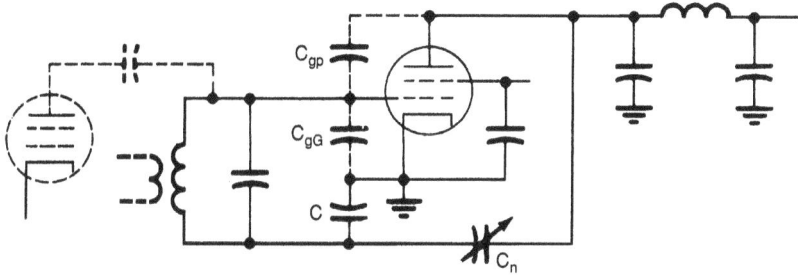

Figure 14.19 Bruene neutralizing circuit for single-ended tetrodes and pentodes.

where C_{gG} is the total capacitance from grid to ground and includes all stray grid-circuit capacitance and output capacitance of the driver stage if capacitive-coupled. The capacitor C may have on the order of 500- to 2000-pF capacitance and also serves to "bypass" the bottom end of the grid circuit to ground. Common values of C_n for this circuit are about 1 to 5 pF.

For broadband operation, it is essential that the common lead inductance in series with C be at an absolute minimum. Best results are achieved by using a feedthrough type of capacitor.

Parasitics

It is hardly necessary to state that all tendencies toward oscillation must be eliminated. Particular care must be used in the design of high-gain linear amplifiers to avoid these tendencies. Low-frequency parasitic circuits are usually caused by RF chokes and bypass and coupling capacitors in the dc feed circuits. Ultrahigh-frequency parasitic resonances almost always occur, and the designer must control them to avoid parasitic oscillation.

The principal UHF parasitic circuit is shown in Figure 14.20. It is always present when a plate tank capacitor C_t is used. The grid-to-cathode impedance at this plate parasitic resonant frequency must be low enough so that oscillation cannot take place, since neutralization is not effective at these parasitic frequencies. The tube-voltage gain can be very high because there is little resistive loading in these parasitic circuits. If the grid-circuit impedance cannot be controlled well enough, the plate parasitic resonance can be lowered in frequency by adding some inductance in the lead from the tube plate to the plate tank capacitor. This resonance must not fall on a second or third harmonic of the fundamental operating frequency, however. If it does, the harmonics of plate current will develop a substantial voltage across the resonant circuit and reduce efficiency, increase distortion, and possibly cause flashovers.

Resistance loading can sometimes be applied successfully to parasitic circuits. Two types of such parasitic suppression circuits are illustrated in

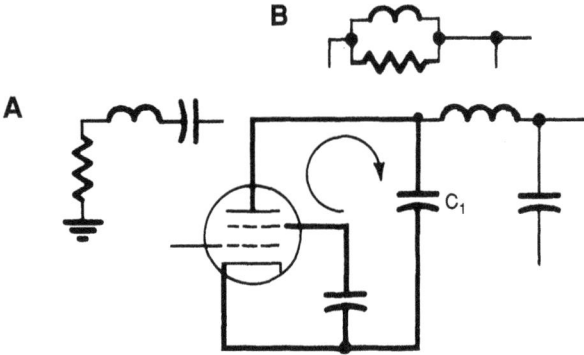

Figure 14.20 Ultrahigh-frequency parasitic resonant circuit and two types of suppressors: (A) shunt type; (B) series type.

Figure 14.20. Figure 14.20A shows a series-resonant trap shunt-type circuit, and Figure 14.20B shows resistance introduced in series with the circuit at the parasitic frequency. The resistance is removed at the operating frequency by the shunting effect of the low-value inductance across it. When the wattage rating of the resistor shown in Figure 14.20B is selected, the dissipation due to harmonics of plate current must be considered.

Parasitic circuits appear in a great variety of unsuspected ways. The UHF resonance described can be quickly found with a grid dip meter. Others must be found by testing. A powerful method of finding oscillation tendencies is to measure tube gain across a wide range of frequencies such as from 100 kHz to 1000 MHz. Response peaks indicate parasitic resonances.

Swamping resistance

High-gain amplifiers can be made more stable by adding swamping resistance across their input or output circuits or both, as shown in Figure 14.21. The gain of pentode or tetrode amplifiers is proportional to their load resis-

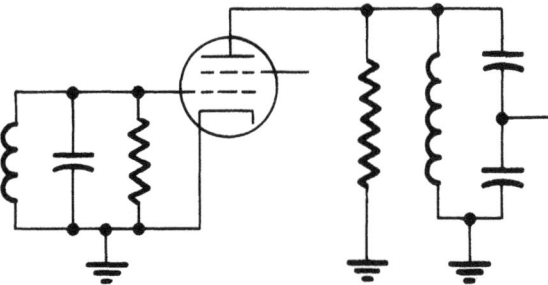

Figure 14.21 Grid- and plate-circuit swamping resistance.

tance, so added plate circuit swamping resistance reduces gain, but this gain reduction improves gain stability. The swamping resistance should be connected in the circuit so that it is effective at parasitic frequencies as well as at operating frequencies.

Swamping across the grid circuit is also effective in reducing the effects of imperfect neutralization. When the grid-to-plate coupling is low, such as in well-screened tubes, neutralization may not be necessary if swamping is sufficient. Grid circuit swamping also helps reduce the effects of reaction upon the input circuit due to impedance in series with the cathode.

Tank circuit losses sometimes vary considerably as the amplifier is tuned over a wide frequency range. The use of swamping resistance will reduce the total effective circuit-impedance variation. In fact, circuits may be added in series with the swamping resistor to equalize an interstage amplifier gain over its frequency range. This is particularly useful for maintaining uniform gain of amplifiers within a feedback loop.

Effects of series cathode impedance

A small amount of filament- (or cathode-) lead inductance will cause a resistive load to appear across the tube input. This inductance may be due to internal filament (or cathode) leads or external leads to ground through bypassing capacitors, as shown in Figure 14.22. When two filament leads are involved, L_k is the equivalent parallel inductance of both of them. The magnitude of this added input resistance is

$$R = \frac{e_g}{\omega^2 L_k C_{gk} I_{pi}} \qquad (14.33)$$

For small-signal class A amplifiers, e_g/I_{p1} can be replaced with $1/g_m$. This source of positive input resistance can be better understood by realizing that it results from driving the cathode a small amount, as in a cathode-driven

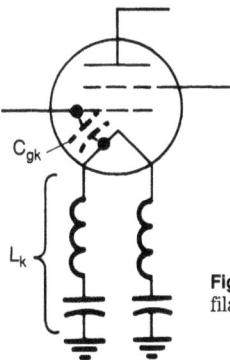

Figure 14.22 Total equivalent filament or cathode lead inductance.

amplifier. Radio frequency grid voltage causes a current to flow through C_{gk} and L_k. The voltage developed across L_k by this current is the amount of cathode drive voltage. The cathode drive power fed through the tube is

$$P_{ft} = \omega^2 L_k C_{gk} e_p I_{p1} \tag{14.34}$$

Note that the amount of grid loading due to this cause increases as the square of the frequency. It can be reduced by keeping L_k to an absolute minimum. Use of cavity-type circuits with ring-seal or coaxial-type tubes at VHF and higher is effective in minimizing input loading.

Very low-impedance filament or cathode bypass capacitors are desirable. An appreciable amount of input loading can increase distortion. Any tube plate-characteristic nonlinearity will cause a variation in this input loading signal level and thereby present a varying load to the driver, thus causing increased distortion in the drive voltage.

Effects of series screen impedance

Imperfect screen bypassing allows coupling from plate to grid in addition to that through C_{gp}. Figure 14.23 illustrates that current through C_{pg2} and Z_s, because of RF plate voltage, produces an RF voltage on the screen. This, in turn, is coupled to the control grid through the grid-to-screen capacitance C_{g1g2}. The resulting voltage on the grid from this coupling depends upon the impedance of the grid-to-cathode circuit. When the screen bypass impedance is capacitive, the coupling from screen to grid is in phase with the plate-to-grid capacitive coupling. It can be neutralized by increasing the neutralizing capacitance in the regular neutralizing circuit.

When the screen bypass impedance is inductive, because of tube leads or capacitor leads, the coupling is out of phase with that through C_{gp}. In fact, if Z_S is inductive by a certain value, the coupling from screen to grid will neutralize the coupling through C_{gp}. With normal bypassing practice, this usually occurs somewhere in the VHF region. The self-neutralizing fre-

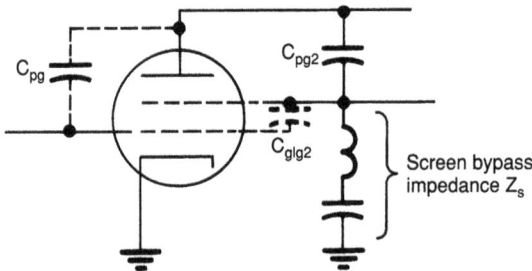

Figure 14.23 Screen bypass impedance permits coupling from plate to grid.

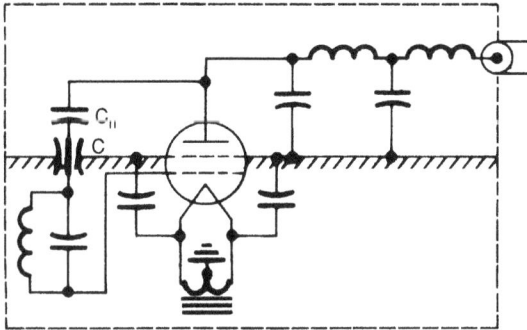

Figure 14.24 Grounded-screen circuit.

quency is sometimes given on tube data sheets. Annular screen bypass capacitors built into the socket that surrounds the tube near the screen terminal are quite effective for small tubes like the 4CX250B. For tubes in the 10-kW category, the capacitor dimensions are so large that resonances within the capacitor become a problem.

Nearly perfect screen bypassing can be achieved by using ring-seal tubes and grounding the screen terminal directly to the chassis deck all around the tube, as shown in Figure 14.24. This eliminates all common screen lead impedance right up to the screen cage within the tube. It has been used very successfully in many PAs with ratings from 1 to 250 kW to make neutralization more independent of frequency and to increase the stability of operation.

14.9 RF Feedback [10]

Radio frequency feedback has been found to be very effective for reducing both amplitude and phase distortion. The SSB wave may be looked upon as a single "sine wave" that varies in amplitude and phase. Radio frequency feedback will tend to restore any errors in amplitude or phase of the RF wave due to amplifier distortion.

It is generally advantageous to include as many stages as practical within the feedback loop. With current techniques it is necessary to limit this to two or three stages, however.

Two-stage feedback

One circuit that has been used a great deal is shown in Figure 14-25. With two high-gain stages the gain is still very respectable even after a gain reduction of 12 to 15 dB by feedback. The feedback voltage is obtained from the PA plate and applied to the cathode of the driver tube. The amount of feedback is determined by the ratio of the voltage from grid to ground and the voltage from grid to cathode.

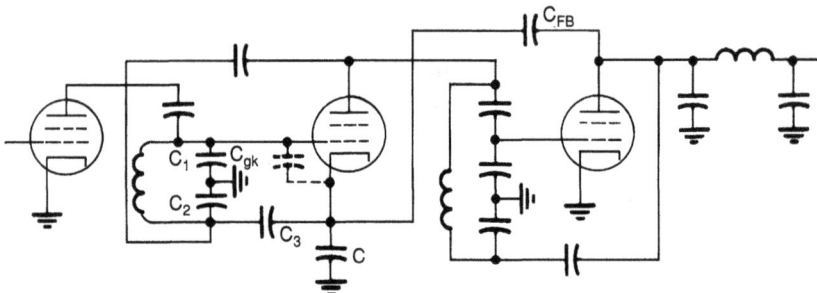

Figure 14.25 Two-stage feedback circuit, including neutralizing circuits.

There are several requirements that must be met to achieve satisfactory operation, however. The feedback voltage-divider capacitance from driver cathode to ground must be high, that is, on the order of 2000 to 5000 pF in the HF range. The feedback voltage drives the cathode, which takes some power. The load resistance that the cathode presents to the feedback voltage divider is on the order of 20 to 200 ohms. The reactance of the cathode capacitor must be much smaller than this to avoid a significant shift in phase of the voltage applied to the cathode. Also, RF amplifiers tend to be unstable when appreciable capacitive reactance appears in series with the cathode.

Large values of C require fairly large values of C_{FB} to develop the necessary feedback voltage. This value of capacitance is across the output plate tank circuit and may raise the circuit capacitance more than desired, so a compromise must be made.

It is necessary to neutralize the cathode-to-grid capacitance to keep the feedback voltage on the cathode from coupling back into the grid tank circuit. The relationship

$$\frac{C_1}{C_2} = \frac{C_{gk}}{C_3} \tag{14.35}$$

must exist for this neutralization. The values of C_1 and C_2 need not be equal, and the capacitance C_2 is usually made about five times C_1. The grid-to-plate capacitance of both stages must also be accurately neutralized. These circuits are included in Figure 14.25.

In medium- and high-power amplifiers, the effect of appreciable voltage on the driver cathode must be considered, as discussed in section 14.6. All possible phase shift between the driver plate and the PA grid is avoided by direct coupling between stages. A voltage or impedance stepdown may be used with tapped tank capacitance as shown. To avoid phase shift, any swamping resistance added to this circuit should be from driver plate to ground rather than from control grid to ground.

The input circuit is not within the feedback loop. It can be capacitively coupled to the plate of a high-output-impedance tube such as a tetrode.

When properly designed and neutralized, this two-stage amplifier is capable of quite high performance. A two-stage amplifier such as this can be made completely stable so it will not oscillate with any degree of mistuning or output loading. Some stray phase shifts always occur, but with careful design the normal losses in the output tank circuit will keep the gain from becoming too high, so stability can be maintained.

Three-stage feedback

Feedback around three stages with tuned coupling circuits between stages has been successfully used [10]. Since three tuned circuits contribute to the phase-gain characteristics of the feedback loop, it is not possible to use as much feedback for a given stability margin. Nevertheless, 10 to 12 dB can be used.

Larger transistors (both bipolar and FET) have become available in recent years, so a better solution in many cases is to employ a broadband solid-state first stage and eliminate the first interstage tuned circuit. The transistor stage output can be loaded with a low value of resistance to minimize phase shift caused by the driver stage input capacitance. The transistors should be operated class A and as linear as possible. A reduced value of collector or drain dc voltage helps to keep dissipation within the device limits.

Details of the transistor amplifier circuit and the means of injecting the feedback signal at its input will be left to the ingenuity of the designer since larger and better devices are continually being introduced.

Adding the third stage for more gain within the feedback loop greatly reduces much of the complication of the two-stage feedback circuit. Also, this stage operates at a lower signal level (by the amount of feedback used) than if feedback were around just the last two stages. The capacitance of the feedback capacitor which shunts the output circuit is much less also.

Stability

The stability of linear amplifiers employing RF feedback is determined by the phase-gain characteristic of the feedback loop. This in turn is determined principally by the resonant coupling circuits. For this to be true, the screen, plate, and cathode RF bypass capacitors must have negligible reactance over the band of interest. Interstage connecting leads must also be kept short to prevent them from acting as transmission lines with resulting phase shift from the plate of one stage to the grid of the next.

The impedance of the output network, as seen from the final tube plate, is within the feedback loop. The input impedance of an output network that couples to the load resistance through one branch of the resonant circuit is not symmetrical about resonant frequency. Examples are low-Q pi or pi-L networks. It is better to use an output network with a more symmetrical passband, such as previously illustrated in Figure 14.12. The properties of the output network are modified in actual circuits by dc blocking capacitors and RF chokes. These must be chosen to have little effect within a few octaves from the operating frequency.

Phase-gain characteristics

The phase-gain characteristic around the feedback loop needs to be carefully controlled in the design and development testing of the RF amplifier. The greatest tendency to oscillate is when the phase shift is 180°. The gain must be down by more than the amount of feedback used where 180° of phase shift occurs in order to avoid oscillation.

Initial testing of a PA is done with the feedback loop opened. When all stages are functioning properly, the gain and phase shift around the feedback loop are measured. This is done by comparing the RF feedback voltage with the input voltage. Ideally, the feedback voltage is in phase with the input voltage. The amount of feedback that would be applied if the feedback loop were closed is

$$\text{Feedback} = 20 \log \frac{e_{\text{in}}}{e_{\text{in}} - e_{\text{fb}}} \text{ dB} \qquad (14.36)$$

For 12 dB of feedback, the feedback voltage e_{fb} would be three-quarters of the input voltage e_{in}.

The gain margin of stability can be found by raising the excitation frequency until the open-loop feedback voltage is 180° out of phase with the input voltage. The gain reduction caused by this frequency change, less the amount of feedback, is the gain margin on the high side of the operating frequency. This procedure is then repeated on the low side of resonance.

The phase margin can be found by raising the excitation frequency until the feedback voltage is down by the amount of feedback that is to be used. The phase angle between e_k and the feedback voltage is measured. This is the phase margin on the high side of resonance. The procedure is then repeated on the low side of resonance.

If the PA employs automatic servo tuning, the servos should be allowed to tune the amplifier, then they should be disabled before shifting the input frequency for the above measurements.

The effect of a high-Q transmitter load impedance upon feedback amplifier stability is easily overlooked, but it can create a serious instability problem. The high-Q load could be a tuned whip or resonant loop antenna, or it could be caused by a high-Q multiplexer or bandpass filter. The problem is that the high-Q load can unload the final amplifier on either side of resonance, causing a large increase in gain. Too much gain increase could cause spurious oscillation, which is indeed a very serious matter. Antenna systems above the 10-kW level seldom have high Q, but high-Q antennas or antenna coupling circuits are frequently encountered up to the 10-kW level.

The final stage phase-gain problem is worst when the load impedance variation is similar to that of a high-Q series-resonant circuit located at an even multiple of half-wavelengths of phase delay from the final amplifier plate.

The above words of caution are not intended to discourage the use of RF feedback, but rather to stress the need for a thorough and careful design of the phase-gain properties of the feedback loop. Ideally, the amplifier should be stable with any load impedance and with any Q. Something less than unconditional stability can be accepted if the transmitter will always work into a low-Q antenna system such as log-periodic or other broadband antennas.

A computer is a very desirable aid in the design of linear amplifiers with feedback around more than two stages. The benefit of a stable amplifier with a substantial amount of RF feedback is superior IMD performance.

PA output (source) impedance

Single-sideband transmitters are typically designed to feed a 50-ohm coaxial transmission line with a specified SWR limit. It must not be assumed that the PA output resistance (looking back into the amplifier) is $50 + j_0$ ohms, like that of most RF signal generators, however. This fact has been of little consequence because the SSB signal bandwidth is usually very small compared to the antenna and output network bandwidth. Therefore the load impedance on the tube anode doesn't vary much across the signal band. On the other hand, very narrow antenna bandwidths such as produced by a tuned whip or a small tuned loop or by a very wide signal bandwidth, such as used for frequency hopping, may cause a very large load impedance variation at the tube anode.

Consider a case where the antenna input impedance varies across a 1.5 SWR circle as the signal frequency is varied from one side of the signal band to the other. For simplicity, let us assume a hypothetical case where the plate load resistance R_L is 1000 ohms and the output network bandwidth is wide enough to be negligible. The plate load impedance will then vary across an arc (which may be of any angular orientation) across a 1.5 SWR circle normalized to 1000 ohms. Increasing the line length to the antenna rotates the arc clockwise (around the center of the chart) 2° for each degree of added line length. When the transmission line to the antenna is many wavelengths long, the ends of the arc come closer together and may even form one or more complete loops. The impedance locus only passes through the $1000 + j_0$ point when the antenna impedance at the end of the coaxial cable passes through $50 + j_0$. Figure 14.26 illustrates three examples. Therefore, the amplitude response of a linear amplifier across the sig-

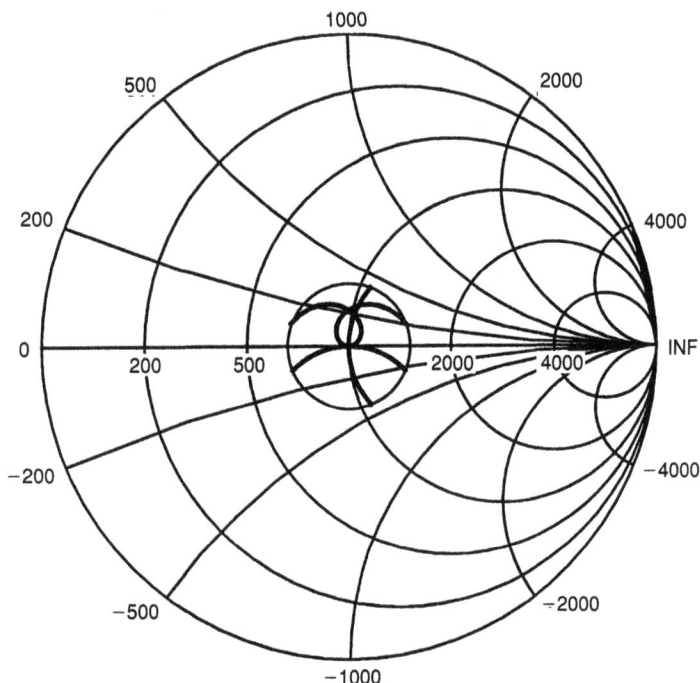

Figure 14.26 Loci of load impedance across a 1.5:1 SWR circle for two short and one long transmission line lengths to antenna.

nal band depends on the effective output (source) resistance of the tube and upon the electrical length of the circuitry and transmission line(s) to the series antenna resistance.

The tube output resistance R_S is determined by the effective plate resistance R_{PE} over an RF cycle and by RF feedback. R_{PE} can be estimated from tube curves such as shown in Figure 14.28. The slope of the constant current curves in the area of point C along the load line is used as follows:

$$R_{PE} = 2\frac{\Delta E}{\Delta I}$$

with constant grid voltage.

The factor of two accounts for the fact that in a class AB amplifier the plate current is effectively cutoff for one-half of the RF cycle. Driving the tube harder so that the load line enters the nonlinear region to the left of point A will cause a reduction in R_{PE}. Operation in that range is undesirable because nonlinearity in this region produces IMD.

R_{PE} is typically at least five times the normal plate load resistance R_L. The actual tube output (source) resistance R_S is < R_{PE} after being modified by

RF feedback, as will be discussed later. This can be measured by injecting a small signal into the output terminal and measuring the SWR caused by a mismatch between R_S and R_L. (Note: R_S is the load for the small test signal.) An SWR of 5 indicates that R_{PE} is five times R_L. (We know that it is higher than R_L, rather than lower, from inspection of the tube curves.)

Figure 14.27 shows a diagram of a test setup for measuring R_S while the PA is delivering a single-tone signal at any power level up to maximum PEP [11]. A test signal on F_1 drives the PA to the desired power output level P_1. This signal passes through two 30-dB directional couplers DC1 and DC2 and to the dummy load DL1. Test signal F_2, derived from a lower-power amplifier, passes through DC2 and injects a signal P_3, 30 dB lower, toward the PA under test. A 50-W signal P_2 on F_2 injects a 50-mW signal into the PA under test. A wide dynamic range (80 dB) spectrum analyzer is used to first measure the level of P_3 being fed into the PA out of one port of DC1. The other port is terminated in 50 ohms. The connections to these two ports are then interchanged to measure the level of the reflected signal P_4 on F_2. The SWR is calculated from the difference in decibels between these two levels. The reflection coefficient is

$$\rho = 10^{-dB/20}$$

$$SWR = \frac{1 + \rho}{1 - \rho}$$

$$R_S = R_L \, (SWR)$$

assuming a lossless output network.

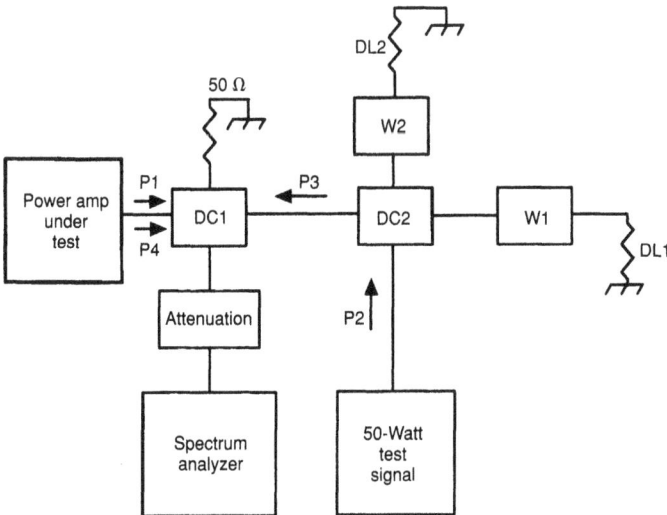

Figure 14.27 Test setup for measuring output impedance.

For example, an amplitude difference of 3.5 dB gives an SWR of 5:1. We can correct for output network loss by reducing the above decibel difference by the loss of two passes through the output network. An output network loss of 5.6% is equivalent to 0.25 dB, which when doubled is 0.5 dB. Subtracting 0.5 dB from the above 3.5 dB leaves 3.0 dB to use for computing R_S.

$$\rho = e^{-3/20} = 0.708$$
$$\text{SWR} = 5.85$$
$$R_S = 5.85 R_L$$

The output impedance Z_{out} looking into the output terminal is then somewhere around a 5:1 SWR circle in the above example. Just where depends upon the electrical length or phase delay of the output network. A Smith chart is used, starting with the voltage node at the plate of the tube. The phase delay of a pi-L network might be on the order of 225°, for example. In this case,

$$Z_{out} = 24.5 + j46.0 \text{ ohms}$$

at the transmitter output terminal.

R_S can be reduced to R_L by the application of the correct amount of RF voltage feedback (RF current feedback would raise R_S). It will always be less than 6 dB. About 4.5 dB would be correct for the above example. In the above test setup, when no F_2 is reflected, $R_S = R_L$, resulting in an infinite decibel difference. Accuracy can be increased by increasing the level of P_2 considerably and reducing P_1 to zero, since R_S does not change much at low power levels.

More RF feedback reduces R_S still more. It can be measured as above, except that R_S is less than R_L by the SWR factor.

The RF power output will vary most when the plate load resistance R_L varies from the low resistance to the high resistance side of a given SWR circle. For the case of $R_S/R_L = 5$, a 1.5:1 SWR and a 1000-W amplifier, the power varies from 390 to 667 W. It only can go to 667 W because the drive had to be backed down to prevent excessive RF plate voltage swing. Incidentally, the plate dissipation for a representative case varies from 827 to 395 W if the load impedance is varied around the 1.5:1 SWR load impedance circle. By using RF feedback to reduce R_S/R_L to 1.00, the RF power output remains constant at 667 W around the SWR circle. The output only rises to 750 W at the center of the circle, therefore the power output remains more nearly constant as the load impedance varies within a given SWR circle when $R_S/R_L = 1$.

Radio frequency feedback in tuned RF amplifiers does not affect the harmonic components of plate current because of the selectivity of the grid and plate-tuned circuits. It just improves the amplitude linearity of the fundamental component versus signal level.

Grounded-grid triode PAs

Grounded-grid (GG) linear amplifiers are seldom used in commercial and military installations because the IMD, when added to driver IM, is often not up to FCC and military standards. The driver is often the principle source of high IM. More IM is tolerated in amateur bands. A typical amateur station consists of a 100-W transceiver and a grounded-grid linear that is switched in when more power is needed. The type of transmitting tubes optimized for grounded-grid use are high-μ zero-bias triodes, and they are operated in class AB_2. Tubes using the focused beam geometry, such as the 8877, typically have a little more gain and lower IMD than the conventional filament-type tubes such as the 3-500Z.

The value of idling plate current for minimum IMD can be found by plotting cathode voltage versus plate current curve, with plate voltage maintained at its dc value (similar to the upper left curve in Figure 14.6). Extend the straight portion of the curve down to the zero plate current line. The plate current at that cathode bias voltage will give minimum IMD. If this amount of idling current causes too much idling plate dissipation, the bias must be increased so that, typically, the idling plate dissipation is no more than two-thirds of rated plate dissipation. This bias voltage is usually less than 10 Vdc. Many designs use a zener diode in series with the dc return from the tube cathode (or filament transformer center tap) instead of applying bias to the control grid. The voltage on the cathode is above ground by the zener voltage, to make the grid negative with respect to the cathode.

Radio frequency drive power is applied to the cathode (filament). About three-quarters of the drive is fed through the tube to the plate circuit, and the other one-quarter is absorbed mostly by the tube grid and a little by the bias circuit. The RF power gain of a typical grounded-grid linear using a high-μ triode is approximately 20.

It is usually best to choose an operating condition close to one of those given by the tube manufacturer on the tube data sheet. Operating parameters for other operating conditions can be computed using the simple equations listed below for approximate results. A selected load line is drawn on a set of constant current curves for grounded-grid operation as shown in Figure 14.28 for an Eimac 3CX1500A/8877 transmitting tube. The following simple equations are used to get the load line located close to where the desired parameters are achieved. These formulas apply for a single-frequency signal representing PEP:

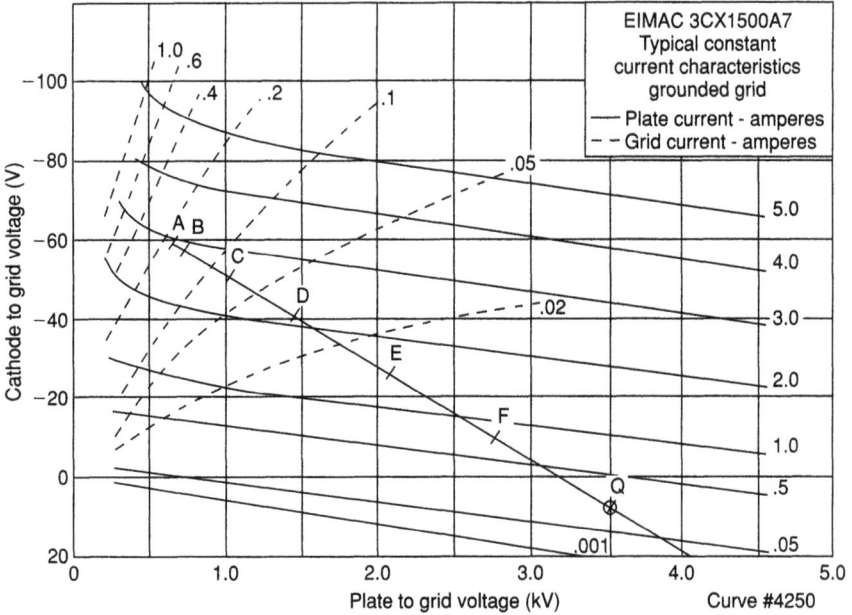

Figure 14.28 Load line for grounded grid 8877 tube. (*With permission of Varian Power Grid Tube Products Co.*)

Watts of average RF output:

$$P_o = \frac{i_b e_p}{4} \tag{14.37}$$

dc plate current:

$$I_b = \frac{i_b}{\pi} \tag{14.38}$$

Watts of RF drive-through:

$$P_{FT} = \frac{i_b e_k}{4} \tag{14.39}$$

Watts of plate dissipation:

$$P_D = E_b I_b - P_o - P_{FT} \tag{14.40}$$

Watts of grid dissipation:

$$P_G = i_g \frac{e_k}{4} \text{ (approx. but on high side)} \tag{14.41}$$

Watts of RF drive power:

$$P_D = P_{FT} + P_G \tag{14.42}$$

Ohms of plate load resistance:

$$R_L = 2\frac{e_p}{i_b} \tag{14.43}$$

Ohms of cathode input resistance:

$$R_K = 2\frac{e_k}{i_b + i_g} \tag{14.44}$$

The output (source) resistance R_S of a cathode-driven tube is higher than for the same tube when grid driven because of the inherent current feedback present in grounded-grid operation. The amount of feedback is determined by the source resistance of the drive at the cathode. A low source resistance reduces the amount of feedback.

The IMD level of a grounded-grid linear amplifier depends not only upon the IMD out of the driving amplifier, but upon the effective driving source impedance Z_{out}. This is the impedance looking back toward the driver from the grounded-grid linear's cathode. The source resistance Z_{out} will equal R_K if the driver $R_S = R_L$. On the other hand, if $R_S/R_L = 5$, for example, Z_{out} will be somewhere on a 5:1 SWR circle centered on R_K (on a Smith chart). If Z_{out} happens to be at the high resistance point on the SWR circle, Z_{out} will equal $5R_K$. The drive will be approximately that of a constant current source. Grounded-grid linear amplifiers inherently driven with a current source will degenerate out much of the plate current nonlinearity (such as might be caused by using a bias higher than the projected cutoff). If Z_{out} is near the low resistance point on the SWR circle, the amount of effective feedback will be very low.

If Z_{out} is near the top or bottom of the SWR circle (between the high and low resistance points), Z_{out} has a large reactive component. In this case, any nonlinear load on the driver caused by the grounded-grid linear amplifier will produce phase nonlinearity. Phase nonlinearity produces IMD components which add to the amplitude components present on one side of the signal and subtract from those on the other side. The total power in the IM products is always increased by the presence of phase distortion. Phase distortion can be eliminated by keeping Z_{out} resistive. This requires that either $Z_{out} = R_k$ or that the phase delay from the driver plate (or collector or drain) to the grounded-grid cathode be some multiple of 90°.

$$Z_o = R_K \frac{R_S}{R_L}$$

for an even multiple of 90° and where R_S and R_L are those of the driver. For an odd multiple of 90°, such as 270°,

$$Z_{\text{out}} = R_K \frac{R_L}{R_S}$$

A low value of Z_{out} has the advantage of reducing the effect of nonlinear grid current loading across the cathode circuit. This leaves the designer with the choice of reducing plate current nonlinearity or of reducing grid current nonlinearity—or of a compromise by using a driver that produces a Z_{out} equal to R_K. In the latter case the phase delay of the drive circuit is unimportant, which simplifies RF coupling circuit design. Another good reason for this compromise is that grid current and plate current nonlinearities are of opposing phase, resulting in partial distortion cancellation, which improves the overall linearity.

Some form of ALC should always be used to prevent driving the grounded-grid linear into the peak flattening range—beyond its linear PEP capability. Excessive drive causes high IM (splatter) on both sides of the signal.

A grounded-grid linear amplifier is a simple, practical, low-cost way to increase power by a factor of 15 to 20 (although analysis of operation is complex). The driver should have considerably lower IMD than what is required out of the grounded-grid linear amplifier. Also, unless the system is designed to take advantage of either a low or high value of Z_{out}, best results will usually be obtained when the driver output (source) impedance at its 50-ohm output terminal is 50 ohms—like that of a test equipment signal generator.

References

1. W. B. Bruene, "Linear Power Amplifier Design," *Proc. of IRE* 48 (December 1956): 1754–1759.

2. E. W. Pappenfus, W. B. Bruene, and E. O. Schoenike, *Single Sideband Principles and Circuits* (New York: McGraw-Hill, 1964).

3. E. L. Chaffee, "Simplified Harmonic Analysis," *Rev. Sci. Instr.* 7 (October 1936): 384.

4. Eimac Tube Performance Computer, Application-Engineering Dept., Eimac Div., Varian, San Carlos, CA.

5. R. I. Sarbacher, "Graphical Determination of PA Performance," *Electron* 15 (December 1942): 62–56.

6. W. B. Bruene, "How to Design R-F Coupling Circuits," *Electron* 25 (May 1952): 134–140.

7. Thomas R. Cuthbert, *Circuit Design Using Personal Computers* (New York: John Wiley & Sons, 1983).

8. W. B. Bruene, Automatically Tuned Coupled Resonant Circuits, U.S. Patent 3,355,667 (November 28, 1967).

9. W. B. Bruene, "How to Neutralize Your Single Ended Final," *CQ* 6 (August 1950): 11.

10. W. B. Bruene, "Distortion Reducing Means for SSB Transmitters," *Proc. of IRE 44* (December 1956): 1760–1765.

11. W. B. Bruene, "RF Power Amplifiers and the Conjugate Math," *QST* (November 1991): 31–32.

15

Antenna Matching Techniques

Glenn R. Snider

The role of the antenna coupler is to provide an impedance matching function in either of two ways: narrowband or broadband. With narrowband matching a near perfect impedance match is achieved at a single frequency. With broadband matching a less than perfect impedance match is accepted in order to get an optimal match over a broad range of frequencies.

Section 15.1 presents the current techniques used for narrowband antenna couplers. Included is the basic approach to be followed, some of the pitfalls, and general engineering advice. It is assumed that the reader has a computer at his/her disposal. Section 15.2 considers broadband matching of antennas in applications where it is advantageous.

15.1 Narrowband Antenna Couplers

The discussion of narrowband couplers will be subdivided into three topics: R/X plane, two examples, and control techniques. The R/X plane is a very useful graphical representation of how individual RF elements affect the input impedance of a coupler. By visualizing the loci of points on a graph with axes of resistive and reactive impedances, one can determine which RF network to choose, as well as how to automatically position the elements to the tune point. Two antenna coupler designs, shunt and whip, will demonstrate the practical application of the R/X plane graph and provide general information for the construction and specification of the RF elements. The last subdivision will deal with impedance measuring devices and the methods used to automatically control the RF tuning elements.

R/X plane

The first problem the engineer faces is, "What RF element configuration best suits the required specifications?" To make this decision, the antenna impedance can be plotted on the R/X plane. Then, by knowing how different types of RF elements, when combined with the antenna impedance, affect the resulting input impedance, logical decisions can be made. Smith has treated this in great detail [1]. The R/X plane is so important in the design of couplers that some of the basic concepts will be repeated here.

Consider the circuit shown in Figure 15.1A, a series circuit consisting of C, L, and R. The input impedance is the vector sum of all the components. When one of the reactive elements is varied, the magnitude of input impedance always falls on a straight line passing through the real axis at R_S. Increasing the inductive component moves the input impedance upward in an inductive direction. Increasing the capacitive component moves the input impedance downward in a capacitive direction.

If the desired input is equal to R_S and the antenna impedance is equal to $R_S + j\omega L_S$, the antenna coupler would consist of a series capacitor of reac-

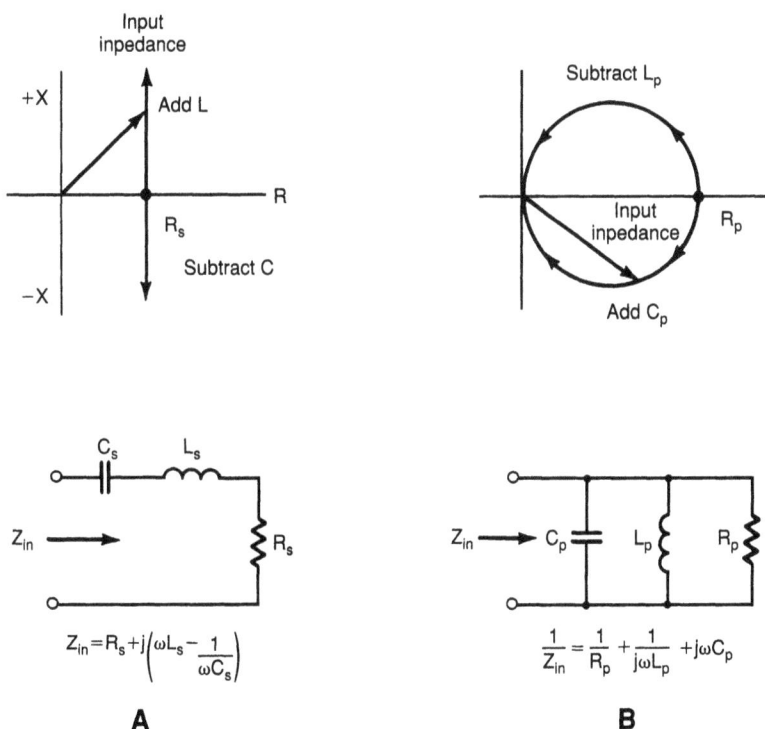

$$Z_{in} = R_s + j\left(\omega L_s - \frac{1}{\omega C_s}\right)$$

A

$$\frac{1}{Z_{in}} = \frac{1}{R_p} + \frac{1}{j\omega L_p} + j\omega C_p$$

B

Figure 15.1   (A) Series and (B) shunt R/X plane.

tance $-j\omega L_S$. To summarize, impedance in series with an antenna translates the input impedance along a line of constant real value.

Figure 15.1B shows the dual of Figure 15.1A. The circuit consists of a parallel C, L, and R. Smith has shown that when one of the reactive elements is varied the magnitude of input impedance always falls on a circle of radius $R_p/2$ and center located on the real axis. Decreasing the inductive impedance moves the magnitude of input impedance in a counterclockwise direction. Decreasing the capacitive impedance moves the magnitude of input impedance in a clockwise direction. To illustrate, if the desired input impedance is R_p and the antenna impedance is $1/(1/R_p + j\omega C_p)$, the antenna coupler would consist of a shunt inductor of impedance $1/(-j\omega C_p)$.

Since antenna impedances are more complicated than the above examples, couplers rarely consist of a single element. It can be shown that with two RF tuning elements any complex impedance can be transformed to a desired resistive impedance. The series and shunt circuits shown in Figure 15.1 can be combined to form a two-element network. Eight possible combinations of two elements are available, of which two will be discussed. The first consists of an input series C followed by an L in shunt with the antenna. The second consists of an input series C followed by a C in shunt with the antenna. These circuits will be developed on the R/X plane to show how each can be used to transform the antenna impedance to a desired resistive input value. In each case, certain impedances cannot be transformed. These areas (no tune) must be avoided either by choosing a different network or by changing the antenna impedance.

Figure 15.2 is the R/X plane of a general-purpose antenna coupler. That is, the circuit formed by the input C and shunt L can transform the maximum area on the R/X plane to the desired input impedance. The load is the antenna impedance, and the shunt L is closest to the load. Referring to Figure 15.1B, the load impedance can then be modified by L. If a very large L is present, the load will be modified very slightly in a circular counterclockwise direction. If the shunt L is near zero, the load impedance will be modified greatly in the same circular direction almost to the origin. The proper value of shunt L is one that places the modified load impedance at a value that has the desired real component and a highly inductive component. The reason for this will soon become apparent. From Figure 15.1B the diameter of the circle on which the transformation takes place is the parallel equivalent resistance of the load. The familiar equation $R_p = R_{ant}[1 + (X_{ant}/R_{ant})^2]$ can be used to find this from the antenna impedance. Take note, the R_p of the antenna must be greater than the desired input impedance. If this is not the case, it is not possible to use this network for the antenna coupler. Because of the circular motion, a shunt element can only transform the real component of input impedance in the range of zero to R_p. Also note that if the antenna impedance is inductive, the antenna resistance must be greater than the desired transformed resistance. If this is not the

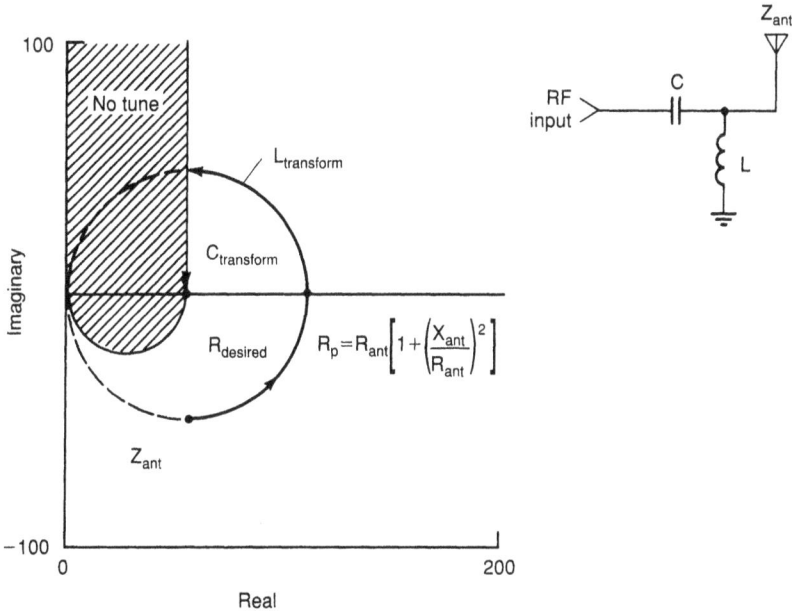

Figure 15.2 Series capacitor shunt coil R/X plane.

case, rotation in a clockwise direction is required. This is not possible with a shunt inductor. These two conditions form the no-tune region on the R/X plane shown in Figure 15.2.

The transformation by the shunt L has placed the modified load impedance at the desired real value. All that remains is to treat this modified load impedance as a new load for the series C. From Figure 15.1A the load impedance is modified on a line of constant real value which in this case has been adjusted by the shunt L to be the desired input value. Adding series C causes the cancellation of the inductive component. The network is tuned to the desired input value. The shunt L could have left the transformed impedance presented to the series element at the desired real value with a capacitive reactive component. However, this requires a series L to reach the tune point, and the circuit has been equipped with a series C.

To summarize this network, the shunt L acts to "load" the input impedance and the series C acts to "phase" the input impedance. *Loading* means adjusting the real component of input impedance to the desired value. *Phasing* means adjusting the reactive component to zero.

The second circuit to be presented on the R/X plane consists of a series C and a C in shunt with the antenna. As can be seen from Figure 15.3, this network transforms a very limited amount of load impedance to a desired real value. Since it is comprised of capacitors, the efficiency is very high for a given volume. Typically, antenna couplers use vacuum or ceramic capacitors that have quality factors Q in the range from 2000 to 5000. An inductor with Q this high requires such a large volume that it is impractical for most applications. The shunt antenna coupler example given below will demonstrate the use of this circuit for a unique antenna type.

On the R/X plane, the shunt C is the nearest element to the antenna. From Figure 15.1B, the load must be rotated clockwise on an R_p circle. As with the first circuit, the clockwise rotation crosses the desired real line in two places. Since the phasing element is a series C, the correct transformation leaves the modified load impedance inductive. This determines one limit for the antenna impedance. The impedance to be tuned must be inductive and have a real value less than the desired input impedance. A second limit can be seen from Figure 15.3. The parallel equivalent resistance R_p of the antenna must be greater than the desired input impedance. Once the antenna has been loaded, the series C can be used to adjust phase, completing the tune. To avoid some confusion, remember that regardless of which type of shunt element is used, the extent of rotation is limited to the origin. In this case, if the antenna impedance were greater than the desired value, the clockwise rotation would first cross the desired real value with a capacitive reactive component and stop at the origin. A second real value crossing is not possible. This circuit cannot tune antenna impedances that allow only one crossover point.

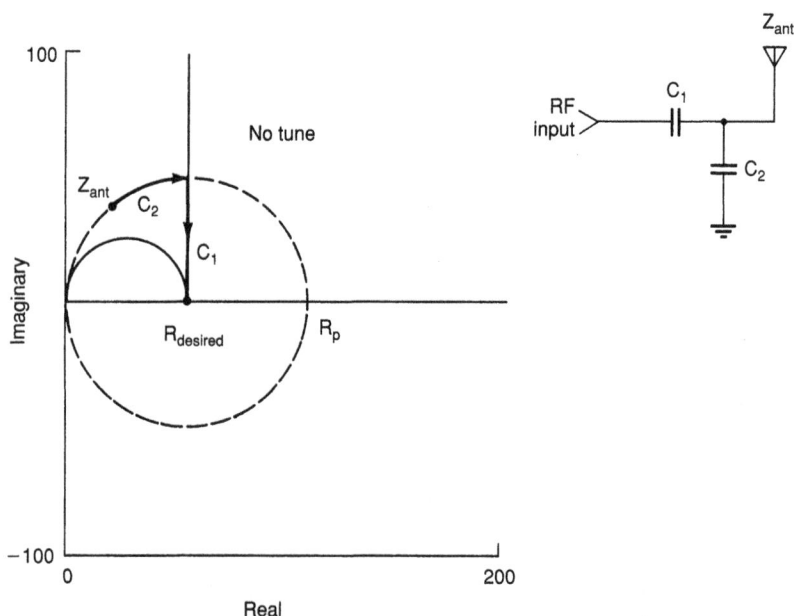

Figure 15.3 Series capacitor shunt capacitor R/X plane.

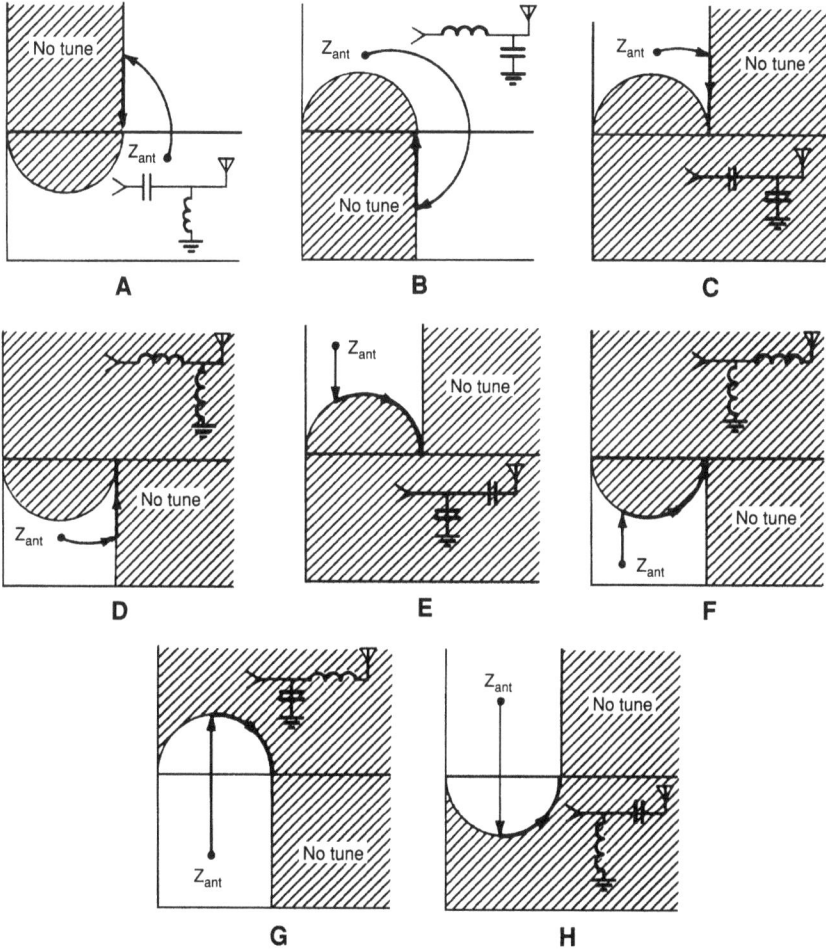

Figure 15.4　R/X plane reference guide.

The previous discussion should allow one to analyze the remaining six possible combinations of two-RF elements. For convenience all eight circuits are
summarized in Figure 15.4. The no-tune regions and element transformations on the R/X plane are shown there. This tabulation is a handy reference
for the designer, since most antenna impedance-versus-frequency characteristics require more than one network configuration. In addition, for all antenna impedances there are two possible RF configurations for tuning.

Example shunt antenna coupler

Figure 15.5 is an R/X plot of an aircraft shunt antenna impedance. For this
example the desired real input impedance is 50 ohms. By inspection of the

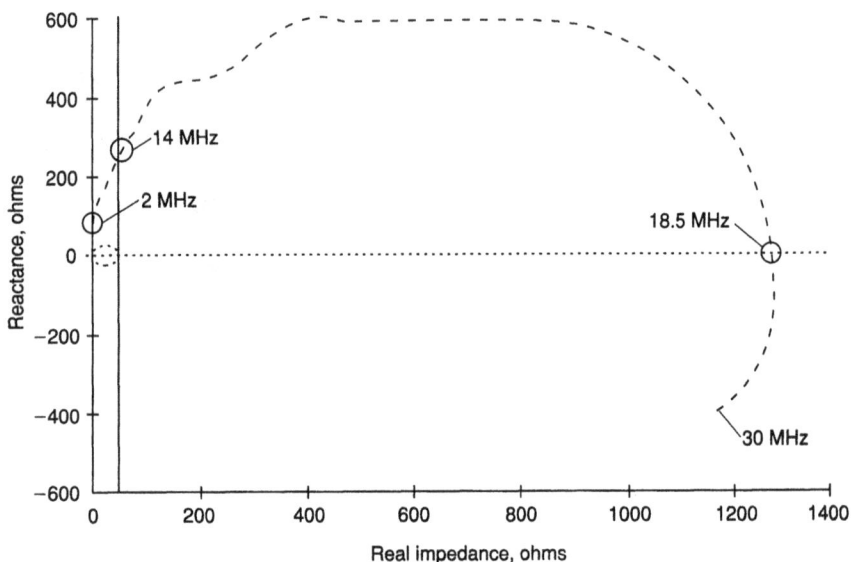

Figure 15.5 Aircraft shunt antenna impedance.

graph the impedance covers all regions of the *R/X* plane with the exception of a circle with diameter of 50 ohms. The conclusion is drawn that to tune the antenna across the entire frequency range more than one RF network will be required. The configurations shown in Figures 15.2 and 15.3 are chosen.

To make the proper choice for the network, more information is needed. The coupler is to be designed with the maximum possible RF efficiency in the low-frequency range. This is necessary because aircraft shunt antennas are poor radiators in that frequency range. Another fact that is extremely important for aircraft design is that the coupler be as small as possible. Usually, aircraft couplers are mounted at the antenna, which is located in the wing or vertical stabilizer where space is at a premium. Examination of vacuum capacitor specifications indicate that *Q*'s of 3000 are readily attainable in compact volumes [100 in.^{3} (1639 cm^3)]. Inductor volume can be estimated with good accuracy using helical resonator techniques [2]. Calculate as follows:

$$Q = 60S\sqrt{F_{\text{MHz}}}$$

$$S = \frac{Q}{60\sqrt{F_{\text{MHz}}}}$$

$$V = 2S^3$$

for

$$Q = 3000 \text{ and } F_{\text{MHz}} = 2$$

$$V = 88388 \text{ in.}^3 \, (1{,}448{,}419 \text{ cm}^3)$$

where Q = quality factor
S = length of side of a square coil cavity (in inches)
F_{MHz} = tune frequency (in MHz)
V = the required volume (in in.^{3})

From the calculation of inductor volume it is obvious that the coupler should consist entirely of vacuum capacitors for minimum size and maximum efficiency. The two all-capacitor networks shown in Figure 15.4 will tune the 2- to 10-MHz range. Since the antenna impedance is very small, its current is very high. The design can be improved if the series capacitor is not placed directly in series with the antenna. If this were done, both of the capacitors would be required to carry high current. Placing the shunt capacitor directly at the antenna requires only the shunt capacitor to carry high current. The series capacitor then carries the smaller 50-ohm input current. The circuit of choice for the low-frequency range is shown in Figure 15.3. Such a circuit has been computer-analyzed, and with 1000 W of input power the shunt capacitor current is greater than 50 A. The series capacitor current is less than 10 A, and the circuit efficiency is greater than 80%.

Referring to Figure 15.5 again, the antenna impedance in the 10- to 30-MHz range falls into the tune range of the circuit shown in Figure 15.2. This has a series capacitor and an inductor in shunt with the antenna. The frequency range could be divided into two segments using two RF networks. However, to minimize size and complexity, this was not done. Since the low-band RF network has a series capacitor, it can be converted to the high-band network by replacing the shunt capacitor with a shunt coil. Fortunately, the antenna in the high-frequency range has better radiation efficiency and, as a result, the antenna Q becomes smaller as frequency increases. This means that the Q of the shunt inductor can be lowered and still maintain reasonable efficiency as well as radiated power. A variable inductor with a Q greater than 100 can be constructed in a volume of 125 in.^{3} (2048 cm^3). Computer analysis indicates with this Q the efficiency will be greater than 60%.

Figure 15.6 shows the configuration for the shunt antenna coupler. The R/X plane is also shown for each frequency range. Notice that the shunt inductor is switched out of the circuit but the shunt capacitor is not. The reason for this can be found from the tuning algorithm. The low band requires the shunt inductor to be switched out of the circuit. The shunt capacitor is varied to cause impedance rotation in a clockwise direction until the 50-ohm line is reached. The series capacitor is varied to cause resonance. However, in the high band the shunt capacitor is placed at minimum capac-

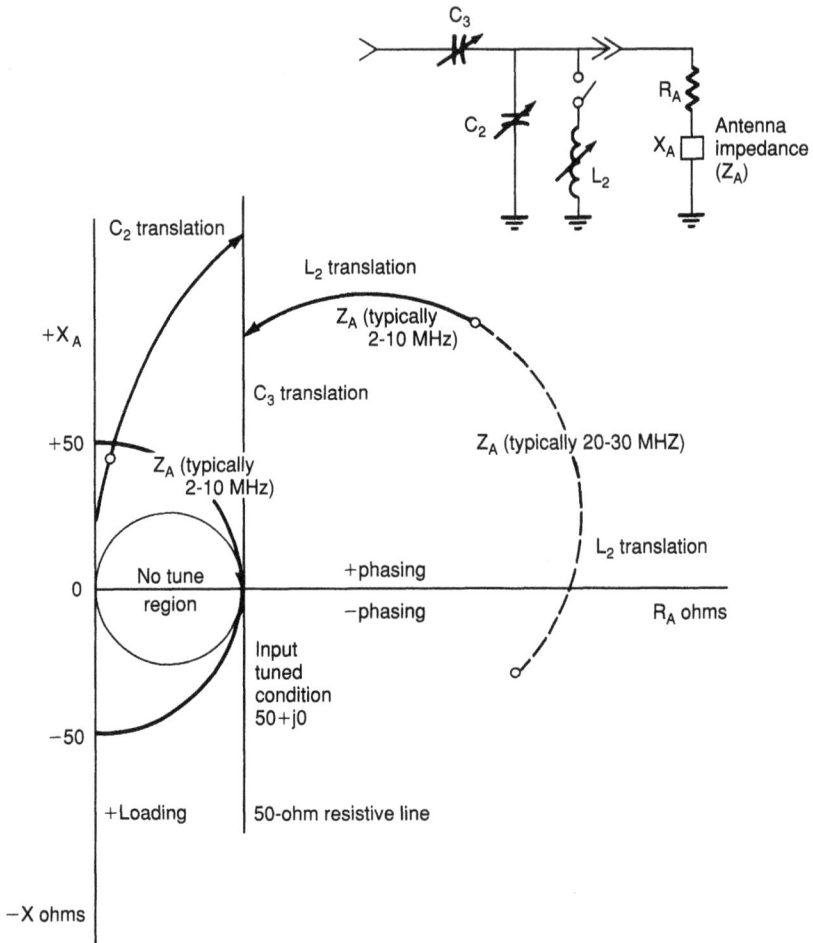

Figure 15.6 Shunt antenna coupler. Note: C_3 translates impedance on a resistive line (vertical). C_2 or L_2 translates impedance on a circle, as shown.

ity and left in the circuit. This results in minimum effect on the antenna impedance. The shunt coil is placed in the circuit and varied in a coarse manner to cause a counterclockwise rotation until the resulting real value of impedance is less than the desired value. This is within the tune range of the low-band network. The shunt capacitor is then servo-tuned to cause the reverse clockwise rotation to achieve the desired value of real impedance. To complete the tune, the series capacitor is again servo-tuned to resonance. The decision at which frequency to switch in the shunt coil has been rendered less critical. In fact, the control logic can make this, based on greater than 50-ohm real and inductive antenna impedance. This technique eliminates one servo system and one high-current switch from the coupler.

Example of a high-power whip coupler

To further illustrate the techniques used for couplers, assume that the antenna to be matched over the 2- to 30-MHz range is a 35-ft (10.67-m) whip. This impedance versus frequency is shown in Figure 15.7. The input power to the network is 20,000 W and the desired input impedance is 50 ohms. The network will be subdivided into three configurations covering 2 to 4 MHz, 4 to 20 MHz, and 20 to 30 MHz. Figure 15.8 shows a block diagram of the coupler. It consists of three modules: transformer, L network, and output coil. The factors which lead to this configuration are determined primarily by the input power.

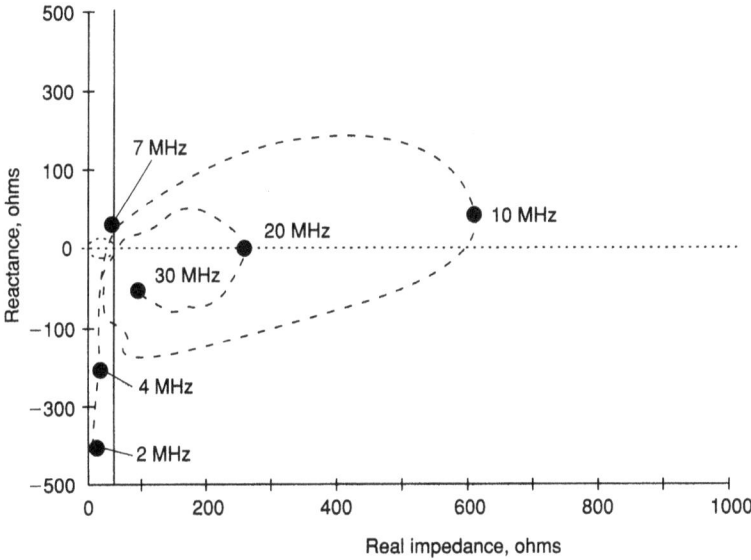

Figure 15.7 A 35-ft (10.67-m) whip antenna impedance.

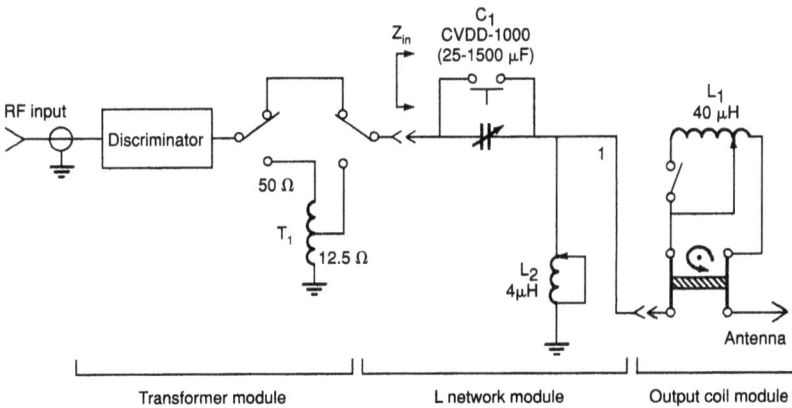

Figure 15.8 Block diagram, 35-ft (10.67-m) whip antenna coupler.

Inspection of the R/X plot of impedance versus frequency in Figure 15.7 indicates that all areas of the R/X plane are covered by the antenna impedance. In addition, the low-frequency range contains a highly reactive component that requires special voltage precautions. From Figure 15.4, several configurations are necessary to reach the desired 50-ohm tune point across the entire frequency range. The circuits shown in Figures 15.4A and 15.4F are used. The three frequency bands will now be discussed.

Three frequency bands

Band 1 (2 to 4 MHz). The antenna impedance in this range can be tuned by the circuits shown in Figures 15.4A, D, F, and G. To help eliminate some of the possibilities, consider that with 20,000 W of RF power the antenna voltage will be extremely high when the reactive component is large. In this case antenna voltages up to 70,000 V are possible. This can be estimated by assuming that all the input power is dissipated in the antenna resistance. A current can then be found that flows into the antenna terminal. From this the antenna voltage can be calculated:

$$V_{\text{peak}} = \left| Z_{\text{ant}} \right| \sqrt{\frac{2P}{R_{\text{ant}}}}$$

$$= 400 \sqrt{\frac{2\,(20{,}000)}{1.31}} = 70{,}000 \text{ V}$$

Clearly the circuits shown in Figures 15.4A and D can be eliminated since these have a common node at the antenna terminal that requires each of the elements to stand off most of the antenna voltage. The circuit shown in Figure 15.4G may be the best choice if this is the only frequency range to be tuned. The use of a vacuum capacitor results in a more efficient design. However, to tune the remaining frequency ranges, the general-purpose network circuit shown in Figure 15.4A is required. This circuit has a shunt coil, and if the circuit shown in Figure 15.4F is chosen for band 1, the shunt coil can be reused.

Tuning the chosen network is accomplished by translating the antenna impedance to a 50-ohm R_p circle using the series load coil and phasing the remaining reactance using the shunt coil. As can be seen from the R/X plane, this network has a limited tuning region and is chosen to prevent extreme high voltage on more than one element. The lossless design equations are presented to aid in estimating coil sizes before a full-scale computer analysis is undertaken.

Shunt coil reactance:

$$X_2 = \sqrt{\frac{2500}{50 - R_{\text{ant}}}}$$

Series coil reactance:

$$X_1 = X_{\text{ant}} - \sqrt{-R_{\text{ant}}^2 + 50 R_{\text{ant}}}$$

where $Q_{\text{ant}} = X_{\text{ant}}/R_{\text{ant}}$
R_p = parallel equivalent antenna resistance
R_i = desired input tune impedance

Band 2 (4 to 20 MHz). For 4 to 20 MHz, the series load coil used in band 1 is switched out. The shunt coil is now located at the antenna and becomes the loading element. To make the network tune all of the impedances, the general tuning circuit shown in Figure 15.4A is chosen. The circuit shown in Figure 15.4B would make an equally good network if it were not for the existing shunt coil availability from band 1.

Tuning is accomplished by matching the antenna to 12.5 ohms and then using a transformer to step up to 50 ohms. A capacitor in series with the antenna would make it possible to tune directly to 50 ohms. This capacitor can be positioned so that the no-tune areas can be avoided. The control logic required is complicated and to keep it simple a transformer which requires no tuning algorithm is used. The lossless ladder network design equations are as follows.

Shunt coil reactance:

$$X_2 = \frac{-X_{\text{ant}} - R_{\text{ant}} \sqrt{R_{\text{ant}}\left(\dfrac{Q_{\text{ant}}^2 + 1}{R_l}\right) - 1}}{1 - R_{\text{ant}}/R_l}$$

Series capacitor reactance:

$$X_{C1} = \sqrt{R_p R_l - R_l^2}$$

where $Q_{\text{ant}} = \dfrac{X_{\text{ant}}}{R_{\text{ant}}}$

R_p = parallel equivalent antenna resistance
R_l = desired input tune impedance

Band 3 (20 to 30 MHz). The tuning network used for band 3 is the same as for band 2 except that the transformer is switched out of the circuit. The antenna will fall in the tune region of 50 ohms. The important design parameter is the minimum value of the series C. If the maximum parallel equivalent resistance of the antenna is less than 900 ohms at 30 MHz, the value calculated from the above equation is

$$X_{C1} = \sqrt{(900)(50) - 2500} = 206$$
$$C_1 = 25.7 \text{ pF}$$

The reader may ask, "Why not tune to 12.5 ohms in band 3 as was done in band 2?" The answer lies in the required values of L and C. Tuning to 12.5 ohms requires a very small shunt L and a very large series C. By removing the transformer, the required range of inductance and capacitance is reduced. The exact reduction is of course determined by the frequency at which the transformer is switched out of the circuit.

Control techniques. Once the RF network has been chosen that meets the requirements, the problem remains to automatically control the element values to achieve a match at all frequencies. This can be done either with digital or analog control circuitry. Both types require some way to know where, on the R/X plane, the transformed antenna impedance is located. The centerpiece of all control schemes is the impedance measuring device. This section will focus primarily on two types. The first is the low-cost conventional HF discriminator which has been used for many years. The second is a tracking impedance measuring system (TIMS) [3].

The HF discriminator is usually used as part of an analog tuning system. This consists of continuously variable RF elements, analog servo amplifiers, an HF discriminator, and logic to turn the servos on and off. The discriminator provides two variable dc voltage output error signals. The phasing output is proportional to the reactive component of impedance. A positive voltage indicates inductive and a negative voltage indicates capacitive. The loading output is a voltage proportional to the magnitude of impedance, and positive voltage indicates an impedance magnitude greater than the desired input value and a negative voltage indicates less. When both outputs are zero, a tune condition exists.

The TIMS is usually used as part of digital tuning system. This system consists of digital variable RF elements, TIMS, and a computer. The TIMS provides an accurate digital representation of the impedance being measured. This allows complicated mathematical network computation to be done by the computer. Direct setting of the digital RF elements is possible without the time-consuming servo tuning associated with the HF discriminator.

Figure 15.9 is a simplified schematic diagram of a loading/phasing dis-criminator. The loading portion consists of a current sample (T_1) and a volt-age sample (C_1, C_2). The current sample formed by T_1 coupling to the RF line produces a voltage across R_1 that is directly proportional to the line current. This voltage is peak detected by CR_1 and capacitor C_3. The voltage sample formed by C_1 and C_2 produces a voltage that is proportional to the line voltage. This voltage sample is peak detected by CR_2 and C_4. Operation of the loading error detector is such that the output error voltage at the cathode of CR_2 is the sum of the two dc peak-detected sample voltages. The sample voltages are chosen to be exactly equal in magnitude and opposite in polarity when the impedance on the RF line is equal to the desired value. When the RF impedance is less than the desired value, the current-sample detected dc voltage increases and the voltage-sample detected dc voltage decreases. This produces a net potential difference on the loading error output. In the circuit shown a negative error signal is produced. When the

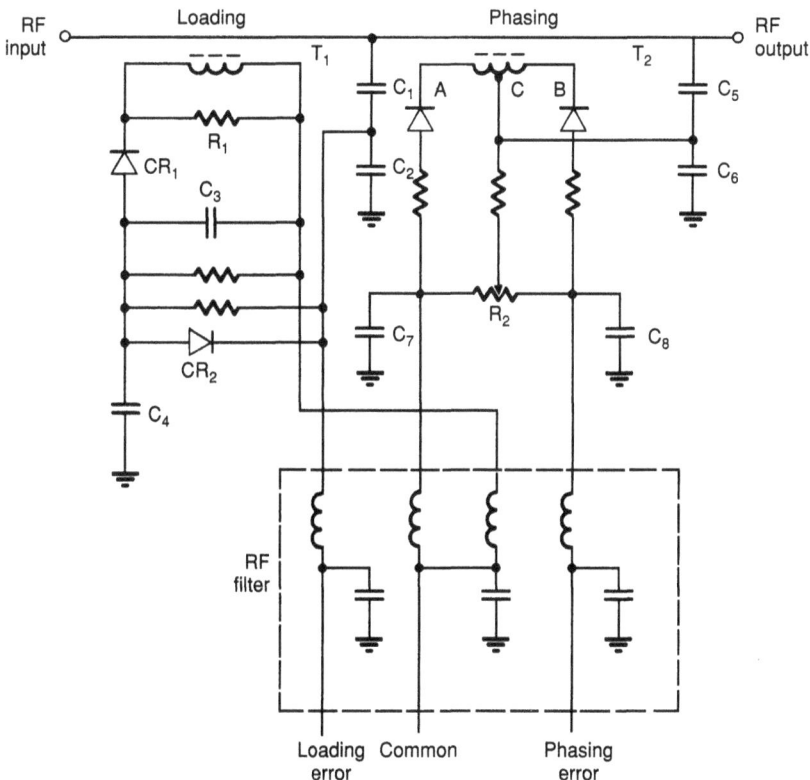

Figure 15.9 Block diagram, HF discriminator.

impedance is greater than the desired value, the exact inverse occurs. Note that zero error signal occurs when the magnitude of the line impedance equals the magnitude of the desired value, not of the real component. This circuit is an impedance magnitude detector.

The phasing circuit is formed by current sample (T_2A, T_2B) and voltage samples (C_5, C_6). With respect to the center tap of T_2, two voltages are produced that are $180°$ out of phase with each other and $90°$ out of phase with the line current. The phase shift with the line current is a result of very light loading of the transformer. The transformer center tap is then connected to the voltage sample. This causes the RF voltage at points A and B to be the vector sum of the voltage sample and the respective current samples:

$$V_a = KI\underline{/+90 + \theta} + K_1 V\underline{/0}$$
$$V_b = KI\underline{/-90 + \theta} + K_1 V\underline{/0}$$

where K and K_1 = sample ratio constants
$\quad I \qquad$ = line current
$\quad V \qquad$ = line voltage
$\quad \theta \qquad$ = phase angle between I and V

When the impedance on the RF line is at zero phase angle, the line voltage and current are in phase $(\theta = 0)$. This means the magnitudes of the sample circuit voltages are equal. When the phase angle of the line impedance is positive $(\theta < 0)$, the magnitude of voltage V_b is greater than voltage V_a. When the phase angle of the line impedance is negative $(\theta > 0)$, the magnitude of V_a is greater than V_b. The circuit then compares the magnitudes of each of the voltage sums. This is done by peak detection using the respective diodes and capacitors. The dc voltage comparison is accomplished by resistor R_2. The output phasing error signal is the voltage across R_2. A positive impedance phase angle results in a positive error signal and a negative angle in a negative error signal. Zero phase angle results in zero error.

The control circuitry that makes use of the loading and phasing discriminator consists of zero crossover detectors to convert the analog error signals to digital. This determines the area on the R/X plane where the impedance is located. Logical decisions can then be made as to which RF element must be changed. The discriminator can be used with either analog or digital elements. In the case of analog elements, the error voltages are fed to a servo motor that changes the variable capacitor or inductor. When used with digital elements, a binary search or some other type of high-speed method is used to position the elements to the correct values. When compared to the TIMS, the loading/phasing discriminator is best suited to analog applications. It does not accurately represent impedances at any other points than zero phasing or proper loading. It is not possible to take full advantage of the speed of digital RF technology without accurate magnitude and phase information.

The detailed design of the TIMS is not within the scope of this chapter; however, an overview will be given here. The TIMS is a true impedance measuring device providing very accurate measurement of the magnitude and phase angle of the voltage coefficient. These measurements are performed under normal operating RF power levels. Figure 15.10 shows the system in block diagram. The basic sensor is a dual directional coupler. The

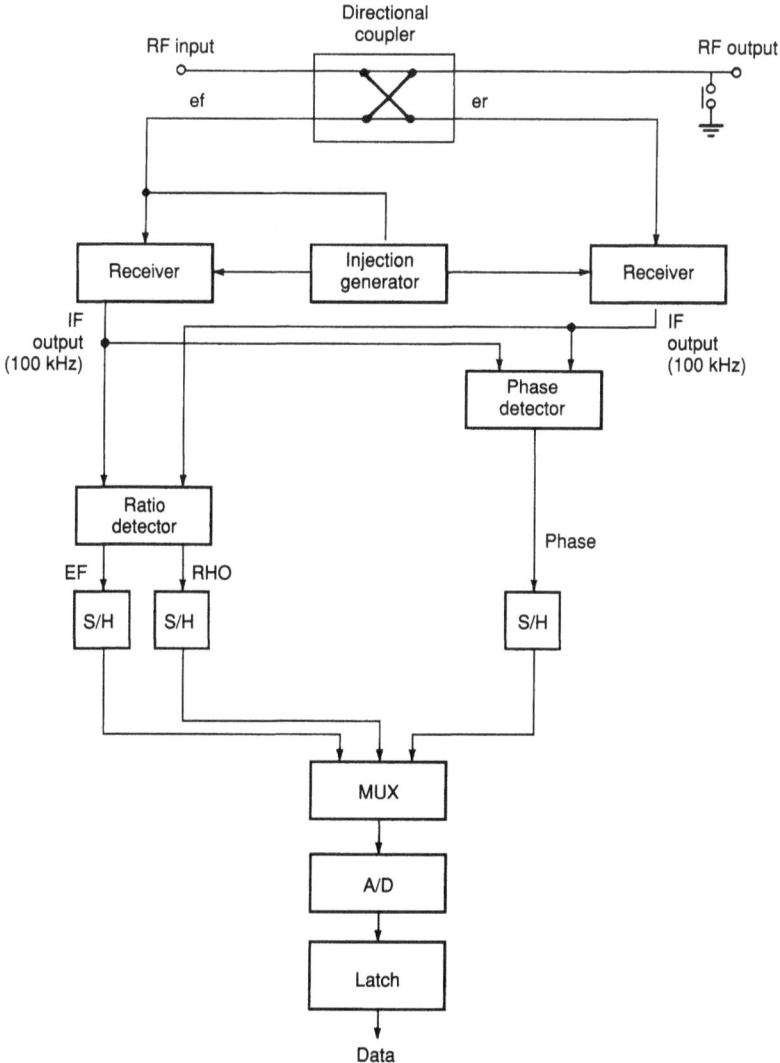

Figure 15.10 Block diagram, tracking impedance measuring system.

directional coupler has RF outputs proportional to the forward and re-flected voltage at its location in the RF path. A shorting switch is provided to calibrate the system and establish the measurement point. Generally the directional coupler is located near the RF elements. A short coaxial cable connects the shorting relay to the directional coupler. Thus, precise location of the measuring point can be accomplished. The remainder of the circuitry determines the phase difference between forward and reflected wave fronts and determines the ratio of reflected to forward voltage. Recall from basic transmission line theory that the ratio of reflected to forward voltage is the magnitude of the reflection coefficient, and the phase difference is the phase angle of the reflection coefficient.

To accurately measure the reflection coefficient, the forward and reflected samples are mixed down to a standard IF of 100 kHz. This is accomplished by using two matched receiver circuits with the IF injection frequency derived directly from the RF input. The downconverted samples are then fed into a phase detector and a ratio detector and converted to dc output samples. The dc outputs are then simultaneously stored in sample-and-hold devices to be processed by the digital circuitry. The digital circuits convert the analog samples to 12-bit information via the multiplexer, A/D converter, and latch. This scheme updates the digital information for magnitude of reflection coefficient and forward power and phase every 50 μs.

One of the greatest advantages of the TIMS is the lack of unwanted signals coupled back onto the RF line. When properly designed receiver circuits are used, the generated noise and spurious signals are well isolated from the RF line. The conventional discriminator couples a considerable amount of noise and spurious signals into the RF path. This may be a problem where multiple receivers and transmitters are located at the same site. The unwanted signals mask the low-level receive signals and interact with other local transmit signals to cause an increase in IM products.

A second advantage of the TIMS can be seen from the application of the exact impedance information made available. When the control for the antenna coupler contains a microprocessor capable of numerical computation, direct calculation of the RF element values is possible. This is done in the same manner that network analysis is done on a mainframe computer. A model of the physical RF network including strays is included in the tune algorithm. The network is analyzed using model impedances and comparisons made with actual impedances measured with the TIMS. In this manner each element in the RF network is positioned. If the computer model is perfect, only one try is required for each element; however, the network strays cannot be measured with sufficient accuracy to achieve this. In practice two or three iterations (impedance measurements) are required. Typically, digital RF networks can be tuned with six iterations. This compares with several hundred required when the conventional discriminator and digital search methods are used. The TIMS dramatically improves the tuning time of a digital antenna coupler.

When the TIMS is used in an antenna coupler containing analog RF elements, the tune time is not greatly improved. This is a result of the nature of the RF elements themselves. Analog RF elements may require several seconds to change values over their entire range. The limiting factor is the speed at which the analog elements can change value. This is not meant to imply the TIMS should not be used in analog systems. Substantial improvement in tune accuracy and reliability is possible.

15.2 Broadband Matching

Broadband matching is a powerful technique that has many design applications. The application of the methods as applied to antenna coupler design will be discussed. Of particular interest is the matching of a complex load impedance to a resistive source. An example of broadband matching a whip antenna, including impedance transformation and resistive padding, follows.

Many authors have presented ways to match impedances over a broad frequency range. One of the first was Fano [4], who showed how to match various combinations of complex or resistive load *and* source impedances. His methods tend to confuse the novice. Cuthbert [5] has taken some of the "black magic" out of the process. To further simplify the application, we will limit the discussion to one type of broadband matching, a complex load to a resistive source. This may seem restrictive, but it will serve as an introduction to the subject and will be useful in many practical antenna matching applications.

Two personal computer programs are supplied with this book. They will enable the user to perform the matching process. CHART.EXE determines the number of reactive elements required, using a graph, and FANO.EXE calculates the elements. Chapter 17 gives detailed instructions for using the programs.

Fano derived the conditions necessary to obtain the optimum broadband match. Levy [6] presented a Chebyshev approximation for these conditions. It can be shown that the maximum reflection coefficient is

$$\rho_{max} = \frac{\cosh{(nB)}}{\cosh{(nA)}}$$

where n is the number of matching elements, and the parameters A and B are found by solving the simultaneous equation set:

a) $\sinh{(A)} - \sinh{(B)} - 2\delta \sin{(\pi/2n)} = 0$

b) $\dfrac{\tanh{(nA)}}{\cosh{(A)}} - \dfrac{\tanh{(nB)}}{\cosh{(B)}} = 0$

where

$$\delta = \frac{Q_{BW}}{Q_{LOAD}}$$

$$Q_{BW} = \frac{F_{high} - F_{low}}{\sqrt{F_{high}F_{low}}}$$

F_{high} and F_{low} are the matching frequencies and $Q_{LOAD} = \dfrac{X_{load}}{R_{load}}$ at $\sqrt{F_{high}F_{low}}$.

Levy then presents the normalized lowpass "g" parameters that are needed to construct the circuit:

$$g_0 = 1$$

$$g_1 = \frac{2\sin\left(\dfrac{\pi}{2n}\right)}{\sinh A - \sinh B}$$

$$g_{i+1} = \frac{4\sin\left[(2i-1)\left(\dfrac{\pi}{2n}\right)\right]\sin\left[(2i+1)\left(\dfrac{\pi}{2n}\right)\right]}{\left[\sinh^2 A + \sinh^2 B + \sin^2\left(\dfrac{i\pi}{n}\right) - 2\sinh A \cdot \sinh B \cdot \cos\left(\dfrac{i\pi}{n}\right)\right]}$$

$$g_{n+1} = \frac{2}{g_n} \cdot \frac{\sin\left(\dfrac{\pi}{2n}\right)}{\sinh A + \sinh B}$$

where $i = 1, \ldots, (n-1)$

A lowpass-to-bandpass transformation is then performed to get the desired circuit. The FANO program calculates the parameters. We then determine the number of poles needed to obtain the desired match. The chart in Figure 15.11 is a plot of decrement (δ) versus input VSWR. From this n can be determined. For $n = 1$ no matching elements are used. For $n = 2$ the load is resonated at the geometric center frequency of the match. As n becomes larger we reach a point where there is little benefit in increasing the circuit complexity. A practical limit is reached when $n = 4$. The CHART program duplicates this plot and allows editing and printing.

Once the Fano matching circuit has been realized in bandpass form, we then deal with the dependent Fano source resistance. This resistance is

VSWR VS decrement

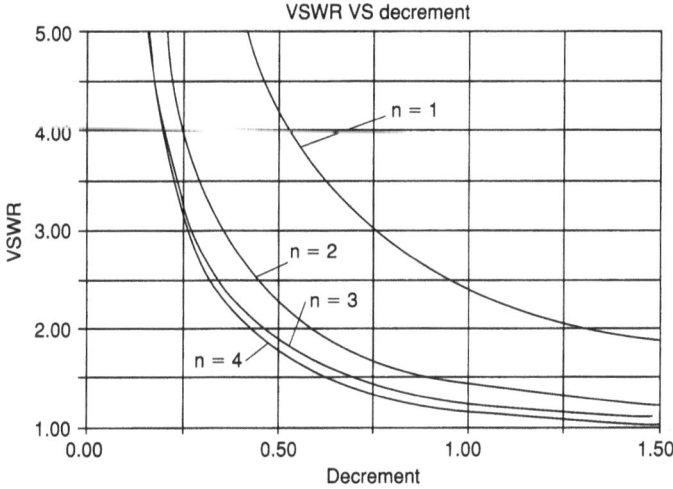

Figure 15.11 VSWR versus decrement for the Fano match.

g_{n+1} times the load resistance. Unfortunately it usually is not the one that is needed. Generally, the impedance match is to be done with some predetermined characteristic impedance in mind. In some cases it may be possible to use a simple transformer to obtain the proper match. However, practical transformers have integer type of transformer ratios (but see Chapter 12) and can sometimes generate distortion when magnetic material is used. There are two practical ways to transform the Fano impedance to the desired characteristic impedance. One is a narrowband approximation: the inverter. The other is a broadband lumped equivalent to a transformer: the Norton transformation.

The inverter [7] has the unique characteristic that the input is always shifted 90° from the output and the input impedance is equal to $Z_{in} = K^2/Z_{out}$ for an impedance inverter and $Y_{in} = J^2/Y_{out}$ for the admittance inverter. A quarter-wave transmission line is an example of an inverter. There are other valuable forms, as shown in Figure 15.12. To illustrate the use of an inverter, refer to Figure 15.13. Inserting an inverter ahead of the shunt elements will 1) transform the input impedance, and 2) the two shunt elements become series elements due to the nature of the inverter. Now the negative inductance of the inverter can be absorbed into the transformed inductor. We see that by adding a single element, the Fano resistance can be transformed to the desired characteristic impedance.

Another way to convert the Fano resistance is to use the Norton transformation [8]. This technique inserts a lumped element transformer into the circuit with a turns ratio equal to the square root of the impedance ratio. As

Admittance inverter J = ωL, 1/ωC

Impedance inverter K = ωL, 1/ωC

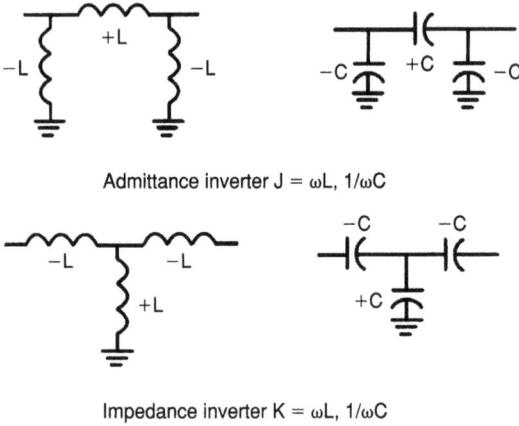

Figure 15.12 Impedance inverters and admittance inverters.

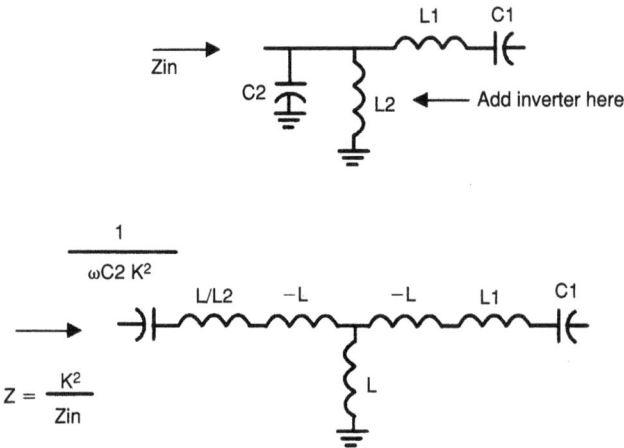

Figure 15.13 Illustrating the use of an impedance inverter.

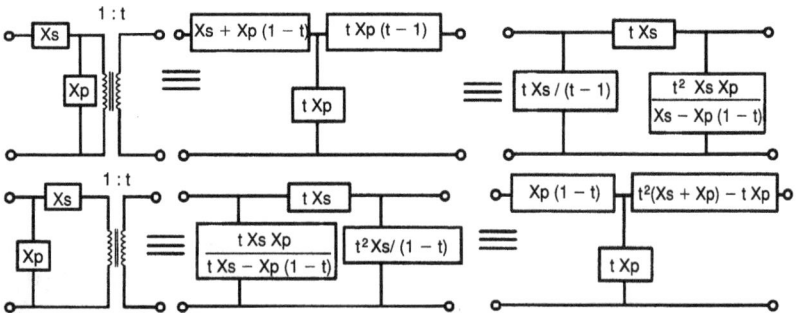

Figure 15.14 Illustrating the use of the Norton transformer.

can be seen from Figure 15.14, two reactances of the Fano circuit can be absorbed into the transformer. With the addition of two elements and by changing the values of two elements the desired result can be obtained. If these reactances are both L's or C's, and if the transform is realized with the same types, the Norton transform becomes broadband and does not affect the matching quality of the Fano network.

An important point is that over a wide band, such as 2.0 to 30 MHz, the broadband matching network for a short (for example, aircraft) antenna must be recalculated frequently. This means that all the elements of the broadband matching network must be servo-tunable. It is important to minimize the number and total volume of these expensive elements. The J or K inverter approach may therefore be more economical than the Norton approach, despite its narrower bandwidth.

Resistive pad matching

There are applications that require a close impedance match which may not be achievable with reactive matching methods. Modern broadband solid-state power amplifiers are sensitive to load impedance. If the load VSWR is greater than 3:1 the output power can approach zero. (See Chapter 12 for more information.) There is a trade-off between power amplifier turndown and power dissipation in a resistive T-pad that achieves an impedance match. This pad has two R1s in series and one R2 in shunt.

The power loss ratio N depends on the pad input SWR (S) and the load SWR (K):

$$N = \frac{(S + 1)(K - 1)}{(S - 1)(K + 1)}$$

The values of R2 and R1 are

$$R2 = R_0 \frac{2\sqrt{N}}{N - 1} \text{ and } R1 = R_0 \frac{N + 1}{N - 1} - R2$$

Antenna matching example

To illustrate the methods we use the example of a 35-ft whip for the frequency range 2.0 to 2.5 MHz. We want the input VSWR to be 1.5:1. On the assumption that this circuit is part of a tunable antenna coupler, only an $n = 2$ Fano match is allowed. Minimum complexity is necessary because each automatically tuned variable RF element is large and very expensive. Inspection of the antenna impedance indicates that it can be represented approximately by a series resistor and capacitor over this frequency range. Entering this data into the FANO program results in the following:

L2 = 1.607 µH	C2 = 3.1515E-10 F	Shunt
L1 = 31.83 µH	C1 = Antenna Reactance	Series

SWR ranges from 7.55 to 10.05:1
Fano Resistance = 100.2 ohms

Transform the Fano impedance to the desired 50-ohm characteristic impedance using an inverter with inductors:

L3 = 10.79 µH	C3 = 3.199E-10F	Series
L2 = 5.045 µH		Shunt
L1 = 26.78 µH	C1 = Antenna Reactance	Series

Since no additional circuit complexity is allowed, the VSWR of 10:1 must be controlled using resistive matching:

$S = 1.5{:}1$ $K = 10.01{:}1$

The power ratio is calculated along with the resistor values of the T-pad:

$N = 4.091$ or a 6.12-dB pad
R2 = 64.45 ohms
R1 = 82.36

A 100-W antenna coupler has been constructed using switched fixed pads of 1 dB, 2 dB, and 3 dB. This coupler also included two digital inductors and two digital capacitors with sufficient range to tune the antenna in either a broadband or conventional narrowband mode. Tuning the broadband matching network requires use of the TIM with impedance measurements at the low, middle, and high frequencies. The tune algorithm uses the calculated parameters from the above example as a starting point for optimization. This results in a final network that accounts for Q losses, as well as strays, and all assumptions about the impedance of the antenna.

References

1. Philip H. Smith, "Charts for L-Type Impedance-Transforming Circuits," in J. Markus and Vin Zeluff (eds.), *Electronics for Engineers*, pp. 272–278 (New York: McGraw-Hill, 1945).

2. A. Zverev, *Handbook of Filter Synthesis*, pp. 499–504 (New York: John Wiley & Sons, 1967).

3. H. L. Landt, "Tracking Impedance Measuring System," U.S. Patent 4,506,209 (March 1985) (assigned to Rockwell Corp.).

4. R. M. Fano, "Theoretical limitations on the broadband matching of arbitrary impedances," *J. Franklin Inst.* (February 1950): 139–154.

5. Thomas R. Cuthbert, Jr., *Circuit Design Using Personal Computers* (New York: John Wiley & Sons, 1983). (See also "Broadband Impedance Matching Methods," *RF Design* (August 1994): 64–70.)

6. R. Levy, "Explicit formulas for Chebyshev impedance matching networks, filters and interstages," *IEE* 6 (1964): 1099–1106.

7. G. L. Matthaei, L. Young, E. M. T. Jones, *Microwave Filters Impedance-Matching Networks and Coupling Structures* (New York: McGraw-Hill, 1964).

8. DeVerl S. Humphreys, *The Analysis, Design, and Synthesis of Electrical Filters* (New York: Prentice-Hall, 1970).

16

Receiver Measurements and EMI Techniques

David H. Church

16.1 Measurements and Design

Performance measurements are an important and necessary part of the design and development of SSB receivers. Since theoretical predictions and analyses are often difficult to achieve, measurements ensure conformance to the design specifications. This chapter discusses many of the measurement techniques used in designing and testing SSB receivers.

16.2 Receiver Measurements

Many of the performance characteristics of receivers are related to two fundamental parameters. These are the internally generated noise and the nonlinearity of the transfer characteristics. Whereas past design emphasis has been on low-noise front ends for the reception of weak signals, today's spectral environment of large transmitter and jammer signals has focused attention on intermodulation (IM) products which may mask small desired signals.

Noise figure

To analyze the noise performance of a receiver (see references 1 and 2), the receiver may be thought of as a set of two-port networks (or two-ports, for short) in cascade. These two-port networks are filters, amplifiers, mixers, and attenuators. In addition to its desired function of amplification or frequency translation, each two-port generates internal noise and distortion that contaminates the desired signal output.

A simple resistor is a one-port network which may be analyzed as a voltage source (thermal noise) in series with source resistance. The thermal noise source voltage, calculated from noise-power spectral density, is given by

$$V_{\text{noise}}^2 = nB = 4kTBR \ [\text{rms noise voltage}^2] \qquad (16.1)$$

where n = noise power density, -174 dBm in a 1-Hz bandwidth
k = Boltzmann's constant, 1.37×10^{-23} J/K
T = absolute temperature, K (293 K normal ambient)
B = bandwidth, Hz
R = source resistance, ohms

The two-port network includes noise sources associated with thermal resistances and junction noise found in diodes and transistors. To describe the effect of internal noise sources, the concept of *noise factor* has been developed. A model of a two-port network with the signal and noise sources is shown in Figure 16.1.

Noise factor is defined as the ratio of the input signal-to-noise ratio to the output signal-to-noise ratio:

$$\text{Noise factor} = \frac{S_{\text{in}}/N_{\text{in}}}{S_{\text{out}}/N_{\text{out}}} \qquad (16.2)$$

$$= \frac{(V_{\text{in}}^2/4\,R_{\text{gen}})/kTB}{(S_{\text{out}}/N_{\text{out}})}$$

$$= \frac{V_{\text{in}}^2/4kTBR_{\text{gen}}}{S_{\text{out}}/N_{\text{out}}}$$

Further,

$$\text{Noise figure} = 10 \log (\text{noise factor}) \qquad (16.3)$$

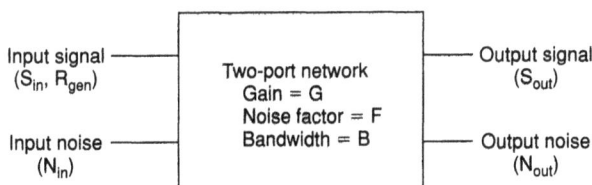

Figure 16.1 Signal and noise analysis of a two-port network.

One method of measuring the noise factor of a receiver is derived from Equation 16.2. If the effective noise bandwidth of the receiver is known, the noise factor may be calculated from the output signal-to-noise ratio resulting from a given input signal level (V_{in}). The effective noise bandwidth of a receiver is a single rectangular bandwidth that would pass as much white-noise power as the actual receiver bandwidth. As an approximation, the 3-dB bandwidth of the receiver may be used as the effective noise bandwidth. This approximation is good if the filter shape factor is small (less than 2:1), which is true of an SSB receiver.

> **Example:** Given a sensitivity measurement of 0.7 μV (open circuit from a 50-ohm generator) for a 10-dB signal plus noise-to-noise ratio (9.5 dB signal-to-noise ratio) in a 2750-Hz bandwidth:
>
> | | |
> |---|---|
> | Signal level | −116.1 dBm |
> | Signal to noise | 9.5 dB |
> | Noise level in 2750-Hz bandwidth | −125.6 dBm |
> | Correction factor for 1-Hz bandwidth | 34.4 dB |
> | Noise level in 1-Hz bandwidth | −160.0 dBm |
> | Thermal noise in 1-Hz bandwidth, 50 ohms | −174.0 dBm |
> | Noise figure of receiver | 14.0 dB |

A preferred method of determining the noise figure of a receiver uses a calibrated noise source. It may be characterized by its excess noise ratio (ENR), which is the ratio of excess noise plus thermal noise (noise source turned on) to thermal noise (noise source turned off). If the noise source is calibrated over frequency (ENR versus frequency), it can be used to quickly measure the noise figure at many frequencies.

The calibrated noise source can be used in the test configuration shown in Figure 16.2. If the noise source has been characterized by its ENR, the Y-method can determine the noise factor of the receiver directly from the value of the ENR. The parameter Y is defined as the ratio of the output noise with the noise source turned on to the output noise with the noise source turned off. The noise factor F is then determined from

$$F = \frac{ENR}{Y - 1} \qquad (16.4)$$

If the noise output doubles (3 dB) when the noise source is turned on then $Y = 2$ and the noise factor of the receiver (plus the noise measuring equipment) is numerically equal to the ENR. Further simplification in the measurement is achieved by using the 3-dB pad to maintain a constant

Figure 16.2 Test configuration for noise figure measurement.

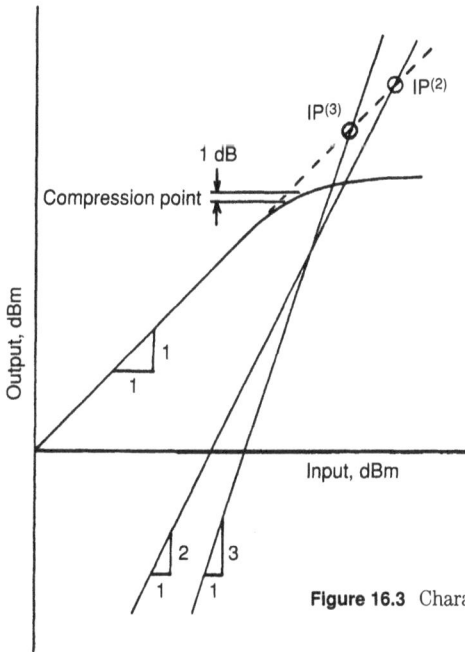

Figure 16.3 Characteristics of IMD.

noise level into the measurement equipment. This results in the following relationship:

$$\text{Noise figure} = \text{ENR} + 1/G_R \tag{16.5}$$

where G_R is the gain of the receiver. The receiver gain is usually very large, thus the noise figure is essentially equal to ENR.

Analysis of the input noise figure of a cascaded system shows the importance of a low noise figure in the first stages. Chapter 4 of this book and reference 1 provide a discussion of noise performance in a cascaded system.

Intermodulation

Two types of IM performance are important in the operation of a receiver: in-band and out-of-band. In-band IM performance is a measure of IM products produced by two tones within the passband of the receiver [3]. These tones may be desired signals, undesired signals, or a combination of both. In-band IM products are produced by nonlinearities within the amplifier stages and filters or by gain modulation from the AGC.

The transfer characteristic shown in Figure 16.3 illustrates the basic parameters used to describe nonlinearities in circuits. Intercept points are the theoretical input levels (extrapolated from actual IM data) at which the IM products are of equal power with the input signals. The compression point describes the point at which the actual and theoretical transfer characteristics differ by 1 dB. Gain compression is often caused by the strong undesired signal affecting the amplifier's bias points and in turn reducing the overall gain. The points of intersection with the fundamental transfer curve are defined as the second-order intercept point and the third-order intercept point. A high intercept point means greater linearity in the transfer characteristic of the front end, resulting in less degradation of the receiver's strong signal performance.

To accurately measure IM performance, a two-tone source must be available that does not generate its own IM products. A typical setup is shown in Figure 16.4. The attenuator is placed between the RF translator and the IF amplifier to prevent the IM product from driving the IF into overload. The AGC is usually turned off. The narrowband filters reduce the harmonics of the source frequencies, eliminating their possible contributions to the IM measurement. In addition, the filters isolate the two generators and thereby eliminate the "back door" phenomenon, by which IM is generated in the output amplifier of each signal generator. Isolation between the two signal sources is improved by the pad networks and the isolation of the hybrid signal combiner. The combiner may be a hybrid transformer, a schematic of which is shown in Figure 16.5A, or a simpler resistive combiner, shown in Figure 16.5B. Chapter 12 provides additional information on hybrid combiners. The transformer provides approximately 20 dB of additional isolation between the signal generators. The resistive combiner has 6 dB of isolation between any two ports and maintains a 50-ohm interface, as does the hybrid transformer.

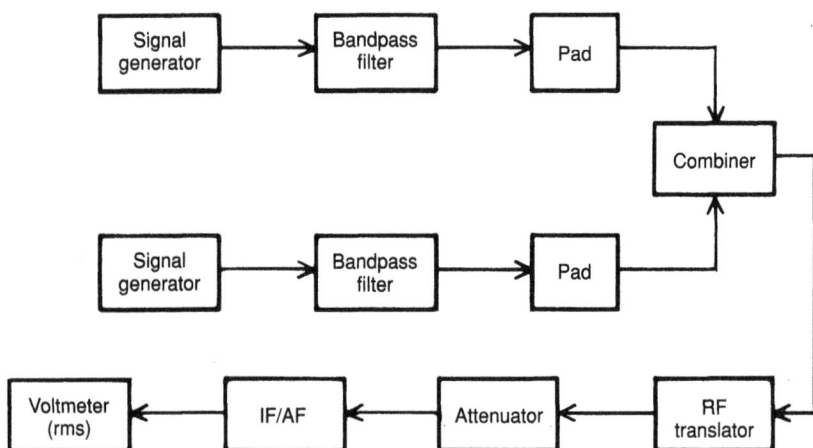

Figure 16.4 Test setup for IMD.

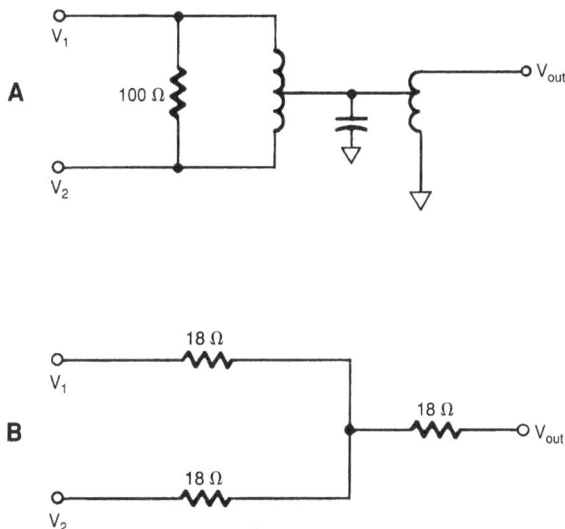

Figure 16.5 RF signal combiner.

For out-of-band IM measurements, the frequencies F_1 and F_2 are outside the receiver's passband, yet will produce IM products that fall in band. As the two tones move further apart, IM improves because of selectivity that may be within the translator. It is always desirable to plot IM versus frequency of separation.

The IM products are related to the out-of-band signal frequencies as follows:

$$\text{Second-order products} = F_1 \pm F_2 \qquad (16.6)$$

$$\text{Third-order products} = 2F_1 \pm F_2 \text{ and } F_1 \pm 2F_2 \qquad (16.7)$$

By determining the level of the IM products in terms of equivalent input signal power, the out-of-band intercept points may be calculated:

$$\text{IP}^{(2)} = \text{second-order intercept point}$$
$$= 2.0 \, (\text{undesired signal}) - 1.0 \, (\text{equivalent product}) \qquad (16.8)$$
$$\text{IP}^{(3)} = \text{third-order intercept point}$$
$$= 1.5 \, (\text{undesired signal}) - 0.5 \, (\text{equivalent product}) \qquad (16.9)$$

The signal levels are given in terms of dBm.

Example: Two out-of-band tones at –10 dBm/tone produce a third-order IM product with an equivalent input level of –80 dBm. So we have

$$\text{IP}^{(3)} = 1.5 \, (-10 \text{ dBm}) - 0.5 \, (-80 \text{ dBm}) = +25 \text{ dBm}$$

Adjacent channel IM is a special case of the out-of-band distortion measurement. The frequencies F_1 and F_2 are chosen such that one of the tones is 30 to 50 kHz from the tuned frequency of the receiver. Typically this IM is produced in the first-mixer network and the first-IF filter. When measuring adjacent channel IM, precautions must be taken to prevent the noise skirts of the signal generators from masking the true IM performance. The effects of reciprocal mixing are discussed later in this chapter.

In-band IM measurements will use frequencies F_1 and F_2, which are within the receiver's passband. The two desired in-band tones are typically chosen to produce audio tones of 900 and 1100 Hz. The IM performance is easily measured with a spectrum analyzer or wave analyzer at the audio output. Intermodulation measured at the audio output is indicative of the total receiver's performance.

Cross-modulation is a type of IMD that is also caused by nonlinearities of the mixers, filters, and amplifiers. This distortion is a measure of the amount of modulation transferred from a strong, undesired, modulated carrier to a desired, weaker, unmodulated carrier at the tuned frequency when both signals are present at the antenna input.

The third-order curvature of an amplifier's transfer characteristic is the major factor in producing cross-modulation products and also third-order IM products. Since third-order curvature is common to both, a relationship may be established between these two distortions such that predictions of cross-modulation performance can be made from IM test results, or vice versa [4, 5].

If a strong undesired AM signal and a desired weaker, unmodulated carrier signal are present, a ratio between the transferred modulation appearing on the desired signal and the modulation of the undesired signal can be established. Similarly, a two-equal-tone signal results in a ratio of the IM products to each of the two tones.

These undesired-to-desired ratios may be expressed as
AM:

$$\frac{3U^2 A_3}{A_1}$$

Two tone:

$$\tfrac{3}{4} D^2 \frac{A_3}{A_1}$$

where U = amplitude of undesired signal
D = amplitude of each fundamental tone
A_1, A_3 = coefficients of transfer function polynomial

Since the coefficients of the transfer function are difficult to measure, they may be eliminated from the final ratio (Equation 16.10) if the peak signal in each test case is equal. This stipulation implies that amplitude D is equal to $0.65U$ when the undesired signal U is 30% modulated. That is, equal peak envelope signals ($2D = 1.3U$) are used to compare IM and cross-modulation. Figure 16.6 shows this relationship.

To compare cross-modulation with two-tone IM, the ratio of the AM and two-tone IM may be calculated:

$$\frac{\text{AM cross-modulation ratio}}{\text{Two-tone intermodulation ratio}} = \frac{3U^2 \left(A_3/A_1\right)}{\tfrac{3}{4}D^2 \left(A_3/A_1\right)} \qquad (16.10)$$

$$= 4\frac{U^2}{D^2}$$

$$= 9.47$$

where $D = 0.65U$ for 30% modulation.

Since the original ratios were defined as undesired signal to desired signal, a larger ratio indicates poorer performance. The final ratio given above shows that for equal peak input signal levels, the AM cross-modulation ratio is 9.47 (19.5 dB) worse than the two-tone IM test ratio. For example, if a two-tone IM test yields IM products 40 dB below the two tones, the cross-modulation test, with equal peak signals, will result in a –20-dB cross-modulation ratio.

Measurement of cross-modulation distortion uses a test procedure similar to that for out-of-band IM. The difference is in the AM of the undesired

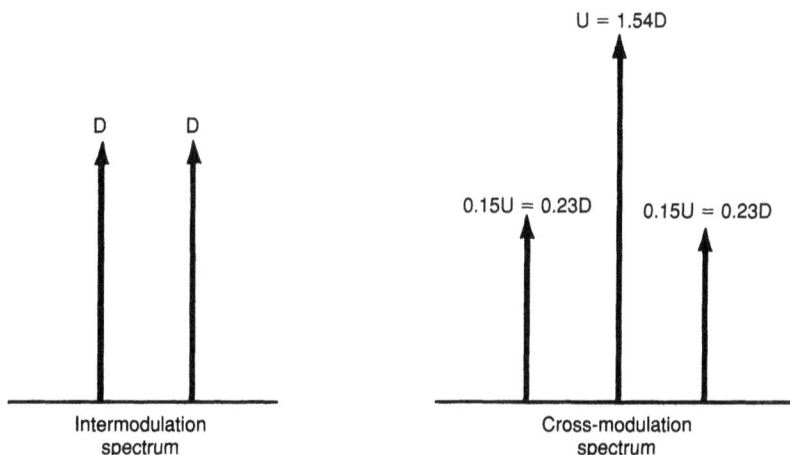

Figure 16.6 Comparison of two-tone and AM signals. The peak signals are equal. Modulation is 30%.

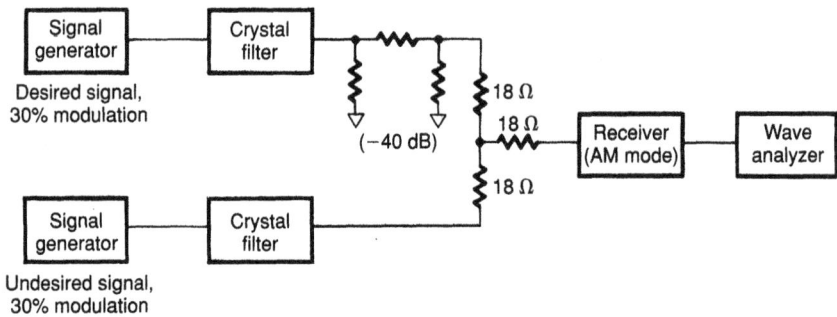

Figure 16.7 Measurement procedure for cross-modulation.

signal and the measurement of the relative difference between the desired and undesired carrier signals. A test setup is shown in Figure 16.7.

A reference measurement is made by modulating the desired signal (1000 Hz at 30% modulation) and measuring the audio output level with the wave analyzer. The modulation of the desired carrier is then turned off. Modulation is then applied to the undesired carrier (1000 Hz, 30% modulation). The modulation frequency is monitored at the audio output with the wave analyzer. Test limits are the undesired carrier level required to produce the specified ratio of undesired to desired audio output. Cross-modulation is usually specified as 10- or 20-dB cross-modulation. That is, the audio caused by the interfering signal is 10 to 20 dB below the modulation caused by the desired modulated signal.

The gain control circuitry located at the front end may contribute to the IM performance. If PIN diodes are used as attenuators, the IM tests should be performed at different bias points of the PIN diodes. The most critical point is the turn-on threshold of the diodes.

Harmonic distortion (audio)

Harmonic distortion can be easily measured with a distortion analyzer (such as the Hewlett-Packard 334) or equivalent test equipment. A known good RF signal is applied to the receiver to produce the desired audio output. The applied signal level must be strong enough to prevent the internal receiver noise from masking the distortion products.

The distortion analyzer uses an internal notch filter to remove the desired signal, allowing the distortion products, hum and noise, to be measured. Total harmonic distortion (THD) is defined as the ratio between the level of the desired signal and the rms level of all the distortion products.

The notch filter used in the distortion analyzer may affect the measurement of THD. In most applications the effect of the notch filter is ignored if the receiver bandwidth is greater than 2 kHz. If narrow bandwidths are used, the THD measurement must be corrected by calculating the noise re-moved by the notch filter.

For applications where the audio signals are to be transmitted over a phone line (or its equivalent), a THD of less than 1% (–40 dB) is desirable. Headphone and speaker outputs do not require the same high-quality re-production, and 3 to 5% THD performance is acceptable.

Ultimate signal-to-noise ratio

The ultimate signal-to-noise ratio is a measurement of the maximum achievable signal-to-noise ratio that may be obtained from a receiver. The sensitivity of a receiver is determined by the noise figure of the receiver and its noise bandwidth. At signal levels 40 dB and greater above the sensitivity level, the signal-to-noise level will be limited by the in-band phase noise of the oscillator injections to the RF mixers and by the internal noise of the de-tector and amplifier circuits.

The graph in Figure 16.8 shows a typical response for output signal-to-noise ratio versus the input signal level. As the close-in noise of the mixer injection begins to predominate, the curve no longer approximates a linear characteristic. Further increases in the input signal level will not increase the output signal-to-noise ratio. This is the ultimate signal-to-noise ratio. Figure 16.9 shows the close-in noise of the first injection and the resultant receiver output spectrum.

The test setup shown in Figure 16.10 may be used to measure the ulti-mate signal-to-noise ratio of an SSB receiver. For each CW signal level ap-plied to the receiver, a reference level is set on the distortion analyzer. The internal notch filter is then tuned to remove the desired audio signal. The remaining noise is the in-band noise plus hum and distortion prod-ucts. By measuring this in-band signal-to-noise ratio for various input sig-

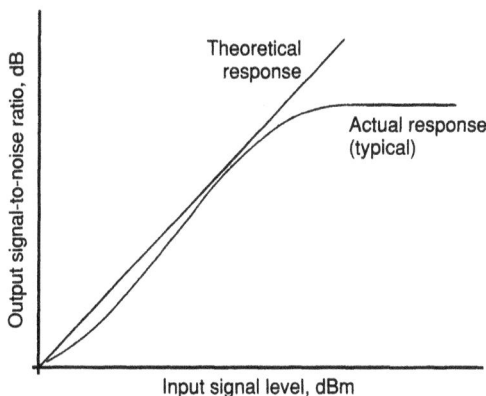

Figure 16.8 Characteristics of ultimate signal-to-noise ratio.

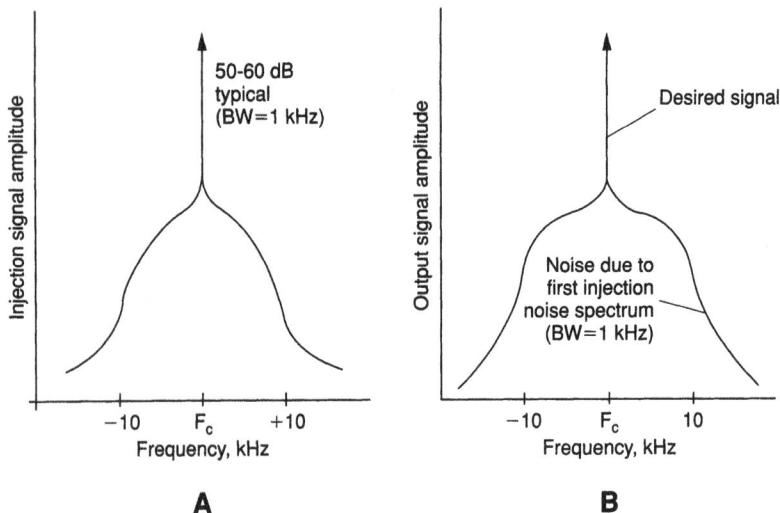

Figure 16.9 Spectral analysis of ultimate signal-to-noise ratio: (A) first injection synthesizer spectrum; (B) output signal spectrum.

nal levels, the graph shown in Figure 16.8 may be obtained. If the noise contains hum and distortion products, these products may be measured using a wave analyzer and their effect subtracted from the measured signal-to-noise ratio.

Since the receiver SSB bandwidth is much greater than the bandwidth of the notch filter used in the distortion analyzer, the amount of in-band noise removed by the notch may be disregarded. The AGC of the receiver is enabled for this test to prevent overload of the various amplifier circuits.

Image and IF rejection

Measurement of the image and IF rejection determines the relative strength of undesired signals at the image frequency and the intermediate frequencies required to produce a reference response at the audio output. A test setup similar to that shown in Figure 16.10 may be used. A reference level is obtained by applying an on-channel signal to produce a 10-dB $(S + N)/N$ ratio at the audio output. The signal generator is then tuned to the image

Figure 16.10 Measurement of ultimate signal-to-noise ratio.

and intermediate frequencies. The amount of signal increase required to obtain the equivalent reference output is the image and IF rejection. Typical rejection levels range from 80 to greater than 100 dB. The optional bandpass filter may be necessary to eliminate noise skirts and spurious responses of the signal generator.

A general discussion of image responses is provided in Chapter 4.

Reciprocal mixing

Performance of the receiver in the presence of strong adjacent channel signals is affected by the noise skirts of the injection signal to the first mixer. These noise skirts can mix with adjacent channel signals to produce noise signals at the IFs. This type of distortion is called *desensitization, reciprocal mixing,* or *noise modulation.*

The mechanism of reciprocal mixing is illustrated in Figure 16.11. An undesired signal located ΔF away from the desired tuned frequency will mix with the noise skirts of the injection signal to produce noise signals at the IFs. In a signal environment containing strong undesired signals located adjacent to the desired signals, the injection noise mixed into the IFs will reduce the signal-to-noise ratio of the desired signal, thus desensitizing the receiver to small signals. Refer to Chapter 2 for a discussion of collocated interference and to Chapter 4 for further discussion of reciprocal mixing.

The level of the noise signals mixed into the IFs depends on the level of the noise skirts (phase noise) relative to the level of the injection signal. A

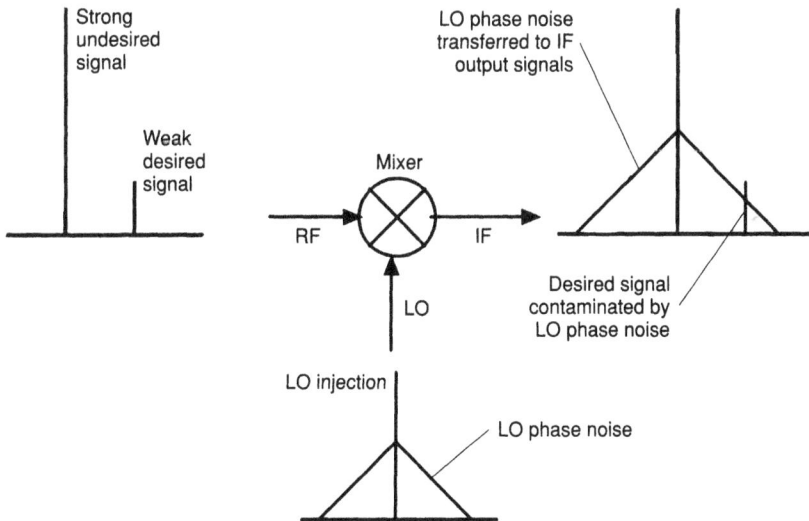

Figure 16.11 Spectral analysis of reciprocal mixing.

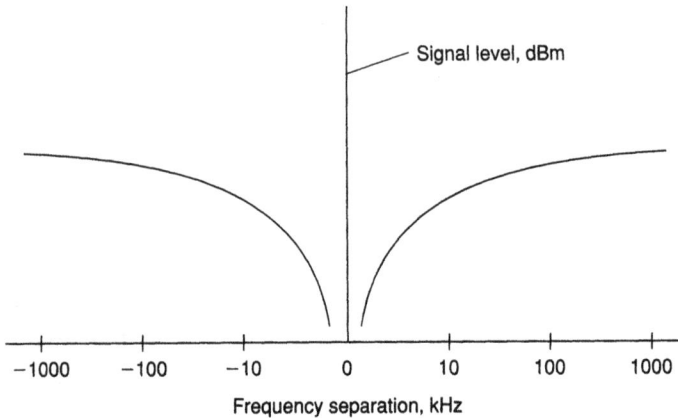

Figure 16.12 Typical reciprocal mixing characteristic. The signal level is that which will produce a 3-dB increase in receiver noise level.

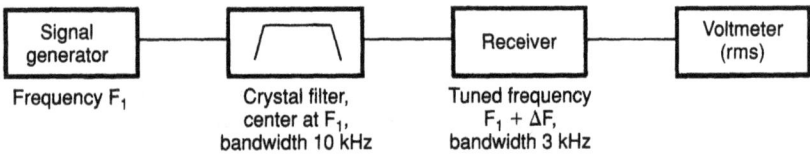

Figure 16.13 Test setup for reciprocal mixing.

plot of the undesired signal level (at ΔF from the tuned frequency) required to produce a 3-dB increase in receiver noise output characterizes the recip-rocal mixing performance of the receiver. Figure 16.12 shows a typical reciprocal mixing response.

It is important to prevent the noise skirts of the undesired signal from affecting the measurements of reciprocal mixing. The test setup shown in Figure 16.13 will produce accurate measurements. The crystal filter located after the signal generator removes the noise skirts of the signal generator. Because the signal generator and crystal filter are at fixed frequencies, the tuned frequency of the receiver is varied by ΔF from the applied signal. For a given tuned frequency, a reference level is measured for a 3-dB signal-to-noise ratio. This equivalent input signal is the noise floor of the receiver. The receiver is now tuned away from the generator frequency. By increasing the signal level until the noise output increases by 3 dB, a relative measurement of the noise skirts of the first injection signal is obtained. This data may be plotted to obtain the typical curve described above (Figure 16.12).

The reciprocal mixing data may be used to determine the effect of adjacent undesired signals on the reception of a desired signal. The following example illustrates this concept.

Example: Given:

1. A desired signal of –107 dBm produces a +20-dB signal-to-noise ratio.
2. An undesired signal of 0 dBm is located 200 kHz from the desired signal.
3. The reciprocal mixing level at 200 kHz is –7 dBm for 3-dB noise increase.

Since the undesired signal is 7 dB above the reciprocal mixing level at 200 kHz away, the undesired signal will produce a 10-dB increase in the noise level of the receiver. This 10-dB noise increase will decrease the normal +20-dB signal-to-noise ratio to about 10 dB.

If the strong undesired signal is close to the weaker desired signal, the desired signal can be totally masked by the reciprocal mixing noise.

Spurious responses

Internally generated spurious responses, sometimes called "birdies" or "tweets," result from injection and other signals leaking between receiver sections to produce on-channel signals. The greater the isolation between the first and second mixers, the less likely that spurious responses will be generated by injection frequencies leaking between the mixer circuits. A second source of spurious responses is discrete signals at the IFs or signal frequencies that are generated internal to the frequency synthesizer. These spurious signals are dependent upon the frequency generation scheme and the tuned frequency of the receiver.

One source of internal spurious signals previously mentioned is the harmonics or fundamentals of the first or second injections leaking into the second or first mixer, respectively. Two examples of these spurious signals are illustrated in Figure 16.14. The isolation provided by electrical shielding around each mixer and the isolation provided by the IF filters will usually reduce and possibly eliminate spurious responses due to the injection frequencies.

Internal spurious responses can be characterized by an equivalent input signal. When a spurious response is detected, the audio output is measured as a reference. The receiver is then tuned off-frequency by 10 kHz and an external signal applied to produce an output equal to the reference level.

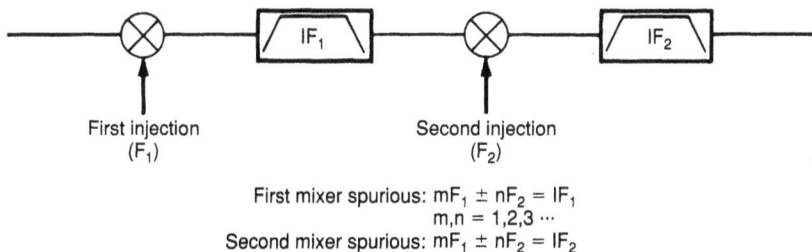

First injection
(F_1)

Second injection
(F_2)

First mixer spurious: $mF_1 \pm nF_2 = IF_1$
$m,n = 1,2,3 \cdots$
Second mixer spurious: $mF_1 \pm nF_2 = IF_2$

Figure 16.14 Internally generated spurious signals.

The strength of the applied signal is the equivalent signal level of the internal spurious response. The AGC network is typically disabled when measuring the equivalent level of internal spurious responses. If the spurious response is of a level to require AGC, the measured equivalent input must be corrected for the AGC attenuation applied.

Oscillator signals or digital clock signals that have frequencies within the operating frequency range of the receiver may also cause spurious responses. These frequencies must be checked during the design and test phases of any receiver development.

Spurious responses may also be the result of crossover points between the PLLs used to generate the LO injection. These crossover points are caused by signal leakage between the PLLs generating discrete signals at or near the IFs of the loop.

If these spurious responses fall in-band to the RF translator (that is, first or second IF) they are classified as internal spurious responses. Those crossover frequencies that do not fall in-band to the translator may produce an in-band response when mixed with a strong out-of-band signal (external spurious responses).

As an example of external spurious responses caused by undesired discrete signals superimposed on the LO injection see Figure 16.15. These discrete signals are typically the synthesizer's loop reference signals appearing as sideband signals on the LO injection. The levels of these discrete signals are measured by a test procedure similar to that used to characterize the spectral response of the LO (see material on reciprocal mixing). External spurious responses are specified in terms of decibels below an on-frequency reference signal.

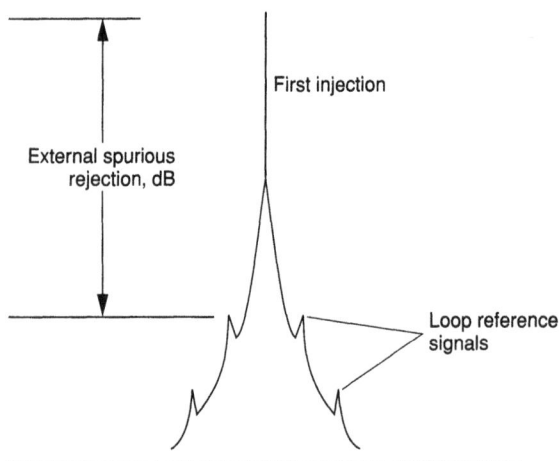

Figure 16.15 Source of externally generated responses.

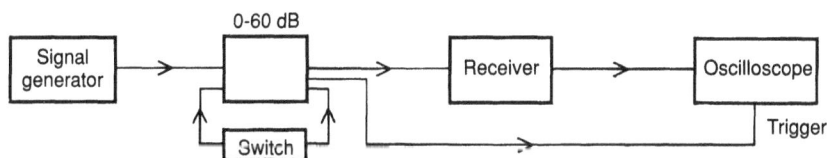

Figure 16.16 Measurement of AGC time constants.

Another source of interference in mixers is harmonic IMD, where harmonics of the RF signal frequency combine with harmonics of the LO frequency to produce outputs that fall in the IF passband. The frequency pairs that produce these spurs can be predicted by using the computer program SPUR.EXE in Chapter 17. They can be tested in production by tuning the receiver and the signal generator to certain specific frequencies and measuring the signal generator level that produces some reference audio output level, say 10 dB above the receiver noise level.

Automatic gain control

An AGC network typically uses a peak detector to determine the strength of the received signal and thus the attenuation required to maintain a constant peak output level. The attack and decay time constants determine how quickly a receiver reacts to changes in the input signal level. To prevent overload of the RF and IF amplifiers when a strong signal is applied, the attack time of the AGC circuit must be relatively fast. The decay time constants determine how fast the RF attenuation is removed when the input signal changes from large to small amplitude.

The test setup shown in Figure 16.16 may be used to measure the attack and decay time constants of a receiver. To simulate the change in signal strength, a 60-dB attenuator is alternately applied to the signal path. Typically the time constants are defined as the time required to reach and stay within 3 dB of the final output level. These definitions are illustrated in Figure 16.17.

Since the AGC circuit is a closed-loop control circuit, careful design must be maintained to prevent instabilities in the loop response. These instabilities can cause the attack characteristic to have an underdamped response. Filter networks within the AGC loop can cause instabilities at discrete frequencies because of differential delay distortion within the filter's passband, especially near the edges of the passband. When examining the characteristics of the AGC network, the stability of the loop should be measured at various frequencies within the passband.

The control range of the AGC circuit and the AGC loop gain will determine the amount of rise allowed in the audio output for a given increase in the input signal. In an analog AGC circuit, a certain amount of rise is neces-

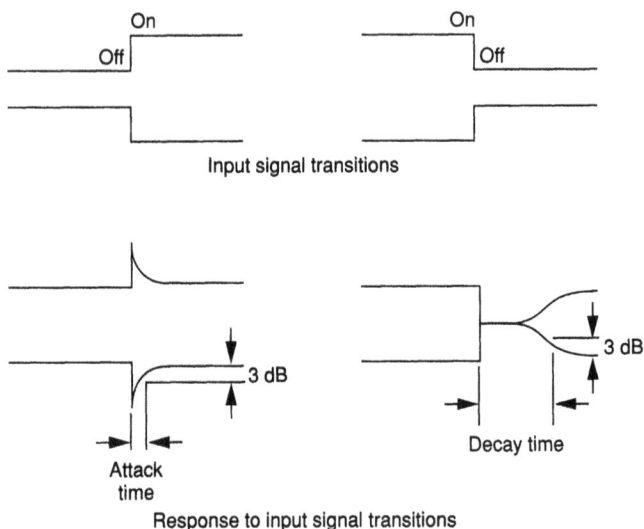

Figure 16.17 Automatic gain control time constants.

sary because of trade-offs between response characteristics and loop stability. In the newer digital receivers, the audio rise is minimized by the use of digital signal processing. Refer to Chapter 8 for further discussion of digital signal processing. The concept of audio rise is illustrated in Figure 16.18.

Blocking

The effects of blocking are due to strong adjacent channel signals that may be present with the received signal. As the signal level of the undesired signal is increased, a level is reached where the input RF amplifiers and mixers be-

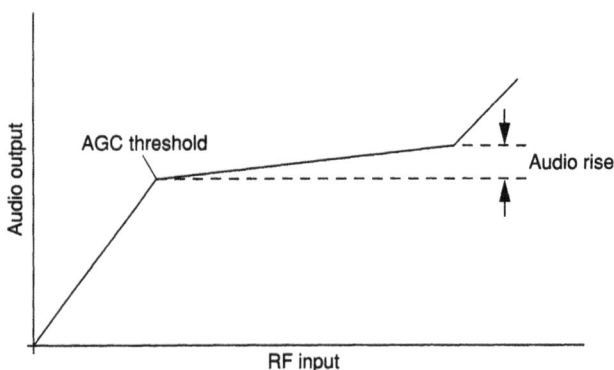

Figure 16.18 Definition of audio rise.

come nonlinear. When this happens, the receiver's gain at the tuned frequency is reduced, that is, the desired signal is blocked by the undesired signal.

When measuring the blocking characteristics, precautions must be taken to prevent reciprocal mixing from degrading the measurement (that is, greater than 500-kHz separation). It may be necessary to use a bandpass filter for each LO to remove the noise skirts of the oscillator signal. A typical measurement test setup is shown in Figure 16.19. The filter located at the output of the "undesired" signal generator prevents the noise skirts of the generator from desensitizing the test procedure.

The frequency of the first signal generator provides an on-channel signal to the receiver. A desired signal level is chosen to provide a 10- to 20-dB ($S + N$)/N ratio at the output. Since the AGC is turned off during this test, a desired signal level is chosen such that the receiver's amplifiers are not overloaded by the desired signal. The level of the adjacent channel signal (greater than 500 kHz from the desired signal) is increased until the signal level of the audio output is decreased by 3 dB. This is known as *3-dB blocking*. The signal levels, frequencies, and gain reductions mentioned here are typical of those used by the communications industry to characterize receiver performance.

Selectivity

The selectivity of a network response is often measured by the 3-dB (half-power frequencies) and the 60-dB attenuation frequencies. The ratio of the 60-dB bandwidth to the 3-dB bandwidth is termed the *shape factor* of the filter.

A second characteristic of the passband is the response ripple. For most applications, the ripple will not exceed 3 dB. In special data transmission filters, the passband ripple will not be more than 1 dB.

Filter selectivity is measured using a test setup similar to that shown in Figure 16.10. The filter under test is inside the receiver. A signal within the passband of the receiver is varied to obtain the maximum output response. Using this point as a reference, the frequency of the input signal or the re-

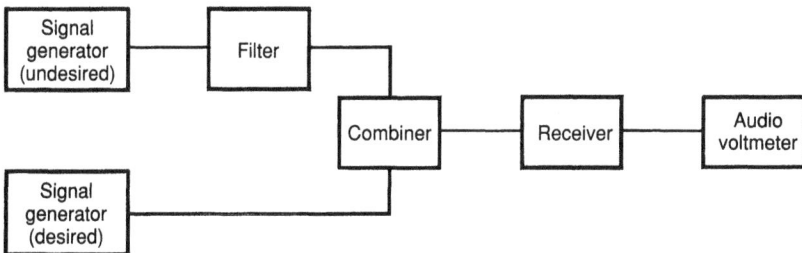

Figure 16.19 Test setup for measurement of blocking characteristics.

ceiver tune frequency is varied until the output response is 3 dB below the reference level. To measure the 60-dB attenuation frequencies, the input signal is increased by 60 dB. The input frequency or the receiver tuning is then varied to obtain the reference output level. For each attenuation point (3 dB, 60 dB) both the upper and lower response frequencies must be measured.

When measuring the selectivity, the applied signal level must not overload the input amplifiers. Automatic gain control is usually turned off to keep the gain control circuits from correcting the passband response. The signal generator must have a low noise level.

Phase-frequency distortion

An ideal receiver will produce a constant time delay for all frequency components with the desired signal. This constant time delay would be produced by a linear phase-frequency relationship. In reality, the phase-frequency response of a receiver is not linear, resulting in phase distortion or time-delay distortion.

Although voice communication can tolerate limited phase distortion of its signal, high-speed data signals rely on precise phase relationships to transfer information. Delay equalization may be added to crystal and mechanical filters to approximate a linear phase-frequency relationship. Developments in the application of digital signal processing for filtering have produced filters with nearly linear phase-frequency relationships.

The time-delay characteristics of a receiver are composed of differential delay and absolute delay. Differential delay measures the variation in delay with respect to a given frequency within the receiver's passband. The total time required for a signal to pass through the receiver is termed the absolute delay.

If the phase-frequency response of the receiver is known, the delay may be calculated as the slope of the phase response at a given frequency. Since the phase-frequency response of a receiver is difficult to measure, the envelope delay can be calculated from the measurements of the delay encountered by a narrowband modulated signal applied to the receiver. The test setup shown in Figure 16.20 is used to measure the envelope delay within the passband of a receiver.

The signal generator is offset from the receiver's tuned frequency to measure the phase difference at each point in the receiver's passband. Typically the offset frequencies range from 200 to 3000 Hz with an increment of 100 Hz. For the test setup shown in Figure 16.20, a phase difference of 1° represents 100 μs of envelope time delay. This measurement is not the true absolute delay of the receiver because of the time delay introduced by the lowpass filter. Data from this measurement can be used to derive the differential delay measurement between adjacent frequencies or with reference to a given frequency in the passband. The response curve shown in Figure 16.21 is typical of an SSB filter.

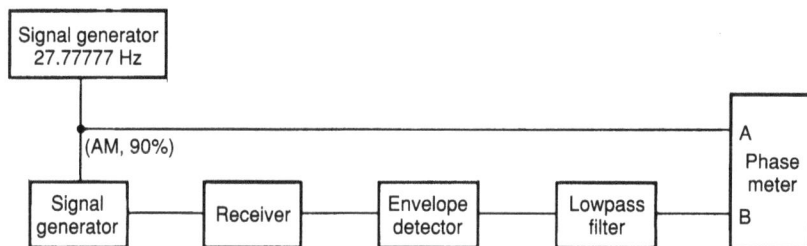

$$\text{Period, T} = \frac{1}{27.77777 \text{ Hz}} = 36 \text{ ms}$$

$$1° = \frac{1}{360} T = 100 \text{ µs}$$

$$\text{Delay (µs)} = (\theta_B - \theta_A) \left(\frac{100 \text{ µs}}{1°} \right)$$

Figure 16.20 Measurement of envelope delay.

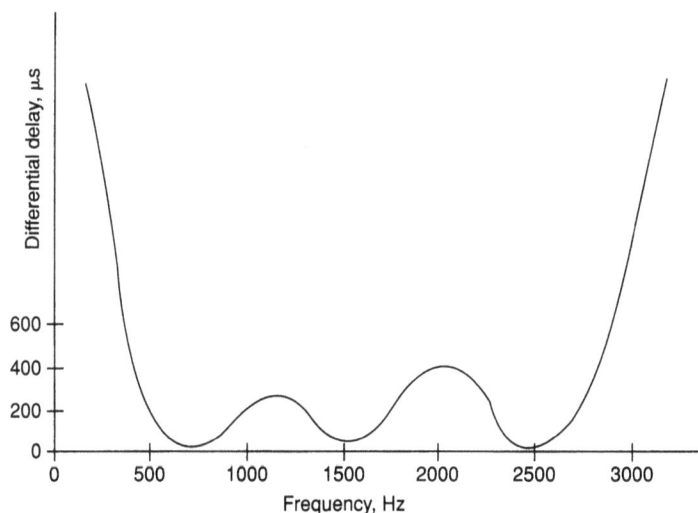

Figure 16.21 Envelope-delay response.

Measurement of the absolute time delay may be obtained using the test setup shown in Figure 16.22. The time difference between the trigger signal initiating the RF pulse signals and the appearance of the output signal is the absolute delay of that frequency. Although various definitions are used to define this time difference, a typical definition is the time required for the signal to be within 3 dB (70%) of its final amplitude. By varying the input signal across the passband of the receiver, the maximum delay time may be measured.

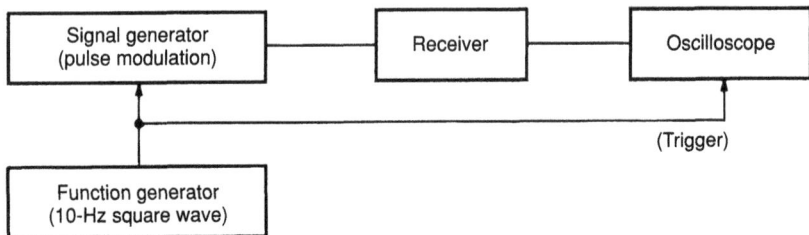

Figure 16.22 Absolute time-delay measurement.

The delay characteristics of a receiver are specified in three parts:

1. Differential delay with respect to a reference frequency within the pass-band

2. Maximum change in differential delay for any 100-Hz increment

3. Maximum absolute delay within the passband

Each of these specifications may be measured using the test setups shown in Figures 16.20 and 16.22.

16.3 EMI Considerations in Receiver and Exciter Design

The importance of designing electronic equipment that is neither suscep-tible to nor a radiator of electromagnetic interference (EMI) has been de-monstrated many times in the history of communications and navigation equipment. Commercial aircraft restrict the operation of portable elec-tronic equipment to prevent interference with the navigation equipment of the aircraft. Extraneous electromagnetic radiation can unsuspectingly pro-vide a homing beacon for weapons systems. Transmitters operating in the HF regions have been known to interfere with TV reception and public ad-dress systems.

The information provided in this section has been gleaned from actual experience in the design and development of HF communication equip-ment and also from the EMI literature.

Overview

The analysis of EMI must begin with the identification of the radiation sources (radiator) and/or the susceptible circuits (receiver). Once identi-fied, design modifications can be implemented to reduce or possibly elimi-nate the source of the interference and the susceptibility of the circuits. Three areas of considerations are presented:

1. Interfaces between functional circuits
2. Layout of the functional circuits
3. Fences and shielding components

The block diagram given in Figure 16.23 shows a generic design of a modern receiver/transmitter. This diagram will be used to illustrate EMI design considerations.

The block diagram consists of four major functional circuits. A double conversion translator is used to translate between the antenna (receive and transmit) and the IF. The IF sampling is then performed by the digital signal processor (DSP) segment. The synthesizer generates the variable injection to the first mixer, the fixed injection to the second mixer, and the clock references required by the DSP circuits. The control circuitry is often digital, therefore also a source of HF noise.

Interfaces between functional circuits

Any degradation caused by interference between circuit areas may be considered as EMI. In a receiver translator, the leakage of injection harmonics between the first and second mixer can generate internal spurious signals that inhibit operation on the affected channels. The generation of undesired transmit responses from the mixing of the injection harmonics degrades the desired signal and may also "jam" equipment operating at the spurious frequency. Electromagnetic interference in the synthesizer circuits may produce spurious responses on the variable and fixed injections that can reduce receive performance in the presence of strong adjacent or out-of-band signals or produce broadband emissions in the transmit mode.

The DSP circuits, by their nature, produce a broad spectrum of digital noise. The major components of this spectrum depend on the clock speeds of the signal processors, the sample clocks for the A/D and D/A converters, and the switching speeds of the supporting circuits. As the processing power increases, so does the frequency and the harmonic content. The energy of this EMI spectrum must be isolated from the sensitive translator and synthesizer circuits.

The isolation of low-level (microvolt-level) circuits from the DSP circuits (+5-V switching signals) is no simple task. The isolation required to prevent the switching signals from affecting the reception of microvolt-level signals is 140 to 160 dB (equivalent to the free space radiation loss between the earth and the moon). Physical separation of the circuits is impractical, therefore EMI reduction and isolation techniques must be applied to the design of the functional circuits and the system integration.

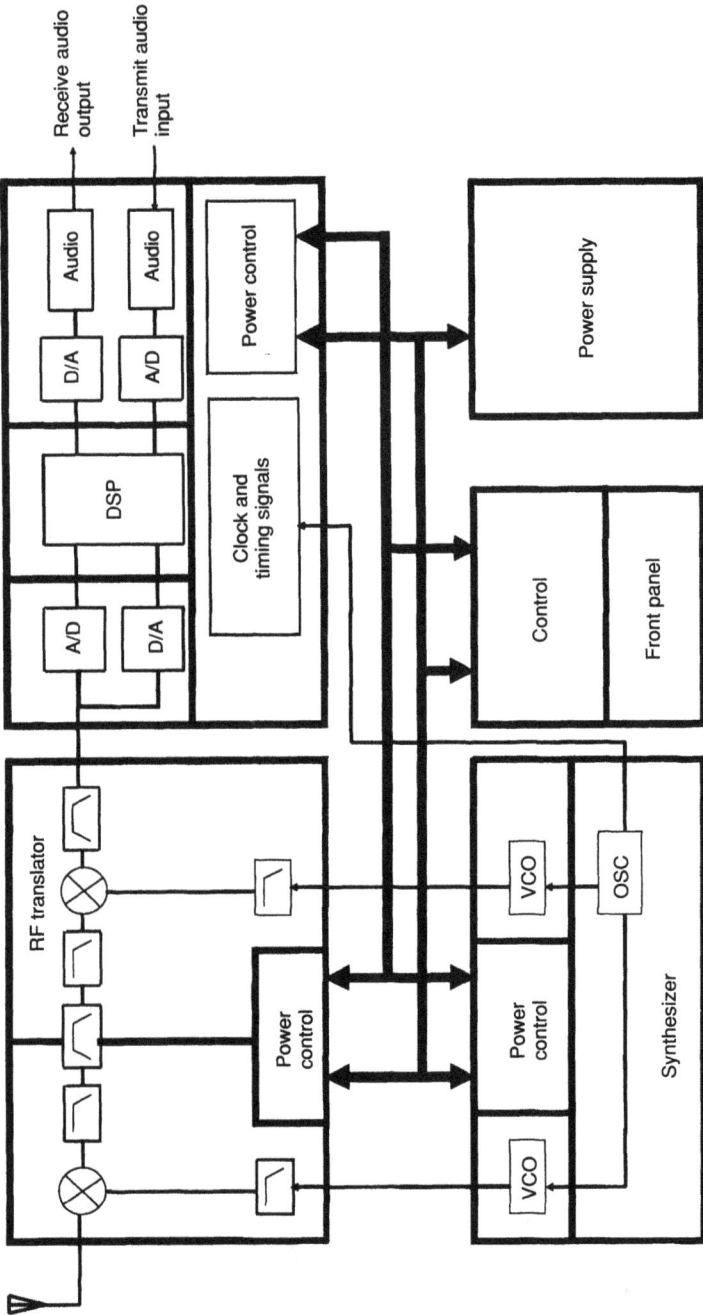

Figure 16.23 Block diagram of a "generic" SSB receiver or exciter that uses DSP and other high-speed digital circuits.

Static and slow-speed interfaces

Control signals between functional circuit areas must be filtered to prevent the conduction of EMI between the circuit areas. Static or slow-speed switching lines are easily filtered by RC or LC networks. The RC network works well for low-current paths where the voltage drop is negligible or will not impact the control function. The LC network is better suited for high-current paths where minimal impact on the primary signal is desired. When using LC filters, ensure that the resonant frequency of the filter does not inadvertently couple EMI *around* the filter. The required filter response should be matched to the expected or measured EMI spectrum carried on the control lines.

High-speed clocks and control signals

Coupling between high-speed clock lines and control lines, as they travel between functional circuits, should be avoided. Interference superimposed on the clock and/or control lines is very difficult to filter from the harmonic-rich waveforms of the desired signal without distorting the desired signal. If the current (energy content) requirements of the clock signal can be reduced, the radiation effects will also be reduced. Shielding and physical isolation of high-speed clocks and control lines provide the best EMI protection.

One successful method of passing high-frequency clock signals involves the creation of the clock from a high-frequency sine wave. The sine wave may then be filtered as it passes between the functional circuits. Narrow filters (crystal bandpass filters or discrete resonators) or lowpass filters, if the EMI is above the sine-wave frequency, may be used to prevent the conduction of EMI. If the frequency of a particular interferer is known, a notch filter may be helpful. This approach is not economical for multiple interferers.

Lumped element line filters

Discrete filters may be used on the interconnects between functional circuits or on interconnects to external equipment. The T-pad or L-pad filter provides a better rejection of EMI because of the series impedance presented to the undesired signals. The pi filter can be used, but must be implemented carefully to prevent EMI from being conducted around the series element by means of the ground path between the input capacitor and the output capacitor.

Multipin connectors are available that contain feedthrough capacitors or pi filters on each pin of the connector. The attenuation characteristics of the filter are limited by the physical constraints of the connector. These filters are typically good for frequencies between 50 MHz and several hundred megahertz. The degree of attenuation will depend on the source and load impedances presented to the filter.

Improving the performance of the IF filter

The isolation between the first and second mixer circuits may be improved by the addition of lowpass filters on the input and output of the IF filter between the circuits. Typically the frequency response of the IF crystal filter will flare back in the stopband regions. This can allow synthesizer spurious products to pass between the mixer circuits. The addition of simple lowpass filters in the signal path can help to maintain the required attenuation. The design of the lowpass filters must not affect the passband or IM performance of the IF filter.

Balanced lines

Balanced lines provide a means for the rejection of common mode signals. The EMI which is coupled onto the balanced lines will cancel at the load.

Return paths

The return path for clock signals and injections is very critical. If the high-order harmonics are forced through the functional circuitry before it returns to the source, spurious signals can be coupled into sensitive circuits. The return paths should be direct and of low impedance to the harmonics of the desired signal.

Circuit layout

The physical layout of the functional circuits is very important in minimizing the effects of EMI. Careful attention to the details of signal flow, parts placement, routing of circuit traces, and grounding will help to isolate the radiators and the sensitive circuits, while providing the most efficient signal flow.

Signal planes, power planes, and ground planes

Multilayer circuit cards are most desirable in the minimization and containment of EMI. A circuit board consisting of two signal planes, a ground plane, and a power plane is a minimum in many of today's operating environments.

Two signal planes will allow the circuit traces on each plane to be orthogonal. This produces the minimum cross-coupling between circuit traces on the two planes and simplifies the interconnect between the functional circuits.

A ground plane will minimize the impedance of return paths between sources and loads. The application of strategically placed cuts in the ground plane can be used to guide the return currents along paths that minimize the cross-coupling into adjacent circuits. Partitioning of the ground plane between analog and digital circuits will reduce the cross-coupling of the broadband digital noise into the sensitive analog circuits.

It is very important that the ground plane consist of large areas of ground plane, not piecemeal grounds or round-about grounds. A central point ground system reduces the common currents of chained grounds. In a chained ground system, the currents of one circuit will flow through the ground path of a second circuit providing a path for the conduction of undesired signals into sensitive circuits.

Power planes are useful for providing low-impedance paths for the power lines. In some applications, it may be possible to combine the power and ground planes on a single circuit layer, while providing an efficient power grid and low-impedance grounds.

Bus bars, which are commercially available for printed circuit cards, are vertically mounted strips of metal that interleave multiple power and ground bars. If the power and ground requirements are not highly critical, bus bars may reduce the need for power and ground planes. In addition, the bus bars increase the stiffness of the circuit boards.

Antenna apertures and bypass capacitors

When locating the elements of a digital circuit, the source and load should be as close to each other as possible. This reduces the effective antenna aperture for radiation and reduces the area of the antenna loop for EMI susceptibility. Bypass capacitors will reduce the antenna aperture of the radiator or the susceptible circuits. The advent of surface mount capacitors allows bypass capacitors to be mounted in close proximity to the source or directly between the power and ground pins with minimum bypass inductance from the circuit traces.

Sensitive circuits

High-speed clock signals or high-power injection signals are potential radiators of EMI. The circuit layouts for these should isolate them from sensitive circuits by as much distance as possible. Additional isolation can be provided by shields and fences. Avoid routing digital signals near A/D inputs, D/A outputs, or near high-gain amplifier networks.

Shields and fences

Shields and fences provide mechanical methods of isolating radiators and sensitive circuits. The design of the shields and fences must take into consideration the type of the radiation (electrical or magnetic) and the frequencies of the signals.

Maximum isolation between circuits can be achieved by mounting each functional circuit in its own shielded box with filtered connectors. This approach to isolation provides the required protection, but it is not the most cost-effective approach in many applications. A good compromise between effectiveness and cost is to isolate the functional circuits with shield fences and covers.

Aluminum is an effective shield to electrical fields and also high-frequency magnetic fields (due to eddy currents). If low-frequency magnetic fields are a problem, steel and special magnetic materials may be required. The shield fences should be mounted to ground traces on the circuit board. This provides the ground potential for the shield. A solder seam between the fence and the ground trace greatly reduces the leakage of high-frequency signals. If tabs are used to mount the shield fences, the following rule-of-thumb provides the spacing between tabs. The minimum distance is one-quarter of the wavelength of 10 times the highest frequency used in the functional circuits. The rule of thumb also applies to the placement of screws used in attaching the circuit card to a shield plate.

Finger stock

An excellent way to provide electrical conductivity for shield fences, covers, and chassis grounds is with finger stock. This material provides multiple points of flexible contact as well as a wiping action to maintain clean contacts, if required. Finger stock comes in many shapes and sizes to meet nearly all applications.

Summary

Electromagnetic interference protection must be considered at the beginning of the design project. The guidelines presented here are the applications of the principles of radiators and receivers at the circuit or system level. The old adage "an ounce of prevention is worth a pound of cure" is applicable to EMI and its prevention. A commonly encountered "economy" is to use a cheap packaging approach that then has to be "bullied" into correct performance, invariably at the very great expense of engineering labor.

References

1. Fitchen and Motchenbacher, *Low-Noise Electronic Design* (New York: John Wiley & Sons, 1973).

2. William A. Rheinfelder, *Design of Low-Noise Transistor Input Circuits* (Hayden, NY: 1964).

3. Wesley Hayward, "Defining and Measuring Receiver Dynamic Range," *QST*, vol. LXIX, July 1975.

4. K. A. Rigoni, *Relationship between AM Cross-Modulation and Third Order Intermodulation Distortion Products*, Collins Radio Co. Tech. Report, August 1961.

5. Harold B. Goldberg, "Predict Intermodulation Distortion from Cross-Modulation Measurements," *Electron. Des.* 18 (May 1970): 76–78.

17

Software for SSB

William E. Sabin

The disk that accompanies this book contains a collection of public domain programs for the IBM-compatible personal computer. These programs will be described in the following sections. They have all been tested over a period of time and are believed to be accurate and bug-free.

17.1 Half-Octave Bandpass Filter Design Program

The program HALFOCT.EXE converts the five-element, all-pole, equally loaded (1-ohm) lowpass prototype filter shown in Figure 17.1A, with a bandwidth of 1.0 rad/sec, into the classical bandpass filter shown in Figure 17.1B, and then into the filter of Figure 17.1C. The program uses the Norton transformation in order to control the values of the inductors. This overcomes a common problem, that these values often vary over a wide range and can be unreasonably large or small for the particular frequency range of the filter. Large values have too much self-capacitance, and small values require very large values of tuning capacitance and tend to have low Q at low frequencies.

For correct operation, the inductor values for L2 and L4 should be the "effective" values, measured at the geometric mean of the passband. These values should not change much due to self-capacitance effects across the passband, therefore coils with large distributed C should be avoided. The self-capacitances of L1, L3, and L5 are part of C1, C6, and C11, respectively. C1, C6, and C11 should therefore be reduced slightly in their hardware values accordingly, and the "true" inductance values of coils L1, L3, and L5 should be the values that the program specifies. This assures that the proper L/C ra-

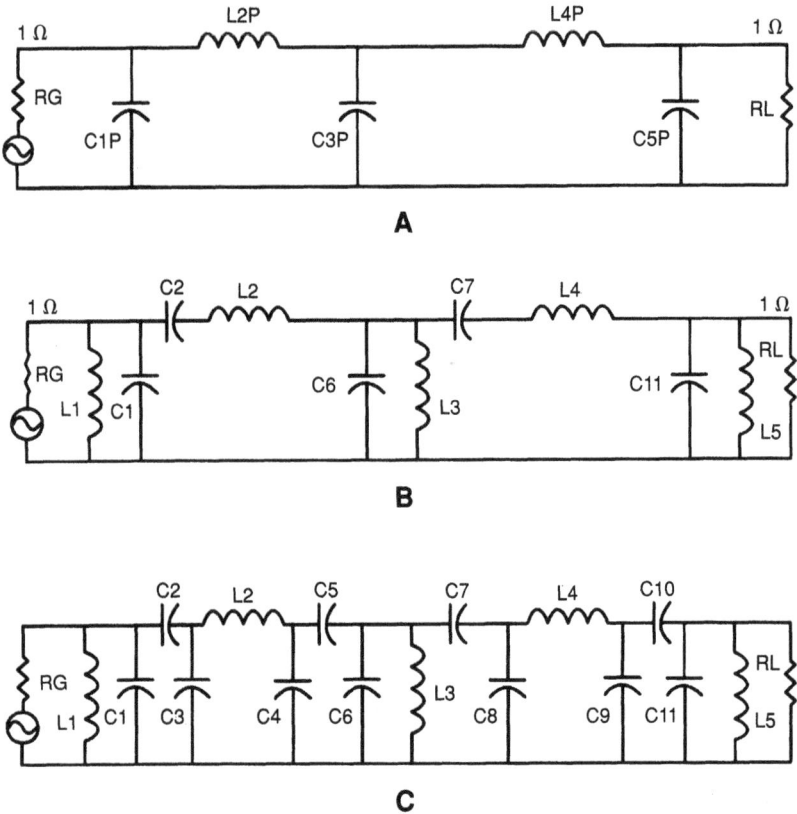

Figure 17.1 The half-octave filter: (A) lowpass prototype; (B) prototype bandpass filter; (C) filter after the Norton transformations.

tios are implemented in the filter. When designing the inductors, make sure that the ferromagnetic cores are not contributing IMD to the system.

If the values of the generator and load resistances are equal, and if the lowpass prototype values are symmetrical around C3P in Figure 17.1A, as they would be for a Butterworth or Chebyshev filter, the program automatically provides five equal values of inductance. This value of generator-load resistance can be varied so that reasonable values of L and C can be realized for the particular frequency range of interest. As R increases, the Ls increase and the Cs decrease.

For the intended preselector application the Butterworth or Chebyshev types would be preferred. In particular, simulations indicate that the Chebyshev has good return loss at the edges of its ripple bandwidth.

Because the filters are wideband filters, wideband impedance matching methods must be used at the input and output. For example, the filter can be matched to the actual generator and load resistances in the equipment by placing taps on L1 and L5, provided that these coils are designed as tightly coupled autotransformers. Transmission line transformers, optimized for the frequency range and with high values of common mode impedance, work quite well also.

If the values of RG and RL (Figure 17.10) are not the same, the program automatically matches these values by assigning inductor values that progress in a constant ratio from input to output. The transformation from RG to RL progresses smoothly along the filter. For example, a 50-ohm antenna could be transformed to 500 ohms at the gate of an FET amplifier.

The automatic mode can be overridden by switching to a manual mode. The user can then fine-tune the four K values to get L and C values that may be more desirable in some sense. These K values are the parameters for the four Norton transformations that take place. It is often possible to get several of the C values close to standard values in this manner, although the process may take some time and experience. It is also possible to get negative (unrealizable) values of C, and these are displayed in red. Very small positive values of C should also be avoided.

If the lowpass prototype is not symmetrical, as for example in Bessel and some other filter types, the design procedure is slightly less straightforward. The usual result is that the five Ls cover a wide range of inductance values. This can be largely offset, for example, by reducing the generator resistance if that end has large L values. The Ls are then more nearly equal. For example, if a generator resistance of 200 ohms and a load of 450 ohms are used, one 4:1 and one 9:1 transmission line transformer can be used to match to 50 ohms.

After the filter is designed a SPICE file can be automatically generated that is ready to run. HALFOCT.EXE requests a name for the SPICE file and a value of coil Q that SPICE uses to get the actual frequency response and return loss. The name of this SPICE file is remembered in an auxiliary file, HALFOCT.TXT, and is suggested to the operator on subsequent occasions. The SPICE file can then be modified by text editing to perform numerous other tasks. Based on the results of the SPICE analysis, the band edge specifications for the filter can be widened slightly to compensate for the narrowing of the passband caused by a finite Q value. The default frequency analysis range is from one-half the upper passband edge frequency to twice the lower passband edge frequency. The SPICE listing lists as comments the four K values and the passband frequency values specified in the program.

The program comes equipped with a set of built-in lowpass prototype values that can be used as is or replaced by the user. The default is a 0.1-dB Chebyshev filter with a ripple bandwidth of 1.0 rad/sec. The return loss for this filter is about 17 dB on peaks (Figure 17.2). The DEFAULT button enters these values and also default values for RGEN, RLOAD, FLO, and FHI. This particular five-element default prototype was chosen because of the compromise between performance (which is excellent) and complexity of

Half-octave filter

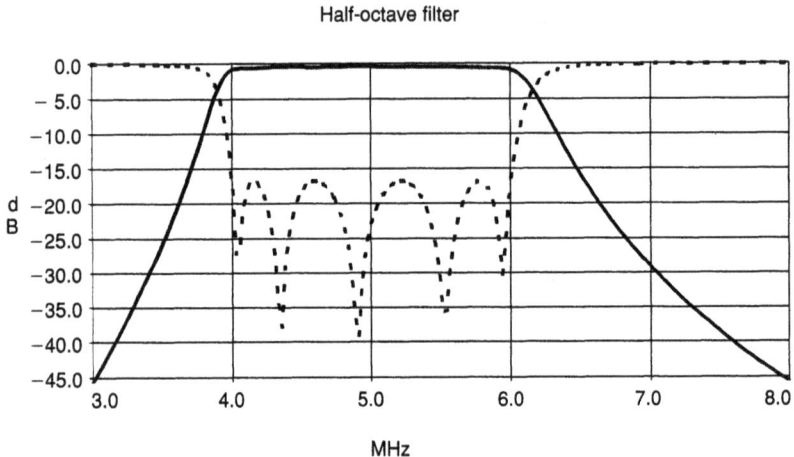

Figure 17.2 Frequency response and return loss of the half-octave bandpass filter.

the final bandpass filter (which is reasonable). For better return loss, use a 0.01-dB prototype or some other type.

The default values for the LPF prototype, RGEN, RLOAD, FLO, FHI, and coil Q can be overridden by entering other values in the various text windows. These new values are then remembered in an auxiliary file named HALFPROT.TXT. The next time the program is run these values are used instead of the default values. But the default values can be restored at any time by clicking the DEFAULT button.

The files HALFOCT.EXE, HALFOCT.TXT, and HALFPROT.TXT must all reside in the same computer directory. The SPICE file can be anywhere.

Although the half-octave filter is the main intention of the program, its usage is not restricted to that frequency ratio. The program is written in Microsoft Visual Basic, version 3.0. It requires Windows, version 3.X and a 286 or better CPU. VGA or better (super-VGA preferred) graphics is a plus. The mouse is used to click the software buttons and select text cells for value entries. An on-screen schematic diagram shows all values of L, C, R, and frequency. Visual Basic also requires the VBRUN300.DLL (or 200) file, which most Windows PCs have, but a public domain copy is provided on the disk. It should reside in the WINDOWS directory. An icon is included that can be used in Program Manager. To speed up the analysis process considerably, a second icon in Program Manager can be set up to perform the simulation with the SPICE program.

17.2 Cascaded Noise Figure and Intercept Program

The program NFIIP.EXE receives input data for the gain, noise figure, and either third-order or second-order intercept points (output or input) of a cascade of stages (1 to 12) in an SSB or CW receiver signal path. The data can be typed in manually or read from a file in memory. It then calculates the cascaded gain, noise figure, and input intercept point. It also calculates the spurious-free dynamic range (SFDR), either third order or second order, based on the bandwidth in the BW window (default – 1 Hz). The stage names, values, and BW can be modified from the keyboard, and stages can be added or deleted. The order of intercept, third or second, must be entered in the IM ORDER window. The calculation assumes two equal-tone levels.

Running the program shows the layout of the screen. Each column is numbered at the top. Next is the NAME of the stage and below it the GAIN, NFIG, and either the IIP or OIP. If input intercept is specified the output intercept is calculated and vice versa. Also required are the noise figure and input intercept (second or third order) for the rest of the receiver following stage 1. The next row of the screen displays the total gain, referred to the input of each stage and calculated from right to left. Next is the cascaded noise figure and the cascaded input intercept point.

To perform the calculation, click the CALC button. To get a correct calculation, there must not be any empty columns between the first stage and the last stage. To get the SFDR for a different BW, enter the new BW and click the CALC button.

To save a file, use FILE and SAVE. Enter the name of the file to save, using the extension .DAT. To retrieve a file, use FILE and GET. A list of files (in the same directory as the program) having the extension .DAT is presented. Click on the desired file.

To delete a column, first erase the contents of the NAME window. Then click the column number above the stage name window.

Suppose column X has a stage name entered. To move column X and those to the left of column X one place to the left, click the column number X. This will not work if column 12 is occupied. If the stage name of column X is empty, clicking the column number will shift the columns that are to the left of column X one place to the right.

To refine the design of the signal path, look for gain, noise figure, and intercept parameters for the various stages that improve the total noise figure, intercept, or SFDR. The sensitivity to various stage parameter values is easily observed.

The PRINT button sends a complete data summary to the printer.

The program is written in Visual Basic, version 3.0. See the section for HALFOCT.EXE for further details on Visual Basic. An icon for the Program Manager screen is also supplied. Clicking the icon loads the program.

17.3 Mechanical Filter SPICE Models

The software disk contains SPICE subcircuit files for the three Collins economy-series mechanical filters (5800 Hz, 2500 Hz, and 500 Hz, part number

526-8636/5/4-010) that are discussed in Chapter 6. These files can be added to the designer's SPICE list to simulate any circuit that uses one or more of these filters. Figure 17.3 shows the equivalent *LCR* circuits for these filters. The generator and load are not included in the SPICE files, but are part of the user-defined circuit to be simulated. That circuit should include the 30-pF shunt capacitors and the 2-kohm source and load resistance values.

These subcircuits are especially useful for studying various methods of impedance matching the filters to the circuit in question. Passband ripple effects and slopes due to termination variations are easily observed. Filter switching circuits can be analyzed. Figure 17.3A is the equivalent circuit for the 2500-/5800-Hz filters, and Fig 17.3B is for the 500-Hz filter.

17.4 Chebyshev Bandpass Filter Program

The program CHEBVALU.EXE calculates the lowpass Chebyshev prototype that has a ripple bandwidth of 1 rad/sec. The number of elements can be 2, 3, . . ., 10. It then calculates the *LCR* values for a classical wide-bandpass filter for a specified generator/load resistance and a lower and upper pair of frequencies for the ripple bandwidth. The value of coil *Q* is requested. The correct value of load resistance is calculated. Figure 17.4 shows the topology of the bandpass filter and the series or shunt *LCR* segments. The first segment can be a series or shunt resonant circuit.

The values of the series and shunt *LCR*s are presented on the screen. A SPICE file can also be generated that is ready to run. A value of coil *Q* is requested. This file must be placed in the same directory that CHEBVALU resides in. The SPICE file returns the frequency response and the return loss. This file can then be modified, using a text editor. For example, a SPICE optimizer can adjust the coil values to accommodate the nearest standard capacitor values. The return loss is a good parameter to optimize.

After the filter is designed at the impedance level that produces the most convenient *LC* values, wideband transformers can be used at input and output to transform to the desired generator and load resistances.

This program can also be used to calculate the lowpass prototype values for five-element filters with various ripple values that can be used in the half-octave filter program described earlier in this chapter.

17.5 Design, Analysis, and Synthesis of PLLs and Their Filters

PLLFILT.EXE is a multipurpose filter design and analysis program. The filter type can be the following:

1. Transfer function: LPF, HPF, or BPF.

2. Ladder network: LPF. Also HPF or BPF if element values are inputted.

3. Active filter: LPF only.

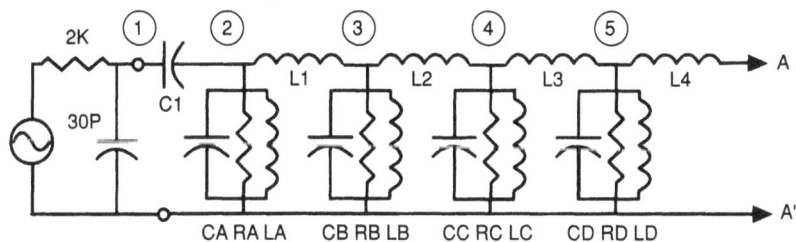

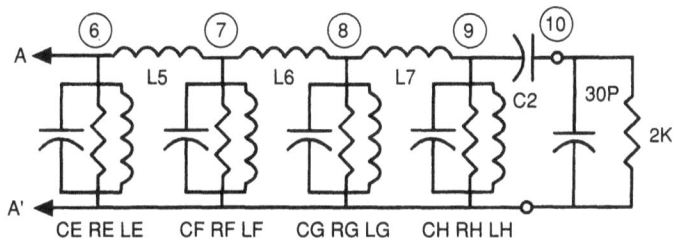

A. SPICE model for 2500-Hz and 5800-Hz filters

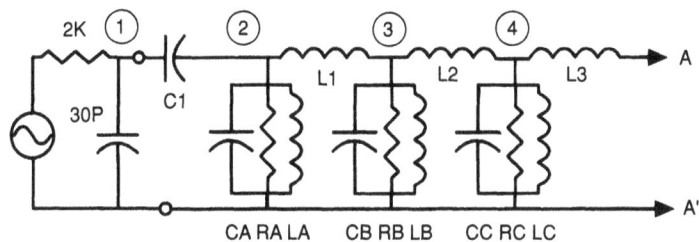

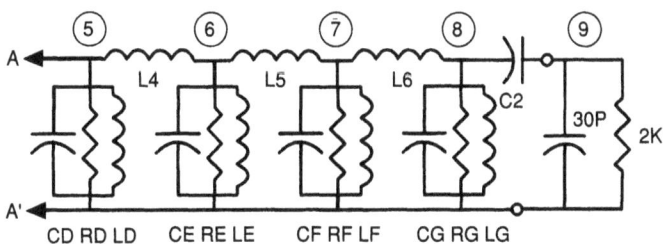

B. SPICE model for 500-Hz filter

Figure 17.3 Equivalent circuits for Rockwell (Collins) economy series mechanical filters: (A) 2500- and 5800-Hz filters; (B) 500-Hz filter.

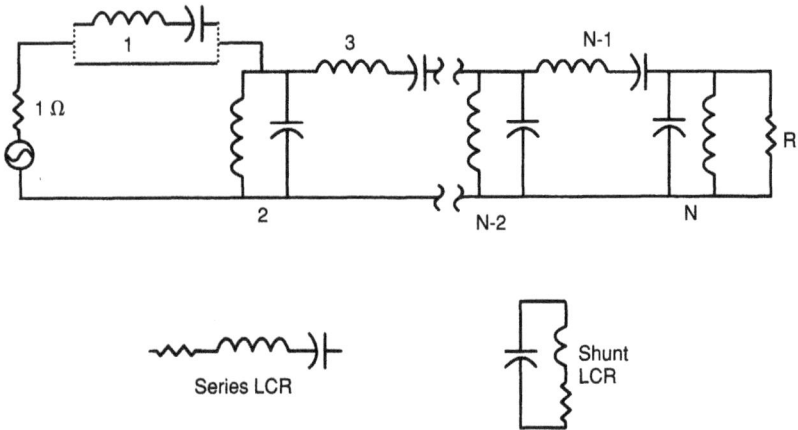

Figure 17.4 Topology of the Chebyshev filter for the CHEBVALU.EXE program.

Input data can be in the following formats:

1. Filter specs: generates an LPF transfer function.
2. Transfer function: used as is or to synthesize an LPF ladder or active filter.
3. Element values for ladder or active filter.

Other options are

1. Sample and hold
2. Real time delay
3. PLL parameters, for loop analysis with filters

Analysis options are

1. Magnitude/phase (Bode) plots of filters alone or PLLs
2. Group delay of any LPF function
3. Step response of any LPF function

A very complete user's guide is included in the program that gives detailed instructions on the usage of this rather complex and very useful program. The user's guide HELPFILT.EXE can be printed using the command line HELPFILT (printer). This program has been used for many years and frequently improved at Rockwell Corp.

17.6 Broadband Matching (Fano Method)

This pair of programs is used to calculate an optimum broadband matching network between a resistive generator and some reactive load such as a random length antenna or whip antenna. FANO.EXE calculates a variety of broadband matching solutions for a given load. CHRT.EXE creates a chart that simplifies the choosing of the correct number of network elements that will allow a desired SWR to be obtained for a given value of the load decrement (defined below). CHRT.EXE requires no inputs; it calculates a chart that can be edited, modified, and printed. FANO.EXE requires several inputs and produces sets of data that represent the circuit values for a classic Fano matching circuit.

Definitions

- Decrement δ is the frequency Q divided by the load Q_l:

$$\delta = \frac{F_h - F_l}{F_0 Q_l}$$

- F_h is the highest matching frequency.
- F_l is the lowest matching frequency.
- F_0 is the geometric mean of F_h and $\bar{F_l}$.

$$F_0 = \sqrt{F_h F_l}$$

- g parameters are the normalized lowpass prototype element values (1.0 ohm, 1.0 rad/sec).

- Bandpass realization is the bandpass matching elements. The load values L_0, C_0 R_1 are specified by the user. R_s is the generator and L_n, C_n are the network elements nearest the generator side of the network.

- The K inverter is a narrowband impedance transformer:

$$K = \sqrt{Z_{in} Z_{out}}$$

- The J inverter is a narrowband admittance transformer:

$$J = \sqrt{Y_{in} Y_{out}}$$

Program execution

Before running the program you must determine the low and high frequencies of the desired match, the number of network poles, and the type of load (series or parallel). Run the CHRT.EXE program. Using the calculated decrement, determine the number of network poles that are required to meet the desired SWR match. The load impedance must be examined, by direct measurement or by calculation, to determine if it is a series or parallel equivalent. This is determined by looking at the loci of resistance versus reactance points plotted versus frequency. If it is a straight line the load is a series equivalent; if it is a circular line the load is a parallel equivalent.

Execute the program. Answer the questions about frequency and load type.

Caution: While entering data when the antenna impedance is most like a parallel equivalent circuit, the following caution is offered. The antenna data can be entered as parallel lumped elements or as an impedance. The impedance value that is entered by the user should be the series equivalent of the lumped elements in the parallel equivalent circuit.

The program then displays the lowpass prototype elements along with the bandpass circuit that forms the broadband match. The input VSWR and source resistance are also displayed. If the VSWR is too high, add another pole and start over. Generally the source resistance is not suitable, and this can be corrected by adding a transformer. Three options are available. The *inverters* are narrowband transformers that work over limited bandwidths but they serve as a good alternative to adding more poles and elements of complexity. The Norton transformation is ideal and transforms without bandwidth limitations. You may pick and choose between these transformers until the best solution is found. When ready to quit, the program asks for a file name in which to store the element values. Give it a name without an extension. The program appends the .DAT extension. Next, simulate the load and network using a network analysis program such as SPICE. The actual performance of the calculated network will be somewhat modified as a result of the lossy elements that are actually used. See also Refs. 1 and 2.

17.7 Digital SSB Receiver Analysis

The program SSBRCVR.XLS is an Excel (version 5.0) spreadsheet program that analyzes the performance of an SSB receiver that uses IF bandpass sampling, as described in Chapter 8 (section 8.10). An example of the program output is shown in Table 8.3. The following points are mentioned in connection with this table:

- This program assumes 50-ohm impedance normalization throughout the analysis.

- The quantities in the upper left corner of the table, from "Analog RF/IF Noise Figure" down through "Analog AGC Range," with the exception of "A/D Converter Quantization Level," are entered by the user.

- The "Analog RF/IF Noise Figure" is the noise figure of the analog front end, as illustrated in Figure 8.36. The overall SSB receiver noise figure will naturally be slightly greater than this because of the addition of quantization noise in the converters and signal processing algorithms (digital filters, etc.). The "Output S/N Ratio in BW2" is accurately calculated assuming A/D quantizing noise only, and thus neglects other DSP algorithm quantization noise sources that may be present. However, in many cases, the A/D noise will be the dominant noise source.

An overall receiver noise figure can certainly be calculated from the spreadsheet output, even though it is not in this example. However, be careful when using cascade noise figure formulas for assigning noise figures to devices such as A/D converters. An "effective" A/D converter noise figure can be calculated, but it is not presented here for two reasons:

1. The A/D converter input impedance is rarely 50 ohms. Many sampling A/D converters have a high impedance (> 5 kohms), while "flash" converters have a very low input impedance that can actually vary with signal level.

2. The A/D converter is a nonlinear device. Unless the signal is large or sufficient noise is present at the input, the noise performance can be quite unpredictable.

Only the peak voltage level of the "A/D Converter Full-Scale Level" must be entered by the user. The level (in dBm) is calculated by the program assuming a 50-ohm normalization. This dBm level can then be used as a reference point for specifying input and output levels and for allocating headroom in the A/D and D/A converter.

For example, if the full-scale level is +24 dBm, and the "Output Signal Level" is entered as +4 dBm, then the D/A converter will be allocated 20 dB of headroom.

The "A/D Converter Quantization Level" is calculated from the other parameters, and is presented so that the user can ensure that sufficient noise is present at the A/D converter to guarantee uniformly distributed quantizing noise, as discussed in section 8.10.

The "Analog AGC Threshold" is the "Antenna Signal Level" at which analog AGC begins to hold the "A/D Signal Level" constant. It is normally set to start front-end AGC action when the ultimate "Output S/N Ratio in BW2" is obtained. Remember that this threshold also determines the amount of headroom allocated at the A/D converter.

The "Analog AGC Range" is the maximum attenuation range of the front-end AGC circuitry.

The first value in the column "Antenna Signal Level" is entered by the user. It is normally set to the minimum receiver signal level which must produce a specified output signal-to-noise ratio. Subsequent values in that column are automatically calculated in 10-dB increments.

The first value in the column "Analog RF/IF Gain" is entered by the user. It is normally set to the lowest gain that will guarantee that the "Noise at A/D Input in BW1" is greater than the "A/D Converter Quantization Level" and still provide an adequate "Output S/N Ratio in BW2." Subsequent values in that column are automatically calculated.

The first value in the column "Output Signal Level" is entered by the user. It is normally set, as described above, to allocate headroom to the D/A converter output. Subsequent values in that column are automatically copied from the entered value.

The values in the last column, "Antenna Overload Level," are calculated and can reach very large theoretical values. Although these values are accurate, this analysis does not consider the practical physical limitations of real-world analog RF/IF components. Obviously only a very special SSB receiver could perform with a +57-dBm interferer at its antenna input port.

When using this program, be sure to operate from a copy of the program so that the original spreadsheet does not become corrupted.

17.8 Mixer Harmonic Intermodulation Program

The program SPUR.EXE shows, in graphical form, the susceptibility of a mixer to harmonic IM caused by harmonics of the RF signal and harmonics of the LO injection. An example is shown in Figure 17.5. The horizontal axis is signal frequency and the vertical axis is ±delta-signal frequency. The plot lines indicate the delta-signal frequency, for a particular signal frequency, at which some harmonic of the interfering RF signal combines with some harmonic of the LO frequency to produce a spurious response that crosses over the mixer output desired signal frequency.

The mixer can be a low-side difference, high-side difference, or summing mixer. The schematic of the mixer type can be viewed and printed. The mixer can be selected as a receive mixer or as a transmit mixer. Chapter 4 gives further discussion of mixer spurious products and how they are calculated.

17.9 Preselector Programs

An Excel spreadsheet program is provided that can be used to calculate the parameters for the min-loss filter, as discussed in Chapter 9. The values for Example 9.2 are provided, but can be modified by the user. A SPICE text file is also provided that provides the frequency response and other data for Example 9.2.

17.10 Automatic Link Evaluation

Chapter 3 describes, briefly, the general methods of HF link evaluation and network design. Because of space limitations in the chapter, more-detailed information is included on the software disk file entitled LINKEVAL.DOC. This file is in a Word for Windows (version 6.0) format, and can be easily translated to other word processor formats or to ASCII text if necessary prior to printout.

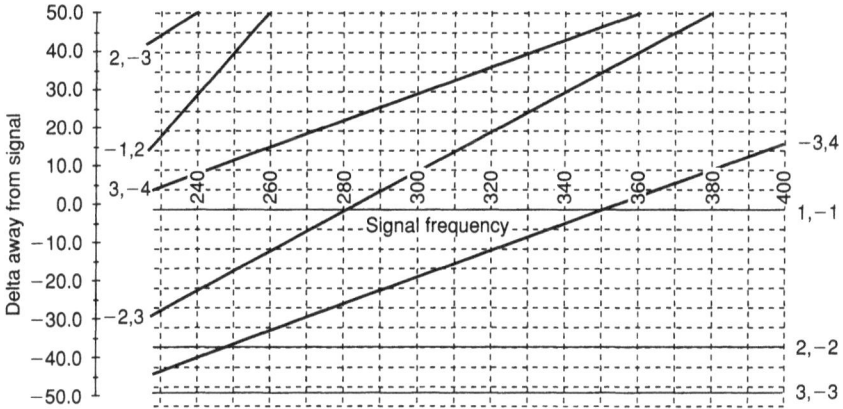

Figure 17.5 Example plot of mixer spurious program SPUR.EXE.

References

1. Cuthbert, T. R., "Broadband Impedance Matching Methods," *RF Design* (August 1994).

2. Cuthbert, T. R., "Broadband Impedance Matching, Fast and Simple," *RF Design* (November 1994).

Index

ABOUT THE EDITORS

WILLIAM E. SABIN is a retired design engineer who was in the advanced technology department, Collins Division, of Rockwell Corporation, where he worked starting in 1964. He received his MSEE degree from the University of Iowa and is widely experienced in HF receiver and exciter design, synthesizer design, power supplies, spread spectrum techniques, and digital design.

EDGAR O. SCHOENIKE is a retired senior technical staff member who was in the advanced technology department at Rockwell Corporation. He holds five U.S. patents. He received his MSEE degree from Iowa State University and has completed all course work for his Ph.D. He is the coauthor of *Single Sideband Principles and Circuits,* also published by McGraw-Hill.

www.ingramcontent.com/pod-product-compliance
Lightning Source LLC
Chambersburg PA
CBHW050452190326
41458CB00005B/1251